Dear Otto,

Your expedition to the islands in 1965 was an important stimulus for our work.

Best regards,

Tod

18 Nov 2017

Plants of Oceanic Islands

Evolution, Biogeography, and Conservation of the Flora of the Juan Fernández (Robinson Crusoe) Archipelago

Bringing together results from over 30 years of research on the Juan Fernández Archipelago off the coast of Chile, this book offers comprehensive coverage of the plants of these special islands. Despite its remote setting in the southeastern Pacific Ocean, the Juan Fernández Archipelago is in many ways an ideal place to ask and attempt to answer basic questions regarding the evolution of vascular plants in an oceanic island environment. By building on a firm taxonomic base for the flora, a new level of understanding regarding evolution, biogeography, and conservation of the plants is presented. This book is an extensive investigation of the origin and evolution of the flora of an oceanic archipelago, and it serves as a valuable resource for researchers and scholars of island biology as well as for conservation biologists worldwide.

Tod F. Stuessy is Professor Emeritus at The Ohio State University and the University of Vienna. His interests include principles of classification, island biology, biogeography, and plant variation and speciation.

Daniel J. Crawford is Adjunct Professor at the University of Kansas. His research centers on the origin, evolution, and reproductive biology of island plants.

Patricio López-Sepúlveda is Assistant Professor of botany at the University of Concepción, Chile. His research interests include island biology, speciation, conservation, and taxonomy of vascular plants, especially using molecular tools.

Carlos M. Baeza is Professor and Chair of the Department of Botany, University of Concepción. His research interests include island biology and cytotaxonomy of Chilean vascular plants.

Eduardo A. Ruiz is Associate Professor and Vice Dean of the Faculty of Natural and Oceanographic Sciences at the University of Concepción. His professional interests include island biology and molecular systematics and population genetics of Chilean endangered species.

Panoramic view of Tres Picos on Robinson Crusoe Island, Juan Fernández Archipelago.

Plants of Oceanic Islands

Evolution, Biogeography, and Conservation of the Flora of the Juan Fernández (Robinson Crusoe) Archipelago

EDITED BY

TOD F. STUESSY
The Ohio State University and University of Vienna

DANIEL J. CRAWFORD
University of Kansas

PATRICIO LÓPEZ-SEPÚLVEDA
University of Concepción

CARLOS M. BAEZA
University of Concepción

EDUARDO A. RUIZ
University of Concepción

CAMBRIDGE
UNIVERSITY PRESS

CAMBRIDGE
UNIVERSITY PRESS

University Printing House, Cambridge CB2 8BS, United Kingdom

One Liberty Plaza, 20th Floor, New York, NY 10006, USA

477 Williamstown Road, Port Melbourne, VIC 3207, Australia

4843/24, 2nd Floor, Ansari Road, Daryaganj, Delhi – 110002, India

79 Anson Road, #06–04/06, Singapore 079906

Cambridge University Press is part of the University of Cambridge.

It furthers the University's mission by disseminating knowledge in the pursuit of education, learning, and research at the highest international levels of excellence.

www.cambridge.org
Information on this title: www.cambridge.org/9781107180079
DOI: 10.1017/9781316841358

First published 2018

Printed in the United Kingdom by TJ International Ltd. Padstow Cornwall

A catalogue record for this publication is available from the British Library.

Library of Congress Cataloging-in-Publication Data
Names: Stuessy, Tod F., editor.
Title: Plants of oceanic islands : evolution, biogeography, and conservation of the flora of the Juan Fernandez (Robinson Crusoe) Archipelago / edited by Tod Stuessy, Ohio State University, University of Vienna, Daniel Crawford, Kansas University, Patricio Lopez-Sepulveda, University of Concepcion, Carlos Baeza, University of Concepcion, Eduardo Ruiz, University of Concepcion.
Description: Cambridge, United Kingdom ; New York, NY : Cambridge University Press, 2017. | Includes bibliographical references and index.
Identifiers: LCCN 2017028541 | ISBN 9781107180079 (hardback)
Subjects: LCSH: Island plants – Juan Fernández Islands. | Island plants – Evolution – Juan Fernández Islands. | Island plants – Conservation – Juan Fernández Islands. | Botany – Juan Fernández Islands – Identification. | Juan Fernández Islands. | Robinson Crusoe Island (Juan Fernández Islands)
Classification: LCC QK473.J8 P53 2017 | DDC 581.983–dc23
LC record available at https://lccn.loc.gov/2017028541

ISBN 978-1-107-18007-9 Hardback

In Memoriam

Professor Clodomiro Marticorena Pairoa
1929–2013

Critical, extremely intelligent, deeply insightful, a true friend and supporter of those he valued, expert in botanical nomenclature and bibliography, dedicated to understanding the Chilean flora, builder of the Herbarium of the University of Concepción, enthusiastic advocate of the Juan Fernández Project, co-leader of our first expedition to the archipelago in 1980, and co-author of many of our publications. We miss him for all these exceptional qualities.

We dedicate this book to past and present Corporación Nacional Forestal (CONAF) administrators, guides, and associates, without whose constructive help and friendship we never would have been able to carry out successful research in the Juan Fernández Islands.

Administrators

Bernardo Ackermann
Gastón González
Iván Leiva Silva

Guides and associates

Alfonso Andauer Schiller
Jorge Angulo Andonie
Guillermo Araya Arredondo
Danilo Arredondo Contreras
Danilo Arredondo Contreras, Jr.
Oscar Chamorro Muena
Esteban Charpentier
Miguel García
Alvis González
Guillermo González Arredondo
Claudio González González
José González de Rodt
Michael González
Bernardo López Rivadeneira
Guillermo López Rivadeneira
José López Rivadeneira
Eduardo Paredes López
Aldo Recabarren Green
Teodoro Rivadeneira Recabarren
Ramón Schiller Recabarren

Representative CONAF guides who have helped us on many research expeditions to the Juan Fernández Archipelago. Left to right: Oscar Chamorro, Ramón Schiller, Guillermo Araya, Esteban Charpentier, and Bernardo López.

Contents

A colored plate section for this book can be found between pages 108 and 109.

Contributors

Gregory J. Anderson
Department of Ecology and Evolutionary Biology, University of Connecticut, USA

Carlos M. Baeza
Departamento de Botánica, Universidad de Concepción, Chile

Gabriel Bernardello
Instituto Multidisciplinario de Biología Vegetal, CONICET, Universidad Nacional de Córdoba, Argentina

Daniel J. Crawford
Department of Ecology and Evolutionary Biology and the Biodiversity Institute, University of Kansas, USA

Josef Greimler
Department of Botany and Biodiversity Research, University of Vienna, Austria

Patricio López-Sepúlveda
Departamento de Botánica, Universidad de Concepción, Chile

Clodomiro Marticorena
Departamento de Botánica, Universidad de Concepción, Chile

Patricia Pacheco
Cristina Barros 653, Las Condes, Santiago, Chile

Karl Reiter
Department of Botany and Biodiversity Research, University of Vienna, Austria

Roberto Rodríguez
Departamento de Botánica, Universidad de Concepción, Chile

Eduardo A. Ruiz
Departamento de Botánica, Universidad de Concepción, Chile

Mario Silva O.
Departamento de Botánica, Universidad de Concepción, Chile

Walter A. Sontag, Jr.
Department of Botany and Biodiversity Research, University of Vienna, Austria

Tod F. Stuessy
Herbarium and Department of Evolution, Ecology, and Organismal Biology, The Ohio State University, USA, and Department of Botany and Biodiversity Research, University of Vienna, Austria

Ulf Swenson
Department of Botany, Swedish Museum of Natural History, Stockholm, Sweden

Koji Takayama
Museum of Natural and Environmental History, Shizuoka, Japan

Hugo Valdebenito
School of Life Sciences, Universidad San Francisco de Quito, Ecuador

Preface

This book has derived from numerous research discussions and expeditions over a period of more than 35 years, beginning in February of 1977. Tod F. Stuessy traveled to Chile for research discussions with Mario Silva O., professor in the Department of Botany of the Universidad de Concepción. The initial idea was to develop a joint research project on patterns and processes of speciation in the genus *Chaetanthera* (Mutisieae, Asteraceae) in the high Andes of southern South America. For a number of practical reasons, it was decided to develop another cooperative research program focused on the Juan Fernández (Robinson Crusoe) Islands, which belong to Chile. Silva was enthusiastic, and the principal taxonomist in the Department of Botany, Professor Clodomiro Marticorena, was also interested in a possible collaboration. A decision was made to develop an evolutionarily oriented project in the archipelago.

From the outset, an important concept for our research in the islands was that it must be internationally collaborative, involving Chilean botanists as well as those from the United States, Europe, and elsewhere. We have viewed this project as a means not only of advancing scientific understanding of the archipelago, justifiable enough in its own right, but also as a mechanism for educational development on all sides. Three Ph.D.'s have been completed in the project, two at The Ohio State University (Patricia Pacheco and Hugo Valdebenito) and one at the Universidad de Concepción (Eduardo A. Ruiz, who also spent research time at Ohio State). Five postdoctoral fellows have participated (Carlos M. Baeza, Gabriel Bernardello, Patricio López-Sepúlveda, Roger Sanders, and Ulf Swenson) plus dozens of students and technicians.

This book now complements several other recent volumes that, taken together, tell much about the Juan Fernández Archipelago. Especially useful for overviews of the islands and their history, culture, and people is the lavishly illustrated *La Isla de Robinson Crusoe* by Patricio Arana (2010). A smaller book, but also useful, is *Les îles de Robinson* by Philippe Danton et al. (1999), which provides a nice introduction and overview of the islands. As for the flora of the archipelago, Philippe Danton and collaborators have contributed an introduction to *Wild Plants of Robinson Crusoe Island* (2004; trilingual, also in Spanish and French) and now the comprehensive *Monografía de las Plantas Vasculares del Archipiélago Juan Fernández, Chile* (Danton and Perrier 2017). Our new book builds on the taxonomic base of understanding of plants of the archipelago by summarizing their evolution, biogeography, and conservation. With these combined new references, a new level of knowledge about the islands and their fascinating plant life has been achieved.

A note on figure citations in this book is in order. Black and white figures are contained within each chapter, but all color figures have been placed between pages 108 and 109. These figures all carry the prefix “C,” for example, Fig. C1, Fig. C2, etc., and they are so cited throughout all chapters of the text.

On a personal level, we have found in these small islands a special world of wild and human nature. People living in isolated corners of our planet tend to reach out easily to visitors, and this is completely the case in the Juan Fernández Islands. We stress the friendship offered by the CONAF guides, to whom we dedicate this book. This volume belongs to them for making it all possible in the first place. We have profited immensely from our involvement with these islands, and we hope that readers will also derive inspiration from them.

Acknowledgments

It is with pleasure, and a profound sense of personal debt, that we acknowledge the help and support of many individuals and institutions over the past thirty-five years. These have been fundamental in aiding our efforts to understand the evolution and biogeography of the plants of the Juan Fernández Islands.

We give thanks to Mario Silva O., professor emeritus at the University of Concepción. It was his strong enthusiasm for the Juan Fernández project that led to its initial development and eventually the production of this book. On numerous occasions, Mario was the person who found the solution to some crisis, whether it be financial, administrative, or political. He was also the head of phytochemical investigations in the islands and was one of the general scientific leaders of the entire project for many years.

We thank the foundations that have provided the financial basis for our 12 expeditions to the islands plus many laboratory investigations: the US National Science Foundation (from International Programs and Systematic Biology), the National Geographic Society, the Österreichische Nationalbank Jubiläumsfond, The Austrian Science Fund (FWF), and the Armand G. Erpf Fund, Inc.

Thanks go to the following institutions for additional financial support: The Ohio State University, the Los Angeles County Museum of Natural History, the University of Vienna (Austria), the University of Concepción (Chile), the Department of Ecology and Evolutionary Biology and Research Foundation of the University of Connecticut, and CONICET and the Universidad Nacional de Córdoba (Argentina).

We very much appreciate the numerous leaves of absence from teaching and administrative duties granted to many of us, which made it possible to carry out the field work required for such an undertaking during 1980 (two trips), 1984, 1986, 1990, 1991, 1995, 1996, 1999, 2000, 2010, and 2011. These research leaves were generously provided by the University of Concepción, The Ohio State University (Columbus), the University of Connecticut (Storrs), the Universidad de Córdoba (Argentina), the Los Angeles Museum of Natural History, and the University of Vienna.

We thank the Corporación Nacional Forestal (CONAF) of Chile, which provided permits to visit, investigate, and collect in the islands. Our research in the Juan Fernández Archipelago would not have been accomplished without the consistent logistic help and administrative support of several chiefs of the Robinson Crusoe National Park: Bernardo Ackermann, Gastón González, and Ivan Leiva. We also thank the park guides (especially Alfonso Andauer, Jorge Angulo, Danilo Arredondo, Guillermo Araya, Oscar Chamorro, Bernardo López, Manuel Recabarren, and Ramón

Schiller), without whose invaluable assistance our work would have been impossible (see Dedication). Our sense of indebtedness to them is very sincere and deep.

We also acknowledge the following Chilean institutions: the Servicio Aeronáutico de Chile for allowing us to use one of their houses to sleep, eat, and carry out research during two expeditions; the Armada de Chile for help with logistic support in getting supplies to the archipelago; and the Servicio Meteorológico (especially Osvaldo Jara and Alex Meneses) for assistance with activities in San Juan Bautista.

The field and laboratory research described in this book has been accomplished with the participation of many students, associates, colleagues, and friends. We are particularly indebted to Pedro Aqueveque, Jorge Arriagada, Luis Letelier, Daniela Bacho, Leonardo Gaete, Alejandro Gatica, Fidelina González, Ana María Humaña, Héctor Ibarra, Paula Kivinen, Gabriele Kottirsch, Richard Jensen, Thomas Lammers, Alejandro Landero, Oscar Matthei, Patricio Novoa, Patricio Peñailillo, Oscar Parra, Richard Roederer, Mauricio Rondanelli, Roger Sanders, Jaime Sepúlveda, Alan Stuessy, Patricia Stuessy, Eric Tepe, Eduardo Ugarte, and Delbert Wiens. Assistance in San Juan Bautista was provided by José M. Gutiérrez, Juanita López, Jorge Palomino, and Valeria Salzmann.

Support was also generously provided by many specialists. The University of Connecticut greenhouse staff, led by Clinton Morse and assisted by Matt Opel, did an excellent job of growing and caring for a number of difficult and unusual plants. Conley McMullen, Warren Wagner, and Steve Weller provided comparative data and many useful comments in the development of Chapter 12. Alicia Marticorena located errors and omissions from the lists of taxa in Chapter 5. John McNeill, with input from Werner Greuter, Kanchi Gandhi, and Paul Rijckevorsel, provided expert nomenclature help in sorting out the proper name of *Erigeron fernandezia*. Mark Carine alerted us to the early collections of George Handisyd, and Charlie Davis aided with interpreting Sloane's handwriting. Andrés Moreira Muñoz directed our attention to the unpublished geological thesis by Astudillo M. John Freudenstein provided assistance in digitization of the figures for the book. John Frederick made helpful critical comments on the manuscript and prepared the indices, for which he received technical guidance from the reference staff of the Upper Arlington (Ohio) Public Library. Also important was the sophisticated electrical expertise provided by Sr. Juan Ghivarello, without whose contributions the isozyme runs on the islands would have been nearly impossible.

Most of the photographs in this book were taken by the editors and authors as follows: Gregory Anderson (frontispiece, dedication, and Fig. C39); Gabriel Bernardello (Figs. 2.3F, C72, and C77); Daniel Crawford (Fig. C41); Josef Greimler (Figs. 7.5A, 8.5, 8.9, C8, C10, C11, C23, C33, C75, C80, C85, C87, C93, C95, C111, C116, C117, and C121; Patricio López-Sepúlveda (Figs. C20, C83, and C90); Roberto Rodríguez (Figs. C7, C12, and C24); and Ulf Swenson (Fig. 9.5B). Patricio Novoa, a participant on the 2011 expedition to Alejandro Selkirk Island (Novoa 2015), also generously contributed images of ferns (Figs. C17, C18, C19, C22, and C25). Alicia Marticorena provided the photograph of her father, Clodomiro Marticorena, and Otto Solbrig allowed us to use photographs of *Ochagavia elegans* and *Robinsonia masafuerae* (Figs. C31, C66, and C67), plus his portrait (Fig. 2.1H), taken on the 1965 US-Chile expedition. All other photos were taken by Tod F. Stuessy. Gilbert Ortiz drafted Figs. 7.1, 7.2, 10.1–10.9, 13.1, 13.2A–X, 14.2, 14.3, and 15.1–15.6 and the maps in Appendix 1, and Patricio López-Sepúlveda prepared Fig. 1.1.

The portraits of botanists in Fig. 2.1 have been cropped or modified from the following sources: A, original oil painting by Sir Thomas Lawrence in the National Portrait Gallery, London; B, Delprete et al. (2002: 625); C, Aguirre de Maino (1981: 332); D, Fürstenberg (1906: frontispiece); E, Moseley (1892: frontispiece); F and G, from photographs in the Department of Botany, Universidad de Concepción, Chile; H, Otto Solbrig, unpublished; I, Paulo de T. Alvim, in Heusser (1977: 248), ©1977, The New York Botanical Garden.

Permissions have been granted to republish figures that originally appeared in journals or books. We gratefully thank the following persons and institutions: Frontispiece, *American Journal of Botany*; Fig. 2.1B, Piero Delprete; Fig. C45, *Systematic Botany* (American Society of Plant Taxonomists); Fig. 3.1, *Science*; Figs. 3.3 and 3.4, the Geological Society; Figs. 3.5 and 3.6, *Opera Botanica* (now *Nordic Journal of Botany*); Fig. 4.4, Hugo Romero; Fig. 4.5, Andrés Moreira-Muñoz and Springer; Figs. 8.2, 8.3, 8.4, and C63, *Aliso*; Figs. 14.1 and 17.1, Taylor & Francis; Fig. 17.2, Robert Whittaker, *Journal of Biogeography*, and John Wiley; Figs. C88 and C104, University of Hawaii Press; Figs. 16.1 and 16.2, *Canadian Journal of Botany* (© 2008, Canadian Science Publishing or its licensors).

Finally, we are very appreciative of the skill and experience of Sr. Santiago Figueroa, seasoned pilot formerly of the Lineas Aéreas Robinson Crusoe, who safely carried us through some not always perfect weather to and from the islands on so many different trips.

Introduction

Tod F. Stuessy, Daniel J. Crawford, Patricio López-Sepúlveda,
Carlos M. Baeza, and Eduardo A. Ruiz

Why on earth would anyone wish to work in the Juan Fernández (Robinson Crusoe) Islands? The lore of the Crusoe name appeals, but these islands seem so distant, so remote, so poorly known – not an ideal place, perhaps, to consider carrying out evolutionary or biogeographical studies. Our selection of this study site was based on a number of considerations.

Despite its remote setting in the southeastern Pacific Ocean, in many ways the Juan Fernández Archipelago is an ideal place to ask and attempt to answer evolutionary and biogeographical questions. First, it consists of only two major small (ca. 50 km^2) islands, with a vascular flora of only 134 endemic species of plants. This would allow, after a period of time, a synthesis of evolutionary aspects for the entire flora. Further, with only two major islands, this greatly reduces the number of plausible evolutionary or biogeographical hypotheses that can be proposed. Second, among these endemics are some strikingly divergent and fascinating taxa, including the monotypic family Lactoridaceae. In a number of groups, much morphological divergence has occurred during evolution from continental progenitors. Third, some groups have obviously radiated within the archipelago, such as the larger genera *Dendroseris* (11 endemic species) and *Robinsonia* (8 endemic species), which offer opportunities to study within-island speciation. Fourth, a good floristic inventory of the vascular flora already existed, based on the earlier works of Hemsley (1884), Johow (1896), and Skottsberg (1921, 1953b; plus Christensen and Skottsberg, 1920). This would allow focus on evolutionary questions without having to do a basic revision of the flora. Fifth, there have been no indigenous peoples in the archipelago (Haberle 2003), which lessens disturbance of the original vegetation; human impact began in 1574 with the discovery of the islands by the Spanish navigator Juan Fernández (Woodward 1969). Sixth, because ties of the endemic flora are obviously close to southern South America (Bernardello et al. 2006), in a biogeographical sense the island closest to the continent, Robinson Crusoe (= Masatierra), has a much higher probability of receiving dispersing immigrants than the second island, Alejandro Selkirk (= Masafuera), 181 km further west into the Pacific Ocean. This sets a hypothesis of biogeographical directionality (and evolution) that is not often encountered. Further, we later learned that Robinson Crusoe Island is older, at 4 million years, in contrast to Alejandro Selkirk Island at 1 to 2 million years (Stuessy et al. 1984), which reinforces this directionality.

It was our belief, then, that the Robinson Crusoe Islands would make an ideal natural system for asking and attempting to answer basic questions regarding the evolution of vascular plants in an oceanic island archipelago. That has, in fact, proven to be the case. From the first publication in 1982 to the present, persons in our group or associated with our group have published more than 60 papers in many different journals. The reports have been widely scattered, however, and one of the objectives of this book is to summarize all these studies in one place for the interested reader.

The impression should not be left, however, that all is simple in working in the Juan Fernández Islands. Most pronounced are practical difficulties with logistics because the islands are isolated and do not have any huge influx of tourists; infrastructure and facilities are not well developed that might allow obtaining all food, beverages, and first aid items for each expedition. These must be brought to the islands either on a boat or by small plane. Hence much planning goes into every expedition, and this is particularly the case on Alejandro Selkirk Island, where no supplies exist – everything must be brought over by boat from Robinson Crusoe Island. Another point is that the collecting work is hard because there are no roads outside of short strips around the airport and in the village. All collecting must be done by hiking with backpacks, and there are few well-developed hiking trails. Being in good physical shape is mandatory for research in the archipelago, and each island has its own special landscape challenges.

Although the primary focus for our studies in the Robinson Crusoe Islands has always been on basic questions of island biology, we have been cognizant of the need for effective conservation. Oceanic islands throughout the world are under intense pressure from human developmental activities. Fortunately, due to the small size and remoteness of the Robinson Crusoe Islands, no high-rise hotels, artificial beaches, casinos, or off-shore banking establishments have been created. Nonetheless, the impact of humans on the islands in the more than four centuries since their discovery in 1574 by Juan Fernández has been substantial, and this continues to place a pressure on the native flora. It is our hope that our studies, and especially this book, will underline the importance of these marvelous islands and the need for their effective conservation.

Part I

Historical Aspects

The Juan Fernández Islands are steeped in history. After their discovery in 1574 by Juan Fernández, many European ships stopped at the archipelago to allow crews to rest and recuperate, obtain fresh water and vegetables, and repair boats. Meat from feral animals (mainly goats) was also gratefully consumed. From these early visits, impacts from humans on the archipelago began to accumulate. To appreciate the present flora and vegetation and to understand the evolutionary and biogeographical dimensions that led to their formation require having explanations of the natural and human-induced historical changes.

Chapter 1 sketches the history of the islands from their discovery to the present day. Very useful have been the numerous historical reports and descriptions of the islands, which have allowed good insights into the changing impacts on vegetation over more than four centuries. Chapter 2 outlines the activities of botanists in the islands over more than 300 years, including our own numerous expeditions from 1980 to 2011. Together these two chapters provide a useful introduction to the archipelago and an update on previous botanical explorations that have culminated in the synthesis of information contained in this book.

1 History of the Islands

Tod F. Stuessy

Discovery

The Juan Fernández (Robinson Crusoe) Islands (Fig. 1.1) were first brought to the attention of Europeans by the Spanish sea captain Juan Fernández, who sighted them on November 22, 1574 (Medina 1974). He sailed from Callao (a port near Lima, Perú) on a voyage south to Chile (Valparaíso or perhaps Concepción) (Woodward 1969). Such a trip normally took three to six months because sailing close to the western coast of South America required fighting the Humboldt Current flowing northward. Seeking a better route, Juan Fernández went first westward and then turned southward, avoiding the negative impact of the coastal current and coincidentally allowing discovery, first, of the Desventuradas Islands (San Félix and San Ambrosio, also owned by Chile) and, further south, of the Juan Fernández Islands. It is unclear whether he observed both the large islands, Robinson Crusoe (Masatierra; originally called Santa Cecilia) (Woodward 1969) and Alejandro Selkirk (Masafuera) or just the former with its small off-shore neighboring island, Santa Clara, but after discovery, he then turned eastward and arrived in Chile in record time. The trip, in fact, was so fast (30 days) that he was viewed with great suspicion, and he may even have been brought before the Inquisition on charges of having used witchcraft for such a fast trip (Woodward 1969). Other ships, however, soon accomplished the same feat, and this opened up a flow of ship traffic past the islands.

A note regarding naming of these islands is in order. The islands were originally, and historically, named after their discoverer, Juan Fernández. The two main islands were called Más a Tierra (literally "closer to the land," often condensed to Masatierra) and Más Afuera ("further away," often Masafuera), reflecting their positions relative to the Chilean continent. In 1962, the Chilean government officially renamed the archipelago as the Robinson Crusoe Islands. Masatierra was renamed "Isla Piloto Robinson Crusoe," and Masafuera was designated "Isla Marinero Alejandro Selkirk," but most people have abbreviated the names to just Robinson Crusoe Island and Alejandro Selkirk Island, which is the approach taken in this book. The reason for these changes was apparently to stimulate tourism, taking advantage of the connection with the famous novel, *Robinson Crusoe*, by Daniel Defoe published to popular acclaim in 1719. Not everyone has been completely enthusiastic about these modifications, however (e.g., Barrera 1963).

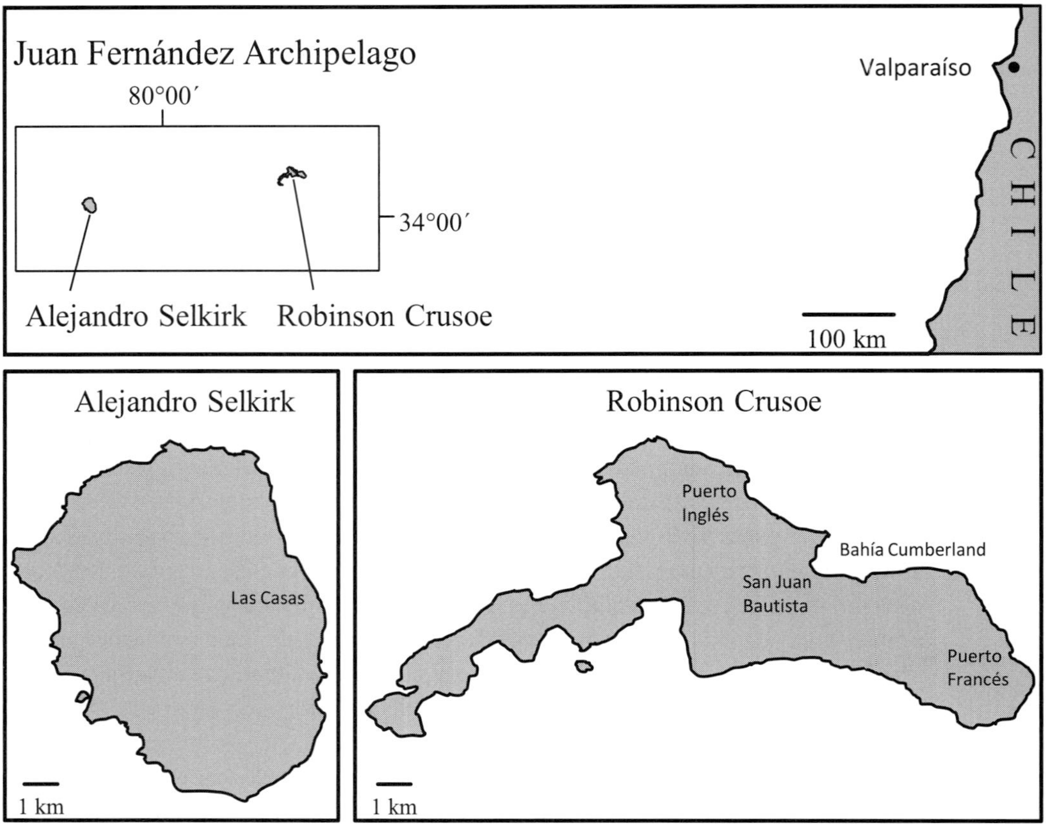

Figure 1.1 Location of the Juan Fernández (Robinson Crusoe) Islands in the eastern Pacific.

There was, in fact, a real Robinson Crusoe, but he bears scant resemblance to the hero in Defoe's famous and engaging novel (Howell 1829). In January of 1704, a Scottish sailing master, Alexander Selkirk, aboard the ship *Cinque Ports*, under command of Captain Thomas Stradling, sailed around Cape Horn and continued northward along the Chilean coast. The ship arrived at Robinson Crusoe Island on February 10, 1704. After a series of incidents involving French ships arriving at the archipelago, a dispute developed between Selkirk and Stradling such that the former insisted on staying ashore. Stradling provided Selkirk with only a few survival provisions, but because the island had drinkable water, vegetated hills with some edible plant species, sea lions in abundance, fish, lobsters, and feral goats that were left from previous voyages, it was possible for Selkirk to live alone on the island (without an assistant, Friday, as occurs in the novel) for more than four years and four months! Many ships arrived at Robinson Crusoe Island during this period, but because of the continuing wars in Europe, it was difficult for Selkirk to know who was friend and who was foe. Eventually, in 1709, he dared to connect with two ships, the *Duke* and the *Duchess*, which, luckily, were British, and seized the opportunity to return home to Europe. This now-bearded man, dressed in goat skins, created quite a sensation, especially when the *Duke* arrived back in England on October 1, 1711. Selkirk was interviewed by many journalists, and reports were published in local outlets. Although Daniel Defoe may or may not have met Selkirk (Woodward 1969), in any event it is certain that he took advantage of this lone mariner's story to write his uplifting and entertaining novel of 1719. It was set, however, somewhere in the Caribbean rather than in the actual Juan Fernández Islands in the eastern Pacific.

Pirates and Early Visitors

Since their discovery, the Juan Fernández Islands in the seventeenth century became a location for ships of many nations to pause to refit the boats and refresh the sailors. The presence of a sheltered natural bay (Bahía Cumberland), a subtropical (or warm-temperate) climate, abundant fresh water, forested slopes, native plants that served as vegetables, and meat in the form of feral goats and native sea lions all combined to attract vessels, especially those that survived the difficult passage around Cape Horn. The sheltered bay also served well as a refuge from which ships from England, France, and The Netherlands could attack colonial Spanish ships and coastal cities. The island became known, therefore, as a stopping place for ships sailing around the world to rest before continuing their journey westward across the Pacific and also to serve as a pirate's lair.

Many of the visitors to the islands published comments on their stays, especially the captains or first mates of ships in the early eighteenth century. It was still of significant interest for readers in continental Europe to learn details of voyages to far-away lands. Reports by George Shelvocke (1726) and Jacob Roggeveen are examples from this period (from the Netherlands; Sharp 1970).

Another important visitor to the Juan Fernández Islands during this period was George Anson, captain of the *Centurion*, which sailed into the main bay of Robinson Crusoe Island on June 9, 1741. The importance of Anson's visit was the detail of the observations that he offered in his logbook (Walter and Robins 1748) on the natural history and geography of the islands. His voyage from England around Cape Horn with three ships was marred by serious illness, only 335 crewmen surviving of a total of 961 at departure from England (Woodward 1969). To cure all remaining crew members, plus to refit the ships needing repair, Anson remained on Robinson Crusoe Island until September 19, 1741. From this position he successfully attacked and captured Spanish ships and pillaged shore communities, finally traversing the Pacific and returning home to England in 1744.

It was due in large measure to Anson's successful disruption of Spanish shipping and commerce that the Spanish government, centered in the Viceroyalty of Perú, in Lima, decided of necessity to officially reclaim the Juan Fernández Islands and establish a permanent settlement there. They dispatched two prominent officers with scientific training, Jorge Juan and Antonio de Ulloa, to reconnoiter the islands and report on the best ways to secure the archipelago for Spain. They visited the islands, principally

Robinson Crusoe Island, on January 7–22, 1743. In their secret report submitted later that year (published and made public only in 1826), they described the natural history of the islands but, more important, recommended the establishment of defenses. The biggest challenge was to arm the natural bay, Bahía Cumberland, so that no marauding ships (especially British) could anchor there.

Spanish Control

Reacting to the report of Juan and Ulloa, the Viceroyalty of Perú in 1750 dispatched a colonization force to the Juan Fernández Islands. This consisted of a ship with 62 soldiers, 171 colonists (including women), and 22 convicts (Orellana R. 1975). Cows, sheep, mules, pigs, and poultry were also transported. The idea was to establish a permanent settlement around Bahía Cumberland. Also dispatched was a ship laden with arms, including rifles, muskets, and gunpowder, plus 18 cannons and 7,400 cannonballs to arm the envisioned fortress (Woodward 1969). The village, named San Juan Bautista, was organized in short order and received more immigrants from Concepción, Chile. Misfortune struck, however, with a severe earthquake and tidal wave on May 25, 1751 (Vicuña Mackenna 1883) that caused substantial damage to the village and fort. Help was immediately provided, however, to allow rebuilding of all structures, and by December of that same year, the fortress (Santa Bárbara) was constructed and nine cannons were placed to challenge and repel uninvited pirates (Morel 1975). Although the small community at this point was not at all self-sufficient, the goal of controlling the islands and reducing pirate action was achieved. The importance of this strategic move was underlined by the reality that the English had successfully colonized the Falkland (Malvinas) Islands, lying off the eastern coast of southern Argentina. With this strategic point under alien control, it emphasized the importance of controlling access to the western coast of southern South America, which the Juan Fernández Islands provided.

The Spanish, however, had settled and fortified only one island, Robinson Crusoe (Fig. C1). Alejandro Selkirk Island (Fig. C2), lacking a natural harbor, was still open for visits by ships of other countries. John Byron, for example, in his ship *Dolphin*, visited Alejandro Selkirk Island in 1765 (Gallagher 1964) to take on water and wood, not realizing that Robinson Crusoe Island was now under Spanish command. Because the Spanish had taken over the eastern island quietly, it came as a surprise to many ships that this former pirate's bay was no longer available for open use. Spain, for its part, continued to fortify the island and to send the worst criminals to this isolated location. The number of settlers, obviously, declined proportionately with increasing numbers of convicts. Some of the prisoners were forced to dig caves along the sea cliffs in which to live. These were documented by Claudio Gay during his visit in 1832 (Muñoz Pizarro 1944) (Fig. 1.2), and they remain of interest to tourists even today. Maintaining a penal colony combined with a settlement, however, became more difficult to sustain over the years, and Robinson Crusoe Island was actually nearly abandoned for a short period in 1814.

Figure 1.2 Bahía Cumberland on Robinson Crusoe Island, showing the small village of San Juan Bautista and the caves (left side of drawing) constructed by convicts in the late eighteenth century. (From Gay 1854.)

While Spanish control succeeded in keeping English, French, and Dutch ships away from the Juan Fernández Islands, the American Revolution of 1776 opened up opportunities for American vessels to visit the archipelago. The whaling industry on the East Coast of the United States, particularly from Massachusetts, flourished by sending ships into different oceans, including around Cape Horn, along Chile, and into the Pacific (Pereira 1971). The large number of seals and sea lions served as a powerful attraction to visit the Juan Fernández Archipelago, especially Alejandro Selkirk Island. It is difficult today to imagine the scale of harvesting of the fur seals; for example, Captain Amasa Delano and his crew alone took 3 million skins to sell in China between 1797 and 1804 (Delano 1817; Woodward 1969)!

Chilean Independence

With the invasion of Spain by Napoleon in 1807 and the abdication of King Fernando VII, this weakness in the Spanish government opened up opportunities for independence in the colonies of Spanish America. Simon Bolívar, José de San Martín, and Bernardo O'Higgins (remarkably the illegitimate son of the Viceroy of Perú, who was the most powerful Spanish authority in the New World), all having spent years in Europe and having been exposed to the liberal thinking of French and American authors, spent a good part of their lives in the liberation of Venezuela, Panamá, Colombia, Ecuador, Perú, Bolivia, Argentina, and Chile. Chile easily won its independence early in 1810, with many *criollos* (persons born in the New World of Spanish descent), including Bernardo O'Higgins, leading the way. This freedom did not last long, however, because Chile was reconquered by royalist troops from Perú. Many republicans (patriots) were rounded up (Vicuña Mackenna 1883), and fifty were sent to Robinson Crusoe Island, where they were kept in the old caves carved earlier by convicts (Fig. 1.2). O'Higgins and others escaped to Argentina, however, and joined with José de San Martín, who was planning the liberation of Chile. This was to be followed by a naval invasion of Perú for conquest of the last bastion of Spanish control. With the final freedom of Chile in 1814, the patriots on Robinson Crusoe Island were now free to return to the mainland and help participate in the formation of a new government.

With the exodus of all prisoners from Robinson Crusoe Island, only a few people elected to remain in San Juan Bautista as entrepreneurs selling fresh water and wood to passing ships. Soon, however, O'Higgins realized the need for a penal colony to remove undesirables from the Chilean mainland, and he therefore reestablished the colony in San Juan Bautista in 1821. This also had the positive effect of once more keeping the archipelago under local control, especially to guard against English or American occupancy.

The early years of the young, independent Chile were filled with the usual amount of political upheavals common to all Latin American republics, and the settlement of San Juan Bautista in the Juan Fernández Islands also followed this trend. The island was rented to Joaquín Larrain in 1829–33, but it continued as a penal colony, receiving all manner of murderers and other criminals. An Englishman, Thomas Sutcliffe, assumed the governorship on November 25, 1834, but this administration was especially affected by the high tidal wave that destroyed the village on February 25, 1835 (Sutcliffe 1839, 1841), exactly three months after Sutcliffe's arrival. Rebuilding ensued, but social instability in the little village was still problematic.

Development of a Stable Community

The gold rush in California in 1848 intensified ship traffic to Robinson Crusoe Island. Many Americans made the long trip around Cape Horn from the East Coast of the United States westward to California (Lewis 1949). This provided helpful income for the islanders in sales of fresh water, wood, meat, and vegetables. These free-spirited visitors also continued the substantial impact on the natural vegetation of the islands, which sadly has been so typical of oceanic islands. As many as fifty ships visited Robinson Crusoe Island in 1849–1850 (Lewis 1949; Woodward 1969), but maritime traffic dropped way down afterward.

The period from 1850 to 1900 on Robinson Crusoe Island was one of relative calm, and European immigrants began to arrive. This allowed a more stable fishing and agriculture community to become established. Sales of fresh water, wood, vegetables, and meat continued to passing ships, but in addition, livestock was encouraged as well as development of the lobster and fishing industries. This phase led to the development of families in the village that are still represented today. For example, Karl Alfred von Rodt, of Swiss origin, after having served in the Austrian Army, emigrated to Chile and eventually to Robinson Crusoe Island in 1877 (von Rodt 1907; Ruh 1975). As with von Rodt, most visitors to the archipelago were taken by its pastoral charm, and many believed that there could be successful economic possibilities, which often turned out to be unrealistic. The population stayed small, therefore, with, for example, only 64 persons in 1877 (only 29 adult males) (Woodward 1969).

World War I led to the very odd and dramatic circumstance of a visit by the German ship *Dresden* and its sinking by British ships in Bahía Cumberland on March 9, 1915 (Parker de Bassi 1987). Chile was officially neutral during the war, so the port commander had permitted the *Dresden* to drop anchor. Two British ships, however, which were tracking the *Dresden*, approached and believed the ship to be making an escape. This led to the British firing on the *Dresden* and eventually sinking her in the bay, where she still rests today at a depth of 60 m (Woodward 1969). Most of the crew of the *Dresden*, however, was able to leave the ship before it was destroyed. A plaque in the cemetery of San Juan Bautista near Punta San Carlos, erected in 1922 by the German community in Valparaíso, commemorates the three German crew members who died in the incident. This monument was damaged, however, during the destructive tsunami that swept the lower portions of the village on February 27, 2010 (Arana 2010).

Awareness of Natural Resources

The early decades of the twentieth century saw continued economic development and social stability plus a growing awareness of the unique natural heritage that the islands possessed. The village continued to grow, with a few hundred persons taking up residence. In response once more for a place to house criminals, Alejandro Selkirk Island was this time used as a penal colony during 1909–30 (Fig. C2). Remains of stone walls from this period can still be seen today.

Most important in these decades were the field studies and publications of Federico Johow, an immigrant to continental Chile from Prussia, and Swedish botanist Carl Skottsberg, who visited the islands in 1906–7 and 1917–18. Johow (1896) summarized information on the flora of the archipelago in Spanish, building on the earlier studies by Hemsley (1884) from the Challenger Expedition (see Chapter 2). Although Skottsberg's publications were in English (1921, 1956) and German (1928), they served well to attract international attention to the amazing diversity of endemic plants.

In realization of the scientific importance of the endemic flora of the islands and in view of the enormous negative impacts perpetrated on the island's ecosystem over nearly 400 years, the archipelago was designated a Chilean National Park on January 16, 1935. Although this was an important step forward in beginning to protect the natural resources of the islands, the lack of enforcement measures resulted in little actual conservation achievements. Furthermore, it was difficult to contain the activities of the villagers, who in some cases were third- or fourth-generation islanders and regarded the islands as virtually their own. The Corporación Nacional Forestal (CONAF) of Chile, which administers the national parks (among other responsibilities), has in recent years become very active in conservation, and many positive steps have been taken. In addition, the archipelago has now been placed on the "tentative" list (an initial step) as a UNESCO World Heritage Site, which has strengthened its international importance.

Contemporary Village Life

The Juan Fernández Archipelago is one of the more isolated civilized locations in the world. Despite being a Chilean national park, the islands are not easy to reach. One can choose flying to the islands in a small five- to eight-seat propeller airplane that leaves from Santiago, or one can take a boat from Valparaíso. The flight lasts 2.5 hours, and the boat trip takes two days. In the small airplane, one lands on the western tip of Robinson Crusoe Island on a single narrow runway. From there one must take a jeep to the coast at Bahía Padre and then a 1.5-hour boat ride around to the eastern side of the island to Bahía Cumberland and the village of San Juan Bautista. The connections, therefore, are fun and filled with some adventure, but they are not designed for tourism with comfort and luxury. Hence many visitors tend to be young, energetic, and not particularly wealthy. To reach Alejandro Selkirk Island (Fig. C2), one must travel by boat for about thirteen hours from San Juan Bautista. There is no airplane connection between the two islands.

San Juan Bautista (Figs. 1.3, C3, and C4), the only permanent village in the islands, is small, with a current population of 885 persons (www.comunajuanfernandez.cl). There are very few streets, one running nearly parallel to the shape of the bay and another (La Pólvora) extending up the island toward the hills. Prior to the destructive tsunami of 2010, there was a municipal building, port captain's office, gymnasium, village plaza, pier, soccer field, several simple restaurants, guest houses, a school, small general stores and souvenir kiosks, a church, houses, a national park information center, and one discotheque (which was open only on Friday and Saturday nights). On February 27, 2010, however, a 3-m-tall tidal wave, originating from a submarine earthquake of level 8.8 near the Chilean coast (Farías et al. 2010), reached Robinson Crusoe Island and destroyed most of the village up to 60 m. Eleven persons perished, and six others were never found. All buildings mentioned earlier were destroyed, and all but the largest trees were also swept out to sea. This area of the island is being rebuilt in attractive style, but the older historical buildings close to the sea have been lost (see Fig. 3.7). Nonetheless, in any architectural configuration then or now, for the typical tourist there is really not much to see and do in the village. Most visitors, therefore, prefer to fly back to the continent after only a few days. For a plant biologist, however, the islands offer endless fascinations for exploration and investigation, as we hope this book successfully communicates.

Figure 1.3 View of the village of San Juan Bautista, Robinson Crusoe Island.

The present economy of the village depends mainly on lobster fishing (Figs. C5 and C6) and tourism. The lobsters have long been fished for sale to continental markets (especially restaurants, where a good lobster dinner can cost US$50). The fishermen of Robinson Crusoe Island belong to a cooperative (Hernández and Monleon 1975), where adequate-sized lobsters from each day's catch are kept in holding pens until transported by plane or boat to the mainland. With heavy fishing, however, the lobster population declined into the 1970s and 1980s (Yañez et al. 1985), and incomes diminished proportionately (Arana 1987). The good news is that with new regulations to promote sustainability, lobster fishing has now made a recovery (Ernst et al. 2013).

Tourism, and especially ecotourism, has now also become important in the economy of the islands. Because the islands are a national park, there are restrictions on land use by the resident population. More important, education in the primary school now has more emphasis on conservation, to prepare the next generation for increasing demands from ecologically minded tourists. A tension had existed between the villagers, some of whom can trace their families in the islands back several generations and who view the islands as their own, and the Chilean government with its conservation mandates. This tension is now dissipating due to the clear need for cooperation to stimulate ecotourism.

In the 1980s, prior to arrival of telephones and televisions, the small village of San Juan Bautista was really isolated from the mainland. There were only two cars in the village because with only two short roads there was little need for automobiles. People walked everywhere and chatted frequently when encountering a friend in passing.

Telephones now make it possible to call from house to house, and this began to change the rhythm of the village. Multichannel television from satellites (now also available on Alejandro Selkirk Island) offered broader entertainment for families at home, which also changed people's social habits. Telephone and Internet connections to the outside world further lessened the villagers' national and international isolation.

Nonetheless, despite modernization, the people of San Juan Bautista continue to be warm and welcoming. They are proud of their island, which is in many ways a modern paradise, where crime is virtually nonexistent, every family can have a house and garden, community support is strong and sincere, and there is a delightful subtropical climate with some of the clearest and most star-strewn night skies on the planet. We were completely captivated by the islands, and this is one reason why we have never hesitated to return to continue our research.

2 Botanists in the Islands

Tod F. Stuessy and Clodomiro Marticorena

To allow an appreciation of the evolutionary data and interpretations presented in this book, we provide in this chapter an outline of the botanical expeditions and collecting that have taken place over the past three centuries. The main objective is to indicate that the floristic inventory of the Juan Fernández Islands is strong. In our own investigations over the past three decades, we have discovered only two new species: *Erigeron stuessyi* (Valdebenito; see Appendix 2) and *Gleichenia lepidota* (Rodríguez-Ríos 1990). There have been only a few additions by other workers, for example, *Robinsonia saxatilis* (Danton 2006a) and *Carex stuessyi* (Wheeler 2007). Few taxonomic surprises remain. This was realized at the outset when we began our investigations in the early 1980s and, in fact, was one of the reasons for focusing on the archipelago, because it allowed a more certain and comprehensive approach to understanding the evolution of the flora. Numerous botanists have collected in the islands (Table 2.1 and Fig. 2.1), but we discuss here only those who have contributed the most to our floristic understanding. Two previously published histories containing much more detail can be found in Johow (1896) and Skottsberg (1953a).

The first recorded collections from the Juan Fernández Archipelago were made by George Handisyd in October of 1690 (Middleton, 1909; Dandy, 1958; Gunckel L., 1971). He served as surgeon aboard the ship Welfare (Gunckel L., 1971) under Captain John Strong. Middleton (1909) implied that the name of the ship was the Modena, but this was apparently the name of a later ship on which Handisyd also sailed (Gunckel L., 1971). The stated objective of the voyage was to trade along the Spanish coast of southern South America, especially on the Chilean side, but the Spanish colonies were suspicious and repulsed their efforts to harbor near cities. Despite this absence of trade and safe harboring, Handisyd was able to go ashore and collect 48 plant specimens (Gunckel L., 1957). The ship also visited Robinson Crusoe Island, and Handisyd obtained two species of ferns (without precise locality): *Adiantum chilense* and *Asplenium dareoides*, both common in the island. At the end of the voyage, these specimens were given to John (Hans) Sloane, who incorporated them in vol. 8 (H.S. 8) of his herbarium, which eventually was donated to the British Museum (Dandy, 1958). Gunckel L. (1971) suggested that a collection by Handisyd of *Solanum* might have come from the Juan Fernández islands, but it may also have been collected on Isla Mocha near the continent or even along the Strait of Magellan. In any case, this specimen appears to be the cultivated *Solanum tuberosum*.

Table 2.1 Principal Botanical Collectors in the Juan Fernández Archipelago

		Island visited	
Year(s) visited	Collector(s)	RC	AS
1690	George Handisyd	X	
1823	Mary Graham (with help from Thomas Cochrane and Mr. Shepard)	X	
1824	David Douglas (with Dr. John Scouler)	X	X
1830	Carlo Bertero	X	
1830	Captain Phillip Parker King	X	
1830	Hugh Cuming	X	X
1832	Claudio Gay	X	X
1851	Thomas Bridges	X	X
1852	Filiberto Germain	X	X
1864	Rodulfo Amando Philippi	X	
1869, 1872	Edwyn C. Reed (with Roberto McSporrman and José Guajardo)	X	
1869	Agustín Guajardo	X	X
1872	Jorge Downton	X	X
1875	Henry Nottidge Moseley (the Challenger expedition)	X	X
1890	Federico Delfin	X	
1891–2, 1895	Federico Johow (with A. Guajardo and F. Germain)	X	X
1901	George Tracy Hastings	X	
1908, 1916–17, 1954–5	Carl Skottsberg (with Inga Skottsberg, Günther Kunkel, Benkt Sparre, and local assistants)	X	X
1922	Otto Tenz	X	
1925, 1927	Gualterio Looser	X	
1938	Thomas Harper Goodspeed (and expedition members)	X	X
1941	P. Aravena	X	X
1942	Edmundo Pisano and Patricio Montaldo	X	
1965	Chilean–US expedition (Otto Solbrig and Carlos Muñoz Pizarro plus many others)	X	X
1980–2011	The Ohio State University (University of Vienna)–Universidad de Concepción expeditions (with many collaborators; see Table 2.2)	X	X
1998–2013	Philippe Danton (with Christophe Perrier and other collaborators; see Table 2.3)	X	X

Note: RC = Robinson Crusoe Island; AS = Alejandro Selkirk Island.
Sources: Based primarily on Johow (1896), Middleton (1909), Looser (1927), Skottsberg (1953, 1958), Gunckel L. (1971), and Martínez (1983).

The second person to collect specimens in the Juan Fernández Archipelago was a woman, Mary Graham (Fig. 2.1A), who published her journal in 1824. At that time, she was the wife of Thomas Graham, a captain in the British Navy, who was to be stationed in the Pacific. He died in her arms in 1822, however, on the voyage around Cape Horn just as they were arriving for him to assume his new post. After this wrenching life trauma, and while waiting for safe passage back to England, Mary

Figure 2.1 Botanists who have collected plants and contributed to floristic understanding in the Juan Fernández Islands: (A) Mary Graham; (B) Carlo Bertero; (C) Claudio Gay; (D) Rodolfo Philippi; (E) Henry Moseley; (F) Federico Johow; (G) Carl Skottsberg; (H) Otto Solbrig; (I) Carlos Muñoz Pizarro. For credits, see Acknowledgments.

Graham renewed her friendship with Thomas Cochrane, who had been hired in early 1818 as Commander of the Chilean Navy in the newly independent country to aid in the transport of troops from Chile to Perú for the invasion of this last royalist stronghold. He accepted this seemingly odd position due to being embroiled in a stock scandal in

England (Cordingly 2007), which made it convenient for him to leave, seeking a new life elsewhere. In 1822, Cochrane had returned from the Peruvian campaign, frustrated by not having been allowed to attack the mainland directly, due to caution from José de San Martín, who wanted the Peruvians to take up independence rather than having it forced on them (Lynch 2009). Because Chile had made the decision to disband its navy, Brazil tendered an offer to Cochrane for him to take command of their new Navy to help gain independence from Portugal. Before Cochrane departed for Brazil, he first sailed on January 18, 1823, in the ship *Colonel Allen* to Alejandro Selkirk (= Masafuera) and Robinson Crusoe (= Masatierra) islands, arriving at the latter on January 22. Mary Graham went ashore on January 25 and 26, and on the 26th made plant collections. These included some interesting endemics, such as *Gunnera peltata* (Gunneraceae) and *Rhaphithamnus venustus* (Verbenaceae), among others (Skottsberg 1953a). The specimens were initially collected by Cochrane and one of the British officers, Mr. Shepard, who brought them back to her in the village. It is likely that the men climbed to the Portezuelo or Cordón Central and back. The entire party also later in the day visited El Pangal, the location of Anson's former campsite in 1740, filled with European fruit trees at the time of his visit. Mary Graham took the specimens under her care, and they are now housed in the herbarium at Kew.

Although botanist David Douglas (with Dr. Scouler) visited Robinson Crusoe Island for three days in 1824 (Mitchell and House 1999), collecting 70 "distinct and exceedingly interesting plants" (Douglas 1914, p. 54), presumably each representing separate species (with specimens now at Kew), a more significant contribution was made by Carlo Guiseppi Bertero (Fig. 2.1B), who stayed on Robinson Crusoe Island for several months in 1830. Originally from Santa Vittoria d'Alba in Piamonte, Italy, Bertero moved to Chile in 1827. Although a physician, he was also an enthusiastic botanist and collected specimens whenever possible, sending material to Europe and publishing articles on his discoveries (Vignolo-Lutati 1955). At the beginning of 1830, in part to leave behind the civil unrest in the young country of Chile, Bertero traveled to the islands and amassed a collection of about 2,000 specimens (Hemsley 1884) containing 300 species, including cryptogams. On return to Valparaíso in the austral winter of 1830, Bertero dispatched a report of his trip (1830) and his specimens to Europe to Delessert in Paris (Delprete et al. 2002), who distributed duplicate sets to other institutions. The large remainder of some 15,000 specimens were eventually auctioned and purchased by the Botanische Reiseverein (Wörz 2007), which sold duplicates to Berlin, Kew, Leiden, Paris, and elsewhere, and these are still easily encountered today. He then sailed off to Tahiti, where he also collected. After this successful botanical adventure, he took a Tahitian boat back to Valparaíso, but it completely disappeared, with all crew and passengers presumed dead.

The next important collector in the Juan Fernández Archipelago was Claudio Gay (Fig. 2.1C). Arriving from Paris in 1828 to teach as professor in a new secondary school in Santiago (Colegio de Santiago), his real interest was to explore and publish on the natural history of this newly independent country. This was the age of world scientific

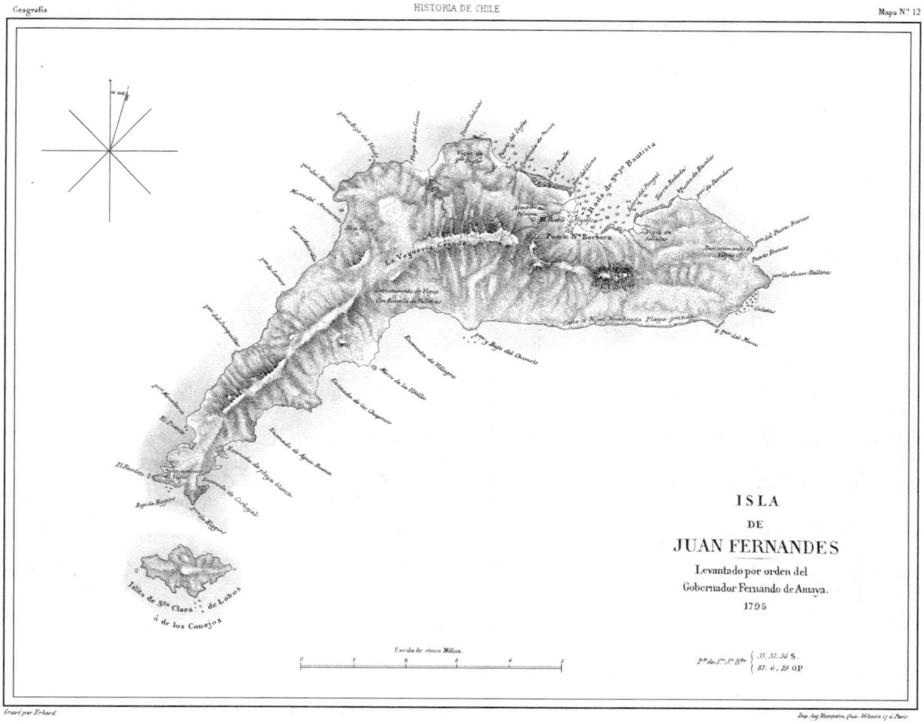

Figure 2.2 Map of Robinson Crusoe Island by Claudio Gay (1854).

exploration, and many young European scientists felt a need to travel, explore, and create reputations for themselves through their accomplishments in natural history (Fuenzalida 1944). Gay traveled widely throughout Chile (Muñoz Pizarro 1944), visiting Robinson Crusoe Island for three weeks in February of 1832 and making numerous collections. On returning to Valparaíso, Gay bundled up most of his specimens, plus others he had collected in Chile, and returned with them to Paris, where he began working on his notes and collections. He returned to Chile in 1834 for more collecting and studies on the continent until 1842, when he returned again to Paris. He soon thereafter began publishing the *Flora Chilena*, which when finished comprised eight volumes (1845–54) of the *Historia Física y Política de Chile*. He also included a useful map of Robinson Crusoe (Masatierra) Island in the *Atlas* (Gay 1854) (Fig. 2.2).

In 1864, another major contributor to understanding the flora of Chile, Rodulfo Amando Philippi (Fig. 2.1D), visited Robinson Crusoe Island (with Antonio Ahrends), but they stayed only four days. Despite this short time, their collections were profitable, resulting in descriptions of six new species. The most interesting was *Lactoris fernandeziana*, sole representative of the ancient family Lactoridaceae (see Chapters 9 and 13 for more comments on this unusual species). Philippi (1856a) was the first to document the occurrence of the noxious invasive species *Aristotelia chilensis* ("maqui") and *Acaena argentea* ("trun"). Species from this short trip were mostly deposited in the Museo de Historia Natural in Santiago (established by Claudio Gay), where Philippi was director. An excellent and very helpful list of the specimens described by him in the nineteenth century, including specimens and types deposited at the Natural History Museum in Santiago (SGO), was compiled by Muñoz Pizarro (1960). Because of the importance of Rodulfo Philippi and his botanist son, Federico, to advancing understanding of the plants of Chile, there are several good biographical studies (Barros Arana 1904; Gotschlich 1904; Fürstenberg 1906; Taylor and Muñoz-Schick 1994).

Henry Nottidge Moseley (Fig. 2.1E) made significant contributions to our knowledge of the flora of the Juan Fernández Islands even though he stayed only three days on Robinson Crusoe Island from November 13–15, 1875. The objective of the voyage, aboard the British ship *H.M.S. Challenger*, was mainly scientific, with a focus on the Atlantic, Pacific, and Indian Oceans (Spry 1877; Linklater 1972). Moseley collected enthusiastically in the small amount of time at his disposal, taking a special interest in the ferns (Moseley 1892), including the endemic *Thyrsopteris elegans*. His specimens contained 105 species and were finally deposited at Kew. Hemsley (1884) wrote up the scientific results, which represented the first comprehensive survey of the entire flora of the islands (focusing, however, on Robinson Crusoe Island). He also categorized the species as endemic or not and attempted to understand their biogeographical affinities. This was the basis for all subsequent floristic studies, especially those of Johow (1896) and Skottsberg (1921, 1953b, 1956).

Federico Richard Adalbert Johow (Fig. 2.1F), originally from Bonn, Germany, and immigrating to Chile in 1889, was a professor at the Instituto Pedagógico (now Escuela Normal) in Santiago. He visited the islands in 1891–2 and again in 1895 for a total of 2.5 months (Skottsberg 1953a). Johow greatly extended knowledge of the flora of the archipelago, presenting the taxa in more or less the same order as in Hemsley (1884) and supplementing the descriptive information from that earlier work. Johow's book is not a comprehensive flora of the islands, but he did a fine job of summarizing available data and added information on the history of botanical collection, previously published works, analyses of the flora, and descriptions and classifications of the vegetation. There is also a series of recommendations to the Chilean government (Johow 1896, pp. 267–74), including a plea for conservation of the spectacular flora. Johow also published separately (1893) a list with discussion of 57 cultivated species he saw on Robinson Crusoe Island, some of which represented danger to the native flora. His collections are scattered in different herbaria, but most are on deposit in the Natural History Museum in Santiago (SGO).

Carl Johan Fredrik Skottsberg (Fig. 2.1G) visited the Juan Fernández Archipelago on three occasions: (1) for six days in 1908 (four days on Robinson Crusoe Island and two days on Alejandro Selkirk Island), (2) for almost five months from December 3, 1916 to April 30, 1917, and (3) for several months in 1954–5. His initial experience in the southern hemisphere was as a student on the Swedish Antarctic Expedition of 1901–3 under the leadership of explorer Otto Nordenskjöld (Peterson 1964), but they did not stop in the Juan Fernández Islands. He continued studying for a Ph.D. at Uppsala University and graduated in 1907, also being appointed lecturer in that same year

(Salisbury 1964). Under the sponsorship of Uppsala University, he led the Swedish Magellanic Expedition to South America in 1907–9 (Peterson 1964), which had a profound impact on his subsequent research directions. On this expedition, although Skottsberg spent only a very limited time in the Juan Fernández Islands, he published several papers, including an account (1910) of the endemic sandalwood (only one living plant remained). His specimens from the voyage are on deposit at the University of Uppsala (UPS).

Skottsberg's second expedition to the islands, for nearly five months in 1916–17, was the principal one for making the many collections that allowed his research and that of his collaborators to be carried out successfully. The timing is interesting because this was in the middle of World War I, which might have been one good reason to get away from the northern hemisphere. This expedition, completed with his wife Inga, resulted in all sorts of natural history collections, rocks, minerals, plants, sea animals, marine and freshwater algae, birds, insects, and analyses of the vegetation. These specimens were deposited in the Göteborg herbarium (GB). For his full description of the expedition, see Skottsberg (1918; unfortunately still available only in Swedish). Research was published in parts (mostly in English and German) beginning in 1920. All articles were distributed together in 1956 (and bound up by libraries) under the title *Natural History of the Juan Fernandez and Easter Island*. This has caused some bibliographical difficulties in that the parts (chapters) were mostly all published as separate pieces prior to 1956 (actual date of printing is given at the end of most chapters).

Skottsberg conducted his third and last expedition to the Robinson Crusoe Islands for several months in the austral summer of 1954–5, completed at the age of 74 (he lived to be 83). Considering the travel challenges of the day, this was a remarkable achievement. Despite his advancing age, he still managed to publish several results from this trip (e.g., Skottsberg 1958).

Skottsberg made major contributions to understanding the natural history of the archipelago. First, he organized a series of collaborators who wrote articles on different aspects of the islands, such as the Myxomycetes (Fries 1920), freshwater algae (Ström 1921), Gasteromycetes (Fries 1921), mosses (Brotherus 1924), lichens (Zahlbruckner 1924), Basidiomycetes (Romell 1926), Ascomycetes (Keissler 1927), thallose Hepaticae (Evans 1930), the genus *Codium* (marine green algal group) (Setchell 1937), marine algae (Levring 1941), leafy liverworts (Herzog 1942), and geology (Quensel 1954), and 61 articles by many authors on zoology (all of vol. 3, 1921–1940).

Second, Skottsberg provided a much more detailed analysis of the vegetation of the flora of the islands (1953a) in comparison with that provided earlier by Johow (1896). Skottsberg used the new (at that time) Braun-Blanquet (1928) method of making "relevés." This involved assessing not only the species associated with each locality but also quantitative estimates of cover, density, and other factors. Each locality was precisely pinpointed, which has allowed recent comparisons of the composition of the flora from the exact same localities (Sanders et al. 1982; Greimler et al. 2002, 2013). Skottsberg (1941) also described the marine algal communities.

Third, Skottsberg added more floristic information (1921, 1953b; Christensen and Skottsberg 1920) to our knowledge of the species, especially the endemic taxa, extending that of Hemsley (1884) and Johow (1896). Skottsberg did not write a complete flora, however. He described new species, wrote keys to some of the larger genera, offered taxonomic opinions and insights, and discussed some nomenclatural aspects, but he never published a comprehensive floristic treatment. This may be due to his extremely broad biological interests, especially reproductive biology, plant organ development, and conservation.

Fourth, Skottsberg made major contributions to understanding the biogeography of the endemic and native flora (1956). Because he was so familiar with the flora of southern South America, plus that of other Pacific islands, particularly Hawaii, it was possible for him to offer hypotheses on relationships and biogeographical connections. He also provided details on pollination and dispersal biology (1928), of fundamental importance for understanding biogeographical and evolutionary patterns.

In recent decades there has been a shift in the nature of expeditions to the Juan Fernández Islands in that science has become more cooperative, even as regards field expeditions. Ease of international travel and increased global wealth have encouraged a change to groups of specialists rather than lone adventurers. This parallels similar changes in laboratory science in biology that have taken place during this same period.

In 1965, a bilateral botanical expedition to the Robinson Crusoe Archipelago was organized by Otto T. Solbrig (Fig. 2.1H) from Harvard University and Carlos Muñoz Pizarro (Fig. 2.1I) of the Museo de Historia Natural in Santiago, Chile (Meyer 1966). This consisted of quite a large group of four Chilean and eight US scientists. The Chilean participants, all from the Museo Natural de Historia Natural, were Carlos Muñoz Pizarro (vascular plant systematist), Nibaldo Bahamonde N. (marine biologist), Fernando Saravia B. (forester), and Eugenio Sierra R. (botanical illustrator). On the US side were Otto T. Solbrig and his assistant James Walker (flowering plants), Harvard University; Raymond Hatcher and his assistant John Engel (bryophytes), University of Wisconsin, Milwaukee; Henry A. Imshaug (lichens and fungi) with assistant Dennis Jackson (who also collected freshwater algae), Michigan State University; Frederick G. Meyer (ferns and introduced plants), US National Arboretum; and Harold E. Moore, Jr. (palms), Bailey Hortorium, Cornell University. The expedition lasted 32 days, from November 26 to December 27 (November 26–December 6, Alejandro Selkirk; December 7–27, Robinson Crusoe). Many valuable collections were made, including 228 samples of freshwater algae, 800 bryophytes, 150 ferns, 670 flowering plants, and 100 wood samples. Few articles were actually published from the expedition, but the studies by Moore and colleagues on the endemic palm *Juania australis* deserve particular mention (Moore 1969; Tomlinson 1969; Uhl 1969). The collections are on deposit at the institutions from which the expedition members came.

The largest series of expeditions to the Robinson Crusoe Islands have been those involving the authors of this book and their collaborators. Beginning in 1980, this has been a continuous research program on the evolution of the flora of the archipelago that has included twelve expeditions (Table 2.2 and Fig. 2.3). The main institutional cooperation of the research program has been between the Universidad de Concepción, Chile, and The Ohio State University, Columbus, OH, USA. From 1997 to 2013, the main collaboration was between Concepción and the University of Vienna, Austria. The focus

Table 2.2 Personnel Associated with the Ohio State–Concepción and Vienna–Concepción Expeditions to the Juan Fernández Archipelago, 1980–2011

Personnel	Institution	Specialty
1980: January 27–February 21 (5,000–5,249)		
Jorge Arriagada	U. Concepción	Flowering plants
Clodomiro Marticorena	U. Concepción	Flowering plants
Oscar Parra	U. Concepción	Algae
Roberto Rodríguez	U. Concepción	Ferns
Roger Sanders	Ohio State U.	Flowering plants
Tod Stuessy	Ohio State U.	Flowering plants
Eduardo Ugarte	U. Concepción	Ecology
1980: November 18–November 30 (5,300–5,524)		
Oscar Matthei	U. Concepción	Monocots (grasses)
Roger Sanders	Ohio State U.	Flowering plants
Tod Stuessy	Ohio State U.	Flowering plants
Hugo Valdebenito	U. Concepción	Flowering plants
1984: January 15–February 13 (6,200–6,680)		
Daniel Crawford	Ohio State U.	Flavonoids
Alejandro Landero	U. Concepción	Flowering plants
Patricia Pacheco	Ohio State U.	Flavonoids
Eduardo Ruiz	U. Concepción	Flavonoids
Tod Stuessy	Ohio State U.	Flowering plants
Hugo Valdebenito	Ohio State U.	Flavonoids
1986: January 16–February 14 (8,000–8,368; 8,380–8,518; 9,000–9,665)		
Michael Doyle	Rancho Santa Ana Bot. Gard.	Cryptogams
Leonardo Gaete	U. Concepción	Flowering plants
Thomas Lammers	Ohio State U.	Flowering plants
Alejandro Landero	U. Concepción	Flowering plants
Eduardo Ruiz	U. Concepción	Flavonoids
Jaime Sepúlveda	U. Concepción	Medical student
Tod Stuessy	Ohio State U.	Flowering plants
Hugo Valdebenito	Ohio State U.	Flowering plants
1990: January 19–February 12 (11,030–11,723)		
Carlos Baeza	U. Concepción	Monocots
Daniel Crawford	Ohio State U.	Isoyzmes
Ana María Humaña	U. Valdivia	Reproductive biology
Patricio López	U. Concepción	Flowering plants
Patricio Peñailillo	U. Concepción	Flowering plants
Mauricio Rondanelli	U. Concepción	Flowering plants
Patricia Stuessy	Ohio State U.	Assistant
Tod Stuessy	Ohio State U.	Flowering plants
Delbert Wiens	U. Utah	*Lactoris* reprod. biol.

Table 2.2 (cont.)

Personnel	Institution	Specialty
1991: January 13–February 3		
(11,728–12,124)		
Gregory Anderson	U. Connecticut	Reproductive biology
Daniel Crawford	Ohio State U.	Isozymes
Patricio López	U. Concepción	Flowering plants
Richard Roederer	Ohio State U.	Assistant
José Soto	U. Concepción	Isozymes
Tod Stuessy	Ohio State U.	Flowering plants
1996: January 10–January 25		
(15,000–15,196)		
Gregory Anderson	U. Connecticut	Reproductive biology
Pedro Aqueveque	U. Concepción	Isozymes
Carlos Baeza	U. Concepción	Monocots
Gabriel Bernardello	U. Cordoba (Argentina)	Reproductive biolgy
Daniel Crawford	Ohio State U.	Isozymes
Héctor Ibarra	U. Concepción	Zoology
Eduardo Ruiz	U. Concepción	Isozymes
Tod Stuessy	Los Angeles Nat. Hist. Mus.	Flowering plants
Ulf Swenson	Los Angeles Nat. Hist. Mus.	Invasive plants
Eric Tepe	Ohio State U.	Isoyzmes
1997: January 11–January 26		
(15,197–15,407)		
Gregory Anderson	U. Connecticut	Reproductive biology
Pedro Aqueveque	U. Concepción	Isozymes, DNA
Marcelo Baeza	U. Concepción	Monocots
Gabriel Bernardello	U. Cordoba (Argentina)	Reproductive biology
Daniel Crawford	Ohio State U.	Isozymes, DNA
Fidelina González	U. Concepción	Isozymes
Gabriele Kottirsch	U. Concepción	Assistant
Patricio López	U. Concepción	Flowering plants
Eduardo Ruiz	U. Concepción	Isozymes, DNA
Tod Stuessy	Los Angeles Nat. Hist. Mus.	Flowering plants
1999: February 1–17		
(No collections; studies of vegetation only)		
Josef Greimler	U. Vienna	Vegetation
Patricio López	U. Concepción	Vegetation
Alan Stuessy	U. Vienna	Assistant
Tod Stuessy	U. Vienna	Flowering plants
2000: February 3–18		
(No collections; studies of vegetation only)		
Josef Greimler	U. Vienna	Vegetation
Patricio López	U. Concepción	Vegetation
Tod Stuessy	U. Vienna	Flowering plants

Table 2.2 (cont.)

Personnel	Institution	Specialty
2010: February 4–24		
(19,100–19,356)		
Daniel Crawford	U. Kansas	Flowering plants
Josef Greimler	U. Vienna	Vegetation
Luis Letelier	U. Talca	Flowering plants
Patricio López	U. Vienna	Flowering plants
Patricio Peñailillo	U. Talca	Flowering plants
Tod Stuessy	U. Vienna	Flowering plants
2011: January 26 –February 24		
(19,401–19,444; 19,603–19,683; 19,800–19,847; 20,000–20,049)		
Carlos Baeza	U. Concepcion	Flowering plants
Alejandro Gatica	U. Talca	Flowering plants
Josef Greimler	U. Vienna	Vegetation
Patricio López	U. Vienna	Flowering plants
Patricio Novoa	Jard. Bot. Nac. (Chile)	Flowering plants
Patricio Peñailillo	U. Talca	Flowering plants
Eduardo Ruiz	U. Concepcion	Flowering plants
Tod Stuessy	U. Vienna	Flowering plants

Note: Numbers in parentheses refer to expedition collection numbers, mostly under Stuessy et al. but sometimes made by other combinations of personnel under the same number series. Total of thirty-nine participants from twelve different institutions; 4,065 collection numbers.

of these expeditions has been on sampling aspects of the flora and vegetation to answer specific evolutionary and biogeographical questions. All told, more than 4,000 collections have been made, many with duplicates. Most of the specimens are on deposit at The Ohio State University (OS) and the University of Vienna (WU), with a full duplicate set at the Universidad de Concepción (CONC). A spirit collection of hundreds of samples (WU, now transferred to W) and a dry wood collection (OS) also exist. More than 60 journal articles have been published from these investigations, including aspects of population divergence, speciation, hybridization, adaptation, biogeography, dispersal, pollination, reproductive biology, vegetation, and invasive species. These are the studies that are now summarized in this book.

More recently, enthusiastic investigations have been carried out on the natural history, flora, and conservation of the plants of the Robinson Crusoe Islands by Philippe Danton and Christophe Perrier from Grenoble, France, with collaborators (Table 2.3). These investigations have led to several very useful publications with excellent drawings and photographs. The first dealt with general aspects of natural history and the cultural life of the islanders (Danton et al. 1999). This was followed by another beautifully illustrated book entitled, *Wild Plants of Robinson Crusoe Island: Identification Guide* (Danton 2004), which provided descriptions, photographs, and illustrations of 45 species. Danton et al. (2006) also published a new catalogue of the flora. More recently, and most important, Danton and Perrier (2017)

Figure 2.3 Members of OSU-CONC expeditions (all left to right): (A) Jan–Feb 1980: Clodomiro Marticorena, Roberto Rodríguez, Oscar Chamorro, Oscar Parra, Alvis González, Roger Sanders, Jorge Arriagada, and Eduardo Ugarte; (B) Jan–Feb 1984: Marcelo González, Patricia Pacheco, Daniel Crawford, Miguel García, Gastón González, Eduardo Ruiz, Hugo Valdebenito, and Alejandro Landero; (C) Jan–Feb 1986 (with Explorers Club flag): Alejandro Landero, Bernardo López, Aldo Recabarren, Leonardo Gaete, Jaime Sepulveda, Jose López, Hugo Valdebenito, and Michael Doyle; (D) Jan–Feb 1990: Patricio López, Patricio Peñailillo, Daniel Crawford, Mauricio Rondanelli, Ana María Humaña, Carlos (Marcelo) Baeza, Tod Stuessy, and Delbert Wiens; (E) January 1996, back row: Gabriel Bernardello, Daniel Crawford, Tod Stuessy, Héctor Ibarra, Gregory Anderson; front row: Ulf Swenson, Eric Tepe, Carlos (Marcelo) Baeza, and Pedro Aqueveque; (F) January 1997: Daniel Crawford, Patricio López, Eduardo Ruiz, Tod Stuessy, Pedro Aqueveque, Gregory Anderson, Gabriel Bernardello, Fidelina González, and Carlos (Marcelo) Baeza.

Table 2.3 List of Botanical Expeditions to the Juan Fernández Archipelago (1997–2013) organized by Philippe Danton, with Christophe Perrier and Collaborators

November 16, 1997–February 10, 1998
(Filming of a documentary for Gaia, Production AED, by Vincent Tardieu)
Michel Baffray
Emmanuel Breteau
Philippe Danton
Jean-Yves Lesouef (National Botanical Conservatory of Brest, CBNB, France)
Franklin Picard (Conservatory of the Specialized Plant Collections, CCVS, France)
November 23, 1998–February 21, 1999
(Filming of reports on the three islands, Chiloe Productions, by William Leroux)
Michel Baffray
Emmanuel Breteau
Philippe Danton
December 14, 2000–March 14, 2001
Philippe Danton
November 8, 2001–January 30, 2002
Philippe Danton
February 14–March 2, 2002
(A group from the Botanical Society of France, SBF)
Michel Boudrie (pteridologist, Guiana and France)
Michel Cambornac (botanist, Soc. Yves Rocher, France)
Philippe Danton
Thierry Delahaye (botanist, Savoy)
René Delépine (phycologist, France)
Bruno de Foucault (phytosociologist, France)
Christiane Gardoux (botanist, France)
Claude Pépin (botanist, Savoy)
Romaric Pierrel (Jard. Bot. de Nancy, France)
Louis Zeltner (botanist, Switzerland)
December 10, 2002–March 11, 2003
Jan Bannister (student, Chile)
Philippe Danton
Francis Hallé (Professor University of Montpellier, France)
Christophe Perrier
Rodrigo Vargas (student, Chile)
November 5, 2003–February 4, 2004
(Filming of a documentary for *Tierra Adentro*, by Paul Landon)
Philippe Danton
Christophe Perrier
December 31, 2004–April 1, 2005
Philippe Danton
Guido Martínez (student, Chile)
Christophe Perrier
September 29–December 29, 2005
(Filming of documentaries for *Ushuaïa Nature*, by Gilles Santantonio, with Nicolas Hulot, and channel *Voyage*, by Marc Mopty)
Philippe Danton
Stéphanie Marcellin
Christophe Perrier

Table 2.3 (cont.)

January 3–April 2. 2008
Philippe Danton
Albert Reif (Prof. University of Freiburg, Germany)
Rodrigo Vargas (student, Freiburg, Germany)
January 7–April 7, 2009
(Filming of a documentary *Faut pas rêver*, by Malick Tialba, with Laurent Bignolas)
Philippe Danton
Cécile Georget (student, France)
Christophe Perrier
May 6–9, 2013
(A group for the Proyecto GEF 83266)
Fernando Baeriswyl (National Coordinator of the Proyecto GEF 83266)
Stephane Buord (National Botanical Conservatory of Brest, CBNB, France)
Hector F. Correa Cepeda (National Botanical Garden, Viña del Mar, Chile)
Philippe Danton (ROBINSONIA Association)
Pedro León Lobos (INIA, Chile)
Miguel Stutzin (Ministerio del Medio Ambiente de Chile)

are in the process of publishing a complete flora of the vascular plants of the archipelago, which is the first complete flora ever presented. This will be particularly useful for attracting attention to conservation of the native and endemic plants as well as offering cautions regarding dangers from invasive species. These descriptive publications on the islands and their flora make excellent companions to the evolutionary and biogeographical studies summarized in this book.

As a general perspective, at the present time, a reasonably good understanding of the plants of the Juan Fernández Islands exists. The flora is now documented, and this book summarizes perspectives on the origin and evolution of the native and endemic plants. The archipelago, because of its geographical and geological simplicity, is ideal for asking and attempting to answer evolutionary and biogeographical questions (Stuessy et al. 2005b). A clear view of the conservation needs for preservation of the flora has also been documented (Biodiversa 2009a, 2009b; see also Chapter 9). What remains to be achieved now is the hard work of finding resources to allow conservation efforts in the archipelago to be successful.

Part II

Physical Setting

Evolutionary and biogeographical phenomena always take place in some specific area of the world, and in the case of the Juan Fernández Archipelago, this means on the two major islands, Robinson Crusoe and Alejandro Selkirk. To understand the environmental context in which immigrant plants arrive, establish, and diverge requires having an appreciation of the physical landscape. It is important not only to reveal the present physical characteristics of both islands but also to attempt to reconstruct their geological history over time. It is only with this information that we can more precisely understand the biogeographical and evolutionary processes that have transpired in the archipelago.

Chapter 3 sketches the geology and soils of the islands. Here we make the important point, so often repeated in other chapters of this book, that the islands are of different geological ages, Robinson Crusoe Island being approximately 4 million years old and Alejandro Selkirk Island being 1 to 2 million years old. Immigrants to the archipelago during the first half of its existence, therefore, could only have become established on the older island. This also happens to be the island closest to the South American continent, which is the major source for propagules to the archipelago. Also most important is that the islands must have been larger at their formation, especially Robinson Crusoe Island, and what is seen now is much reduced from its original size. This reduction would have substantially affected surface area, environmental conditions, and population and specific diversity.

Chapter 4 provides data describing the climate of both islands. The location of the archipelago at 33°S latitude in the southeastern portion of the Pacific, westward from the north-flowing, cool Humboldt current, provides conditions for the development of subtropical vegetation. Daytime temperatures fluctuate from 11 to 20°C during the year, and nighttime freezing temperatures are extremely rare. Precipitation varies currently from 550 to 1,650 mm/year, but if the geological reconstructions of island ontogeny are accurate, especially for Robinson Crusoe Island, the elevational profile would initially have been much higher, with significant impact on patterns of rainfall. As for soils, they derive from volcanic lava and ash, both of which erode relatively quickly, one of the factors leading to loss of surface area over time. Basaltic dikes underpin the major ridges of the islands, which are more resistant to erosional forces and serve today to give structure to the landscape, especially on the more weathered Robinson Crusoe Island.

3 Geology and Soils

Walter A. Sontag, Jr., and Tod F. Stuessy

To understand processes and patterns of evolution in any area of the world requires having a clear context of place and time. Oceanic islands have the great advantage in that they are delimited spatially on all sides by water, and in this way, they are more clearly defined than most continental regions. Nevertheless, to interpret evolutionary events, it is essential to obtain more details about any island system. It is important to know the age of an island (or archipelago); what minerals, rocks, and soils it contains; and how it might have become modified through time. All these components have influenced evolution of the native and endemic flora. This chapter focuses on the age of the islands, the nature of volcanic activity that occurred to form them, and the substrates (minerals, rocks, and soils) that were deposited and developed further. Also very significant is the pattern of erosion and subsidence of the islands during their geological history.

Geology

To set the geological context, the Juan Fernández Archipelago is located between 667 and 848 km west of continental Chile in the southeastern Pacific Ocean on the Nazca Plate (Devey et al. 2000) (Fig. 3.1). It comprises two large islands of comparable size, Robinson Crusoe (= Masatierra, 33°37'S, 78°50'W) and Alejandro Selkirk (= Masafuera, 33°45'S, 80°46'W), which are 48 and 50 km², respectively (Stuessy 1995). There is also a very small island, Santa Clara (33°42'S, 79°01'W) (Baker et al. 1987; Castilla and Oliva 1987), which is separated southwest approximately 1 km off the coast of Robinson Crusoe by a shallow strait. Santa Clara encompasses an area of merely 2.2 km². The islands are mainly characterized by assemblages of valleys, ridges, ravines, and gorges (i.e., quebradas) described in considerable detail by Skottsberg (1953a, 1954). The weathered Robinson Crusoe Island is dominated by its highest peak, El Yunque, rising to 915 m, whereas the dome-like Alejandro Selkirk Island has a much higher summit, Los Inocentes, reaching to 1,319 m (Baker et al. 1987; Stuessy 1995). Santa Clara is only 350 m high (Stuessy 1995).

Age of the Islands

The geological ages of the islands of the archipelago, based on their origins through volcanic activity, have been determined radiometrically in several studies using

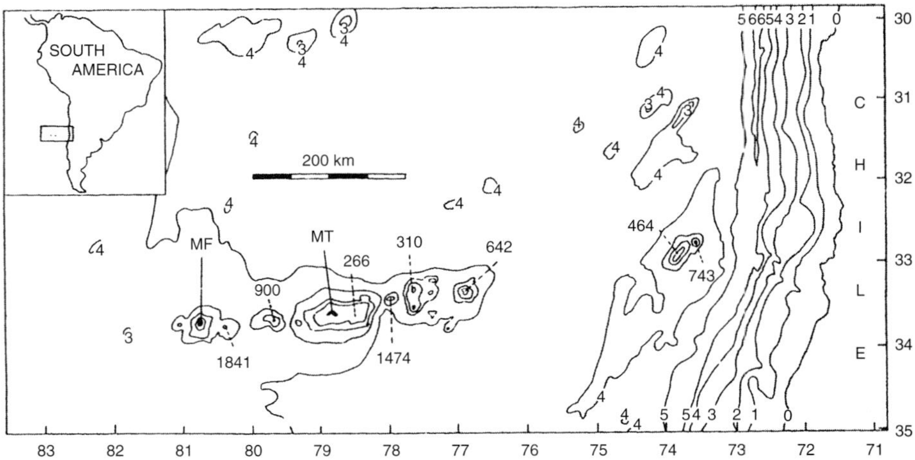

Figure 3.1 Bathymetric map of the South Pacific Ocean surrounding the Juan Fernández Islands. Single-digit depths given in thousands of meters. MF, Masafuera (= Alejandro Selkirk); MT, Masatierra (= Robinson Crusoe). (Data from Mammerickx and Smith 1978 and Prince et al. 1980. From Stuessy et al. 1984.)

potassium-argon dating techniques. Booker et al. (1967) provided ages from several samples in the range of 3.1 to 3.5 Ma for Robinson Crusoe and 0.85 to 1.3 Ma for Alejandro Selkirk. Ferrara et al. (1969, abstract) gave a range of dates for Robinson Crusoe Island of 2.0 to 3.9 Ma and for Alejandro Selkirk of 0.87 to 1.3 Ma. Stuessy et al. (1984), also from a limited collection of rock samples, derived ages of 1 to 2.4 Ma for the younger Alejandro Selkirk, 3.8 to 4.2 Ma for Robinson Crusoe, and 5.8 ± 2.1 Ma for Santa Clara. Baker et al. (1987) collected samples (analyzed by D. C. Rex) from Robinson Crusoe and obtained 4.0 ± 0.2 Ma. More recently, Lara et al. (in preparation, cited in Astudillo M. 2014) reported 3.85 ± 0.15 Ma for Robinson Crusoe Island and 0.93 ± 0.02 Ma for Alejandro Selkirk Island. All radiometrical evaluations of the islands in the archipelago therefore are consistent in indicating that Robinson Crusoe (and its closely associated Santa Clara) is the older island at about 4 Ma, and Alejandro Selkirk is much younger at about 1 Ma. The radiometrical attribution of different ages for the two major islands corresponds well to their obvious geomorphological appearance (Stuessy et al. 1984; Baker et al. 1987) (Fig. 3.2). Robinson Crusoe Island is clearly much more eroded and has broader valleys (Fig. 3.2A), and in contrast, Alejandro Selkirk Island is more dome shaped and characterized by deep ravines (Fig. 3.2B). These data are extremely important because they provide an absolute time frame within which all evolutionary phenomena in the islands can be interpreted. The islands, therefore, are geologically, and hence evolutionarily, youthful.

Much earlier, Brüggen (1950) suggested that the islands were of Eocene age, in part to help explain the existence of the endemic ancient angiosperm *Lactoris fernandeziana*. This species is the sole representative of the family Lactoridaceae, which is known from fossil pollen records in the Late Cretaceous of Africa (Zavada and Benson 1987), Late

Figure 3.2 Geomorphological comparison between Robinson Crusoe Island (A), showing broad, eroded valleys, and Alejandro Selkirk Island (B), revealing younger, deep amphitheater-headed valleys.

Cretaceous to Oligocene of Australia (MacPhail et al. 1999), and Early Miocene of southern South America (Gamerro and Barreda 2008). Despite the obvious occurrence of *Lactoris*-type pollen in the southern hemisphere long before the Juan Fernández Archipelago was formed geologically, the data do not support an ancient origin for the islands or their endemic flora. It is true that these two recently originated volcanic islands are built on 32- to 38-Ma-old oceanic crust (Gerlach et al. 1986), but this has little import for interpretation of the evolution of the terrestrial biota.

Formation of the Islands

Robinson Crusoe and Alejandro Selkirk Islands represent intraplate volcanoes thought to have originated from a single hotspot on the Nazca Plate (Baker et al. 1987; Huene et al. 1997). The islands are entirely volcanic and show no evidence of having been connected to the South American mainland (Stuessy et al. 1984). The islands lie along an east-west-trending bathymetric ridge probably associated with the Challenger Fracture Zone (Gerlach et al. 1986) (Fig. 3.3). This ridge is thought to represent a hotspot trace that can be seen up to 500 km eastward. Westward 100 km from Alejandro Selkirk Island lie two seamounts (i.e., submarine volcanoes), Friday and Domingo (Farley et al. 1993; Devey et al. 2000), that arise from an abyssal plain approximately 3,450 m deep. The recovery of fresh volcanic samples from the Friday seamount, which is situated on 28-Ma-old crust (Devey et al. 2000), suggests that the active hotspot underlies this area. Previously, Baker et al. (1987) had hypothesized that the active hotspot is currently situated beneath Alejandro Selkirk, but this does not appear to be the case, particularly because we now know this island to be 1 Ma old.

The Nazca Plate, on which the islands have emerged, has continuously been moving from west to east. A plate motion of 6 cm/year (Minster and Jordan 1978) is consistent both with estimates of sea-floor spreading from the East Pacific Rise in this part of the plate and the observed distance of 181 km separating the older Robinson Crusoe Island from the younger Alejandro Selkirk Island

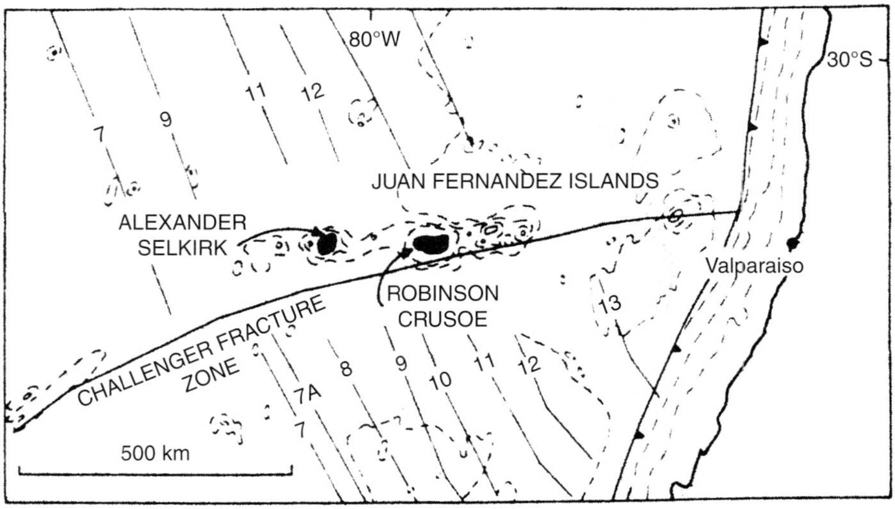

Figure 3.3 Part of the Nazca Plate with the location of the Juan Fernández Islands and magnetic lineations. The Challenger Fracture Zone seems to be associated with the east-west-trending bathymetric ridge, probably a hot spot trace (cf. Gerlach et al. 1986) on which the islands are situated. Numbers refer to the geomagnetic polarity time scale. (From Baker et al. 1987, after Corvalan 1981.)

(Minster and Jordan 1978; Baker et al. 1987; Sanders et al. 1987). Dividing the distance between the two islands by the rate of plate motion yields 3.02 Ma, essentially the difference between the two islands dated radiometrically as 4 and 1 Ma, respectively. Yelles-Chaouche et al. (1987) pointed to different rates of 4.0 to 17.1 cm/year depending on which part of the plate was being examined.

Some easterly seamounts exist that could possibly have been situated above sea level at some point in the past (Stuessy et al. 1984; Vergara and Morales, 1985; Yañez et al. 2009), but no evidence of terrestrial deposits or organisms has yet been found. The most likely candidate for being above the sea might have been an island 135 km to the east of Robinson Crusoe Island and presently 341 m under the ocean (Fig. 3.1, as 464) called Guyot O'Higgins (Vergara and Valenzuela 1982; Kopp et al. 2004). A radiometric date of 9.26 ± 0.28 Ma has been reported for this guyot (Lara et al., in preparation, in Astudillo M. 2014, p. 53). It might not have been impossible, therefore, for these two islands to have been above the sea at the same time. The easternmost island, however, would have to have been above the surface of the water initially, and there is no evidence that this was the case. Furthermore, it is also possible that it might have been covered by the sea due to rapid subsidence and erosion before the Juan Fernández islands were formed, and hence island-hopping would have been prohibited. A point in favor of this interpretation is that molecular divergence data among congeneric endemic Juan Fernández Island species so far analyzed do not support older ages of their formation. Astudillo M. (2014, p. 52) provides a useful summary of these and other guyots in and around the Juan Fernández Archipelago.

Nature of Substrates

Chemical and isotopic compositions of the rocks of the Juan Fernández Archipelago are similar to the lavas characterizing oceanic islands such as the Galápagos Archipelago (Stuessy et al. 1984). Although the Juan Fernández Archipelago is characterized by basaltic lava lying close to the Earth's mantle (Gerlach et al. 1986; Baker et al. 1987), there are distinctive differences in topography and structural background of the two main islands. The older Robinson Crusoe Island is characterized by an abruptly rising anvil-shaped mountain, Cerro Yunque, which is thought to be in the middle of at least four volcanic centers (Fig. 3.4). The largest of these is the Cumberland Bay area (diameter ca. 4 km); the other three formations, La Vaquería, Puerto Inglés, and Puerto Francés, are distinctly smaller (Baker et al. 1987). The existence of these separate centers is suggested by the opposing dip of lavas on either side of the valleys. These major features are surrounded by several additional petrographical groupings, which provide the basic geological structural pattern of the island.

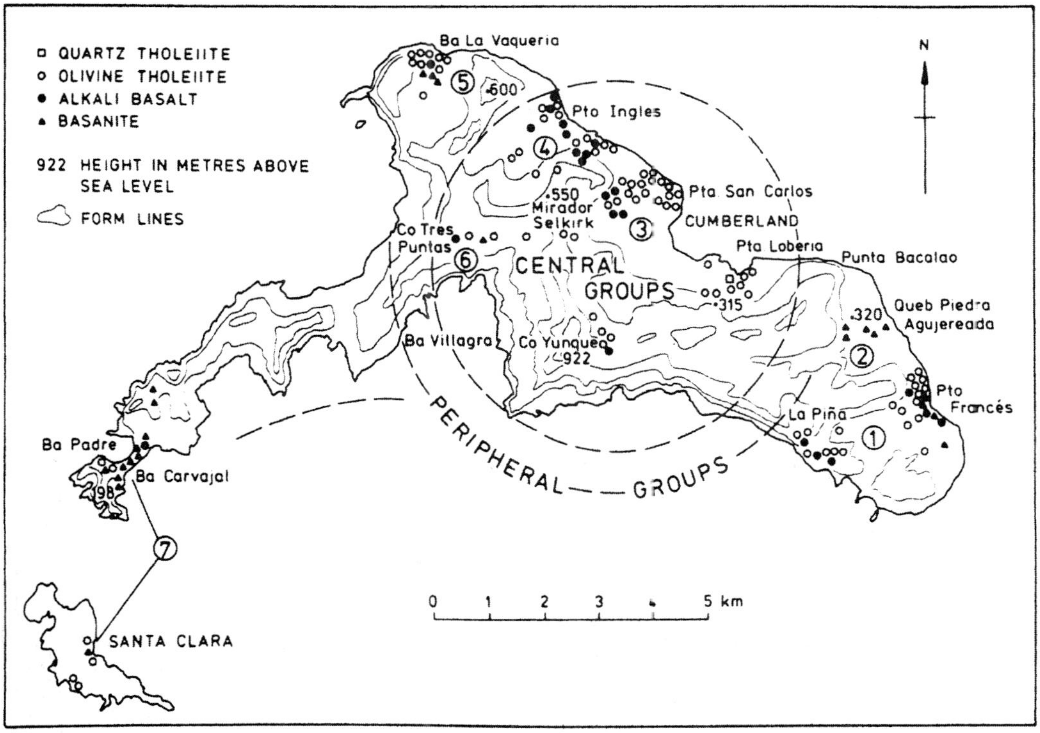

Figure 3.4 Map of Robinson Crusoe Island with principal localities, volcanic centers, and geopetrographical groupings of volcanic rocks. Central and peripheral groups (given as numbers) were delineated on a geographical basis and thereafter classified on geochemical grounds. (From Baker et al. 1987.)

On Robinson Crusoe Island, the strata layers vary in composition from highly normative nepheline to alkali basalts, olivine and quartz tholeiites, and hawaiites (Baker and Keyvan-Scocouhi 1982). A more detailed analysis is presented by Gerlach et al. (1986) and Farley et al. (1993). An extensive volcanic sequence more than 2,200 m thick is present. The lowest unit, at Punta Larga, is more than 800 m thick, made up of extensive basalt flows with rare interbedded pyroclastics. The middle unit, at Puerto Inglés, about 1,200 m thick, was found to consist of olivine-phyric to picritic basalts interbedded with a relatively greater proportion (ca. 30%) of pyroclastics. The youngest unit, at Bahía del Padre (ca. 250 m thick), mainly consists of pyroclastics with minor olivine basalt flows.

The younger Alejandro Selkirk Island, measuring 10.5 km (N–S) by approximately 6 km (E–W), is topped by Los Inocentes, which is much higher (1,319 m) than its dominating counterpart, Cerro Yunque (915 m), on the neighboring Robinson Crusoe Island (Baker et al. 1987). Los Innocentes, whose configuration is generally well preserved, appears to be the single volcano on the island. Olivine-rich dike rocks are common (Natland 2003). Owing to their abundance, the new name *masafuerite* was proposed by Johannsen (1937). Emergences of the basaltic cones and structures are the dominating components of the islands' history.

The geology of the Juan Fernández Archipelago therefore consists mainly of rocks and minerals of volcanic origin. A number of the crystalline minerals no doubt formed during the cooling phase of each island. The major impact of these differences is seen in the two environments of lava-derived slopes and soils and the open basaltic ridges. These two main environments have selected for different species of plants. For example, in *Robinsonia* (Asteraceae), *R. gayana* is confined to the wind-swept ridges, whereas the other species are adapted to the more protected moist forest.

Ontogeny of the Islands

Very important for interpretation of evolutionary and biogeographical events in the Juan Fernández Islands (as well as within any oceanic island) is an understanding of how they have changed over geological time. Islands are very dynamic ecosystems, constantly changing, and they erode and reduce in surface area, rarely adding new land due to volcanic activity, but all eventually disappearing under the ocean with complete extinction of their terrestrial biota (Ramalho et al. 2013). Understanding this island ontogeny is fundamental for attempting comparisons between islands in the same or different archipelagos that may be at different stages of geological aging (Stuessy 2007; Whittaker et al. 2010). To understand the dynamics of these changes requires examination of physiographical features above sea level as well as underwater (bathymetric) contours. Present geological and geomorphological features of the Juan Fernández Islands have derived from the long-term eroding power of wind, precipitation, and stream water, as well as island subsidence and marine erosion (e.g., Baker et al. 1987) over a period of 4 million years.

In general, it has been estimated that an island volcano might erode at the rate of about 8 cm every 1,000 years (Ziegler 2002), and in this way, the volcanic slopes rapidly become furrowed by newly developing stream channels. These sorts of phenomena have been described from Hawaii by Stearns (1966). Radiometric measurements obtained from the island of Oahu suggest that a period of 1 to 2 million years is required for deep erosional impacts on the terrestrial parts. This time frame, in fact, fits well with the eroded appearance of the higher parts of Robinson Crusoe Island (Fig. 3.2A) and the "pali"-like shear cliffs created on the windward slopes. In contrast, corresponding processes on the much younger Alejandro Selkirk Island appear to be in their initial stages (Sanders et al. 1987). Trachyte basalt still dominates in the upper 500 m of this younger island (Quensel 1954), and the erosional valleys are still long and narrow (Fig. 3.2B), with modest merging of the upper tributaries and only small amphitheater termini (Sanders et al. 1987).

Submergence processes have also exerted another major impact on island development. As the Nazca Plate moves eastward and eventually subducts at the western margin of South America, it slowly dips downward, taking the emergent Juan Fernández Islands lower. Ziegler (2002) has inferred for Hawaii an average tectonic lowering for all the main islands of about 2 cm (0.8 inch) per 1,000 years (Ziegler 2002). If this rate also occurred with the Juan Fernández Archipelago, Robinson Crusoe Island would have lowered approximately 80 m and Alejandro Selkirk about 20 m during their existence. Sanders et al. (1987) addressed geological reconstruction of the Juan Fernández Islands by using bathymetric maps for submarine contours (Armada de Chile 1965; Mammerickx and Smith 1978; Prince et al. 1980) and extrapolating island outline and surface in a way that had previously been applied for other Pacific islands, particularly the Hawaiian Archipelago (see Stearns 1966). The available bathymetric maps of the Juan Fernández Archipelago appear to show submarine canyon-like structures extending to −2,500 m below both islands, but there is no evidence as to what portion of these extents might have been above water in the past (Sanders et al. 1987). Robinson Crusoe Island is located on a definite platform between −200 and −500 m, and a similar situation also prevails for Alejandro Selkirk Island (with much narrower bathymetric margins), which suggests these as past erosional platforms and hence also provides the initial diameters of the islands. According to this reconstruction, Robinson Crusoe Island originally would have been much larger than at present and would also have included the currently separated island of Santa Clara (Fig. 3.5E). Alejandro Selkirk Island now rests on an only slightly smaller sea-covered base than when the island was formed (Fig. 3.6). For newer bathymetric data and more precise geological reconstructions of these ontogenetic changes, refer to the thesis by Astudillo M. (2014). This new information correlates well with earlier hypotheses of Sanders et al. (1987).

The sea level of both islands would have been affected somewhat from Pleistocene glaciation, as was the case elsewhere on Earth (Weigelt et al. 2016) (Fig. 3.5F). Due to the vertical nature of these volcanic islands, however, the lowering of sea level would not have had a significant impact on the territory of the two Juan Fernández Islands. It is presumed that plant populations would have migrated downward in response to available new land but that they would also have retreated upward at the end of glaciation. No evidence exists for permanent snow or ice on the two islands, presumably because both islands were already too low by the time Pleistocene glaciation ensued.

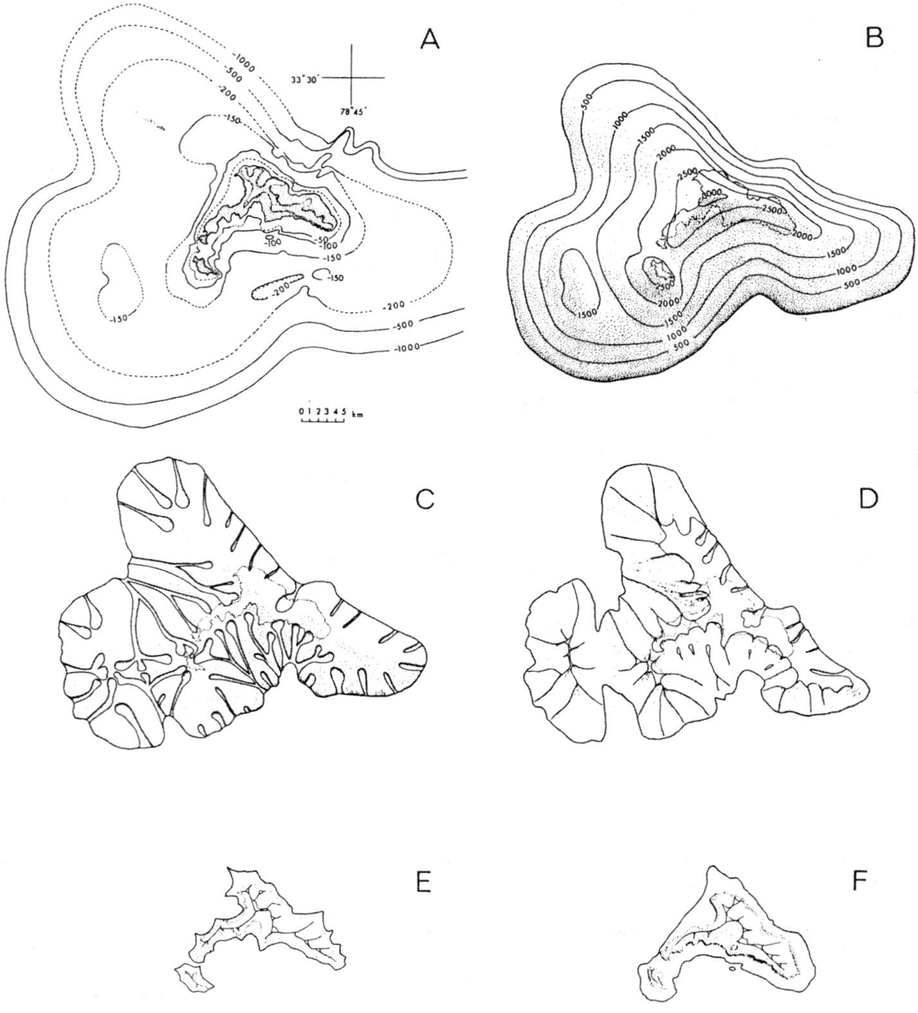

Figure 3.5 Reconstruction of the geological history of Robinson Crusoe Island including Santa Clara. (A) Present configuration of the island showing bathymetric contours to −1,000 m. (From Armada de Chile 1965; Mammerickx and Smith 1978; Prince et al. 1980.) Dashed lines here represent interpolated levels between known data points. (B) Shape of original island, 4 mybp. Dotted lines here and in C—F show present island outline. (C) Erosional patterns showing amphitheater-headed valleys (compare Fig. 3.2B of Alejandro Selkirk Island at corresponding erosional stage), 3 mybp. (D) Further erosion showing coalescence of adjacent valleys, 2 mybp. (E) Following subsidence of island and further erosion by wave action, 1 mybp. (F) Lowering of sea level during Pleistocene glaciation, 10,000 ybp. (After Sanders et al. 1987.)

Taking subsidence and erosion into account, therefore, and attempting to reconstruct the historical development of the islands outlined earlier, one can estimate the reduction from their sizes since their emergences. This application results in a loss of terrestrial area of 95% for the older and originally much larger Robinson Crusoe Island and 28% for Alejandro Selkirk Island (Stuessy et al. 1998a). Conversely, this corresponds to an

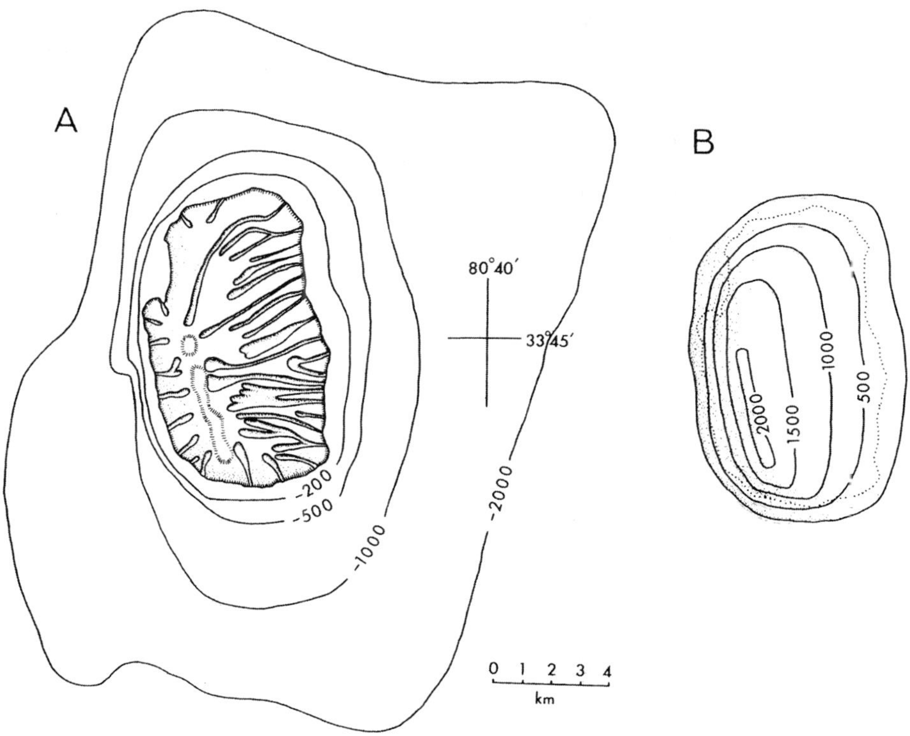

Figure 3.6 Reconstruction of geological history of Alejandro Selkirk Island. (A) Configuration of the island showing bathymetric contours to −2,000 m. (From Armada de Chile 1965, Mammerickx and Smith 1978, and Prince et al. 1980.) (B) Shape of original island, 1 mybp. Dotted line shows present island outline. (After Sanders et al. 1987.)

original size of 1,092.5 km² for Robinson Crusoe and 69.2 km² for Alejandro Selkirk (Figs. 3.5 and 3.6). These facts are very important in interpreting the biogeography of the archipelago (see Chapter 17).

Historical Record of Earthquakes and Tsunamis

Despite the volcanic origin and history clearly indicated by the geological sequences of the Juan Fernández Islands, no volcanic activity has been recorded in historical times in the islands themselves. However, in 1835, submarine volcanic eruptions took place next to Robinson Crusoe Island (Sutcliffe 1839), with tectonic activity registered at a depth of 90 m below sea level. This event resulted in a tidal wave that destroyed much of the village of San Juan Bautista.

More recently, in February of 2010, another tsunami severely affected the archipelago, destroying about half the village on Robinson Crusoe Island (Pérez 2010). On Saturday morning, February 27, 2010, a tsunami occurred in the archipelago at approximately 4:23 am. This was caused by a submarine earthquake at a depth of 35 km near mainland Chile (115 km northeast of Concepción and Talcahuano) at level 8.8 (Richter), resulting in great

Figure 3.7 Main street of the settlement of San Juan Bautista on Robinson Crusoe Island one year after the tsunami incident of February 27, 2010. The houses that existed at this site, which were available as tourist lodgings, were occupied by Sr. Green and his family. Compare with Fig. C3.

damage to coastal towns. On Robinson Crusoe Island, the downtown area of San Juan Bautista was completely destroyed, from the cemetery south to El Palillo. This included the post office, city hall, port captain's office, church, souvenir shops, restaurants, small markets, bread store, school, cultural center and museum, information kiosk (CONAF), and historical homes. All fishermen's lockers were destroyed, as also were any boats on land. The plaza was totally razed, and nearly all plants, except the large trees, were destroyed (Hahn et al. 2014). The wave washed upward to the level of the Santa Bárbara Fort (or the caves of the patriots). The wave was 2.34 m tall (AFP 2010), but the force of the wave sent water up the slopes much higher (ca. 60 m). Eleven persons were killed, including a student in marine biology from Chile who was working on the island. The impact on the village was enormous (Fig. 3.7).

Soils

Diversity of Soils on the Islands

The most detailed examination of the soils of the Juan Fernández Archipelago was done by Ortiz (1982, pp. 100–34 and Anexo 1), but these dealt only with examination of sites

Table 3.1 Soil Properties at Several Sites on Robinson Crusoe Island

Site	Depth	pH	Organic carbon	Available phosphorus	Ca^{2+}	Mg^{2+}	Na^{+}	K^{+}
PI 1	0–3	5.50	9.74	86	16.94	18.04	0.86	1.43
PI 2	3–14	5.80	1.51	42	12.95	16.34	0.72	1.27
PI 3	14–58	5.75	2.36	41	9.57	12.49	0.87	0.56
PI 4	58–87	5.60	1.68	10	10.80	12.03	0.97	0.17
PF 1	1–8	5.60	5.68	4	8.84	2.62	0.64	0.83
PF 2	8–24	5.80	3.89	8	9.00	2.85	0.68	0.91
PF 3	24–34	5.90	3.02	5	11.30	3.47	0.79	0.78
PF 4	34–60	6.00	2.38	3	11.17	3.31	0.84	0.83
EY 1	0–7	6.00	14.62	55	29.75	17.48	0.52	2.55
EY 2	7–18	6.20	11.02	47	31.75	15.62	0.65	2.31
EY 3	18–50	6.80	9.92	24	36.00	16.45	0.73	2.76
LV 1	3–25	6.30	4.76	180	34.00	8.22	0.57	4.80
LV 2	25–40	6.70	3.31	125	32.25	7.81	0.63	4.05
LV 3	40–70	7.10	1.97	48	26.50	7.81	0.66	3.30
CH 4	0–28	3.70	43.45	58	9.37	15.22	2.00	1.27
CH 5	28–42	4.10	36.89	7	0.87	1.33	0.70	0.34

Note: Depth in centimeters; organic carbon in percent; available phosphorus in parts per million; cations in milliequivalents (meq) per 100 grams; PI, Puerto Inglés; PF, Puerto Francés; EY, El Yunque; LV, La Vaquería; CH, Chifladores.
Source: From Ortiz (1982); data taken from table 5 in the Anexos.

on Robinson Crusoe and Santa Clara Islands. On Robinson Crusoe Island, Ortiz collected three to nine samples from each of six localities (San Juan Bautista, Puerto Inglés, Puerto Francés, El Yunque, La Vaquería, and Los Chifladores). Because of serious erosion of the original volcanic soils of the island, much of the study focused on patterns of erosion, which has obvious importance with reference to conservation of the native and endemic plants. Nonetheless, details of soil quality are provided for a number of these localities. Soils are characterized by the rocky volcanic underground. Thus organic substrate usually is poor, which is particularly conspicuous on the sea cliffs and high basaltic ridges. The variation in soil types ranges over a number of described types (based on Skottsberg 1953a; Quensel, 1954), such as dark brown forest soil, gravelly humus soil, weathered basalt fragments and humus particles mixed together, and yellowish-reddish sand with rock fragments.

A reasonable, biologically relevant characterization of the properties of the upper sediment, including amounts of cations and some other analytical chemical parameters, can be derived from Table 3.1. These data reveal that considerable variation in soil types does exist on Robinson Crusoe Island. Unsurprisingly, as one goes higher up into the moist forest zone, the percent of organic carbon increases (e.g., on the top of El Yunque and in the upper portions of the Cordón de Chifladores). Availability of different cations also varies from locality to locality, which no doubt reflects the patterns of erosion and weathering of the exposed volcanic rocks.

Table 3.2 Different States of Erosion Found on Robinson Crusoe Island

State of erosion	Area (hectares)	Percent of total area
No apparent erosion	1,775.20	37.68
Light erosion	120.80	2.57
Moderate erosion	563.20	11.96
Considerable erosion	750.40	15.93
Most severe erosion	1,084.40	23.01
Erosion along cliffs or small offshore islets	417.26	8.85
Totals	4,711.26	100

Modified after Ortiz 1982; table 3 in Anexos.

Although no soils have been analyzed yet from Alejandro Selkirk Island, we have observed a remarkable form of substrate in the "alpine" zone on Cordón Inocentes. Toward the south near the upper part of Quebrada Varadero at 1,050 m, there occurs a thick organic layer dominated by the thallose liverwort *Marchantia berteroana* (Hepaticae) (Greimler et al. 2013; documented by collection *Stuessy & Sepúlveda 9552*). It appears, in fact, that this may be a mat several meters thick on which very little else seems able to grow. One might be justified in calling this a "peat zone," although it is composed entirely of liverworts rather than mosses.

Changes in Soil Composition

Erosion in the islands caused both by natural sources and human impact has played a major role in soil development and change (Table 3.2). The continuous natural effects of wind represent a strong influence on the upper substrate layer, especially where vegetation is lacking or poor (see Chapters 6 and 7). Since European visitations started in the sixteenth century, the islands' appearance has been modified most severely through human impact. Direct deforestation by logging activities has been one crucial factor (cf. Skottsberg 1954; Stuessy et al. 1998b; Haberle 2003). Moreover, the ongoing browsing activity of goats, introduced shortly after 1574, strongly affected vegetation and, as a consequence, soils. From a conservation standpoint, therefore, the present state and conditions of erosion have to be judged as highly critical (Ortiz 1982). Only about 38% of the total surface of Robinson Crusoe Island has no apparent erosion, which coincides reasonably well with the percentage of native, unaltered vegetation that still remains on the island (Greimler et al. 2002a).

Soils and Relevance for Adaptive Radiation

Data are not abundant on the possibility of edaphic factors being a stimulus leading to reproductive isolation and eventual adaptive radiation of plants in the archipelago. Investigations (Sanders et al. 1987) on evolution in the adaptively radiated genera *Robinsonia* and *Dendroseris* (both Asteraceae) have revealed only limited soil features that correlate with the different endemic species. The data collected were organic matter, pH, P, K, Ca, Mg, Fe, cation exchange capacity (CEC), and extractable K, Ca, and Mg (following standard procedures, e.g., North Dakota A.E.S. 1975). These data were analyzed with principal component analysis, but they were first combined with other "abiotic" data involving slope and terrain. The combined data for *Dendroseris* showed much overlap in factors associated with five species, with *D. neriifolia* being somewhat distinct. With *Robinsonia*, most species shared similar patterns except for *R. gayana*, which is confined to open ridges. The results demonstrate that the substrate, slope, and terrain aspects do differ to some extent within the islands, but they have apparently not been major factors in driving adaptive radiation in these two genera on the islands. It must be cautioned that much environmental change due to subsidence and erosion has occurred in the archipelago, and it is impossible to know if these factors might have been more important when the species were actively diverging. Note that in the only other detailed soils analyses done on Robinson Crusoe Island (Ortiz 1982) (see Table 3.1), quite different values of organic matter and available cations can be seen in the different localities sampled. However, many of these sites (such as Puerto Inglés, Puerto Francés, and La Vaquería) do not contain endemic species of plants. These valleys have been substantially altered over time by natural erosion aided by human impact, and they consist now mostly of introduced weeds. Very relevant would be comparable studies of soils on Alejandro Selkirk Island, which has been less altered by the impacts of subsidence and erosion, particularly in relation to the adaptive radiation of *Erigeron* (Asteraceae) (López-Sepúlveda et al. 2015a).

4 Climate and Weather

Walter A. Sontag, Jr., and Tod F. Stuessy

To understand the evolution of the plants of the Juan Fernandez Archipelago, it is important to know the formation of the islands, their resulting substrates, and their changes over the past 4 million years, all of which were considered in Chapter 3. These factors play a major role in shaping the environment for the kinds of plant species that are now seen in the archipelago. The formation of the islands obviously sets the stage for the arrival and establishment of plant colonists from diverse parts of the globe. After arrival, the success of establishment of immigrants to the islands, and their subsequent assembly into vegetation zones, also depend on climatic factors. Climate is the accumulation of patterns of weather that occur daily over the archipelago. The type of climate, involving temperature, rainfall, and winds, over all parts of the islands of the archipelago has much to do with the nature of the plant species and vegetation that result over evolutionary time. This chapter chronicles these factors. The main objective is not to present all available data but instead to offer a summary of environmental conditions that give a sense of the basic climate in the archipelago within which the flora has evolved.

As an overview, the climate of the Juan Fernández Islands is usually classified as a warm-temperate climate, characterized by equivalent dry and moist seasons (Fuenzalida 1966) or, alternatively, as a Mediterranean-type climate with a strong oceanic influence (Hajek and Espinoza 1987). Novoa and Villaseca (1989) define the climate as a warm-marine climate with mild winters, and Johow (1896) regarded it as subtropical.

Sources of Data

Data measurements in the Juan Fernández Archipelago are not numerous or comprehensive, but they suffice to give a general picture of climate. Skottsberg (1953a) was the first to present precise weather data to the scientific community, although these were limited, incomplete, and localized. These data included air temperature, precipitation, light conditions, and winds. The data came mainly from a meteorological station in San Juan Bautista plus original observations and measurements at different places and times on both islands. These latter data are especially interesting because they represent the only relevant data available from areas outside the village of San Juan Bautista.

More recently, Neshyba and Silva (1985) presented and discussed climatological data from Robinson Crusoe and Santa Clara Islands from 1974 to 1984. Observations included sea surface and air temperatures, atmospheric pressure, and rainfall. The latter three measurements were obtained from the meteorological station at San Juan Bautista monitored by the Instituto Hidrográfico of the Armada de Chile. Results were given in useful tables and timeline graphs over the period of years, which demonstrates a basically stable climate in the archipelago.

Hajek and Espinoza (1987) presented a more comprehensive collection of data from the archipelago, including temperature and precipitation as well as atmospheric pressure, cloudiness, sunshine values, humidity, and evapotranspiration. These authors took into account meteorological data from the period between 1924 and 1980, much of them from 1965 to 1971, from the Dirección Meteorológica de Chile. Novoa and Villaseca (1989) present some of the same data in useful summary form. Another helpful summary is available online from climatemps.com. It is important to remember that nearly all these data have been collected at only one weather station on Robinson Crusoe Island at about 15 m above sea level and situated in San Juan Bautista (cf. Greimler et al. 2002; P. López-Sepúlveda, personal communication), and therefore, available annual mean data do not represent the climatic situation in the native forest at higher elevations.

Temperature and Precipitation

The data on temperature in the Juan Fernández Archipelago are of three basic types: (1) occasional data recorded by visiting ships and early colonists, of historical interest only; (2) measurements taken consistently at 15 m at a meteorological station on Robinson Crusoe Island beginning first in 1901 (Skottsberg 1953a) and continuing with some interruptions to the present day; and (3) many data points at different elevations gathered by Carl Skottsberg from both islands of the archipelago during 1916 and 1917 (Skottsberg 1953a). The basic temperatures obtained from the village give the lowest mean minimum air temperature as 11.6°C, the highest maximum as 20.4°C, and the mean annual temperature as 15.6°C (Table 4.1). The temperature varies with the seasons (warmest in January and coolest in July) but not dramatically so (Fig. 4.1), showing a difference of only approximately 5.5°C on average (Hajek & Espinoza 1987). Temperatures keep to this basic pattern with little deviation, as seen in a 20-year profile.

Perhaps more interesting temperature data were collected by Skottsberg (1953a) from many parts of both islands, 94 recordings from Robinson Crusoe Island from 5 to 780 m and 55 measurements from Alejandro Selkirk Island ranging from 215 to 1,500 m (his high-altitude calculation was incorrect; this island is only 1,319 m high). For Robinson Crusoe Island, Skottsberg's recordings were taken from December 3, 1916 to April 24, 1917, that is, the austral summer. These recordings are useful because they began each day at sea level and ended each day at the same place; these measurements were always similar, varying by only a few degrees (Celsius). Most field data were collected near midday (11:00–15:00), with some exceptions. The results of these measurements are that it is cooler up on the slopes, about 3°C more or less. So much of daily

Table 4.1 Average Yearly Air Temperatures and Ranges on Robinson Crusoe Island (at San Juan Bautista), 1974–84

Year	Mean temperature (°C)	Standard deviation	Observed range	Months of observations
1974	[a]	[a]	11.7–[a]	9
1975	14.4	2.1	11.6–17.9	11
1976	15.3	2.6	11.8–19.3	10
1977	15.6	2.3	11.8–18.4	12
1978	16.0	1.8	13.2–18.3	10
1979	15.4	2.0	12.8–18.4	11
1980	15.7	2.9	12.6–20.4	12
1981	15.3	2.4	12.4–18.4	12
1982	15.6	2.3	12.7–19.0	11
1983	15.3	2.8	11.9–19.0	12
1984	[a]	[a]	[a] –19.2	6
Annual mean	15.6			

[a] Insufficient information.
Source: From Neshyba and Silva (1985, p. 47).

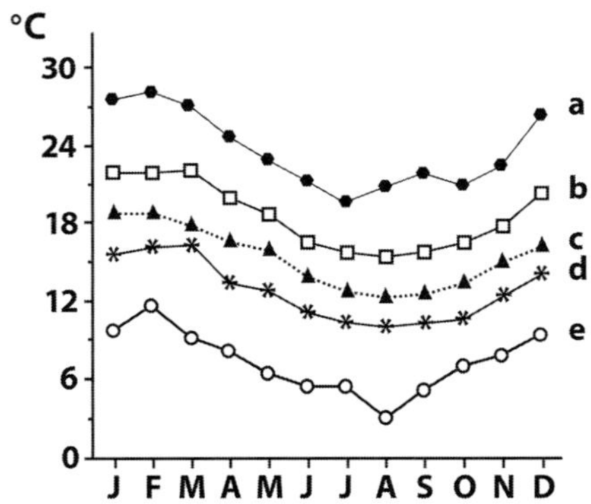

Figure 4.1 Monthly mean temperatures on Robinson Crusoe Island: (A) absolute maximum temperature; (B) mean maximum temperature; (C) mean temperature; (D) mean minimum temperature; (E) absolute minimum temperature. (From Hajek and Espinoza 1987. p. 75.)

temperature in the islands depends on weather, especially cloud patterns and fog, which makes it impossible to document the relationship of elevation and temperature in a linear fashion. In general, as one goes upward on Pacific Ocean islands, a decrease in temperature of 2.8°C per 500 m may be a realistic assessment (Carlquist 1980). Because elevations do not exceed heights of more than 1,319 m in the Juan Fernández Archipelago, no inversion layer is to be expected (Sanders et al. 1987). From seashore to the highest ridges, a maximum temperature gradient of only 5.6°C might be anticipated, but this does not take into account other factors, such as wind and shading effects, as already mentioned.

Table 4.2 Annual Precipitation on Robinson Crusoe Island (at San Juan Bautista) 1974–84

Year	Precipitation (mm)	Months of observations	Percent in comparison with historical annual mean
1974	626	9	[a]
1975	1,003	11	112
1976	820	11	96
1977	1,066	12	116
1978	731	11	91
1979	977	12	106
1980	1,658	12	180
1981	1,136	12	123
1982	1,166	12	126
1983	853	12	93
1984	553	6	[a]

[a] Insufficient information.
Source: From Neshyba and Silva (1985, p. 48).

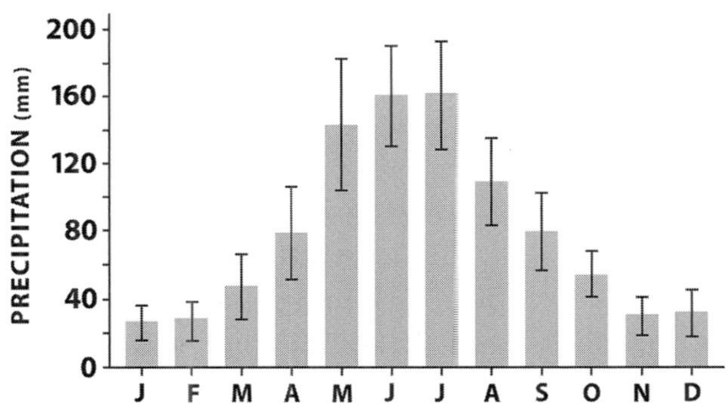

Figure 4.2 Monthly precipitation (average; bars show standard deviation) for Robinson Crusoe Island from the meteorological station at San Juan Bautista. (From Hajek and Espinoza 1987, p. 71.)

The mean annual precipitation on Robinson Crusoe Island is 956 mm (Hajek and Espinoza 1987). Novoa and Villaseca (1989) give total precipitation as 922.10 mm. Again, these data were collected in the village. Skottsberg (1953a) provides a measurement of 1,081.2 mm at 345 m. Data collected between 1974 and 1984 show yearly rainfall conditions as being rather steady, but with some variations (Table 4.2). However, amounts per month clearly differ, and there is a strong seasonality in rainfall (Fig. 4.2). The seasonal comparison shows winter precipitation to be by far the highest. Maximum precipitation within 24 hours varies between 15 and 86 mm (Hajek and Espinoza 1987). Strong rainfall differences should be expected owing to the altitudinal effects and the orientation of the prevailing winds. Cereceda et al. (1994) sampled rainfall at five locations on Robinson Crusoe Island during the winter of 1992, and they recorded twice the rainfall at the Mirador de Selkirk (Portezuelo) as in the village of San Juan

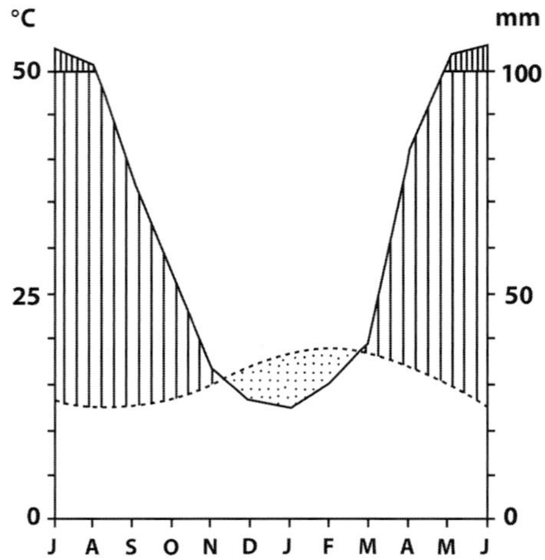

Figure 4.3 Climatic diagram (following Gaussen-Walter; e.g., Walter 1973) for the Juan Fernández Islands based on monthly averages. The dark, narrowly lined tips refer to precipitation over 100 mm/month. The stippled region shows the arid summer months with precipitation less than 50 mm, and the broadly lined regions indicate higher humidity during the austral fall and spring. The dashed line indicates the average monthly temperature. (From Hajek and Spinoza 1987, p. 82.)

Bautista. Kunkel (1957), in describing the quebradas (ravines) leading from the summit of El Yunque, the highest point on Robinson Crusoe Island (915 m) (Anonymous 1978), reported high humidity, very moist soils, and water dripping off leaves and small stems. Periods of extended drought or flooding are uncommon in the islands (Haberle 2003). Rain frequently falls on both sides of the main ridges, with a more slender zone existing on Robinson Crusoe Island (due to its rather narrow physical characteristics) and a broader zone on the more dome-shaped Alejandro Selkirk Island.

The much higher younger island is exposed to abundant orographical rain, primarily from southwesterly winds, in addition to precipitation supplied by the western storm track (Haberle 2003). The high altitudes of this island above 600 m are frequently covered by clouds, most probably receiving much more precipitation and having longer periods of humidity, which along with winter frosts and snowfalls aid in maintenance of an upper forest limit between 700 and 750 m. Stronger habitat differentiation is more noticeable on Alejandro Selkirk Island than on Robinson Crusoe Island (Greimler et al. 2002a, 2013; see also Chapter 6). In general, environmental differences within the islands are not very striking, certainly nothing in comparison with the Hawaiian or Galápagos Archipelagos (T. Stuessy, personal observation), which have very marked environmental zonation. In the Juan Fernández Archipelago, a low-elevation dry zone around the islands gradually merges into a wetter zone along the slopes.

The two climatic variables of temperature and precipitation can be combined into a climate diagram (Fig. 4.3), which provides a useful summary of the principal climatic factors. During the winter months, rainfall is ample, being over 100 mm/month. In the austral summer, however, rainfall drops markedly to less than 50 mm/month. Temperature is relatively stable throughout the year, increasing slightly during the summer months.

Ocean Currents and Winds

The more regional climate factors surrounding the archipelago are sea and airborne currents. Ocean currents along the western coast of South America run predominantly south to north (Frakes 1979; Schopf 1980; Rahmstorf 2002) (Fig. 4.4). The cold subantarctic Humboldt ocean current has strongly influenced climate in the islands (Santibañez 1945). While its impact on the environment of the archipelago is less direct than other factors that influence the landscape, ocean currents are responsible for maintaining a cooler warm-temperate (or subtropical) climate overall.

As for winds, southern and southeastern winds dominate (Fig. 4.5). In January, winds come from the south and southeast, calms are about 21%, and wind speeds average 17 knots (Hajek and Espinoza 1987). In July, calms amount to about 32%; the average wind speeds are 12 knots for southern and southeastern winds and, to a lesser degree, from southwestern, western, northwestern, and northern winds. During the austral winter, the winds become more variable. In the eastern Pacific, westerlies predominate to the south and easterlies to the north. In the austral summer, the high-pressure cell can be shifted a few degrees south such that the winds blow toward the islands from the east or southeast nearly one-third of the time (Skottsberg 1954; Newell et al. 1972; Van Loon 1972). These winds are rather mild, variable, and short lived.

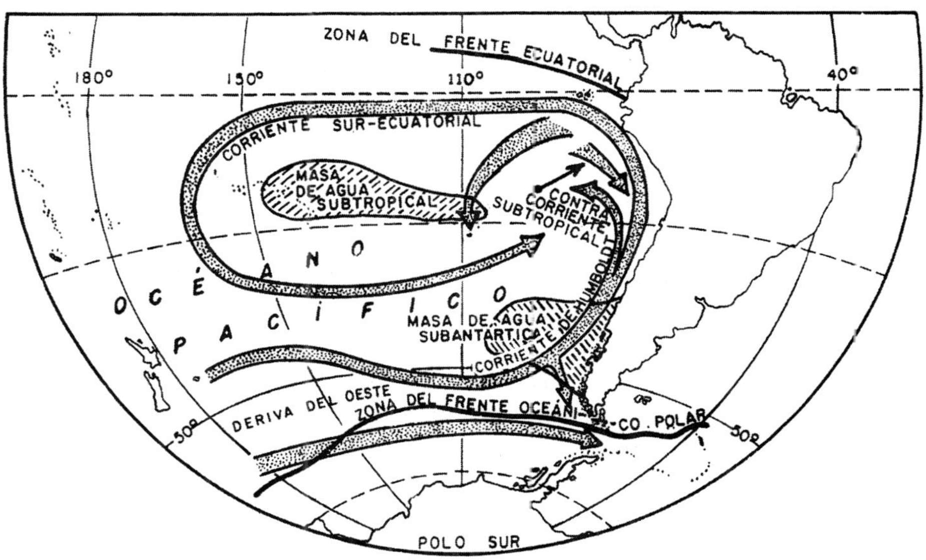

Figure 4.4 Major ocean currents in the southeastern Pacific. (From Romero A. 1985, p. 129.)

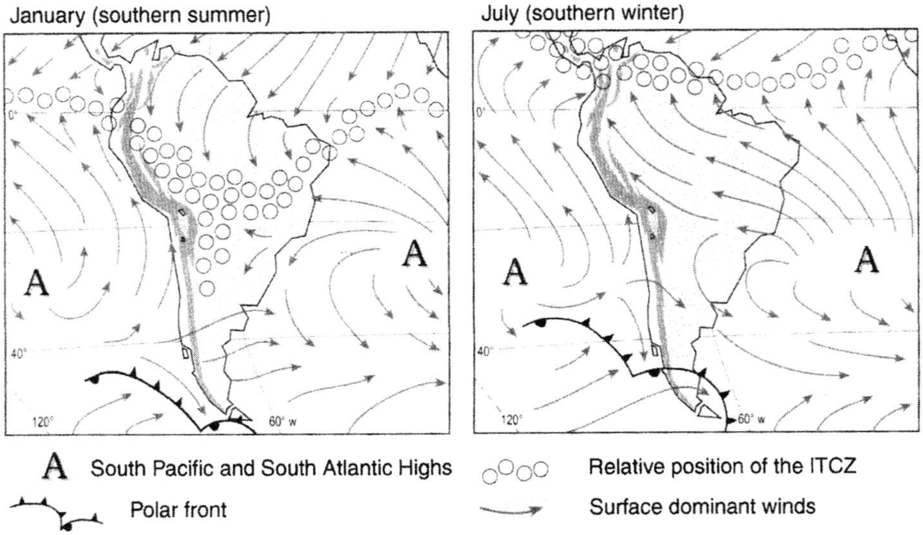

Figure 4.5 Diagrams of atmospheric circulation over South America in January and July. ITCZ, Intertropical Convergence Zone. (From Moreira-Muñoz 2011, p. 32.)

Other Climatic Factors

Other climatic parameters include atmospheric pressure, sunshine, humidity, and evapotranspiration. The following data come mainly from Hajek and Espinoza (1987).

Subtropical high-pressure cells are located just north of the archipelago at approximately 30°S and 90–120°W (Kendrew 1961; Van Loon 1972; Schopf 1980). Generally speaking, the atmospheric pressure is usually high, with values around 1,020 mbar. Regarding sunshine, the ratio of highest actual to astronomical maximum sunshine is in December, at 44%. The relative humidity has a mean value of 75%, with an annual range between 70% and 78%. The yearly transevapotranspiration value is 544 mm.

Skottsberg (1953a) carried out a series of experiments in 1916–17 in which he exposed photographic paper in different environments to assess the available light. For calibration, he set full exposure in open sun at a value of 1. Other situations have much less light and provide interesting contrasts: on the edge of the forest near the ridges, ⅓–⅛; in the deep quebradas of Alejandro Selkirk Island, ½–1/10; in the typical forest (e.g., Villagra on Robinson Crusoe Island), 1/20–1/45; in *Boehmeria* groves (trees; Urticaceae), 1/50; in *Dicksonia* (tree fern) forest, 1/420; and under dense maqui (*Aristotelia chilensis*), 1/2000. These are very useful data because they show how such differences in light intensity add to the mosaic of environmental heterogeneity in the archipelago. It also stresses that once the invasive maqui takes hold and forms dense stands, no other species can grow underneath, emphasizing the danger this species poses for the native and endemic flora of the islands.

El Niño–Southern Oscillation (ENSO)

A special cyclical climatic factor is the El Niño–Southern Oscillation (ENSO) phenomenon. This occurs when a band of warmer water develops in the central and eastern tropical Pacific that also affects the coast of South America. This phenomenon may have strongly influenced the vegetation on Alejandro Selkirk Island after 4,500 years before present (BP)(Haberle 2003, p. 251), as suggested by charcoal records collected at high altitude (i.e., subalpine to alpine environments). These findings point to the onset of protracted drought events from 4,500 ^{14}C yr BP and even stronger occurrences of that kind between 3,500 and 2,500 ^{14}C yr BP. This is in accord with the reported general stronger effects of ENSO in the southern hemisphere after 5,000 ^{14}C yr BP (Diaz and Markgraf 1992; Liu et al. 2014).

The impact of ENSO, however, might have been much more complicated than just by direct meteorological effects. The inferred extension of bird migration patterns in the Pacific during the middle to late Holocene may also have been a response to changing climate and oceanographical conditions caused by the itensified El Niño–related climate dynamics (Thompson and Ollason 2001). In turn, the occupation of the then-existing vegetation zone on Alejandro Selkirk Island by migratory petrel species (*Pterodroma longirostris* and *P. externa*) may have had a significant impact on development of the treefern-shrubland mosaic vegetation pattern through burrowing activity (Haberle 2003). Considering the enormous numbers of birds (about 1 million breeding pairs estimated in 1985–6 in the islands; Brooke 1987a), their possible historical impact might indeed have played a tremendous role.

Low rainfall often occurs during strong La Niña years, which is the cool phase of ENSO, when cooler sea surface temperatures prevail (McPhaden et al. 2006). This is associated with an increased influence of the subtropical high-pressure cell (cf. Allan et al. 1996; Haberle 2003). The importance of this occurrence may be found in the greater impact of the ENSO phenomenon on vegetation patterns in the southern hemisphere after 5,000 ^{14}C yr BP (McGlone et al. 1992; Haberle 2003). Haberle (2003) also points out that because there was no human occupation on Alejandro Selkirk Island prior to its discovery in the late sixteenth century, climate change and environmental variability, allied with an increase in ENSO activity during the middle to late Holocene, apparently led to more burning events, which did have an influence on the vegetation. Specifically, evidence so far suggests that the ENSO phenomenon overall seems to have played only a modest role in the Juan Fernández Archipelago (Neshyba and Silva 1985).

Importance of Climate to Biogeography

Although biogeographical concepts and discussion form much of Chapters 16 and 17, it is obvious that climatic factors have played an important role in the establishment and formation of the flora and vegetation of the Juan Fernández Archipelago. Fundamental to these concerns are aspects of dispersal of propagules, and wind and ocean currents must be considered. It is quite likely that many of the ferns and some of the flowering plants have arrived to the islands through the air. Fern spores travel easily with the wind and can survive for months or even years (Lloyd and Klekowski 1970; Windham et al. 1986). The evidence that this has happened can be seen by the low level of endemism among the ferns in the archipelago and, perhaps even more important, that once an endemic fern species originates on one island, it frequently disperses and also successfully colonizes the other island (Stuessy et al. 1990; see Chapter 13). Although prevailing winds usually come from the west, there are also occasional cyclonic winds that blow across the southern part of South America and out into the Pacific Ocean (Sanders et al. 1987). Furthermore, it is known that the patterns of wind circulation over the southeastern Pacific have changed during the past 4 million years (Habicht 1979). The conclusion, therefore, is that a number of the presently occurring endemic species of flowering plants and ferns arrived in the archipelago by wind dispersal. Ocean currents follow this same basic pattern, although the evidence for water dispersal in the native and endemic flora is scant (see Chapter 16). In general, the Humboldt current now flows south to north along the continental Chilean coast, but it earlier trended further westward (Habicht 1979), hence reaching the Juan Fernández Islands.

Weather

In the Juan Fernández Archipelago, the most important consideration for productive research is weather, not climate. The data on temperature and precipitation (Tables 4.1 and 4.2) provide some indication of the likely weather patterns in the islands, but how these factors interact on a daily basis is even more pertinent. The weather in both islands is changeable and somewhat unpredictable. With modern satellite weather data, cell phones, and Internet connections now available on Robinson Crusoe Island, it is much easier to plan daily trips over the islands. Alejandro Selkirk Island, although more isolated, has good radio contact with the park service (CONAF) on Robinson Crusoe Island, so daily weather reports are always available on both islands.

Because no roads exist on the islands, other than in the village of San Juan Bautista and near the airstrip, all collecting must be done on foot with backpacks. Many of the paths are over soil that becomes muddy and difficult during a rain. Worse are the basalt ridges that become dangerously slippery when wet. Hence, when it rains hard in the islands, no field work is possible. Many short-lived sprinkles also occasionally occur, but they usually do not produce enough moisture to stop work activities, particularly in the dense forest. On Alejandro Selkirk Island, thick fog often completely covers the top of the island over 500 m, a condition that makes collecting extremely dangerous and completely inadvisable.

For planning research expeditions in the archipelago, therefore, extra days need to be added to compensate for lost time due to inclement weather. This can also be an issue with regard to arriving and leaving the islands. All passengers to the archipelago, either by boat or by small airplane, arrive first on Robinson Crusoe Island. If the weather is bad, no planes can fly to or from the island. Landing and takeoff in the six- to nine-seat propeller planes are done mostly by visual cues, and cloud cover will close the small airstrip. The sea between Robinson Crusoe Island and Valparaíso on the continent can also become dangerous, which may delay sailing departures by several days. Broad oceanic storm fronts can also come into the region, restricting travel to and from the islands. On one unusual occasion several years ago, Japanese visitors came to Robinson Crusoe Island, intending to stay for three days, but due to bad weather, they ended up staying two weeks (Clarke 2001, p. 33)! These considerations all have to be built into the planning of research expeditions.

Part III

The Green Landscape

A fundamental understanding of the evolutionary and biogeographical processes in the Juan Fernández Archipelago, as well as proper management of conservation priorities, requires having a solid inventory of taxa and their assembly into vegetational patterns. The good news for this archipelago is that, over the course of more than 150 years, a comprehensive inventory of the species of the islands has been developed. The islands are not large, each only approximately 50 km^2. Although there are no roads for convenient access to all points of the islands, with diligence, most of the island surface can be visited at least to some extent. Due to the many botanical expeditions that have already been conducted in the archipelago, few taxonomic surprises remain; only a few new species have been described in recent decades. More challenging in developing our revised list of endemic and native taxa of the archipelago has been the need to take into account results of recent research, especially from modern molecular phylogenetic studies. This often results in transfers of species from one genus or family to another rather than reduction or addition of taxa.

Chapter 5 provides lists of native and endemic taxa, introduced species, and taxa currently under cultivation in the islands (in gardens and public spaces). The numbers deriving from the list of endemic and native species have been used for our calculations of endemism, geographical distribution within the archipelago, and summaries of life forms, as well as serving as the basis for biogeographical inferences. Chapter 6 presents an overview of vegetation types in the islands. These descriptions and maps of vegetation have been published previously, but this book provides a comparison of the patterns that occur on the two major islands. Because they are of different geological ages, there is an opportunity to understand the changes that have taken place through time, especially on the older Robinson Crusoe Island.

5 Taxonomic Inventory

Tod F. Stuessy, Roberto Rodríguez, Carlos M. Baeza, and Patricio López-Sepúlveda

This chapter presents a list of ferns, gymnosperms, and angiosperms that reside in the Juan Fernández Archipelago. Having an inventory of existing species is fundamental to all other data and concepts presented in this book. At a general level, a list of included species provides the reader with a concept of what the flora entails. This is particularly significant in island floras because the composition is often very different, even "disharmonious" (Carlquist 1974), in comparison with continental source regions. A list of species is also needed to allow phylogenetic relationships to be assessed critically and to gain insights into modes of speciation. Likewise, no biogeographical hypotheses can be developed unless a clear view of existing species is available. Such information also allows statistics to be developed for the flora, particularly assessments of the biological characters of the island flora and subdivisions into endemic, native, and introduced categories. For effective conservation, having these categories well delimited is essential. This also helps focus attention on the most dangerous invasive species that need to be monitored over time.

Previous Inventories of the Flora

Because the flora of the Juan Fernández Archipelago is relatively small and the islands themselves relatively tiny, the challenge of providing an inventory of the islands has not been insurmountable. Although collecting activity in the archipelago is not easy, largely due to the rugged terrain and the absence of roads, it is possible with diligence to gain access to most areas of the islands. As a result, by the time of the inventories of Christensen and Skottsberg (1920) and Skottsberg (1921), most of the vascular flora was reasonably well documented. As evidence, in the past three decades, only five new species have been described: *Gleichenia lepidota* (Rodríguez-Ríos 1990), *Robinsonia saxatilis* (Danton 2006a), *Carex stuessyi* (Wheeler 2007), *Erigeron corrales-molinensis* (Danton 2014, not accepted in this book), and *Erigeron stuessyi* (Valdebenito, see Appendix 2).

Details of previous expeditions to the Juan Fernández Archipelago and publications on the flora have been presented already in Chapter 2, and these details will not be repeated here. The first attempt to comprehensively chronicle the flora, however, was done by Hemsley (1884), who wrote up the results of collections brought back from the Challenger Expedition. He recognized 105 vascular plant species. This initial treatment was added to substantially by Johow (1896), who greatly increased our understanding of the flora and recognized 236 species. Skottsberg followed with his several expeditions and comprehensive listing and discussions on the flora (1921, 1956), in which he lists 330 species. More recently, the catalogue of Marticorena et al. (1998) presented 423 species for the flora, and Danton et al. (2006) followed with a listing of 519 species (excluding those cultivated in gardens). The increase in numbers of species from the archipelago in recent decades has been due largely to documentation of the introduced species in part because these keep arriving in the islands (especially on Robinson Crusoe Island, where the permanent village is located) and also because they were somewhat ignored in previous floristic treatments. The number of introduced species reported has increased from less than 50 in the mid-nineteenth century (Philippi 1856a, b) to more than 227 at the end of the twentieth century (Swenson et al. 1997).

Other resources for understanding the flora of the islands are those dealing with the entire country of Chile, of which the Juan Fernández Archipelago is part. For the ferns and fern allies, there is the useful volume on the ferns of Chile by Gunckel (1984) and, more recently, a book by Rodríguez (1995), which forms much of volume 1 of the new series *Flora de Chile*. The gymnosperms have also been treated in this same first volume of the *Flora*. Some families of angiosperms now have also been published in additional fascicles, but the series is still incomplete.

Concepts of Taxa

Important for presentation of any taxonomic inventory is a brief discussion of the taxonomic concepts being used. Fundamental in the taxonomic hierarchy and the level to which all other categories directly relate is the species (Stuessy 2009). We regard a species as a series of populations that potentially or actually exchange genes at some minimum level and that are largely reproductively isolated from other such population systems. This is essentially the biological species concept proposed by Dobzhansky (1937) and developed more thoroughly by Mayr (1942, 1963). Because species are reproductively isolated, in higher plants they are normally also morphologically and genetically distinct. For purposes of any practical work in the Juan Fernández Islands, if a population is not morphologically distinct from another population, it makes no sense to regard it as a distinct species. One might imagine a case of a cryptic species, which for some reason diverged genetically but not morphologically. Although one might be tempted to recognize such an interesting population formally at the specific level, it would be difficult to work with in the field and unhelpful for nearly all conservation initiatives. More attention has recently been given to this issue of cryptic variation in oceanic island floras (Crawford and Stuessy 2016).

Deciding what species exist on an oceanic island is most important for several reasons. A standard list needs to be provided so that all other investigations and calculations can be done in a consistent fashion. It is impossible to calculate levels of endemism, for example, if we do not know the number of species. One cannot calculate biogeographical affinities with continental source regions if we have no idea of what the species on the islands are, nor can one infer phylogenetic relationships or modes of speciation. At a very practical level, because nearly all endangered species legislation focuses on the species, it would be impossible to understand the conservation imperative in an archipelago without a consistent species list. An example of the problems that inconsistent species concepts can provide for an oceanic island system is the Hawaiian Archipelago, in which many populational variants or interspecific hybrids were treated as good species by Harold St. John (e.g., St. John and Takeuchi 1988, in *Cyrtandra*) and others over many decades. It was only with publication of the comprehensive flora of the Hawaiian Archipelago (Wagner et al. 1990) that a consistent and comprehensive species concept was applied to the flora that allowed further interpretations to be made.

In the Juan Fernández Archipelago, a too-narrow species concept that formally recognizes all sorts of populational variation would result in a long list of species. However, an overly broad species concept would submerge important morphologically and genetically distinct populations invisibly under a broad umbrella, with resulting loss of information regarding biodiversity in the archipelago. A consistent and realistic species concept, therefore, is required for dealing with the vascular flora of the archipelago. Fortunately, no worker inventorying the plants of these islands over the past 130 years has presented an extreme view of species. As a result, nearly all species recognized have morphological distinctness such that they can be recognized in the field and herbarium. There are some complexes, for example, the *Erigeron ingae* complex, that are more difficult and apparently reveal populational divergence that is still actively taking place (López-Sepúlveda et al. 2015b). Some species have also been synonymized, but these are rare occurrences in the flora.

Going downward from the specific level are the categories of subspecies, variety, and form. A few workers have used these levels to describe the morphological variations found among populations in the Juan Fernández Archipelago, and we have maintained some of these taxa. Appropriate application of these infraspecific concepts depends on having a large series of populations to assess relative similarities and differences across a larger geographical area. Because the islands are very small (ca. 50 km^2 for each of the major islands) and the populations of plants generally small, these concepts are not very applicable in the flora. It must be remembered that approximately 20% of the endemic species of flowering plants are known from no more than about 25 or fewer individuals (Stuessy et al. 1998b; see Chapter 9), which impedes application of population concepts and spatial relationships.

Going upward from the species level is the genus. In the Juan Fernández Islands, there has been more taxonomic uncertainty and change in generic concepts than with species. A genus is a lineage of one or more closely related species that is morphologically and genetically distinct from other such groups (Stuessy et al. 2014a). Genera are usually reproductively isolated from each other, but not always. In particular, rapid morphological divergence in genera endemic to oceanic islands is not always accompanied by strong genetic divergence. Genera may also be geographically isolated on an island and hence spatially reproductively isolated but easily crossed artificially in a common research garden. A good example of this is among the genera of the silverswords (Madiinae) in the Hawaiian Archipelago (Carr 2003). In nature, due to ecological divergence and isolation, these genera rarely hybridize, but all can be crossed artificially in the common garden. In the Juan Fernández Islands, ×*Margyricaena* (Rosaceae; Skottsberg 1921) represents a natural intergeneric hybrid between two genera, *Margyricarpus digynus*, a native species, and *Acaena argentea*, an introduced one (Crawford et al. 1993a).

Even without genetic data for comparison, for generic delimitation one relies on morphological similarities and differences among species. Because of rather dramatic divergences that can accrue among species of particular lineages, different opinions on generic limits can prevail. The most dramatic example of this in the Juan Fernández Archipelago is with the genus *Dendroseris*. Even Skottsberg, who applied a very consistent view of taxa at all levels, first (1921) regarded this genus in a broad sense, consisting of three subgenera, and then (1953b) changed his mind and elevated each of these to generic level. More recent workers, for example, Marticorena et al. (1998) and Crawford et al. (1992a), have once again treated all species of this lineage as belonging to a single genus because they are all derived from a single colonization of the islands and form a monophyletic group (Kim et al. 2007; see Chapter 13).

More problematical with regard to defining generic limits, however, has been the application of cladistic classification by some workers with emphasis on strict holophyly (= monophyly in a cladistic context). With cladistic classification, groups cannot be accepted that do not include *all* descendants from a common ancestor (Hennig 1966). At first glance, this may seem reasonable enough, and normally, no difficulties arise. But in oceanic islands quite a number of problems result from this perspective, and some are evident in the Juan Fernández Archipelago. The issue derives from new molecular data, that is, nucleotide sequences, that demonstrate that some island lineages have evolved out of a larger parental ancestral complex in the continent. In a cladistic context, to recognize the derived island species as a distinct genus, no matter how divergent morphologically or at the molecular level, would render the parental complex of species "paraphyletic," which is defined as a monophyletic group deriving from a single ancestor that does *not* contain *all* descendents from that ancestor. This is unacceptable to cladists (Hennig 1966). There are two cladistic solutions to this dilemma. The first is to not recognize the island species as a distinct genus and keep them taxonomically within the parental genus. The second is to recognize the island lineage as a genus along with the specific continental relatives (progenitors) and break up the larger continental group into smaller comparable genera that would be coordinate with the island one.

Two conspicuous examples of the submergence of endemic genera for cladistic reasons occur in the Juan Fernández Islands: *Dendroseris* and *Robinsonia* (both Asteraceae). The former is the largest endemic genus in the archipelago, and the latter is the next largest, with 11 and 8 endemic species, respectively. Collectively, they contain 18% of the endemic species of angiosperms in the archipelago. Kim et al. (2007), based on molecular phylogenetic studies of the genus *Sonchus*, have suggested that *Dendroseris* appears nested within it, and they have submerged all species of the latter into the former (Mejías and Kim 2012). Regarding *Robinsonia*, Pelser et al. (2007, 2010a) have done similar molecular phylogenetic studies with it and the large genus *Senecio*, once again finding the former nested within and originating out of the latter. These workers (Pelser 2010b) have also formally made the combinations into *Senecio*, and *Robinsonia* has also disappeared as an endemic genus within the islands. Stuessy et al. (2014b) have demonstrated that if concepts of strict holophyly were applied worldwide in oceanic island archipelagos, approximately 32% of endemic genera would disappear.

For the flora of the Juan Fernández Islands, we do not accept the application of strict concepts of holophyly in the classification of genera for several reasons. First, we regard the delimitation of genera to be a case of maximizing character information for achieving high predictive quality. Because island taxa often undergo dramatic morphological change as they evolve into the new island environment, this degree of difference should be recognized at the generic, rather than only the specific, level. This is why these groups have been called genera in the first place prior to the appearance of cladistic concepts. Second, we accept that species and genera can evolve from out of each other rather than assuming that all such groups can only diverge in parallel from a common ancestor. This is progenitor-derivative evolution, particularly well documented at the specific level (Crawford 2010). Third, the application of strict concepts of holophyly that results in loss of endemic genera in islands lowers the conservation imperative for many archipelagos. Such taxonomic changes are acceptable if errors of judgment in the past need to be corrected, but to do so solely to enforce strict adherence to cladistic classification seems to us misleading.

In the cases of *Dendroseris* and *Robinsonia*, therefore, we maintain both as distinct endemic genera in the Juan Fernández Archipelago, and we do not accept the corresponding name changes for species as proposed by Mejías and Kim (2012) and Pelser et al. (2010b). *Dendroseris* is a group of species originating from a single introduction (Crawford et al. 1992a) that are rosette trees with dichotomous branching, very distinct from their herbaceous relatives within *Sonchus*. *Robinsonia* is also a rosette tree and dioecious. This combination of features is unknown within *Senecio* of South America. Mejías and Kim (2012) have argued that because these distinctive characters are often a result of adaptations to new oceanic island habitats, which have occurred in parallel in many oceanic archipelagos, they should not be accorded taxonomic import. This misses the point that these distinctive features in *Robinsonia* and *Dendroseris* are apparently under genetic control and are not present in known progenitors. That this tendency has occurred in different archipelagos does not vitiate generic recognition in these two cases. Inferior ovaries, for example, hypothesized as adaptations for protecting ovules (Grant 1950), occur in many different groups of flowering plants, but this does not devalue their taxonomic utility among genera in different families.

Another example of application of strict holophyly in the Juan Fernández Archipelago involves endemic species of *Uncinia* and *Carex*. A number of molecular phylogenetic investigations using parsimony, maximum likelihood, and Bayesian inference with both nuclear and chloroplast loci have now been conducted in tribe Cariceae of Cyperaceae (Yen and Olmstead 2000a, 2000b, *ndhF*, *trnL-F*; Roalson et al. 2001, ITS, *trnT-L-F*; Starr et al. 2004, ITS, ETS 1; Waterway and Starr 2007, ITS, ETS 1f, *trnL-F*; Starr et al. 2008, ITS, ETS 1f; Starr and Ford 2009, ITS, ETS 1f). Because of the large size and morphological complexity of this group, there has been much focus on determining generic limits and relationships (Muasya et al. 2009). All studies reveal that *Uncinia* is a holophyletic group that nests within *Carex*, presumably having evolved from out of the larger parental genus (Saville and Calder 1953). In the summary provided by Starr and Ford (2009), *Uncinia* is supported by 98% bootstrap with *U. kingii* included, and 100% if this species is excluded (see also Starr et al. 2004). Reznicek (1990, p. 1419) earlier transferred this species to *Carex* because the rachilla is only weakly hooked, hence making "*Uncinia* a much more uniform and presumably natural genus." Another species, *U. microglochin*, has also been at times transferred into *Carex* because of its exserted but unhooked rachilla (Starr et al. 2004). It should also be mentioned that based on a comprehensive morphological cladistic analysis of all genera of Cyperaceae (Bruhl 1995), *Uncinia* was maintained as a good genus with closest ties to *Cymophyllus*. Starr et al. (2004), based on molecular phylogenetic assessments, revealed *Uncinia* to have 99% bootstrap support (with ITS and parsimony) and a long-separated branch from species of *Carex* (with maximum likelihood phylogram with ITS and ETS 1f data). They also recommended separation of *Carex* and *Uncinia* because (p. 540) "… the completely closed utricle, the only unambiguous character that unites *Carex* and *Uncinia*, is homoplastic." Despite these previous viewpoints, all species of *Uncinia* have recently been submerged into *Carex* in order to maintain holophyly of the latter in the context of strict cladistic classification (Global Carex Group 2015). If this approach were to be adopted for the species of the Juan Fernández Archipelago, three of the six names would have to be changed completely due to the epithets already being occupied within *Carex*. *Uncinia costata* would become *Carex plurinervata, U. douglasii* changes to *C. fernandesiana* (an inconvenient name because there already exists the quite similar but not strictly homonymous *C. fernandezensis* Mackenzie ex G. A. Wheeler) (Wheeler 2007, p. 127), and the native *U. tenuis* becomes *C. firmula*. The other three species, *U. aspericaulis* (endemic), *U. macloviformis* (endemic), and *U. phleoides* (native), retain their epithets within *Carex*. In this book we do not follow transfer of *Uncinia* into *Carex* but retain the names in the former genus. *Uncinia* is distinct from *Carex* by the former having hooked rachillae exserted from the utricle, which, in our view, is sufficient for generic recognition (see agreement by Reznicek 1990, p. 1421).

Another case worth mentioning is the family Lactoridaceae, containing the single genus and species, *Lactoris fernandeziana*. Molecular phylogenetic studies (Qiu et al. 1993; Wanke et al. 2007) have sometimes shown this taxon to be joined with other genera within Aristolochiaceae, although sequence divergence has occurred between it and other members of this family. From the perspective of cladistic classification, and hence to avoid an unacceptable paraphyletic Aristolochiaceae, *Lactoris* should not be placed in its own family but rather submerged into Aristolochiaceae, where it resides in the latest Angiosperm Phylogeny Group classification (APG IV 2016). Lactoridaceae is the only family in the world that is restricted to an oceanic island. It also is of great conservation interest because there are no more than approximately 1,000 individuals left on Robinson Crusoe Island, fortunately in remote areas (Bernardello et al. 1999;

Crawford et al. 2001a; Ricci 2001a). Paleopalynological studies (Zavada and Benson 1987; Macphail et al. 1999; Gamerro and Barreda 2008) have revealed that the species is ancient, with fossil pollen documented from Cretaceous to Miocene, and has a broad southern hemisphere distribution (known from Africa, Australia, and southern South America), but now it is reduced to confinement (relictual) only on Robinson Crusoe Island. We reject submergence of *Lactoris* into Aristolochiaceae because of its very different morphology (Carlquist 1964; Tucker and Douglas 1996; González and Rudall 2001; González et al. 2001; Kelly and Gonzalez 2003), palynology (Zavada and Taylor 1986; Sampson 1995), anatomy (Carlquist 1990; Wagner et al. 2014), and embryology (Tobe et al. 1993; González et al. 2001) and because of the long nucleotide-based branches that separate it from other families. In this book, *Lactoris* is maintained within its own family, Lactoridaceae (for additional data and arguments, see Stuessy et al. 1998c).

Lists of Species

The species that are found in the Juan Fernandez Archipelago have been grouped into the three categories of native and endemic, introduced, and cultivated (Tables. 5.1, 5.2, and 5.3). For evolutionary and biogeographical purposes, the species of importance are those that have arrived in the islands and/or speciated there, that is, the native and endemic ones. All our discussions in this book on biogeographical and evolutionary principles and calculations deal with these species. Introduced species have come to the archipelago via aid from humans, either purposely or inadvertently. These are also of historical interest, as well as serious conservation concern, because many of them are invasives that have already become serious pests that threaten the native and endemic species (e.g., *Rubus ulmifolius* and *Aristotelia chilensis*). Cultivated plants are those that exist now in gardens or common areas of the two islands, essentially in the village of San Juan Bautista (Robinson Crusoe Island) and the settlement of Las Casas (Alejandro Selkirk Island). It is useful to keep a watchful conservation eye on these species, however, because the harmless garden plant of today, if it escapes, could become a troublesome or even dangerous pest tomorrow. There is a long history of reporting cultivated plants in the archipelago, both for cultural and horticultural objectives. Johow (1893), for example, made a detailed analysis of the species found on the islands at that time. We do not provide here an analysis of all species reported historically because this is of no particular relevance for present conservation purposes. The data listed in Table 5.3 have been summarized from Swenson et al. (1997), Cuevas (2004), Danton et al. (2006), and López-Sepúlveda et al. (2013a). In addition to the list of scientific names of species given in Table 5.1, we also present here a list of the local (or vernacular) names used by people in the Juan Fernández Archipelago (Table 5.4). These labels are particularly helpful to use when working with guides from the village. The list is based primarily on the summary provided by Gunckel (1968), which involved compilation from previous floristic treatments plus his own experience in the islands.

Table 5.1 List of Native and Endemic Taxa of Ferns, Fern Allies, and Angiosperms of the Juan Fernández Archipelago

Taxon	Distribution	Life form	Biogeographical status	Conservation status
FERNS				
Aspleniaceae				
Asplenium dareoides Desv.	AS, RC	Herb	Native	Vulnerable
Asplenium macrosorum Bertero ex Colla	AS, RC	Herb	Endemic	Endangered
Asplenium obtusatum G.Forst. var. *sphenoides* (Kunze) C.Chr. ex Skottsb. (Fig. C8)	AS, RC, SC	Herb	Native	Vulnerable
Asplenium stellatum Colla	AS, RC	Herb	Endemic	Endangered
Blechnaceae				
Blechnum chilense (Kaulf.) Mett.[a] (Fig. C9)	AS, RC	Herb	Native	LC
Blechnum cordatum (Desv.) Hieron.	AS, RC	Herb	Native	LC
Blechnum cycadifolium (Colla) Sturm (Fig. C10)	AS, RC	Tree fern	Endemic	Vulnerable
Blechnum hastatum Kaulf.	AS, RC, SC	Herb	Native	LC
Blechnum longicauda C.Chr. (Fig. C11)	AS	Herb	Endemic	Cr endangered
Blechnum mochaenum G.Kunkel var. *fernandezianum* (Looser) de la Sota	AS, RC	Herb	Endemic	Vulnerable
Blechnum schottii (Colla) C.Chr. (Fig. C12)	AS, RC	Herb	Endemic	Vulnerable
Cystopteridaceae				
Cystopteris fragilis (L.) Bernh. var. *apiiformis* (Gand.) C.Chr.	AS	Herb	Native	Endangered
Dennstaedtiaceae				
Histiopteris incisa (Thunb.) J.Sm.	AS, RC	Herb	Native	LC
Hypolepis poeppigii (Kunze) R.Rodr.	AS, RC	Herb	Native	Endangered
Dicksoniaceae				
Dicksonia berteroana (Colla) Hook. (Figs. C13 and C14)	RC	Tree fern	Endemic	Vulnerable
Dicksonia externa Skottsb. (Figs. C15 and C16)	AS	Tree fern	Endemic	Vulnerable
Lophosoria quadripinnata (J.F.Gmel.) C.Chr. (Fig. C17)	AS, RC	Herb	Native	LC
Dryopteridaceae				
Elaphoglossum lindenii (Bory) Moore	AS, RC	Herb	Native	Endangered
Megalastrum glabrius (C.Chr. & Skottsb.) Sundue, Rouhan & R.C.Moran	AS	Herb	Endemic	Cr endangered
Megalastrum inaequalifolium (Colla) A.R.Sm. & R.C.Moran (Fig. C18)	RC, SC	Herb	Endemic	Vulnerable
Megalastrum masafuerae Sundue, Rouhan & R.C.Moran	AS	Herb	Endemic	Cr endangered
Polystichum tetragonum Fée (Fig. C19)	AS, RC	Herb	Endemic	Vulnerable
Rumohra berteroana (Colla) R.Rodr.	AS, RC	Herb	Endemic	Vulnerable
Gleicheniaceae				
Sticherus lepidotus (R.A.Rodr.) R.A. Rodr. & Ponce	AS	Herb	Endemic	Endangered

Table 5.1 (cont.)

Taxon	Distribution	Life form	Biogeographical status	Conservation status
Sticherus quadripartitus (Poir. in Lam.) Ching (Fig. C20)	AS	Herb	Native	DD
Sticherus squamulosus (Desv.) Nakai var. *squamulosus*	RC	Herb	Native	Vulnerable
Hymenophyllaceae				
Hymenophyllum caespitosum Gaudich.	AS, RC	Herb	Native	Vulnerable
Hymenophyllum caudiculatum Mart. var. *productum* (C.Presl) C.Chr.	AS, RC	Herb	Native	Endangered
Hymenophyllum cruentum Cav. (Fig. C21)	AS, RC	Herb	Native	Vulnerable
Hymenophyllum cuneatum Kunze (Fig. C22)	AS, RC	Herb	Native	LC
Hymenophyllum falklandicum Baker var. *falklandicum*	AS	Herb	Native	Endangered
Hymenophyllum ferrugineum Colla var. *ferrugineum*	AS, RC	Herb	Native	Vulnerable
Hymenophyllum fuciforme Sw.	AS, RC	Herb	Native	Endangered
Hymenophyllum pectinatum Cav.	AS, RC	Herb	Native	Endangered
Hymenophyllum plicatum Kaulf.	AS, RC	Herb	Native	LC
Hymenophyllum rugosum C.Chr. & Skottsb.	AS, RC	Herb	Endemic	Vulnerable
Hymenophyllum secundum Hook. & Grev.	AS	Herb	Native	Endangered
Hymenophyllum tortuosum Hook. & Grev. var. *tortuosum*	AS	Herb	Native	Vulnerable
Polyphlebium exsectum (Kunze) Ebihara & Dubuisson	AS, RC	Herb	Native	Vulnerable
Polyphlebium ingae (C.Chr. & Skottsb.) Ebihara & Dubuisson	RC	Herb	Endemic	Endangered
Polyphlebium philippianum (Sturm) Ebihara & Dubuisson	RC	Herb	Endemic	Endangered
Ophioglossaceae				
Ophioglossum fernandezianum C.Chr.	RC	Herb	Endemic	DD
Polypodiaceae				
Grammitis magellanica Desv.	AS, RC	Herb	Native	Vulnerable
Pleopeltis ×cerro-altoensis Danton & Boudrie	RC	Herb	Endemic	Cr endangered
Pleopeltis macrocarpa (Bory ex Willd.) Kaulf. (Fig. C23)	AS, RC	Herb	Native	LC
Polypodium intermedium Colla subsp. *intermedium* (Fig. C24)	RC	Herb	Endemic	Vulnerable
Polypodium intermedium Colla subsp. *masafueranum* C.Chr. & Skottsb.	AS	Herb	Endemic	Endangered
Polypodium masafuerae Phil.[b]	AS	Herb	Endemic	Endangered

Table 5.1 (cont.)

Taxon	Distribution	Life form	Biogeographical status	Conservation status
Pteridaceae				
Adiantum chilense Kaulf. var. *chilense*	AS, RC, SC	Herb	Native	LC
Argyrochosma chilensis (Fée & Remy) Windham	AS, RC	Herb	Endemic	Vulnerable
Pteris berteroana J.Agardh (Fig. C25)	AS, RC	Herb	Endemic	Vulnerable
Pteris chilensis Desv.	AS, RC	Herb	Native	LC
Pteris semiadnata Phil.	AS, RC	Herb	Native	Endangered
Tectariaceae				
Arthropteris altescandens (Colla) J.Sm. (Fig. C27)	AS, RC	Herb	Endemic	Vulnerable
Thyrsopteridaceae				
Thyrsopteris elegans Kunze (Fig. 9.6)	AS, RC	Tree fern	Endemic	Endangered
FERN ALLIES				
Lycopodiaceae				
Lycopodium gayanum J.Remy	AS	Herb	Native	Vulnerable
Lycopodium magellanicum (P. Beauv.) Sw. var. *magellanicum*	AS	Herb	Native	Endangered
ANGIOSPERMS – ARCHAEANGIOSPERMAE				
Lactoridaceae				
Lactoris fernandeziana Phil. (Figs. C26 and 9.7)	RC	Subshrub	Endemic	Endangered
Piperaceae				
Peperomia berteroana Miq. subsp. *berteroana* (Fig. C27)	AS, RC	Herb	Endemic	Vulnerable
Peperomia fernandeziana Miq. (Fig. C28)	AS, RC	Herb	Native	Vulnerable
Peperomia margaritifera Bertero ex Hook.	RC	Herb	Endemic	Endangered
Peperomia skottsbergii C.DC.	AS	Herb	Endemic	Endangered
Winteraceae				
Drimys confertifolia Phil. (Fig. C29)	AS, RC	Tree	Endemic	Vulnerable
ANGIOSPERMS – MONOCOTYLEDONAE				
Arecaceae (Palmae)				
Juania australis (Mart.) Drude ex Hook.f. (Fig. C30)	RC	Tree	Endemic	Endangered
Bromeliaceae				
Greigia berteroi Skottsb.	RC	Herb	Endemic	Endangered
Ochagavia elegans Phil. (Fig. C31)	RC	Herb	Endemic	Vulnerable
Cyperaceae				
Carex berteroniana Steud.	AS, RC	Herb	Endemic	Vulnerable
Carex fernandezensis Mackenzie ex G.A. Wheeler	RC	Herb	Endemic	Endangered
Carex phalaroides Kunth	RC	Herb	Native	Vulnerable

Table 5.1 (cont.)

Taxon	Distribution	Life form	Biogeographical status	Conservation status
Carex stuessyi G.A. Wheeler	AS	Herb	Endemic	Endangered
Cyperus eragrostis Lam.	AS, RC	Herb	Native	Vulnerable
Cyperus reflexus Vahl	RC	Herb	Native	LC
Eleocharis fuscopurpurea (Steud.) H.Pfeiff.	RC	Herb	Native	LC
Machaerina scirpoidea (Steud.) Koyama ex M.T.Strong	RC	Herb	Endemic	Vulnerable
Oreobolus obtusangulus Gaudich.	AS	Herb	Native	DD
Scirpus cernuus Vahl	AS, RC, SC	Herb	Native	LC
Scirpus nodosus Rottb.	AS, RC	Herb	Native	LC
Uncinia aspericaulis G.A.Wheeler	AS	Herb	Endemic	Endangered
Uncinia costata Kük.	AS	Herb	Endemic	Endangered
Uncinia douglasii Boott	AS, RC	Herb	Endemic	Vulnerable
Uncinia macloviformis G.A.Wheeler	AS	Herb	Endemic	Endangered
Uncinia phleoides (Cav.) Pers.	AS	Herb	Native	Vulnerable
Uncinia tenuis Poepp. ex Kunth	AS	Herb	Native	Vulnerable
Iridaceae				
Herbertia lahue (Molina) Goldb.	AS, RC	Herb	Native	Vulnerable
Libertia chilensis (Molina) Gunckel (Fig. C32)	AS, RC	Herb	Native	LC
Juncaceae				
Juncus capillaceus Lam.	AS, RC	Herb	Native	LC
Juncus imbricatus Laharpe	AS, RC	Herb	Native	Vulnerable
Juncus pallescens Lam.	RC	Herb	Native	Vulnerable
Juncus planifolius R.Br.	RC	Herb	Native	Vulnerable
Juncus procerus E.Mey (Fig. C33)	AS, RC	Herb	Native	Vulnerable
Luzula masafuerana Skottsb.	AS	Herb	Endemic	Endangered
Orchidaceae				
Gavilea insularis M.N. Correa	AS	Herb	Endemic	Endangered
Poaceae				
Agrostis masafuerana Pilger	AS	Herb	Endemic	Endangered
Chusquea fernandeziana Phil.	RC	Shrub (bamboo)	Endemic	Vulnerable
Danthonia chilensis E.Desv. var. *chilensis*	RC	Herb	Native	LC
Danthonia malacantha (Steud.) Pilg.	RC	Herb	Native	DD
Megalachne berteroana Steud.	RC	Herb	Endemic	Vulnerable
Megalachne masafuerana (Skottsb. & Pilg. ex Pilg.) Matthei	AS	Herb	Endemic	Endangered
Megalachne robinsoniana C.Peña	RC	Herb	Endemic	Endangered
Piptochaetium bicolor (Vahl) E.Desv.	RC	Herb	Native	LC
Podophorus bromoides Phil.	RC	Herb	Endemic	Extinct?
ANGIOSPERMS – DICOTYLEDONAE				
Apiaceae (Umbelliferae)				
Apium australe Thouars	RC	Herb	Native	LC
Apium fernandezianum Johow	RC, SC	Herb	Endemic	Endangered

Table 5.1 (cont.)

Taxon	Distribution	Life form	Biogeographical status	Conservation status
Centella asiatica (L.) Urb.	RC	Herb	Native	Vulnerable
Eryngium bupleuroides Hook. & Arn. (Fig. C34)	RC	Rosette tree	Endemic	Endangered
Eryngium ×fernandezianum Skottsb.	RC	Rosette tree	Endemic	Endangered
Eryngium inaccessum Skottsb.	RC	Rosette tree	Endemic	Endangered
Eryngium sarcophyllum Hook. et Arn.	AS	Rosette tree	Endemic	Extinct
Asteraceae (Compositae)				
Abrotanella linearifolia A. Gray	AS	Herb	Native	Endangered
Centaurodendron dracaenoides Johow (Fig. C35)	RC	Rosette tree	Endemic	Endangered
Centaurodendron palmiforme Skottsb.	RC	Rosette tree	Endemic	Endangered
Dendroseris berteroana (Decne.) Hook. & Arn. (Figs. C36, C37, and C38)	RC	Rosette tree	Endemic	Endangered
Dendroseris gigantea Johow	AS	Rosette tree	Endemic	Cr endangered
Dendroseris litoralis Skottsb. (Fig. C41)	RC, SC	Rosette tree	Endemic	Cr endangered
Dendroseris macrantha (Bertero ex Decne.) Skottsb.	RC	Rosette tree	Endemic	Cr endangered
Dendroseris macrophylla D.Don	AS	Rosette tree	Endemic	Cr endangered
Dendroseris marginata (Bertero ex Decne.) Hook. & Arn. (Figs. C42, C43, and C44)	RC	Rosette tree	Endemic	Endangered
Dendroseris micrantha (Bertero ex Decne.) Hook. & Arn. (Figs. C47, C48, and C49)	RC	Rosette tree	Endemic	Endangered
Dendroseris neriifolia (Decne.) Hook. & Arn. (Figs. C50, C51, and C52)	RC	Rosette tree	Endemic	Cr endangered
Dendroseris pinnata (Bertero ex Decne.) Hook. & Arn. (Figs. C39 and C40)	RC	Rosette tree	Endemic	Endangered
Dendroseris pruinata (Johow) Skottsb. (Figs. C45 and C46)	RC, SC	Rosette tree	Endemic	Endangered
Dendroseris regia Skottsb.	AS	Rosette tree	Endemic	Cr endangered
Erigeron fernandezia (Colla) Harling (Figs. C53 and C54)	AS, RC	Herb	Endemic	Vulnerable
Erigeron ingae Skottsb. (Fig. C55)	AS	Herb	Endemic	Endangered
Erigeron luteoviridis Skottsb.	AS	Herb	Endemic	Endangered
Erigeron rupicola Phil. (Fig. C56)	AS	Herb	Endemic	Vulnerable
Erigeron stuessyi Valdebenito (see Appendix 2 and Fig. C57)	AS	Herb	Endemic	Endangered
Gamochaeta chamissonis (DC.)[d] Cabrera	AS, RC	Herb	Native	Endangered
Lagenophora hariotii Franch.	AS	Herb	Native	Vulnerable
Robinsonia berteroi (DC.) Sanders, Stuessy & Marticorena (Fig. C58)	RC	Rosette tree	Endemic	Extinct[e]
Robinsonia evenia Phil. (Figs. C64 and C65)	RC	Rosette tree	Endemic	Endangered
Robinsonia gayana Decne. (Fig. C61)	RC	Rosette tree	Endemic	Endangered

Table 5.1 (cont.)

Taxon	Distribution	Life form	Biogeographical status	Conservation status
Robinsonia gracilis Decne. (Figs. C62 and C63)	RC	Rosette tree	Endemic	Endangered
Robinsonia macrocephala Decne.	RC	Rosette tree	Endemic	Extinct
Robinsonia masafuerae Skottsb. (Figs. C66 and C67)	AS	Rosette tree	Endemic	Endangered
Robinsonia saxatilis Danton	RC	Rosette tree	Endemic	Cr endangered
Robinsonia thurifera Decne. (Figs. C59 and C60)	RC	Rosette tree	Endemic	Cr endangered
Taraxacum fernandezianum Dahlst.	AS, RC	Herb	Native	DD
Yunquea tenzii Skottsb.	RC	Rosette tree	Endemic	Cr endangered
Berberidaceae				
Berberis corymbosa Hook. & Arn.	RC	Shrub	Endemic	Endangered
Berberis masafuerana Skottsb.	AS	Shrub	Endemic	Endangered
Boraginaceae				
Selkirkia berteroi (Colla) Hemsl.	RC	Shrub	Endemic	Cr endangered
Brassicaceae (Cruciferae)				
Cardamine chenopodiifolia Pers.	RC	Herb (annual)	Native	Vulnerable
Cardamine flaccida Cham. & Schltdl.	RC	Herb	Native	Vulnerable
Cardamine kruesselii Johow ex Reiche	AS	Herb	Endemic	Endangered
Campanulaceae				
Lobelia anceps L.f.	AS, RC	Herb	Native	Vulnerable
Wahlenbergia berteroi Hook. & Arn. (Fig. C68)	RC, SC	Shrub	Endemic	Vulnerable
Wahlenbergia fernandeziana A.DC. (Fig. C69)	RC	Shrub	Endemic	Vulnerable
Wahlenbergia grahamiae Hemsl.	RC	Shrub	Endemic	Vulnerable
Wahlenbergia masafuerae (Phil.) Skottsb.	AS	Shrub	Endemic	Endangered
Wahlenbergia tuberosa Hook.f.	AS	Shrub	Endemic	Endangered
Caryophyllaceae				
Spergularia confertiflora Steud. var. *confertiflora*	AS, RC, SC	Herb	Endemic	Vulnerable
Spergularia confertiflora Steud. var. *polyphylla* (Phil.) Skottsb.	AS, RC	Herb	Endemic	Vulnerable
Spergularia masafuerana Skottsb.	AS	Herb	Endemic	Endangered
Chenopodiaceae				
Chenopodium crusoeanum Skottsb.	RC	Shrub	Endemic	Cr endangered
Chenopodium nesodendron Skottsb.	AS	Shrub	Endemic	Cr endangered
Chenopodium sanctae-clarae Johow (Fig. C70)	SC (RC, cult.)	Shrub	Endemic	Cr endangered
Convolvulaceae				
Calystegia tuguriorum (G.Forst.) R.Br. ex Hook.f.	AS	Herb	Native	Endangered
Dichondra sericea Sw.	AS, RC	Herb	Native	Vulnerable
Ericaceae				
Empetrum rubrum Vahl	AS	Shrub	Native	Extirpated
Pernettya rigida (Bertero ex Colla) DC. (Fig. C71)	AS, RC	Shrub	Endemic	Vulnerable

Table 5.1 (cont.)

Taxon	Distribution	Life form	Biogeographical status	Conservation status
Escalloniaceae				
Escallonia callcottiae Hook. & Arn. (Fig. C72)	RC	Shrub	Endemic	Vulnerable
Euphorbiaceae				
Dysopsis hirsuta (Mull.Arg.) Skottsb. (Fig. C73)	RC	Herb	Endemic	Vulnerable
Fabaceae (Leguminosae)				
Sophora fernandeziana (Phil.) Skottsb. var. *fernandeziana* (Fig. C74)	RC	Tree	Endemic	Endangered
Sophora fernandeziana (Phil.) Skottsb. var. *reedeana* (Phil.) Skottsb.	RC	Tree	Endemic	Endangered
Sophora masafuerana (Phil.) Skottsb.	AS	Tree	Endemic	Endangered
Gunneraceae				
Gunnera bracteata Steud. ex Bennett	RC	Large herb	Endemic	Vulnerable
Gunnera masafuerae Skottsb. (Fig. C75)	AS	Large herb	Endemic	Vulnerable
Gunnera peltata Phil. (Fig. C76)	RC	Large herb	Endemic	Vulnerable
Haloragaceae				
Haloragis masafuerana Skottsb. var. *masafuerana*	AS	Herb	Endemic	Endangered
Haloragis masafuerana Skottsb. var. *asperrima* (Skottsb.) Orchard	AS	Herb	Endemic	Endangered
Haloragis masatierrana Skottsb. (Fig. C78)	RC	Herb	Endemic	Endangered
Lamiaceae (Labiatae)				
Cuminia eriantha (Benth.) Benth. (Fig. C77)	RC	Tree	Endemic	Endangered
Cuminia fernandezia Colla	RC	Tree	Endemic	Endangered
Loranthaceae				
Notanthera heterophylla (Ruiz & Pav.) G.Don	RC	Shrub	Native	Extirpated[f]
Myrtaceae				
Myrceugenia schulzei Johow (Fig. C79)	AS	Tree	Endemic	Vulnerable
Myrteola nummularia (Poir.) O.Berg	AS	Shrub	Native	LC
Nothomyrcia fernandeziana (Hook. & Arn.) Kausel (Fig. C80)	RC	Tree	Endemic	Vulnerable
Ugni selkirkii (Hook. & Arn.) O.Berg (Fig. C81)	RC	Shrub	Endemic	Endangered
Orobanchaceae				
Euphrasia formosissima Skottsb.	AS	Herb	Endemic	Endangered
Phrymaceae				
Mimulus glabratus Kunth	AS, RC[g]	Herb	Native	Vulnerable
Plantaginaceae				
Plantago australis Lam.	RC	Herb	Native	LC
Plantago fernandezia Bertero ex Barnéoud	RC	Rosette tree	Endemic	Cr endangered
Plantago firma Kunze ex Walp.	RC, SC	Herb (annual)	Native	DD

Table 5.1 (cont.)

Taxon	Distribution	Life form	Biogeographical status	Conservation status
Ranunculaceae				
Ranunculus caprarum Skottsb. (Fig. C82)	AS	Herb	Endemic	Cr endangered
Rhamnaceae				
Colletia spartioides Bertero ex Colla	RC	Shrub	Endemic	Endangered
Rosaceae				
Acaena masafuerana Bitter (Fig. C83)	AS	Herb	Endemic	Vulnerable
×*Margyracaena skottsbergii* Bitter	RC	Herb	Endemic	Extinct (in wild)
Margyricarpus digynus (Bitter) Skottsb.	RC	Herb	Endemic	Endangered
Rubus geoides Sm. (Fig. C83)	AS	Low shrub	Native	Vulnerable
Rubiaceae				
Coprosma oliveri Fosberg	RC	Tree	Endemic	Endangered
Coprosma pyrifolia (Hook. & Arn.) Skottsb.	AS, RC	Tree	Endemic	Endangered
Galium masafueranum Skottsb.	AS	Herb	Endemic	Endangered
Hedyotis salzmannii (DC.) Steud.	RC	Herb	Native	Vulnerable
Nertera granadensis (Mutis ex L.f.) Druce	AS	Herb	Native	LC
Rutaceae				
Zanthoxylum externum (Skottsb.) Stuessy (see Appendix 2)	AS	Tree	Endemic	Endangered
Zanthoxylum mayu Bert. (Fig. C84)	RC	Tree	Endemic	Vulnerable
Salicaceae				
Azara serrata Ruiz & Pavón var. *fernandeziana* (Gay) Reiche (Fig. C85)	RC	Tree	Endemic	Endangered
Santalaceae				
Santalum fernandezianum F.Phil.	RC	Tree	Endemic	Extinct
Solanaceae				
Nicotiana cordifolia Phil. subsp. *cordifolia*	AS	Subshrub	Endemic	Endangered
Nicotiana cordifolia Phil. subsp. *sanctaclarae* Danton	SC (RC^h)	Subshrub	Endemic	Endangered
Solanum fernandezianum Phil.	RC	Herb	Endemic	Endangered
Solanum pentlandii Dunal subsp. *interandinum* (Bitter) Edmonds	AS	Herb	Native	Endangered
Urticaceae				
Bohemeria excelsa (Bertero ex Steud.) Wedd. (Fig. C86)	RC	Tree	Endemic	Endangered
Parietaria debilis G.Forst.	AS, RC, SC	Herb	Native	LC
Urtica glomeruliflora Steud.	AS, RC	Herb	Endemic	Endangered
Urtica masafuerae Phil.	AS	Herb	Endemic	Cr endangered
Verbenaceae				
Rhaphithamnus venustus (Phil.) B.L.Rob. (Fig. C87)	AS, RC	Tree	Endemic	Vulnerable

Note: Placement of genera into families of angiosperms follows Mabberley (2008) and APG (1998, 2003, 2009, 2016) with some modifications and for ferns from Windham (1987), Kramer and Green (1990), Smith et al.

(2006), de la Sota et al. (2007), Christenhusz et al. (2011), and Liu et al. (2013). For more information on the species listed as "endangered," see Chapter 9. This list is based primarily on herbarium collections at CONC, OS, and WU, plus citations in Johow (1896), Skottsberg (1921), Marticorena et al. (1998), Danton et al. (2006), Baeza et al. (2007), Wheeler (2007), Rodríguez (2015), Freire et al. (2016), and C. Taylor (unpublished manuscript, Urticaceae). Herb, perennial herb; LC, least concern; Cr endangered, critically endangered; DD, data deficient (i.e., status uncertain); RC, Robinson Crusoe Island; AS, Alejandro Selkirk Island; SC, Santa Clara Island.

[a] Tryon and Stolze (1993) and Prada et al. (2008) have placed *Blechnum chilense* in synonymy under *B. cordatum*, but we follow Rodríguez (2015) in maintaining them as distinct.

[b] de la Sota (2007), Smith and Tejero-Díez (2014), and Danton et al. (2015) have treated this species as belonging to *Pleopeltis*, but we consider it better left in *Polypodium* due to a lack of peltate scales (R. Rodríguez, personal communication).

[c] Baeza et al. (2002, 2007) indicate that this species appears to be extinct.

[d] Freire et al. (2016) have recently submerged the formerly endemic *Gamochaeta fernandeziana* (Phil.) Anderb. into *G. chamissonis*, which is known also in Chile and adjacent areas of Argentina.

[e] In April of 2016, there was an online report by Mauricio Silva that the guides of CONAF have located one (male) plant of *Robinsonia berteroi* on the summit of El Yunque on Robinson Crusoe Island. This is encouraging news, but confirmation of this report is needed.

[f] Danton et al. (2006) have suggested that this species may be extirpated.

[g] von Bohlen (1995) questions whether *Mimulus glabratus* occurs on Robinson Crusoe Island.

[h] Growing cultivated on Robinson Crusoe Island only in the CONAF garden in San Juan Bautista.

Table 5.2 List of Introduced Taxa of Gymnosperms and Angiosperms of the Juan Fernández Archipelago

Taxon	Distribution	Life form
GYMNOSPERMS		
Cupressaceae		
Cupressus macrocarpa Hartw.[a]	AS, RC	Tree
Cupressus sempervirens L.	RC	Tree
Pinaceae		
Pinus radiata D.Don	AS, RC	Tree
ANGIOSPERMS – MONOCOTYLEDONAE		
Amaryllidaceae		
Amaryllis belladona L.	RC	Herb
Leucojum vernum L.	RC	Herb
Araceae		
Zantedeschia aethiopica (L.) Spreng.	AS, RC	Herb
Commelinaceae		
Tradescantia fluminensis Vell.	RC, SC	Herb
Cyperaceae		
Isolepis cf. *cernua* (Vahl) Roem. & Schult.	AS	Herb
Iridaceae		
Crocosmia ×*crocosmiiflora* (G.Nicolson) ex N.E.Br.	AS, RC	Herb
Iris ×*germanica* L.	AS, RC	Herb
Juncaceae		
Juncus bufonius L.	AS, RC	Herb

Table 5.2 (cont.)

Taxon	Distribution	Life form
Poaceae		
Agrostis stolonifera L.	AS, RC	Herb
Aira caryophyllea L.	AS, RC	Herb
Aira praecox L.	AS, RC	Herb
Anthoxanthum odoratum L.	AS, RC	Herb
Avena barbata Pott ex Link	AS, RC, SC	Herb
Brachypodium distachyon (L.) P.Beauv.	AS	Herb
Briza maxima L.	RC	Herb
Briza minor L.	AS, RC, SC	Herb
Briza subaristata Lam.	RC	Herb
Bromus berteroanus Colla	RC, SC	Herb
Bromus catharticus Vahl	AS, RC	Herb
Bromus cebadilla Steud.	AS, RC	Herb
Bromus diandrus Roth	AS, RC	Herb
Bromus hordeaceus L.	AS, RC, SC	Herb
Bromus lithobius Trin.	AS, RC	Herb
Bromus stamineus E.Desv.	AS, RC	Herb
Chaetotropis chilensis Kunth	RC	Herb
Chaetotropis imberbis (Phil.) Björkman	AS, RC	Herb
Chascolytrum subaristatum Desv.	RC	Herb
Cynosorus echinatus L.	AS, RC	Herb
Dactylis glomerata L.	RC	Herb
Digitaria sanguinalis (L.) Scop.	RC	Herb
Echinochloa colona (L.) Link	AS	Herb
Elusine tristachya (Lam.) Lam.	AS, RC	Herb
Festuca arundinacea Schreber	RC	Herb
Gastridium ventricosum (Gouan) Schinz & Thell.	RC	Herb
Hordeum chilense Roem. & Schult.	RC	Herb
Hordeum murinum L. subp. *murinum*	AS, RC, SC	Herb
Hordeum secalinum Schreb.	RC	Herb
Leptophyllochloa micrathera (E.Desv.) C.Calderón ex Nicora	RC	Herb
Lolium multiflorum Lam.	AS, RC	Herb
Lolium perenne L.	RC	Herb
Nassella laevissima (Phil.) Barkworth	AS, RC	Herb
Nassella neesiana (Trin. & Rupr.) Barkworth	AS, RC	Herb
Paspalum dasypleurum Kunze ex E.Desv.	RC	Herb
Paspalum distichum L.	RC	Herb
Phalaris amethystina Trin.	RC	Herb
Phalaris angusta Nees ex Trin.	RC	Herb
Poa annua L.	AS, RC	Herb
Poa pratensis L.	AS, RC	Herb
Polypogon australis Brongn.	AS, RC	Herb
Setaria parviflora (Poir.) Kerguélen	AS, RC	Herb
Setaria viridis (L.) P.Beauv.[b]	RC	Herb
Trisetum caudulatum Trin.	AS, RC	Herb
Vulpia bromoides (L.) Gray	AS, RC, SC	Herb

Table 5.2 (cont.)

Taxon	Distribution	Life form
Vulpia muralis (Kunth) Henrard	RC	Herb
Vulpia myuros (L.) C.C.Gmel. var. *megaleura* (Nutt.) Auq.	AS, RC	Herb
ANGIOSPERMS – DICOTYLEDONAE		
Aizoaceae		
Carpobrotus aequilaterus (Haw.) N.E.Br.	AS, RC	Herb
Tetragonia tetragonoides (Pall.) Kuntze	AS, RC, SC	Herb
Amaranthaceae		
Amaranthus deflexus L.	AS, RC	Herb
Amaranthus hybridus L.	RC	Herb
Apiaceae (Umbelliferae)		
Ammi visnaga (L.) Lam.	RC	Herb
Anethum graveolens L.	AS	Herb
Apium australe Thouars	RC	Herb
Apium laciniatum (DC.) Urb.	RC	Herb
Conium maculatum L.	AS, RC	Herb
Coriandrum sativum L.	AS, RC	Herb
Cyclospermum leptophyllum (Pers.) Sprague var. *leptophyllum*	RC	Herb
Daucus montanus Humb. & Bonpl. ex Schult.	RC	Herb
Foeniculum vulgare Mill.	AS, RC	Herb
Petroselinum crispum (Mill.) A.W.Hill	AS, RC	Herb
Sanicula crassicaulis Poepp. ex DC.	AS, RC	Herb
Torilis nodosa (L.) Gaertn.	RC	Herb
Apocynaceae		
Vinca major L.	RC	Herb
Asteraceae (Compositae)		
Amblyopappus pusillus Hook. & Arn.	RC, SC	Herb
Anthemis cotula L.	RC	Herb
Artemisia absinthium L.	RC	Herb
Bahia ambrosioides Lag.	AS	Herb
Bidens pilosa L.	AS, RC	Herb
Calendula officinalis L.	RC	Herb
Carduus pycnocephalus L.	RC	Herb
Carthamus lanatus L.	AS, RC	Herb
Centaurea melitensis L.	AS, RC, SC	Herb
Chamomilla recutita (L.) Rauschert	RC	Herb
Chrysanthemum coronarium L.	RC	Herb
Cichorium intybus L.	AS, RC	Herb
Cirsium vulgare (Savi) Ten.	AS	Herb
Conyza bonariensis (L.) Cronquist	AS, RC, SC	Herb
Cotula australis (Spreng.) Hook.f	AS, RC, SC	Herb
Cotula coronopifolia L.	RC, SC	Herb
Crepis capillaris (L.) Wallr.	RC	Herb
Cynara cardunculus L.	AS, RC	Herb
Delairea odorata Lem.	RC	Herb

Table 5.2 (cont.)

Taxon	Distribution	Life form
Galinsoga parviflora Cav.	AS, RC	Herb
Gamochaeta americana (Mill.) Wedd.	AS, RC	Herb
Gamochaeta coarctata (Willd.) Kerguélen	AS, RC	Herb
Gamochaeta spicata (Lam.) Cabrera	AS, RC	Herb
Gamochaeta stachydifolia (Lam.) Cabrera	AS, RC, SC	Herb
Gnaphalium aldunateoides J.Remy	AS	Herb
Hypochaeris glabra L.	AS, RC, SC	Herb
Hypochaeris radicata L.	AS, RC, SC	Herb
Lapsana communis L.	AS, RC	Herb
Madia sativa Molina	AS	Herb
Matricaria recutita L.	RC	Herb
Micropsis nana DC.	RC	Herb
Pseudognaphalium cheiranthifolium (Lam.) Hilliard & B.L.Burtt	AS, RC, SC	Herb
Senecio mikanioides Otto ex Walp.	RC	Herb
Senecio vulgaris L.	AS, RC	Herb
Silybum marianum (L.) Gaertn.	AS, RC, SC	Herb
Solidago chilensis Meyen	RC	Herb
Soliva stolonifera (Brot.) Loudon	RC	Herb
Sonchus asper Hill	RC	Herb
Sonchus oleraceus L.	AS, RC, SC	Herb
Sonchus tenerrimus L.	AS, RC, SC	Herb
Taraxacum officinale F.H.Wigg.	RC	Herb
Xanthium spinosum L.	AS, RC	Herb
Boraginaceae		
Cynoglossum creticum Mill.	RC	Herb
Myosotis laxa Lehm.	RC	Herb
Myosotis sylvatica Hoffm.	RC	Herb
Brassicaceae (Cruciferae)		
Brassica napus L.	AS, RC, SC	Herb
Brassica nigra (L.) W.D.J.Koch	RC	Herb
Brassica oleracea L.	AS, RC	Herb
Brassica rapa L.	RC	Herb
Cardamine hirsuta L.	RC	Herb
Hirschfeldia incana (L.) Lagr.-Foss.	RC	Herb
Lepidium bonariense L.	AS, RC	Herb
Matthiola incana (L.) R.Br.	AS, RC, SC	Herb
Nasturtium officinale W.T.Aiton	RC	Herb
Rhaphanus sativus L.	AS, RC	Herb
Rorippa nasturtium-aquaticum (L.) Hayek	AS, RC	Herb
Sisymbrium officinale (L.) Scop.	RC	Herb
Campanulaceae		
Campanula rapunculoides L.	RC	Herb
Lobelia tupa L.	RC	Herb
Caryophyllaceae		
Cerastium fontanum Baumg. subsp. *vulgare* (Hartm.) Greuter & Burdet	AS	Herb

Table 5.2 (cont.)

Taxon	Distribution	Life form
Cerastium glomeratum Thuill.	AS, RC, SC	Herb
Paronychia franciscana Eastw.	AS, RC	Herb
Polycarpon tetraphyllum (L.) L.	AS, RC, SC	Herb
Sagina chilensis Gay	AS, RC, SC	Herb
Silene gallica L.	AS, RC, SC	Herb
Spergularia rubra J.Presl ex C.Presl	AS, RC	Herb
Stellaria chilensis Pedersen	AS, RC	Herb
Stellaria media (L.) Vill.	AS, RC	Herb
Chenopodiaceae		
Beta vulgaris L. subsp. *maritima* (L.) Arcang.	AS, RC	Herb
Chenopodiastrum murale (L.) S.Fuentes, Uotila & Borsch	AS, RC, SC	Herb
Chenopodium album L.	RC, SC	Herb
Dysphania ambrosioides (L.) Mosyakin & Clemants	AS, RC	Herb
Dysphania multifidum (L.) Mosyakin & Clements	AS, RC	Herb
Sarcocornia fruticosa (L.) A.J.Scott	AS, RC, SC	Herb
Convolvulaceae		
Convolvulus arvensis L.	AS, RC	Herb
Ipomoea indica (Burm.) Merr.	RC	Herb
Dipsacaceae		
Dipsacus sativus (L.) Honck.	RC	Herb
Scabiosa atropurpurea L.	RC	Herb
Elaeocarpaceae		
Aristotelia chilensis Stuntz	AS, RC	Tree
Euphorbiaceae		
Euphorbia lathyris L.	AS	Herb
Euphorbia peplus L.	AS, RC	Herb
Ricinus communis L.	AS, RC	Shrub
Fabaceae (Leguminosae)		
Acacia (*Racosperma*) *dealbata* Link	AS, RC	Tree
Acacia (*Racosperma*) *melanoxylon* R.Br.	AS, RC	Tree
Albizia lophantha (Willd.) Benth.	RC	Tree
Galega officinalis L.	AS, RC	Herb
Medicago arabica (L.) Huds.	AS, RC	Herb
Medicago lupulina L.	RC	Herb
Medicago polymorpha L. var. *brevispina* (Benth.) Heyn	RC	Herb
Medicago polymorpha L. var. *polymorpha*	AS, RC, SC	Herb
Medicago polymorpha L. var. *vulgaris* (Benth.) Shinners	RC	Herb
Medicago sativa L.	AS, RC	Herb
Melilotus indicus (L.) All.	AS, RC, SC	Herb
Robinia pseudoacacia L.	RC	Tree
Teline monspessulana (L.) K. Koch	RC	Shrub
Trifolium dubium Sibth.	AS	Herb
Trifolium pratense L.	RC	Herb
Trifolium repens L.	AS, RC	Herb
Fagaceae		
Quercus robur L.	RC	Tree

Table 5.2 (cont.)

Taxon	Distribution	Life form
Gentianaceae		
Centaurium cachanlahuen B.L.Rob.	RC, SC	Herb
Geraniaceae		
Erodium cicutarium (L.) L'Hér. ex Aiton	AS, RC, SC	Herb
Erodium moschatum (L.) Aiton	RC	Herb
Geranium core-core Steud.	AS, RC	Herb
Geranium dissectum L.	RC	Herb
Geranium pusillum L.	AS, RC, SC	Herb
Geranium robertianum L.	RC	Herb
Hydrangeaceae		
Hydrangea macrophylla (Thunb.) Ser.	RC	Shrub
Hypericaceae		
Hypericum perforatum L.	AS, RC	Herb
Lamiaceae (Labiatae)		
Marrubium vulgare L.	AS, RC, SC	Herb
Melissa officinalis L.	AS, RC	Herb
Mentha aquatica L.	RC	Herb
Mentha pulegium L.	AS, RC	Herb
Mentha suaveolens Ehrh.	RC	Herb
Origanum majorana L.	AS, RC	Herb
Plectranthus major L.	RC	Herb
Prunella vulgaris L.	RC	Herb
Stachys arvensis L.	RC	Herb
Lardizabalaceae		
Lardizabala biternata Ruiz & Pavón	RC	Twining vine
Linaceae		
Linum usitatissimum L.	RC	Herb
Loranthaceae		
Tristerix corymbosus (L.) Kuijt	RC	Shrub
Lythraceae		
Lythrum hyssopifolia L.	AS, RC	Herb
Malvaceae		
Anoda cristata (L.) Schltdl.	RC	Herb
Malva nicaeensis All.	RC	Herb
Malva parviflora L.	RC, SC	Herb
Modiola caroliniana (L.) G.Don	AS, RC	Herb
Moraceae		
Ficus carica L.	AS, RC	Tree
Myrtaceae		
Eucalyptus globulus Labill.	AS, RC	Tree
Ugni molinae Turcz.	AS, RC	Shrub
Oleaceae		
Ligustrum ovalifolium Hassk.	RC	Shrub
Onagraceae		
Fuchsia magellanica Lam.	RC	Subshrub
Oenothera affinis Cambess.	AS, RC	Herb

Table 5.2 (cont.)

Taxon	Distribution	Life form
Oenothera picensis Phil.	AS, RC	Herb
Oenothera rosea Aiton	AS, RC	Herb
Oxalidaceae		
Oxalis corniculata L.	AS, RC	Herb
Oxalis debilis Kunth	RC	Herb
Oxalis micrantha Bertero ex Colla	AS, RC, SC	Herb
Oxalis pes-caprae L.	RC	Herb
Oxalis purpurea L.	RC	Herb
Papaveraceae		
Fumaria capreolata L.	RC	Herb
Papaver somniferum L.	RC, SC	Herb
Pittosporaceae		
Pittosporum crassifolium Banks & Sol. ex A.Cunn.	RC	Shrub
Pittosporum eugenioides A.Cunn.	RC	Shrub
Plantaginaceae		
Callitriche lechleri (Hegelm.) Fassett	RC	Herb
Cymbalaria muralis G.Gaertn., B.Mey. & Scherb.	RC	Herb
Digitalis purpurea L.	AS	Herb
Kickxia elatine (L.) Dumort.	AS	Herb
Plantago lanceolata L.	AS, RC	Herb
Plantago major L.	AS, RC	Herb
Veronica anagallis-aquatica L.	RC	Herb
Veronica arvensis L.	RC	Herb
Veronica persica Poir.	AS, RC	Herb
Polemoniaceae		
Gilia valdiviensis Griseb.	AS	Herb
Microsteris gracilis (Hook.) Greene	RC	Herb
Polygonaceae		
Polygonum aviculare L.	AS, RC	Herb
Polygonum hydropiperoides Michx.	RC	Herb
Polygonum lapathifolium L.	RC	Herb
Rumex acetosella L.	AS, RC	Herb
Rumex conglomeratus Murray	AS, RC	Herb
Rumex crispus L.	AS, RC, SC	Herb
Rumex foliosus Rech.f.	AS	Herb
Rumex pulcher L.	AS, RC, SC	Herb
Portulacaceae		
Calandrinia monandra DC.	AS	Herb
Portulaca oleracea L.	AS	Herb
Primulaceae		
Anagallis arvensis L.	AS, RC	Herb
Anagallis minima (L.) E.H.L.Krause	AS	Herb
Ranunculaceae		
Anemone decapetala Ard.	RC	Herb
Ranunculus muricatus L.	AS, RC	Herb

Table 5.2 (cont.)

Taxon	Distribution	Life form
Rosaceae		
Acaena argentea Ruiz & Pavón	AS, RC, SC	Herb
Acaena ovalifolia Ruiz & Pavón	AS, RC	Herb
Fragaria chiloensis (L.) Duchesne	AS, RC	Herb
Prunus cerasus L.	RC	Tree
Rubus ulmifolius Schott	AS, RC	Shrub
Rubiaceae		
Galium aparine L.	AS, RC	Herb
Sherardia arvensis L.	RC, SC	Herb
Rutaceae		
Ruta chalepensis L.	AS, RC	Shrub
Salicaceae		
Populus pyramidalis Salisb.	RC	Tree
Salix babylonica L.	AS, RC	Tree
Scrophulariaceae		
Verbascum thapsus L.	AS	Herb
Verbascum virgatum Stokes	AS, RC	Herb
Simaroubaceae		
Ailanthus altissima (Mill.) Swingle	AS	Tree
Solanaceae		
Cestrum parqui L'Hér.	AS, RC	Shrub
Datura stramonium L.	RC	Herb
Nicotiana tabacum L.	AS, RC	Herb
Physalis peruviana L.	AS, RC	Herb
Solanum argenteum Dunal ex Poir.	AS, RC	Small tree
Solanum furcatum Dunal in Poir. var. *furcatum*	AS, RC, SC	Herb
Solanum marginatum L.f.	RC	Shrub
Tropaeolaceae		
Tropaeolum majus L.	AS, RC	Herb
Urticaceae		
Urtica urens L.	RC, SC	Herb
Valerianaceae		
Centranthus ruber (L.) DC.	RC	Herb
Verbenaceae		
Verbena litoralis Kunth	AS, RC	Herb
Vitaceae		
Vitis vinifera L.	AS, RC	Vine

Note: Placement of genera into families of angiosperms follows Mabberley (2008) and APG (1998, 2003, 2009, 2016) with some modifications. This list is based primarily on herbarium collections at CONC, OS, and WU plus citations in Johow (1896), Skottsberg (1921), Pedley (1986, 1987), Matthei et al. (1993), Swenson et al. (1997), Marticorena et al. (1998), Cuevas et al. (2004), Danton et al. (2006), Baeza et al. (2007), Wheeler (2007), López-Sepúlveda et al. (2013), and C. Taylor (unpublished manuscripts, Aizoaceae, Urticaceae). AS, Alejandro Selkirk Island; RC, Robinson Crusoe Island; SC, Santa Clara Island.

[a] Due to taxonomic difficulties in the genus *Cupressus*, Swenson et al. (1997) question whether this species exists in the archipelago.

[b] Baeza et al. (2007) suggest that this species may not be present in the archipelago.

Table 5.3 List of Cultivated Species of Gymnosperms and Angiosperms of the Juan Fernández Archipelago Cited by Swenson et al. (1997), Cuevas (2002), Danton et al. (2006), and López-Sepúlveda et al. (2013a) from the Village of San Juan Bautista on Robinson Crusoe Island (Containing the Majority of the Cultivated Species) and the Settlement at the Mouth of Quebrada Casas on Alejandro Selkirk Island

GYMNOSPERMS
Araucariaceae
Araucaria angustifolia (Bertol.) Kunze
Araucaria excelsa (Lamb.) W.T.Aiton
Cupressaceae
Cupressus goveniana Gordon
Cupressus macnabiana A.Murray
Platycladus orientalis (L.) Franco
Taxodium distichum (L.) Rich.
Pinaceae
Pinus pinaster Aiton

ANGIOSPERMS – ARCHAEANGIOSPERMAE
Lauraceae
Cryptocarya alba (Molina) Looser
Laurus nobilis L.
Persea americana Mill.
Magnoliaceae
Magnolia grandiflora L.
Magnolia quinquepeta (Buc'hoz) Dandy
Monimiaceae
Peumus boldus Molina
Piperaceae
Peperomia nummularioides Griseb.

ANGIOSPERMS – MONOCOTYLEDONAE
Agavaceae
Agave americana L.
Yucca aloifolia L.
Alliaceae
Allium cepa L.
Allium porrum L.
Allium sativum L.
Allium schoenoprasum L.
Nothoscordum gramineum Beauv.
Amaryllidaceae
Clivia miniata (Lindl.) Bosse
Crinum ×*powellii* hort. ex Baker
Hippeastrum puniceum (Lam.) Urb.
Leucojum aestivum L.
Araceae
Arum elongatum Steven
Arum italicum Mill. subsp. *italicum*
Colocasia esculenta (L.) Schott
Monstera deliciosa Liebm.
Philodendron bipinnatifidum Schott ex Endl.
Zantedeschia elliottiana Engl.

Table 5.3 (cont.)

Arecaceae (Palmae)
Phoenix canariensis hort. ex Chabaud
Trachycarpus fortunei (Hook.) H.Wendl.
Washingtonia filifera (Linden ex André) H.Wendl.
Asphodelaceae
Aloe arborescens Mill.
Aloe ciliaris Haw.
Aloe saponaria Haw.
Aloe vera L.
Asparagaceae
Chlorophytum comosum (Thunb.) Jacques
Cordyline australis Endl.
Cordyline stricta Endl.
Sansevieria trifasciata Prain
Bromeliaceae
Ananas comosus (L.) Merr.
Tillandsia aeranthos (Loisel.) L.B.Sm.
Cannaceae
Canna indica L.
Cyperaceae
Cyperus involucratus Rottb.
Cyperus papyrus L.
Hemerocallidaceae
Hemerocallis fulva L.
Hyacinthaceae
Ornithogalum caudatum Jacq.
Scilla peruviana L.
Iridaceae
Chasmanthe aethiopica (L.) N.E.Br.
Dietes iridioides (L.) Sweet ex Klatt
Freesia refracta (Jacq.) Klatt
Gladiolus hortulanus L.H.Bailey
Iris orientalis Mill.
Ixia flexuosa L.
Sparaxis tricolor (Schneev.) Ker-Gawl.
Tigridia pavonia (L.f.) DC.
Watsonia borbonica (Pourr.) Goldblatt
Watsonia meriana (L.) Mill.
Liliaceae
Lilium longiflorum Thunb.
Marantaceae
Ctenanthe oppenheimiana (E.Morren) K.Schum.
Maranta leuconeura E.Morren
Musaceae
Ensete ventricosum (Welw.) Cheesman
Musa acuminata Colla
Strelitzia reginae Aiton
Poaceae (Gramineae)
Arundo donax L.

Table 5.3 (cont.)

Chusquea culeou É.Desv. in Gay
Cortaderia selloana (Schult. & Schult.f.) Asch. & Graebn.
Cynodon dactylon (L.) Pers.
Koeleria micrathera Griseb.
Sorghum halepense (L.) Pers.
Sporobolus indicus (L.) R.Br.
Zea mays L.
Xanthorrhoeaceae
Phormium tenax J.F.Forst. & G.Forst.
Zingiberaceae
Hedychium flavescens Carey ex Roscoe

ANGIOSPERMS – DICOTYLEDONEAE
Acanthaceae
Acanthus mollis L.
Justicia magnifica Pohl. ex Nees
Aceraceae
Acer negundo L.
Actinidiaceae
Actinidia deliciosa (A. Chev.) C.F.Liang & A.R.Ferguson
Aizoaceae
Carpobrotus edulis (L.) N.E.Br.
Lampranthus blandus (Haw.) Schwantes
Lampranthus laxifolius N.E.Br.
Lampranthus roseus (Willd.) Schwantes
Amaranthaceae
Gomphrena elegans Mart.
Annonaceae
Annona cherimola Mill.
Apiaceae (Umbelliferae)
Daucus carota L.
Apocynaceae
Nerium oleander L.
Araliaceae
Fatsia japonica Decne & Planch.
Hedera algeriensis Hibberd
Hedera helix L.
Schefflera arboricola (Hayata) Merr.
Asteraceae (Compositae)
Anthemis arvensis L.
Arctotis acaulis L.
Argyranthemum frutescens (L.) Sch.Bip.
Aster novae-angliae L.
Bellis perennis L.
Cosmos bipinnatus Cav.
Cynara scolymus L.
Dahlia excelsa Benth.
Dahlia ×hortensis Guillaumin
Dahlia ×pinnata Cav.
Delairea odorata Lem.

Table 5.3 (cont.)

Dendranthema grandiflora (Ramat.) Kitam.
Erigeron karwinskianus DC.
Euryops chrysanthemoides (DC.) B.Nord.
Farfugium tussilagineum (Burm.) Kitam.
Felicia amelloides (L.) Voss
Gazania rigens (L.) Gaertn.
Helianthus annuus L.
Lactuca sativa L.
Leucanthemum ×superbum (J.Ingram) Kent
Matricaria chamomilla L.
Osteospermum ecklonis (DC.) Norl.
Pericallis ×hybrida B.Nord.
Petasites fragrans (Vill.) C.Presl.
Rhodanthe manglesii Lindl.
Senecio bicolor Ces., Pass. & Gibelli subsp. *cineraria* (DC.) Chater
Soliva sessilis Ruiz & Pav.
Tagetes patula L.
Tanacetum parthenium Sch.Bip.
Zinnia violacea Cav.
Balsaminaceae
Impatiens sodenii Engl. & Warb. ex Engl.
Impatiens walleriana Hook.f.
Basellaceae
Anredera cordifolia (Ten.) Steenis
Begoniaceae
Begonia albopicta Bull
Begonia cucullata Willd.
Begonia fuchsioides Hook.
Bignoniaceae
Jacaranda mimosifolia D.Don
Podranea ricasoliana Sprague
Boraginaceae
Echium vulgare L.
Brassicaceae (Cruciferae)
Capsella bursa-pastoris (L.) Medik.
Coronopus didymus (L.) Sm.
Lobularia maritima (L.) Desv.
Cactaceae
Hyalocereus undatus (Haw.) Britton & Rose
Opuntia ficus-indica (L.) Mill.
Campanulaceae
Lobelia erinus L.
Cannabaceae
Cannabis sativa L.
Caprifoliaceae
Lonicera japonica Thunb.
Sambucus mexicana C.Presl. ex DC.
Viburnum opulus L.

Table 5.3 (cont.)

Caricaceae
Carica pubescens Lenné & K.Koch
Vasconcellea cundinamarcensis V.M.Badillo
Caryophyllaceae
Dianthus barbatus L.
Dianthus caryophyllus L.
Silene coronaria (L.) Clairv.
Chenopodiaceae
Beta vulgaris L. subsp. *vulgaris*
Crassulaceae
Aeonium arboretum Webb & Berthel.
Aeonium haworthii (Salm-Dyk) Webb & Berthel.
Aeonium holochrysum Webb & Berthel.
Crassula multicava Lem.
Echeveria glauca (Baker) E.Morren
Kalanchoe blossfeldiana Poelln.
Sedum prealtum A.DC.
Sedum spectabile Boreau
Cucurbitaceae
Cucumis sativus L.
Cucurbita ficifolia Bouché
Cucurbita maxima Lam.
Cucurbita pepo L.
Elaeocarpaceae
Crinodendron patagua Molina
Ericaceae
Rhododendron L. subg. *Pentanthera* G.Don f.
Euphorbiaceae
Euphorbia helioscopia L.
Euphorbia milii Des Moul. var. *milii*
Euphorbia pulcherrima Willd. ex Klotzsch
Fabaceae (Leguminosae)
Bauhinia candicans Benth.
Cytisus scoparius (L.) Link
Lathyrus latifolius L.
Lathyrus odoratus L.
Lathyrus tingitanus L.
Phaseolus coccineus L.
Phaseolus vulgaris L.
Pisum sativum L.
Trifolium glomeratum L.
Trifolium suffocatum L.
Vicia faba L.
Vicia sativa L.
Fagaceae
Castanea sativa Mill.
Geraniaceae
Pelargonium × *asperum* Willd.
Pelargonium graveolens L'Her.

Table 5.3 (cont.)

Pelargonium peltatum (L.) L'Her.
Pelargonium domesticum L.H.Bailey
Pelargonium hortorum L.H.Bailey
Hydrangeaceae
Philadelphus latifolius Schrad.
Hypericaceae
Hypericum calycinum L.
Lamiaceae (Labiatae)
Ajuga reptans L.
Lavandula angustifolia Mill.
Ocimum basilicum L.
Plectranthus fruticosus L'Her.
Rosmarinus officinalis L.
Salvia involucrata Cav.
Salvia leucantha Cav.
Salvia microphylla Kunth
Salvia officinalis L.
Thymus vulgaris L.
Linaceae
Linum perenne L.
Lythraceae
Cuphea ignea A.DC.
Lagerstroemia indica L.
Malvaceae
Abutilon striatum G.F.Dicks. ex Lindl.
Alcea rosea L.
Hibiscus rosa-sinensis L.
Lavatera arborea L.
Sparmannia africana L.f.
Moraceae
Ficus elastica Roxb.
Myrtaceae
Psidium cattleianum Sabine
Nyctaginaceae
Bougainvillea glabra Choisy
Mirabilis jalapa L.
Oleaceae
Fraxinus ornus L.
Jasminum mesnyi Hance
Olea europaea L.
Onagraceae
Fuchsia ×hybrida hort. ex Siebold & Voss
Orobanchaceae
Bartsia trixago L.
Papaveraceae
Eschscholzia californica Cham.
Papaver rhoeas L.
Passifloraceae
Passiflora caerulea L.

Table 5.3 (cont.)

Phytolaccaceae
Phytolacca dioica L.
Pittosporaceae
Pittosporum tenuifolium Gaertn.
Plantaginaceae
Antirrhinum majus L.
Hebe ×andersonii (Lindl. & Paxt.) Ckn.
Hebe ×franciscana (Eastw.) Souster
Polygonaceae
*Rheum hybrid*um Murr.
Portulacaceae
Portulacaria afra Jacq.
Primulaceae
Primula vulgaris Huds.
Primula ×polyantha Mill.
Proteaceae
Gevuina avellana Molina
Punicaceae
Punica granatum L.
Ranunculaceae
Aquilegia vulgaris L.
Consolida ajacis (L.) Schrödinger
Rosaceae
Cydonia oblonga Mill.
Cydonia vulgaris Pers.
Eriobotrya japonica (Thunb.) Lindl.
Geum chiloense Balbis ex Ser.
Malus domestica Borkh.
Malus pumila Mill.
Potentilla (*Fragaria*) *chiloensis* (L.) Mabb.
Prunus armeniaca L.
Prunus cerasifera Ehrh.
Prunus domestica L.
Prunus dulcis (Mill.) D.Webb
Prunus persica (L.) Batsch
Pyrus communis L.
Rosa L. spp.
Rubus loganobaccus L.H.Bailey
Spiraea chamaedryfolia L.
Rutaceae
Citrus limonum Risso
Citrus sinensis Pers.
Salicaceae
Salix caprea Boiss. & Buhse
Santalaceae
Santalum album L.
Sapindaceae
Koelreuteria paniculata Laxm.
Saxifragaceae
Bergenia crassifolia (L.) Fritsch
Saxifraga stolonifera Meerb.

Table 5.3 (cont.)

Scrophulariaceae
Buddleja globosa Hope
Myoporum laetum G.Forst.
Solanaceae
Brugmansia sanguinea D.Don
Brugmansia suaveolens (Willd.) Bercht. & J.Presl
Capsicum annuum L.
Petunia ×hybrida E.Vilm.
Solanum jasminoides Paxton
Solanum ligustrinum Lodd., G.Lodd. & W.Lodd.
Solanum lycopersicum L.
Solanum nigrum L.
Solanum pseudocapsicum L.
Solanum tuberosum L.
Theaceae
Camellia japonica L.
Ulmaceae
Ulmus thomasii Sarg.
Urticaceae
Parietaria judaica L.
Soleirolia soleirolii (Req.) Dandy
Verbenaceae
Aloysia salviefolia (Hook. & Arn.) Moldenke
Lantana camara L.
Violaceae
Viola odorata L.

Note: Most of these species are in gardens, but some of them have escaped, and they may be encountered growing in the streets or paths, near houses, or even on the outskirts of the village or settlement. Other species have been deliberately planted within common areas. *None* of these species have been included in our lists of native and endemic or introduced species used for the statistics of the flora of the islands, but they are included here for cultural and conservation interests.

The lists of species in the Juan Fernández Archipelago have been taken primarily from Marticorena et al. (1998) but with additions from Danton et al. (2006). These two catalogues are, in fact, very similar. Differences lie primarily in the inclusion by Danton et al. of a number of introduced species that have escaped from private gardens in the village of San Juan Bautista and remain there feral along the streets and paths. Whether they should be included or not is a matter of taste, to some degree, but inclusion does have the beneficial effect of helping to alert one to potential dangers from highly competitive invasive species that could possibly, if not contained, spread into the native vegetation. Because our focus has been on the evolution and biogeography of the flora, we have not included all these in our list. To include them, now confined only to the village, also distorts calculations of meaningful statistics within the established flora. For a more comprehensive overview of these introduced taxa, we refer the reader to Danton et al. (2006) and Danton and Perrier (2017).

Table 5.4 List of Local Names that Have Been Applied to the Vascular Flora of the Juan Fernández Archipelago

Species	Family	Common name
Acaena argentea	Rosaceae	Trun
Amblyopappus pusillus	Asteraceae	Manzanilla
Apium fernandezianum	Apiaceae	Panul
Aristotelia chilensis	Elaeocarpaceae	Maqui
Berberis corymbosa	Berberidaceae	Michay
Boehmeria excelsa	Urticaceae	Manzano
Coprosma hookeri	Rubiaceae	Olivo, olivillo
Coprosma pyrifolia	Rubiaceae	Peralillo
Chusquea fernandeziana	Poaceae	Colihue
Dendroseris spp.	Asteraceae	Col, colecillo
Drimys confertifolia	Winteraceae	Canelo
Gunnera spp.	Gunneraceae	Pangue
Juania australis	Arecaceae	Chonta, col de palma
Margyricarpus digynus	Rosaceae	Sabinilla
Myrceugenia schulzei	Myrtaceae	Luma
Nothomyrcia fernandeziana	Myrtaceae	Luma, lumilla, pimiento, temu
Ochagavia elegans	Bromeliaceae	Chupón
Pernettya rigida	Ericaceae	Murtillo
Phrygilanthus berteroi	Loranthaceae	Quintral
Rhaphithamnus venustus	Verbenaceae	Juanbueno, arrayán macho, espinillo
Robinsonia spp.	Asteraceae	Resino
Robinsonia gracilis	Asteraceae	Resinillo
Robinsonia macrocephala	Asteraceae	Incienso
Rubus ulmifolius	Rosaceae	Zarzamora
Santalum fernandezianum	Santalaceae	Sándalo
Solanum fernandezianum	Solanaceae	Papa silvestre
Sophora fernandeziana	Fabaceae	Mayu-monte (mayu del monte), guayacán
Sophora masafuerana	Fabaceae	Leña dura, madera dura
Ugni molinae	Ericaceae	Murtilla
Zanthoxylum externum	Rutaceae	Naranjillo
Zanthoxylum mayu	Rutaceae	Mayu, naranjillo, palo amarillo

Source: Adapted from Gunckel (1968).

Table 5.1 provides a comprehensive list of species of ferns, fern allies, gymnosperms, and angiosperms that are known within the Juan Fernández Archipelago. The initial data for ferns and fern allies were provided by R. Rodríguez, derived from his treatment of these groups in the *Flora of Chile* (Rodríguez 1995). Some species were collected many years ago and have not been seen recently. One suspects that these may no longer reside in the islands, but there are still unexplored niches on both islands, and it is not impossible that some of these might still be present. We have designated species as extinct that are with some high degree of certainty no longer occurring in the islands. In addition to giving familial placement, data on distribution in the archipelago, life form, biogeographical status, and conservation status are also provided for each taxon. In nearly all cases there is little doubt regarding these data, but occasionally a question arises about a species being native or introduced. We have given our best judgment on these cases, taking into account as much historical evidence as possible to help guide our decisions. The conservation categories follow the IUCN definitions, and these are explained in more detail in Chapter 9. Referral of species into these categories is based on data from our own collections, our field notes, and categorizations in Rodríguez (1995) and Danton et al. (2006).

Table 5.2 lists the introduced species that have been reported to grow in the archipelago. These are listed separately from the endemic and native species because adventives come and go in different areas, especially around the human settlements. We have elected to chronicle only species that have been included in the works of Matthei et al. (1993), Swenson et al. (1997), Marticorena et al. (1998), Cuevas et al. (2004), Danton et al. (2006), and López-Sepúlveda et al. (2013a), which span the past two decades. Listing species cited in earlier works is of historical interest, but this does not give a useful picture of the adventive species currently in the archipelago, which are obviously of conservation import. Even with this approach, it is probable that not all of the listed species are still found in the archipelago. A number of the reports are of garden escapees, which may or may not ever become permanently established in the islands. Some of these could possibly become serious pests, and therefore, it is useful to bring attention to their occurrence. For more details on adventive species, see Chapter 8.

We also list in Table 5.3 the species that have been reported as cultivated in the archipelago. Although these are still confined to gardens or common areas and under human care, there exists the potential that some of these might escape and pose additional threats to the native vegetation. The data come primarily from Danton et al. (2006) for Robinson Crusoe Island and López-Sepúlveda et al. (2013a) for Alejandro Selkirk Island, with additional specimen data from both islands from Matthei et al. (1993). Species seen in private gardens have not been collected and hence are included in the table based on observation only.

An important point to stress is that these lists of species are based strongly on existing herbarium specimens, particularly on deposit at the University of Concepción (CONC), The Ohio State University (OS), and the University of Vienna (WU). These are the largest collections of herbarium materials from the Juan Fernández Islands available anywhere in the world. Many other institutions have specimens from the archipelago, particularly the Herbarium and Botanical Museum, Göteborg, Sweden (GB), Harvard University Herbaria (GH), Herbarium of the Royal Botanic Gardens, Kew (K), Herbarium of the Museo Nacional de Historia Natural, Santiago, Chile (SGO), and the United States National Herbarium (US), and these have also been consulted in the course of development of the list of species. The important point is that the existence of a herbarium specimen is concrete evidence of the occurrence of a species in the islands. This can be consulted and reevaluated at any time. Observational reports, not supported by specimens, are just too uncertain to be relied on for precise evaluations of the entire flora. Despite having taken this specimen-oriented stance, we nonetheless have consulted the lists of species and their distributions cited by other authors. In a few instances we have included these, particularly regarding aggressive introduced species cited by Danton et al. (2006)

The designation of family concepts and included genera within the flora has followed several basic references. For ferns, fern allies, and gymnosperms, we have followed Kramer and Green (1990) with updates from Smith et al. (2006). Fern classification is now in a very active phase due to the gathering of new molecular data and fresh phylogenetic interpretations (e.g., Schuettpelz and Pryer 2007; Christenhusz and Chase 2014). Some of the changes are based on discovering that some families (and genera) are polyphyletic, and other instances are more a reaction to preserve holophyly. From our perspective, the changes recommended by Smith et al. (2006) seem reasonable, and therefore, we have followed them. A few genera have been moved to new family positions, and this would be in contrast to their placements in other lists, for example, Skottsberg (1921). *Arthropteris* was placed in Oleandraceae but is now in Tectariaceae (Zhang et al. 2016). *Cystopteris* was previously in Dryopteridaceae, but it is now placed in its own family Cystopteridaceae (Rothfels et al. 2012, 2013). *Elaphoglossum* is now in Dryopteridaceae (from Lomariopsidaceae), *Grammitis* is now lodged in Polypodiaceae (from Grammitidaceae), and *Lophosoria* is now in Dicksoniaceae (from Lophosoriaceae) (Wolf et al. 1999).

The taxonomy of the fern family Hymenophyllaceae has been considerably modified in recent years based on new molecular phylogenetic data and analyses, and some comments are needed (see also Chapter 13). Traditionally, two genera have been recognized, *Hymenophyllum* and *Trichomanes*. These distinctions were maintained by Morton (1968) in a comprehensive classification of the family, but he added four additional monotypic genera: *Hymenoglossum, Serpyllopsis, Rosenstockia*, and *Cardiomanes*. He also provided subgeneric and sectional subdivisions within the two larger genera. Tryon and Tryon (1982) recognized the two principal genera, and they submerged *Serpyllopsis* and *Hymenoglossum* into *Hymenophyllum* as subgenera (subg. *Cycloglossum* and subg. *Hymenoglossum*, respectively) but made no mention of the other two genera. Iwatsuki (1990) recognized two subfamilies: Cardiomanoideae, with the single genus *Cardiomanes* (monospecific and endemic to New Zealand), and Hymenophylloideae, containing *Hymenophyllum* and *Trichomanes*, plus *Serpyllopsis, Hymenoglossum, Cardiomanes*, and *Sphaerocionium*, which previously had resided as a section within *Hymenophyllum* in the system of Morton (1968).

The first modern approach to relationships in Hymenophyllaceae focused on *Trichomanes*. Dubuisson (1997a) carried out an ambitious cladistic analysis of many species of the genus based on morphological and anatomical data, but he did not include the two island endemics, *T. ingae* and *T. philippianum*. Morton (1968) had placed both these species in *Trichomanes* subg. *Trichomanes* sect. *Lacosteopsis* (= sect. *Vandenboschia* Copel. in Dubuisson 1997a). Dubuisson (1997b) continued with *rbcL* sequences within *Trichomanes*, but again, neither of the Juan Fernández endemics was sampled. The overall results from *rbcL* were generally positive, and it opened the way for additional molecular work on the family. Pryer et al. (2001) applied *rbcL* to the entire family, and although the sampling was limited to only 14 species of *Trichomanes* (of more than 325) and nine of *Hymenophyllum* (of more than 300), plus *Serpyllopsis*, the basic division into *Trichomanes* and *Hymenophyllum* (including *Serpyllopsis*) was substantiated. Other studies using molecular data with *Trichomanes* have now been completed (Dubuisson et al. 2003; Ebihara et al. 2007), but most have not included the endemic species of Juan Fernández, and none has included both of them for analysis.

Cladistic morphological and molecular phylogenetic studies have also been conducted among species of *Hymenophyllum*. Hennequin (2003) completed a morphological analysis in a cladistic context using Morton's sections as the taxonomic units for analysis, but the results showed more agreement with Iwatsuki's (1990) generic arrangement. The sequences *rbcL* plus *rps4* and intergenic spacer *rps4-trnS* were used by Hennequin et al. (2003), but neither of the island endemics was included (i.e., *H. rugosum* and *H. caespitosum* [= *Serpyllopsis caespitosa*]). Hennequin et al. (2006) also completed another survey, using *rbcL, rbcL-accD*, and *rps4-trnS*, that was focused on species within the subgenus *Mecodium* and therefore did not include the island endemics. The only study that included one of the Juan Fernández endemic species, *H. rugosum*, was Hennequin et al. (2010), based on *rbcL, rbcL-accD* IGS, *rps4*, and *rps4-trnS*. These results showed a tie with *H. tunbridgense* from different localities (France, Scotland, Tanzania, and Madeira, but also Chile).

Based on the new *rbcL* data and analyses, a new classification of Hymenophyllaceae has now been proposed (Ebihara et al. 2006). *Hymenophyllum rugosum* is placed in subg. *Hymenophyllum* along with *H. caespitosum* (formerly *Serpyllopsis caespitosa*), and *H. cuneatum* is referred to subg. *Mecodium*, none of which causes disruption in our list of species for the flora of the islands. *Hymenoglossum cruentum*, however, is now placed also in *Hymenophyllum* with two other species (*H. heimii* of Madagascar and *H. asplenioides* from Central and South America) as subg. *Hymenoglossum*. In *Trichomanes* s.l., eight different genera are now recognized, each of which corresponds to a clade or lineage resulting from analysis of the molecular data. *Trichomanes ingae* and *T. phillipianum*, the two island endemics, along with 13 other species, are now placed in the genus *Polyphlebium*. Whether these evolutionary units might better be recognized at the subgeneric level within the monophyletic *Trichomanes* s.l. remains to be determined in the future, but the evolutionary groupings do seem clear, and hence, in this book we follow the new classification of Ebihara et al. (2006). All these combinations into *Hymenophyllum* have also been followed by Larsen et al. (2013). These authors have also decided not to recognize *H. cruentum* var. *rariforme* on the grounds that it falls within the range of variation within this relatively common species, a perspective with which we agree. It was treated as an endemic variety on Robinson Crusoe Island by Christensen and Skottsberg (1920) based primarily on the small size of the fronds.

Christenhusz and Chase (2014) have provided a new comprehensive classification of ferns based on available DNA data and application of cladistic rules. Changes this system would bring from the concepts used in this book are the downgrading of a number of families to subfamilial level and these then placed into enlarged familial concepts. For the Juan Fernández ferns, these changes would entail the following: Blechnaceae would become a subfamily under Aspleniaceae, Dryopteridaceae would become a subfamily under Polypodiaceae, Dicksoniaceae and Thysopteridaceae would become subfamilies under Cyatheaceae, and Tectariaceae would become a subfamily under Polypodiaceae. Whether these concepts will be adopted by the pteridological community remains to be seen.

For angiosperms, we have followed basically Mabberley (2008) and APG (1998, 2003, 2009, 2016), with some deviations. We recognize three taxonomic classes (Stuessy 2009): Archaeangiospermae (= basal angiosperms), Monocotyledonae, and Dicotyledonae. The availability of nucleotide data throughout the angiosperms has led to several changes in family composition, which we have adopted for the lists of taxa in the Juan Fernández Archipelago. This approach will, quite naturally, cause changes in the statistics of numbers of families in comparison with the previous assessments of Johow (1896), Skottsberg (1921), Marticorena et al. (1998), and Danton et al. (2006), even though there has been little actual change in the plant diversity present over the last century or more, with the exception of many more invasive plants, as mentioned earlier.

Of all the families of angiosperms that have been affected by molecular phylogenetic analyses, Scrophulariaceae figures prominently. Long a family characterized by generalized floral and fruit features (e.g., Cronquist 1981), it has now been shown to be a polyphyletic assemblage consisting of different and unrelated evolutionary lineages (Olmstead and Reeves 1995; Olmstead et al. 2001; Oxelman et al. 2005; Tank et al. 2006). The genera of the Juan Fernández Archipelago placed in the previous family circumscription in Skottsberg (1921) are *Cymbalaria, Digitalis, Euphrasia, Kickxia, Mimulus, Verbascum*, and *Veronica*. These genera have now been separated and placed in different families as follows: Orobanchaceae, *Euphrasia*; Plantaginaceae, *Cymbalaria, Digitalis, Kickxia*, and *Veronica*; and Phyrmaceae, *Mimulus*. The only genus in the islands that remains in Scrophulariaceae is the introduced *Verbascum*. Although we are not experts by any means on this group of genera, we have some sympathy with Brummitt (2014) in that the traditional family may now have been oversplit, whereas at the same time the concept of Plantaginaceae has been enlarged morphologically beyond useful informational limits. The molecular data leave little doubt that *Plantago* is close in nucleotide affinities to *Veronica* and *Digitalis*, the last two genera having been traditionally closely associated. The question is whether inclusion of all these genera (and others, including *Cymbalaria* and *Kickxia*) increases the comprehensive evolutionary information content of the resulting group or dilutes it. It is particularly hard to comprehend the evolutionary relatedness, or lack thereof, as revealed by morphology between *Plantago* and *Veronica*. Nonetheless, because we are not prepared to provide a comprehensive alternative solution to relationships among all these genera, we follow the familial placements as suggested by Mabberley (2008), which are based on Olmstead et al. (2001) and Oxelman et al. (2005). That *Euphrasia* belongs in Orobanchaceae seems abundantly clear from molecular data (Bennett and Mathews 2006; Gussarova et al. 2008; Těšital et al. 2010). This also seems reasonable because the genus is hemiparasitic and fits well with the mode of life of other genera in this family. The placement of *Mimulus* has been more difficult (e.g., Olmstead et al. 2001), but we follow Mabberley (2008), who follows molecular analyses of Beardsley and Olmstead (2002) and Beardsley et al. (2004) in suggesting referral of this genus to Phyrmaceae.

Although we accept the aforementioned changes with respect to Scrophulariaceae based on new molecular evidence, we do not think it meaningful to submerge Chenopodiaceae into Amaranthaceae. There is no question that these two families are closely related. Molecular phylogenetic investigations have revealed (Kadereit et al. 2003; Müller and Borsch, 2005) that the former is nested within the latter. No one has disputed that molecular and morphological characters delimit the clade called Chenopodiaceae. The only reason to synonymize these two families appears to be so that in a cladistic context Amaranthaceae does not become paraphyletic, unacceptable by the rules of strict cladistic classification (Hennig 1966). It seems very clear that Chenopodiaceae have evolved from out of Amaranthaceae, but such an evolutionary phenomenon, although surely commonplace in angiosperms, is incompatible with cladistic methodology. We see no benefit of submerging Chenopodiaceae into Amaranthaceae for that reason alone. Sufficient distinctive character complexes exist to clearly delimit the two as coordinate families and hence preserve higher predictive evolutionary informational value.

As for placements of genera within families and species within genera, some changes have been proposed in recent times. We accept that *Azara* seems better positioned in Salicaceae (Chase et al. 2002). Regarding generic concepts, we agree that the data seem convincing that *Fagara* should be submerged into *Zanthoxylum* (Rutaceae). The morphological data (Brizicky 1962; Hartley 1966; Waterman 1975; Beurton 1994) reveal that *Fagara* is polyphyletic and also falls well within the limits of *Zanthoxylum*. The available molecular data (Chase et al. 1999) are not helpful in this particular situation because the two species in the islands have not yet been sampled. *Fagara mayu* has previously been in *Zanthoxylum*, which makes available a name for use. *Fagara externa*, however, apparently has not yet been combined, which we therefore accomplish here (see Appendix 2). Susanna and García-Jacas (2000) have submerged the previously distinct endemic genus *Yunquea* (Asteraceae, Cardueae) into *Centaurodendron* (also an endemic genus), and it may well belong there. *Yunquea*, however, has not been studied thoroughly at either the morphological or molecular level. It grows only on the highest and nearly inaccessible peak on Robinson Crusoe Island, El Yunque, and very little material has ever been collected. It seems premature, therefore, for this submergence, and therefore we continue to list it as a distinct genus pending further investigations.

Pernettya (Ericaceae) is a particularly interesting case that merits more comment. There is one endemic species in the archipelago, *P. rigida*, and this occurs on both islands. The genus was originally treated as distinct by the previous monographer, Sleumer (1936, 1985), and followed by some other authors (Kausel 1949). Other workers, however, have combined the genus into *Gaultheria* (Stevens 1971; Middleton and Wilcock 1990; Middleton 1990, 1991; Teillier and Escobar 2013) on grounds that some species appear to provide bridging character features and that some intergeneric hybrids have been detected. Luteyn (1995) has discussed these distinctions between the two genera and keeps them separate, stressing the fruit difference of berry in *Pernettya* and capsule in *Gaultheria*. Intergeneric hybridization between genera certainly suggests that they are closely related (which no one disputes in this case), but this by itself is no obligatory reason for combining genera (Stuessy 2009). Recent molecular phylogenetic studies using ITS and chloroplast regions (Bush et al. 2009; Fritsch et al. 2011) show that *Pernettya* is a holophyletic group but nested within the broader *Gaultheria.* From a cladistic perspective, therefore, *Pernettya* should disappear as a distinct genus. As explained earlier, we do not follow the approach of submergence of taxa simply to avoid topological paraphyly (i.e., to preserve holophyly). We favor taking all aspects of phylogeny into account, including evolutionary and genetic divergence, for maximally informative classification. The fruit differences emphasized by Luteyn (1995) seem to us reasonable, and therefore, in this book we continue to recognize *Pernettya* as distinct from *Gaultheria*. As a note for the reader, if one wishes to place *P. rigida* into *Gaultheria*, the proper name must be *Gaultheria racemulosa* (DC.) Middleton (cf. Middleton and Wilcock 1990).

Comments on *Taraxacum fernandezianum* are also pertinent. It was recognized as an endemic species by Skottsberg (1921). Recent cytogenetic studies by Baeza et al. (2013) reveal that the karyotypes of two individuals of putative *T. fernandezianum*, one from Robinson Crusoe Island and the other from Alejandro Selkirk Island, appear identical to known cytogenetic configurations of *T. officinale*, the common dandelion. Considering the known morphological variation within the worldwide weed, *T. officinale*, we suspect that this may simply be a case of introduction and rapid morphological divergence through drift over many generations. It is known that *T. officinale* traveled from Europe to other continents as early as the seventeenth century (Solbrig 1971), and it also may have arrived with ships in the Juan Fernández Islands during this time.

Statistics of the Flora

The basic data on the vascular taxa of the Juan Fernández Archipelago are contained in Tables 5.1 (native and endemic species), 5.2 (introduced taxa), and 5.3 (cultivated species). In this section of the chapter, we synthesize points from these data and present further interpretations. We also discuss the distributions of all taxa on the three major islands of the archipelago. Endemic species are of particular interest for understanding processes of evolution in the archipelago, especially populational divergence and speciation. The native species, in combination with the endemics, are of interest for interpreting biogeographical sources for the flora and for understanding patterns of distribution within the archipelago. The endemic species are also of great interest from the standpoint of uniqueness within the world's flora, which can be assessed by determining percentages of endemism at the specific and generic levels. The endemics are also the main focus for conservation initiatives within oceanic islands. More discussion on the evolutionary and biogeographical significance of these data and patterns is presented in Chapter 13.

Table 5.5 presents statistics of the basic data of all endemic, native, and introduced vascular taxa in the archipelago. The total flora consists of 475 taxa, including 135 endemics, 73 natives, and 267 introduced taxa. Among the endemic and native *taxa*

Table 5.5 Numbers of Endemic, Native, and Introduced Taxa (Species, Subspecies, and Varieties) of Vascular Plants in the Flora of the Juan Fernández Archipelago

Biogeographical category	Total number	Ferns and fern allies	Gymnosperms	Archae-angiospermae	Monocots	Dicots
Endemic	135	26	0	5	19	85
Native	73	31	0	1	19	22
Subtotals	208	57	0	6	38	107
Introduced	267	0	3	0	55	209
Totals	475	57	3	6	93	316

Table 5.6 Numbers (Over Total Numbers of Natives and Endemics) and Percentages of Endemic Families, Genera, and Species (*not* Including Infraspecific Taxa) in the Vascular Flora of the Juan Fernández Archipelago

Taxon	No. endemic families/ total no.	Percent endemic families	No. endemic genera/total no.	Percent endemic genera	No. endemic species/total no.	Percent endemic species
Ferns and fern allies	1/14	7	1/24	4	25/56	45
Archaeangiospermae	1/3	33	1/3	33	5/6	83
Monocotyledonae	0/7	0	3/21	14	19/38	50
Dicotyledonae	0/31	0	8/56	14	80/103	78
Totals	2/55	4	13/104	13	129/203	64
Angiosperms	1/41	2	12/80	15	104/147	71

Note: The total numbers do not include introduced taxa, only those that are native and endemic.

(not species), 65% are endemics and 35% are natives. It is unsurprising that the number of introduced taxa (267; 56% of total) is higher than the combination of native and endemic taxa (209; 44%). The taxa of ferns and fern allies show 46% endemism in contrast to the angiosperms with 72%. No native or endemic gymnosperms exist in the flora, only three introduced taxa. Regarding angiosperms, the dicots show a high percentage of endemism (79%) in contrast to the monocots, with only 50% endemism. The largest families of native and endemic taxa are Hymenophyllaceae in the ferns with fifteen taxa and Asteraceae in angiosperms with thirty-one taxa.

Table 5.6 gives the numbers and percentages of endemic families, genera, and species (not including infraspecific taxa) in the vascular flora of the Juan Fernández Archipelago. There are two monospecific endemic families, one from the ferns (Thyrsopteridaceae) and another in the Archaeangiospermae (Lactoridaceae), yielding 4% (2/55) for the flora. At generic level, there are 12 endemic genera of angiosperms and one (*Thyrsopteris*) of ferns, or 13 total for 13% (13/104) of the flora. Most important, 64% of the species are endemic, 71% for angiosperms and 45% for the ferns.

As for variation in life form among the endemic and native taxa of vascular plants (Table 5.7), the preponderance of diversity lies with the herbs (66%), with the truly woody shrubs (not including subshrubs) and trees summing to 18%. Tree ferns make up

Table 5.7 Summary of Life Forms among the Native and Endemic Vascular Plants (Taxa) of the Juan Fernández Archipelago

Taxon	Herb	Subshrub	Shrub	Tree	Tree fern	Rosette tree
Ferns and fern allies	53				4	
Archaeangiospermae	4	1		1		
Monocotyledonae	36		1 (bamboo)	1 (palm)		
Dicotyledonae	44	2	19	15		27
Totals	137	3	20	17	4	27

Table 5.8 Distributions of Specific and Infraspecific Taxa of Endemic, Native, and Introduced Vascular Plants on the Major Islands of the Juan Fernández Archipelago

Biogeographical category	Alejandro Selkirk Island	Robinson Crusoe Island	Santa Clara Island
Endemic			
Ferns and fern allies	19	19	1
Archaeangiospermae	3	4	0
Monocotyledonae	10	11	0
Dicotyledonae	34	56	7
Angiospermae subtotals	47	71	7
Subtotals	66	90	8
Native			
Ferns and fern allies	30	24	3
Archaeangiospermae	1	1	0
Monocotyledonae	11	16	1
Dicotyledonae	13	14	2
Angiospermae subtotals	25	31	3
Subtotals	55	55	6
Introduced			
Gymnosperms	2	3	0
Monocotyledonae	33	52	7
Dicotyledonae	121	191	37
Angiospermae subtotals	154	243	44
Subtotals	156	246	44
Totals	277	391	58

Note: Some taxa are found on both islands.

only 2% of the flora; rosette trees are more abundant (13%), as is often the case in island archipelagos.

Table 5.8 summarizes the distributions of all taxa among the three major islands of the Juan Fernández Archipelago. The total distributions indicate that Robinson Crusoe Island harbors 391 taxa, Alejandro Selkirk Island has 277, and Santa Clara Island has only 58. As for the endemics, 90 taxa occur on Robinson Crusoe Island, 66 on Alejandro Selkirk Island, and only 8 on Santa Clara Island. Although the two major islands are nearly the same size, the much higher level of endemics on Robinson Crusoe Island most probably relates to its original larger surface area and elevation and its greater geological age, both of which would allow more speciation to take place with accumulating diversity over time. It is interesting that the distribution of native species is the same on these two islands, both fifty-five (plus six on Santa Clara), although Robinson Crusoe Island is much nearer to the major source area (South America). Regarding the introduced species, it comes as no surprise that a much higher level of taxa has been chronicled for Robinson Crusoe Island (391), which historically has received most of the human visits and activities, and this continues to the present because of the village of San Juan Bautista. No ferns are listed as being introduced, being only natives or endemics.

With regard to the cultivated plants, the total number of recorded species for the Juan Fernández Archipelago is 269 (Table 5.3). Because these are not presently in the natural areas of the islands, they pose no immediate threat to the native vegetation. Some of these species, however, such as *Lonicera japonica*, could become aggressive pests in the future. A prohibition on cultivating possibly threatening species in gardens would be a desirable conservation precaution.

6 Vegetation

Josef Greimler, Patricio López-Sepúlveda, and Karl Reiter

Vegetation Inventory of the Juan Fernández Islands

A number of studies on vegetation of the Robinson Crusoe Archipelago have been completed over the past 120 years. A first summary of the general vegetation of the archipelago was provided by Johow (1896) in addition to his treatment of the flora of the archipelago. In the years 1908, 1916, and 1917, Swedish botanist Carl Skottsberg visited the islands and collected data for a profound analysis of the vegetation, which was, however, published nearly half a century later (Skottsberg 1953a). Besides the many informative and detailed descriptions of the vegetation, he also provided graphical summaries of the vegetation on both islands by very simple sketch maps. These maps were very helpful, but they were based on inaccurate topographical maps and had no support from comprehensive aerial photographs. Two further analyses focused on Robinson Crusoe Island: a detailed report of the vegetation on and around the summit of El Yunque by Kunkel (1957) and a transect analysis by Schwaar (1979). A revised map based on Skottsberg's (1953a) map, including new observations from 1976 and 1979, was provided by Nishida and Nishida (1981). This map, however, is of limited resolution, displaying only six vegetation units. Another mapping effort of the vegetation of the entire archipelago was completed by Ortiz R. (1982), who used aerial photographs taken in 1980 by the Chilean Air Force. This analysis yielded more than 120 different categories of vegetation, many based on dominance of single species, and together providing an overly complex picture of the spatial distribution of the vegetation. In investigating bird communities, Hahn (1998) showed a rough distribution of habitat types on both islands based on Skottsberg's maps and data, updated by detailed actual observations. Modified versions of Skottsberg's maps are also included in a short summary in Mueller-Dombois and Fosberg (1998). Cuevas (2002) investigated the proportions of native and alien plants on Alejandro Selkirk Island (AS) in some relevés taken from different habitats. Danton et al. (2006) gave short descriptions of the vegetation on the islands in their catalogue of vascular plants in the archipelago, whereas Danton (2006b) focused on the *Myrceugenia* (*Nothomyrcia*) forests on both islands.

We collected data for updated and detailed analyses of the vegetation on both islands in the years 1999 and 2000 on Robinson Crusoe Island and in 2011 on Alejandro Selkirk Island, resulting in descriptions of the actual vegetation supported by classified vegetation tables and colored vegetation maps (Greimler et al. 2002a, 2013). For generating the vegetation map of Robinson Crusoe Island, we used the aerial photographs taken in 1980 by the Chilean Air Force as field maps. Because of the lack of orthophotographs, delineation of the field maps was transferred to a geocoded digital elevation model developed by us. For the vegetation map of Alejandro Selkirk Island, we relied essentially on the ESRI basemaps and Bing images (Web-based mapping service provided by Microsoft) taken between 2001 and 2004 and a digital elevation model provided by CONAF. For spatial data handling and management, as well as map drawings, we used the geographical information system ARC-GIS (version 10 in 2012–13). For estimating abundance or dominance (cover) of each species recorded in a plot (relevé), we used the cover-abundance scale of the Zürich-Montpellier approach (Braun-Blanquet 1964). One-hundred and six relevés were completed on Robinson Crusoe Island and 90 on Alejandro Selkirk Island. The relevés were preprocessed and classified using the TWINSPAN algorithm (Hill 1979) of the program JUICE 7 (Tichy 2002). Further details on the methodology are given in Greimler et al. (2002a, 2013).

The vegetation maps show vegetation units (represented by polygons) initially drawn in the field on field maps and revised later in the laboratory including all available information from photographs, sketch drawings, additional notes, and the vegetation relevés. The groups of dominant species in those relevés define the plant communities, which in many cases correspond to the vegetation units on the maps.

Not every detail of vegetation can be shown on the maps. Classification of vegetation for the purpose of mapping is always a compromise between scale, accuracy, and practicability depending on the degree of resolution of the desired map. Any reasonably resolving vegetation map of a natural setting contains some generalizations. On both islands, but especially on Robinson Crusoe Island, we had to combine several plant communities into larger vegetation units. The actual distribution of the plant communities and the vegetation units, therefore, cannot be entirely congruent. The plant communities, or phytocoenoses, that we have identified on the islands are dominated by the growth forms (or synusia) of trees, shrubs (including tall ferns), herbs, and grasses. For the synusia of bryophytes, we have recorded only the overall cover as a rule. Except for a few patches dominated by liverworts, the vascular plants are the best indicators of plant communities on the islands.

The Vegetation of Robinson Crusoe Island

The following 16 vegetation units are shown on the map (Fig. C88; see also Figs. C89–C103 for different views of landscapes and vegetation on this island).

1 Upper Montane Forest

The endemic trees *Drimys confertifolia* and *Nothomyrcia fernandeziana* are of high abundance and dominance; most tree ferns and *Gunnera bracteata* are confined to this essentially upper montane forest (Fig. C97). This unit consists of several different communities, mostly above 400 m along the main ridges, around and on El Yunque and in the high regions of the eastern part of the island. These communities often replace each other depending on microrelief, slope, and aspect and thus often form a mosaic pattern that is not practicable for resolution on the vegetation map. In the cloud zone of the highest altitudes above 500 m, the soil is rich in organic components. Moisture is constantly maintained in the litter and in old leaf bases/scales of the trunks, which are often covered by epiphytes, especially mosses.

We use the term "upper montane forest" in a broader sense, deviating from Skottsberg (1953a), who restricted it only to communities of highest elevations. Our criteria for delimiting upper from lower montane forest are high frequency and abundance of the tall ferns *Thyrsopteris elegans* and *Dicksonia berteroana* and other taxa typical for the assemblages of endemic plants in higher elevations.

1.1 In the ***Thyrsopteris elegans–Dicksonia berteroana*** **community**, the dominant tree ferns *Dicksonia berteroana* and *Thyrsopteris elegans*, either solitary or in combination (occasionally together with *Blechnum cycadifolium*), form a dense cover usually reaching 3 m in height, while the trees such as *Drimys confertifolia* and *Nothomyrcia fernandeziana* in most cases play a minor role on the very steep slopes. Other ferns, including mostly epiphytic species of *Hymenophyllum*, add to the high fern diversity. *Gunnera bracteata* has its optimal distribution in this community, as do *Coprosma oliveri*, *Cuminia eriantha*, *Robinsonia gracilis*, *R. evenia*, and *Asplenium macrosorum*. According to Kunkel (1957) and Danton (2000), similar assemblages do occur in the highest location on the island, the summit region of El Yunque (915 m).

1.2 In the ***Drimys confertifolia–Nothomyrcia*** **(*Myrceugenia*)** ***fernandeziana*** **community**, occurring above approximately 350 m, the trees *Drimys, Nothomyrcia, Fagara*, and *Juania* often form a dense forest. Single trees reach heights up to 15 m on considerably steep slopes and even up to 25 m on moderate slopes. The tree ferns also reach up to 5 m. Nearly all the ferns present in the *Thyrsopteris-Dicksonia* community can be found in this forest, but additional ones are more frequent here: *Lophosoria quadripinnata, Megalastrum inequalifolium*, and *Polystichum tetragonum*, to mention just the tallest ones. *Rhaphithamnus venustus*, a small tree, and *Dysopsis hirsuta*, an herb, are found mainly in this forest. Altogether this community contains the highest species diversity among native island plant communities, except for some high-elevational variants of the *Blechnum cycadifolium* community. Coverage of herbs is usually greater than in the dense tree fern community, and mosses again play an important role.

This upper montane forest community corresponds to Skottsberg's (1953a) *Drimys-Nothomyrcia* associations (including *Nothomyrcia-Dicksonia*, *Drimys–Blechnum cycadifolium*, and *Drimys-Dicksonia* associations) and occurs in many variants. Kunkel (1957) found similar assemblages in the canyons of the El Yunque summit but without *Nothomyrcia* and *Fagara* (missing from the highest elevations). Impact by the introduced shrubs *Aristotelia chilensis* and *Rubus ulmifolius* is higher than in the tree fern community. On the lowest sites, for example, in sector Villagra, this forest is highly threatened by these aliens; even though they are not yet present within the forest, they occur abundantly nearby in the canyons (quebradas), forest clearings, and landslide areas and along the central path.

1.3 ***Gunnera peltata* community**. In the bottom of the canyons, *Gunnera peltata* often forms a monodominant shrub layer nearly excluding the trees. Below the very broad leaves of *Gunnera* are found ferns such as *Megalastrum inaequalifolium*, *Blechnum chilense*, *B. schottii*, and *Thyrsopteris elegans*. Due to the high amount of litter of *Gunnera* and the tall ferns, the herb layer is very poor. Mosses can be abundant on the rocks and trunks. We observed this community at between 350 and 450 m, whereas Skottsberg (1953a) found *Gunnera peltata* associations from lower elevations. Kunkel (1957) found *G. peltata* on El Yunque exclusively in the bottoms of the canyons, whereas *G. bracteata* also occurs in the forest on the slopes. The community can be seen as a linear extension of the upper montane floristic elements following the water courses downward into the lower montane zone and even lower, where they meet the aliens *Aristotelia chilensis* and *Rubus ulmifolius*.

1.4 ***Blechnum cycadifolium* community**. *Blechnum cycadifolium* is the tree fern with the highest abundance and the broadest altitudinal and ecological amplitude on the island. It forms huge and rather pure stands on the slopes of higher elevations. The very rare small endemic shrub *Lactoris fernandeziana* is found in this community. On very extreme slopes we found *B. cycadifolium*, often very short in stature, together with *Nothomyrcia* and *Pernettya*. This fern is also very abundant on dry, wind-exposed sites and often a codominant in the *Pernettya* and *Ugni* shrub areas on the higher ridges. Skottsberg (1953a) described three associations with different combinations of trees, shrubs, and *B. cycadifolium* on Robinson Crusoe Island.

2 Lower Montane Forest

This is essentially the *Nothomyrcia fernandeziana* forest because this is the dominant tree, often accompanied by *Drimys confertifolia* and *Fagara mayu*. This forest is found mostly below 400 m, especially in the dry northwestern and southeastern regions of Robinson Crusoe Island. In general, there is less moisture due to lower precipitation and higher incoming radiation. On slopes adjacent to the bottoms of the ravines, *Boehmeria excelsa* is a common tree. Disturbance by cattle and alien plants is rather high; for this reason, a high proportion of the lower montane forest can be found as a component of the Mixed Unit 12 (forest with alien plants).

2.1 In the ***Nothomyrcia* (*Myrceugenia*) *fernandeziana* community**, the trees with dominant *Nothomyrcia* cover between 50% and 100%. Tree ferns, with the exception of scattered *Blechnum cycadifolium*, are missing, and shrubs usually play a minor role. The herb layer usually is poor, but smaller ferns can be abundant. Cryptogams also play a minor role, but in some sites lichens and liverworts form a crust covering the soil. Species diversity in this forest is much lower than in the upper montane *Drimys-Nothomyrcia* forest. *Nothomyrcia*, *Drimys*, and *Fagara mayu* form the tree layer, occasionally accompanied by *Aristotelia chilensis* (as tree and/or shrub). In the understory, the more drought-resistant ferns *Rumohra berteroana*,

Adiantum chilense, and *Blechnum hastatum* are very common, with *Histiopteris incisa* occurring occasionally, as well as the aggressive alien *Acaena argentea*. This community comprises more or less all the lower montane *Nothomyrcia* associations reported by Skottsberg (1953a). Of the several rare plants (*Robinsonia macrocephala*, *Dendroseris micrantha*, and *Sophora fernandeziana*) found in this forest in earlier times, we saw only one *D. micrantha*. This may be to some extent a random-sampling effect, but it may also be an effect of shrinking population sizes of these taxa since the times of Skottsberg's field observations. In general, an increased disturbance from aliens is indicated by *Aristotelia chilensis*, *Acaena argentea*, *Ugni molinae*, and *Rumex acetosella*.

2.2 The ***Boehmeria excelsa* community** is found in the valley bottoms and canyons of the lower altitudes. *Boehmeria* often becomes dominant with no other trees besides *Nothomyrcia*. The understory is very poor due to the dense cover of *Boehmeria* and in some places due to the feeding and trampling by cows. Skottsberg (1953a) observed very similar *Boehmeria* associations but also mixed with other trees, many ferns, and some grasses. These mixed assemblages are included in the *Nothomyrcia fernandeziana* community in our analysis.

3 *Ugni molinae* Scrub

Often-dense scrub (shrubland) occurs along the ridges and wind-exposed slopes (see Fig. 8.6) and especially on the eroded slopes near the village of San Juan Bautista, where the introduced *Ugni molinae* started its invasion of the island. *Ugni molinae* is dominant in lower elevations, along with the endemic *Pernettya rigida* (this shrub fading out at the lowest elevations). At higher elevations, *Blechnum cycadifolium* and the endemic shrub *U. selkirkii* become abundant.

3.1 ***Pernettya rigida–Ugni selkirkii* community.** The endemic *Ugni selkirkii* is confined to higher elevations. Together with the endemics *Pernettya rigida* and *Blechnum cycadifolium*, it forms an often-dense scrub along the ridges up to the highest summits (Kunkel 1957). Trees, when present, do not grow beyond shrub size. In the relatively lower accessible elevations (below 700 m) we found rather dense stands with only very few gaps for herbs, although mosses are occasionally abundant. Up to approximately 650 m we also found the alien *Ugni molinae* mixed with the endemics, which was not mentioned by Kunkel (1957) on his way up from Cordón Camote (ca. 600 m) to El Yunque (916 m). *Ugni molinae* has certainly replaced *U. selkirkii* in lower elevations. This community most likely also appears in Unit 13 (scattered native plants among rocks) in the vegetation map. However, we lack detailed observations in higher elevations.

3.2 ***Pernettya rigida–Ugni molinae* community.** The two shrubs *Pernettya rigida* and *Ugni molinae* together with *Blechnum cycadifolium* are the dominants of this typical scrub on the ridges and wind-exposed slopes in higher elevations. Some smaller ferns and *Gunnera bracteata* of the upper montane forest can still be found there. In lower elevations, *U. molinae* often becomes dominant with some alien herbs such as *Rumex acetosella*, *Aira caryophyllea*, and *Anthoxanthum odoratum* in the ground layer, whereas the alien shrubs *Aristotelia chilensis* and *Rubus ulmifolius* occur frequently at any elevation. *Ugni molinae* was found around the village by Johow (1896) and Skottsberg (1953a: observed 1917). Skottsberg (1921, p. 221) also gave a detailed report: "... lower slopes of Cordón Central, and along the path to Portezuelo, to about 200 m; two small shrubs near the Selkirk memorial, 590 m; Q. [Quebrada] Villagra rare." The situation has now changed dramatically. Obviously, the introduced *U. molinae* is rapidly invading native communities, especially the *Pernettya–Ugni selkirkii* scrub, replacing the latter endemic congener and perhaps also limiting *Pernettya*, which is nearly absent in some relevés at lower altitudes. Anderson et al. (2000b) suggested that these two lithophytic and xerophytic species, *U. molinae* and *Pernettya rigida*, respectively, are proliferating due to soil degradations in recent decades.

4 *Rubus ulmifolius–Aristotelia chilensis* Scrub

This mostly very dense thicket comprises the introduced spiny raspberry *Rubus ulmifolius* (Fig. 8.3) and the multistemmed arching tree *Aristotelia chilensis* (Fig. 8.1). *Rubus* (zarzamora) displays highest density around the village, along the path (Fig. 8.4) up to Mirador de Selkirk (Portezuelo, with the Selkirk memorial plaque), and down the other side of the ridge into Villagra valley. This noxious *Rubus ulmifolius–Aristotelia chilensis* community is also frequent in the Plazoleta del Yunque. In more remote and higher elevations, larger native trees may be included, but as a rule, *Aristotelia* is highly abundant, forming a low tree layer of about 5 m (Fig. 8.2), overtopping *Rubus* shrubs. *Aristotelia chilensis* was obviously introduced earlier, as can be concluded from Johow (1896) and (Skottsberg 1953b). Obviously, *Rubus ulmifolius* is a similarly rapid and dangerous invader, as are *Ugni molinae* and *Aristotelia chilensis*. While Kunkel (1957) did not record *Rubus* and *Aristotelia* on El Yunque, this has now changed for the worse because a French expedition to the summit of El Yunque in 1999 recorded single individuals of *Aristotelia* and *Rubus* (Philippe Danton, personal communication 1999; Danton 2000; see also Chapter 8).

5 Fern Assemblage

The dominant fern *Histiopteris incisa* forms large clonal and very dense patches through rhizomes. This assemblage is found on edges of forests and on drier or eroded slopes. Occasionally, a few shrubs are included; lower herbs and mosses play a minor role. The huge clonal systems built up by *Histiopteris* often cover 100 or even 1,000 m^2. This fern, as well as the occasionally accompanying *Rumohra berteroana*, *Adiantum chilense*, and *Blechnum hastatum*, seems more drought resistant than *Blechnum chilense* and *Lophosoria quadripinnata*, which are usually found in the forests. The shrub *Ugni molinae* also invades this community, together with *Acaena argentea*, another alien of the ground layer.

6 *Libertia chilensis* Assemblage

This assemblage of introduced and native plants is found on eroded slopes close to the village. *Libertia chilensis* is locally abundant either in small groups or covering larger areas of eroded slopes close to San Juan Bautista. Although this is not a very distinct community, we keep it separate because we suspect it to be a remnant of a native floristic assemblage on eroded slopes prior to massive invasion of these areas by *Ugni*, *Acaena*, and other introduced herbs and grasses.

7 *Acaena argentea* Assemblage

The stolons of *Acaena* form a very dense carpet-like herbaceous thicket (Fig. 8.8). This almost monodominant herb community comprises only a few other species with low abundance and is present everywhere from the lower to medium elevations except for the very dry and windy westernmost part of the island. Only a few introduced grasses and herbs are scattered within these dense carpets of *A. argentea*. Of the native and endemic taxa, only *Blechnum hastatum* and *Haloragis masatierrana* can be found frequently at low abundance. These taxa, as well as occasional native grass species of *Nassella* and *Piptochaetium*, may point to very dramatic vegetational changes in lower elevations.

According to Johow (1896), *Acaena argentea* had already been "la maleza mas comun de la isla" and was a serious pest in pastureland even more than 100 years ago. Skottsberg (1921) judged this taxon as "one of the most widespread and noxious weeds," having been introduced doubtless inadvertently from mainland South America (Matthei 1995). The spread of *A. argentea* represents a main act in the tragedy of the island flora. On the one hand, this taxon builds large and rather dense "carpets" covering deforested and eroded slopes, protecting these areas from further erosion. On the other hand, it not only outcompetes native taxa but also invades natural communities such as the *Nassella* grassland.

8 *Nassella* Grassland

The tussock-forming grasses *Nassella laevissima* and *Piptochaetium bicolor* are dominants in this native "steppe-like" grassland, extending from dry lower areas up to medium elevations (Fig. C102). This grassland is absent in the very dry westernmost part of the island.

8.1 The ***Piptochaetium bicolor–Nassella laevissima* community** occurs from lowest elevations up to approximately 500 m. *N. laevissima* is the most common species of the native tussock-forming grasses. They usually build the matrix of the steppe-like grasslands, which thousands or millions of years ago must have covered most of the dry western part and, more recently, the lower eastern parts of the island. In a few relevés the taller *Piptochaetium bicolor* is more abundant than *N. laevissima. Nassella neesiana* (= *Stipa fernandeziana*) occurs occasionally but is never very abundant. This native grassland is highly disturbed by aliens, the most serious pressure coming from *Acaena argentea*, which appears to be invading the few remaining "pure" steppes. But many of the Euro-Mediterranean weeds (e.g., *Rumex acetosella*, *Anthoxanthum odoratum*, *Bromus hordeaceus*, *Aira caryophyllea*, *Briza minor*, *B. maxima*, *Vulpia* spp., *Avena barbata*, *Dipsacus sativus*, and *Anagallis arvensis*) are present in many relevés and also must be considered a threat. Skottsberg (1953a), who did not pay much attention to the grasslands, did observe *Acaena argentea* and most of the above-mentioned weeds in this community back in 1917. The native grassland is highly impacted by the large rabbit populations and other introduced foraging animals. In addition to foraging pressure, these animals help disperse weedy species even more broadly over the island (e.g., as with the hooked fruits of *Acaena*). In maps prepared by Ortiz R. et al. (1982), nearly all grassland areas were classified as degraded or severely degraded.

8.2 ***Juncus capillaceus* community.** This community never covers large areas, but it locally replaces the *Piptochaetium bicolor–Nassella laevissima* grassland on concave slopes, flat depressions, and so on, where moisture can accumulate. Disturbance in this community is also high due to invasion by the aliens *Aira caryophyllea*, *Rumex acetosella*, and *Anthoxanthum odoratum*.

9 Weed Assemblages

These include grasslands dominated by introduced (mostly European) taxa, as well as several other herbaceous weedy communities in the drier and lower parts of the island.

9.1 The ***Rumex acetosella–Aira caryophyllea* community** with a low cover of grasses and herbs usually below 50% is often found on eroded slopes, which seem to be devoid of vegetation when first observed from a distance. In some relevés we found very dense crusts of lichens and liverworts protecting the powder-like soil from surface erosion.

9.2 We found the ***Briza maxima* community** with a cover of grasses and herbs between 50% and 95% close to the village, where it seems to represent a transition from native *Nassella* grassland to alien communities because both species of *Nassella* are found there. It is rich in other introduced herbs and grasses and might have been used as pasture in earlier historical periods, when more cattle were present near the village. *Acaena argentea* is present there, too.

9.3 The ***Vulpia* community** with a cover of grasses and herbs between 60% and 100% is found in the dry lower and western parts of the island. Due to recording in late summer, it was difficult to discriminate among the three species. As far as we could estimate, especially in the western part of the island, *Vulpia bromoides* is the dominant taxon. In this same dry western region, many of the ruderals are absent, but *Hordeum murinum* is abundant.

9.4 In the ***Avena barbata* community,** the grasses cover between 50% and 100%. This is the typical steppe-like grassland of the small island, Santa Clara, dominated by *Avena barbata* and *Bromus hordeaceus* in the large tableland area. These two taxa produce a high amount of litter, which covers the gaps between the grasses. Due to recording in the dry late summer, no dicot herbs were found except *Rumex crispus*. On the more open slopes we observed a higher abundance of *Hordeum murinum* and *H. secalinum*. Already in 1917 Skottsberg (1953a) noticed the high abundance of *A. barbata* and that there was "little or nothing" left of the native flora of Santa Clara.

9.5 **Ruderal communities** and **mixed rock and coastal communities** occur near the village, near huts and landing places, and very often on steep, rocky slopes, especially on the coast. Species composition can vary. *Ruta chalepensis* appears to form communities on steep faces of the coastal rocks in the northeastern part of the island, as could be judged from boat observations.

10 Tall Ruderals

Tall-growing (1–2 m) ruderal plants occasionally form large patches (*Conium maculatum*, *Silybum marianum*, *Centaurea melitensis*, and *Papaver somniferum*) that occur in flat areas and valley bottoms at lower elevations. We identified a **mixed tall ruderal community** with the tall annuals or biannuals presented earlier. This community is found mostly at the bases of slopes, in depressions, and in flattened areas. In addition to *Conium*, *Silybum*, and *Centaurea*, *Papaver somniferum* was found to be highly abundant near the airfield. Other introduced tall weeds, for example, the perennial *Lobelia tupa*, often form large groups on eroded slopes.

11 Cultivated *Eucalyptus, Pinus*, and *Cupressus*

Cultivated areas of *Eucalyptus globulus*, *Pinus radiata*, and *Cupressus goveniana* (Fig. C101) cover the eroded slopes around the village of San Juan Bautista. The trees were deliberately planted to help stem soil erosion and to provide needed wood for boats, homes, cooking, and heating. These aggressive trees, however, are now maintaining themselves through natural seedling establishment.

These forest plantations harbor only a poorly developed herb layer. The introduced trees, *Eucalyptus globulus, Cupressus goveniana*, and *Pinus radiata*, form either pure or mixed stands depending mostly on initial planting. But they also regenerate abundantly, in some places climbing up adjacent slopes. Groups of *Eucalyptus* (Cordón Salsipuedes) and *Cupressus* (ridge to Damajuana) were found up to an elevation of 400 m. *Eucalyptus* has some features (allelopathy, dense stands with much litter) that can affect or exclude other taxa (Swenson et al. 1997). At least two taxa of *Cupressus* are present on Robinson Crusoe Island (Swenson et al. 1997), where *C. goveniana* is more abundant (*C. macnabiana* is not recorded in the vegetation analyses). *Acacia dealbata* and *A. melanoxylon* apparently have been introduced more recently because they were not mentioned by Johow (1896) or Skottsberg (1921, 1953a, 1953b).

12 Forest with Alien Plants

In this mixed assemblage of forest composed mainly of endemics, but with high alien impact, we could not distinguish between the upper and lower montane forest components in every part of the island, although it is possible to do so in certain areas. This unit therefore contains both the upper montane forest, highly disturbed, and the lower montane forest, highly disturbed by the invasive shrubs *Aristotelia chilensis*, *Rubus ulmifolius*, and *Ugni molinae* and in the herbaceous layer by *Acaena argentea*. Often a small-scale mixture can be found of pure native forest with the invading shrubs on landslides and eroded patches.

13 Scattered Native Plants among Rocks

Found at higher elevations, on steep slopes, this unit comprises essentially elements of the upper montane forest. This unit also includes some patches of single *Nothomyrcia* trees in eroded areas above Puerto Francés.

14 Scattered Grassland and Weeds among Rocks

This unit comprises a mixture of native and alien grasses plus different weedy assemblages occurring in lower areas, on eroded slopes, on coastal rocks, and so on.

15 Grassland with *Acaena*

These mosaic patterns of mostly native grasses and dense patches of *Acaena argentea* are found in the drier areas, especially on the westernmost parts of the island.

The following two categories are not plant communities, but they are shown on the vegetation map because they have high conservation importance.

16 Rocks and Eroded Areas

This unit denotes bare ground lacking any vegetation or only scattered plants covering less than 10% of the surface (Fig. C103). Because our data were recorded in late summer (as well as the aerial photographs from the Chilean Air Force), it is possible that overestimation of the proportion of these areas has occurred. Nonetheless, the great extent of these eroded, bare areas cannot be overemphasized.

17 Settlement, Airfield

This unit comprises the area of the village (San Juan Bautista) with buildings, construction sites, dirt roads, footpaths, horticultural areas, and the airfield environs in the westernmost part of the island.

The Vegetation of Alejandro Selkirk Island

The following nine vegetation units are shown on the map (Fig. C104; see also Figs. C105 through C124 for different views of landscapes and vegetation on this island).

1 *Dicksonia externa* Tree Fern Community

Dicksonia externa is very common in higher elevations (700–1200 m) (Figs. C15 and C111), forming sort of an upper montane "forest" on the southern part of the island that is often hidden by clouds and obviously receives substantially more moisture than the lower northern part. Soil conditions are characterized by an often-thick organic layer that together with the often abundant bryophytes collects much moisture. The tree fern often forms monodominant dense and dark stands, usually without other plants below. However, in most parts of the island, *Dicksonia* occurs mixed with *Lophosoria quadripinnata. Drimys confertifolia* is often found in the shallow valleys of those highest elevations and on the southern slopes. More open variants of the community reveal a high diversity with *Blechnum cycadifolium*, *Myrteola nummularia*, the rare *Oreobolus obtusangulus* (seen only once), and several species of *Hymenophyllum*, in addition to other more widespread taxa. *Histiopteris incisa* is often dominant in gaps from a few square meters up to approximately 100 m^2.

2 *Lophosoria quadripinnata* Fern Community

The tall fern *Lophosoria quadripinnata* (Figs. C17, C117, and C120) often becomes dominant in elevations above 500 m, except for the southern part and the dry northern table land. This community grows on silty-sandy soil or solid rock, which is often covered by a dark organic layer and much coarse litter in very dense stands. This fern scrub is best developed in aspects toward the south and west between 900 and 1,200 m, while it becomes more open with a high abundance of *Pernettya rigida* and mixed with grassland when facing north and east. In higher elevations on the westward slopes and in the shallow origins of the valleys, we occasionally found small trees of *Drimys confertifolia* hidden in the ferns as well as small groups of tall trees of this species. *Lycopodium magellanicum* is often abundant, whereas other subantarctic elements, such as *Rubus geoides*, *Nertera granadensis*, and *Uncinia tenuis*, as well as the endemics *Acaena masafuerana* and *Luzula masafuerana*, as a rule prefer the highest elevations. This fern scrub was termed "estepa de helechos" by Johow (1896), although it has nothing in common with a steppe, as Skottsberg (1953a) argues. We have only limited observation of the inaccessible western cliffs, where *Gunnera* is obviously more common in patchy vegetation with *Lophosoria* and grasses called "rock face carpets" (Skottsberg 1953a).

3 Fern-Grassland Mosaic

In the grassland occurring between the patches of *Lophosoria* and *Pernettya*, *Anthoxanthum odoratum* is the dominant species in higher elevations (Fig. C120). This mosaic of fern scrub and grassland occurs in any physical setting, especially along the ridges and summits, on the steep slopes of the canyons, and on the southern and western declivities toward the sea at between 500 m and the summit area around 1,300 m. Patches of erosion or rocks bare of vegetation are often found in this setting

Figure C1. San Juan Bautista, on Bahía Cumberland, Robinson Crusoe Island.

Figure C2. Las Casas settlement, Alejandro Selkirk Island.

Figure C4. Central plaza of San Juan Bautista (1991).

Figure C5. Lobster boat arriving at Santa Clara Island.

Figure C3. Main street of San Juan Bautista (1991, before tsunami).

Figure C6. Endemic lobster (*Jasus frontalis*).

Figure C7. *Arthopteris altescandens*.

Figure C8. *Asplenium obtusatum* var. *sphenoides*.

Figure C9. *Blechnum chilense*.

Figure C10. *Blechnum cycadifolium*.

Figure C11. *Blechnum longicauda*.

Figure C12. *Blechnum schottii*.

Figure C13. *Dicksonia berteroana* and other ferns.

Figure C14. *Dicksonia berteroana*.

Figure C15. *Dicksonia externa* and *Lophosoria quadripinnata*.

Figure C16. *Dicksonia externa*.

Figure C17. *Lophosoria quadripinnata*.

Figure C18. *Megalastrum inaequalifolium*.

Figure C19. *Polystichum tetragonum.*

Figure C20. *Sticherus quadripartitus.*

Figure C21. *Hymenophyllum cruentum.*

Figure C22. *Hymenophyllum cuneatum.*

Figure C23. *Pleopeltis macrocarpa.*

Figure C24. *Polypodium intermedium* subsp. *intermedium.*

Figure C25. *Pteris berteroana.*

Figure C26. *Lactoris fernandeziana.*

Figure C28. *Peperomia fernandeziana.*

Figure C29. *Drimys confertifolia.*

Figure C27. *Peperomia berteroana* subsp. *berteroana.*

Figure C30. *Juania australis.*

Figure C31. *Ochagavia elegans.*

Figure C32. *Libertia chilensis.*

Figure C33. *Juncus procerus.*

Figure C34. *Eryngium bupleuroides.*

Figure C35. *Centaurodendron dracaenoides.*

Figure C36. *Dendroseris berteroana.*

Figure C37. *Dendroseris berteroana.*

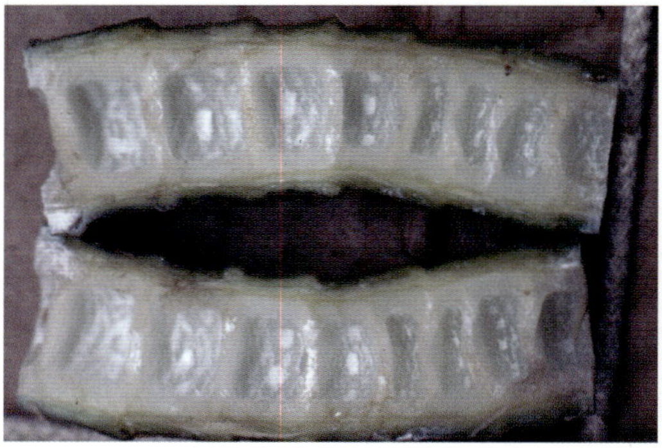

Figure C38. *Dendroseris berteroana*, stem lx.

Figure C39. *Dendroseris pinnata.*

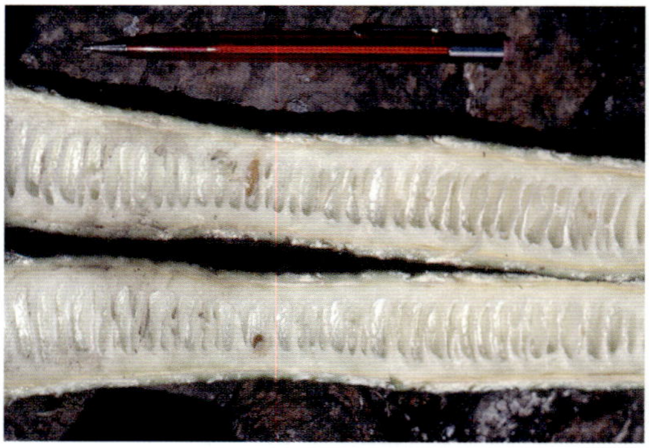

Figure C40. *Dendroseris pinnata*, stem lx.

Figure C41. *Dendroseris litoralis.*

Figure C42. *Dendroseris marginata.*

Figure C43. *Dendroseris marginata.*

Figure C44. *Dendroseris marginata*, stem lx.

Figure C45. *Dendroseris pruinata.*

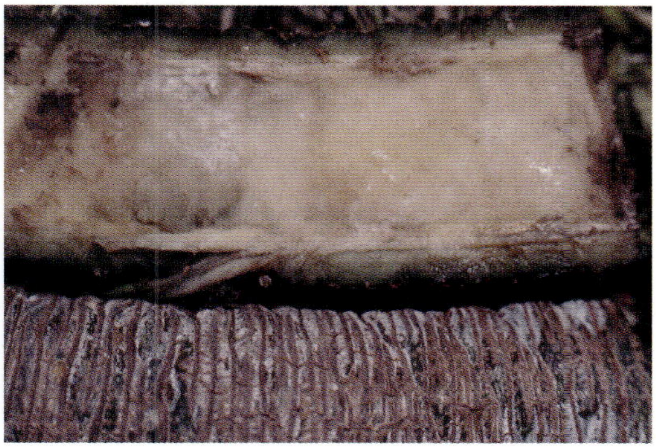

Figure C46. *Dendroseris pruinata*, stem lx.

Figure C47. *Dendroseris micrantha.*

Figure C48. *Dendroseris micrantha.*

Figure C49. *Dendroseris micrantha*, stem lx.

Figure C50. *Dendroseris neriifolia.*

Figure C51. *Dendroseris neriifolia.*

Figure C52. *Dendroseris neriifolia*, stem lx.

Figure C53. *Erigeron fernandezia.*

Figure C54. *Erigeron fernandezia.*

Figure C55. *Erigeron ingae.*

Figure C56. *Erigeron rupicola.*

Figure C57. *Erigeron stuessyi.*

Figure C58. *Robinsonia berteroi.*

Figure C59. *Robinsonia thurifera.*

Figure C60. *Robinsonia thurifera.*

Figure C61. *Robinsonia gayana.*

Figure C62. *Robinsonia gracilis*.

Figure C63. *Robinsonia gracilis* (left, female; right, male).

Figure C64. *Robinsonia evenia*.

Figure C65. *Robinsonia evenia* (left, female; right, male).

Figure C66. *Robinsonia masafuerae*.

Figure C67. *Robinsonia masafuerae* (left, female; right, male).

Figure C68. *Wahlenbergia berteroi.*

Figure C69. *Wahlenbergia fernandeziana.*

Figure C70. *Chenopodium sanctae-clarae.*

Figure C71. *Pernettya rigida.*

Figure C72. *Escallonia callcottiae.*

Figure C73. *Dysopsis hirsuta*.

Figure C74. *Sophora fernandeziana* var. *fernandeziana*.

Figure C75. *Gunnera masafuerae*.

Figure C76. *Gunnera peltata*.

Figure C77. *Cuminia eriantha*.

Figure C78. *Haloragis masatierrana*.

Figure C79. *Myrceugenia schulzei*.

Figure C80. *Nothomyrcia fernandeziana*.

Figure C81. *Ugni selkirkii*.

Figure C82. *Ranunculus caprarum.*

Figure C83. *Rubus geoides.*

Figure C84. *Zanthoxylum mayu.*

Figure C85. *Azara serrata* var. *fernandeziana.*

Figure C86. *Bohemeria excelsa.*

Figure C87. *Rhaphithamnus venustus.*

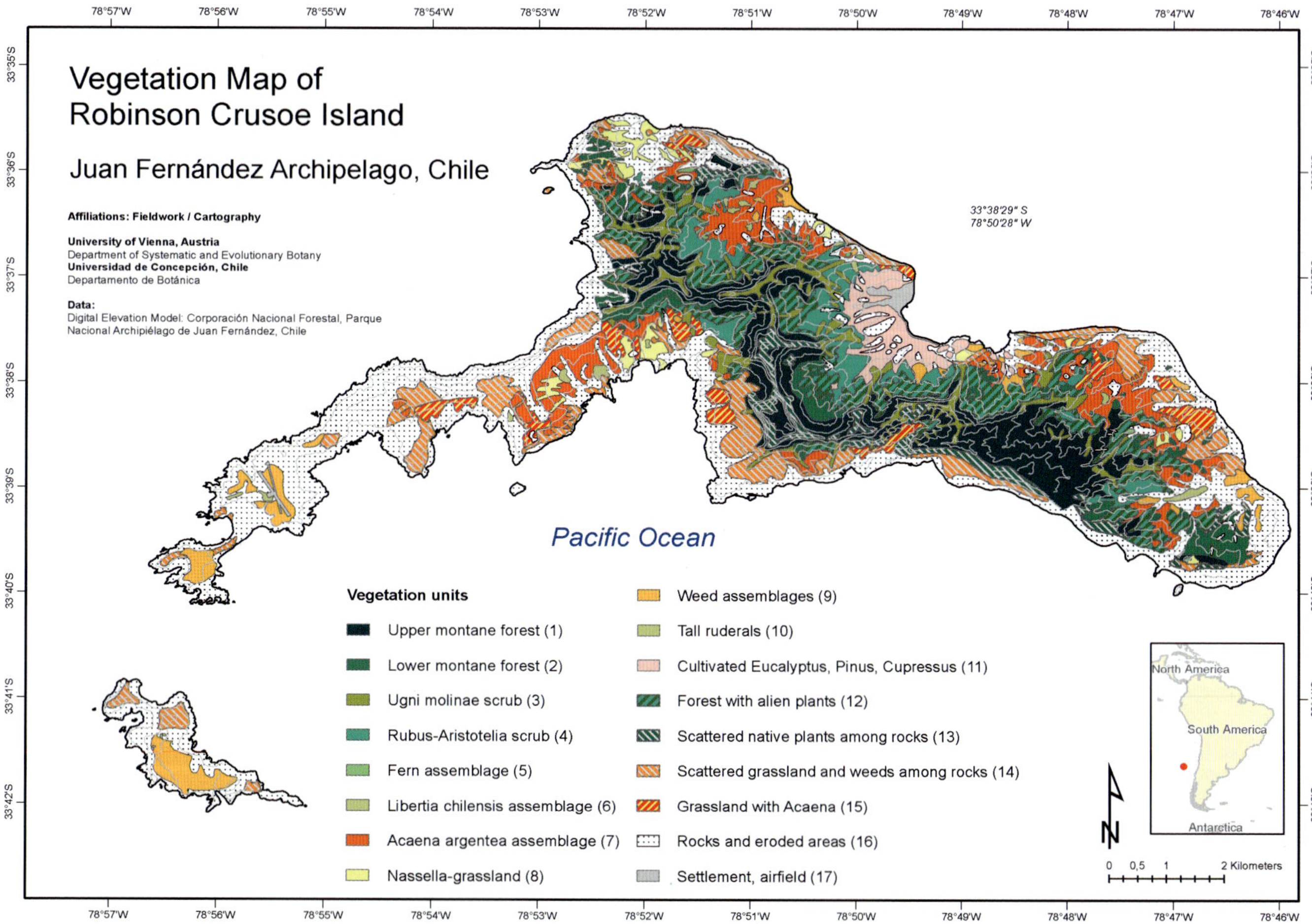

Figure C88. Vegetation map of Robinson Crusoe Island. From Greimler et al. (2002, modified).

Figure C89. Panorama of area surrounding San Juan Bautista, showing the peaks Damajuana (left) and El Yunque (right).

Figure C90. Southwestern side of Robinson Crusoe Island, with Tres Picos in the center.

Figure C91. Coastal cliffs along upper ridge of La Vaquería looking westward toward Tres Picos on the northern side of Robinson Crusoe Island.

Figure C92. La Pascua, Puerto Francés, looking toward southwest side of Robinson Crusoe Island.

Figure C93. La Pascua, upper ridge of Puerto Francés, Robinson Crusoe Island, forest dominated by *Nothomyrcia fernandeziana* and *Drimys confertifolia*.

Figure C94. Upper reaches of La Vaquería, Robinson Crusoe Island, showing scattered trees of *Nothomyrcia fernandeziana*.

Figure C95. El Camote, Robinson Crusoe Island, looking up toward El Yunque (center right).

Figure C96. Main ridge of Cordón Central, Robinson Crusoe Island.

Figure C97. Upper montane forest in Valle Villagra, Robinson Crusoe Island.

Figure C98. Valle Villagra, Robinson Crusoe Island, showing effect of wind-shear on vegetation.

Figure C99. Path through forest to Selkirk's Lookout, Robinson Crusoe Island.

Figure C100. Robinson Crusoe Island, Cerro Centinela in middle ground, with El Yunque in background; lower montane forest, with *Nothomyrcia fernandeziana* dominant in lower parts of valley.

Figure C101. Cultivated *Cupressus goveniana* in area surrounding San Juan Bautista.

Figure C102. Grassland in La Vaquería, Robinson Crusoe Island.

Figure C103. Remnants of forest and eroded region in Puerto Francés on trail up toward La Pascua, Robinson Crusoe Island.

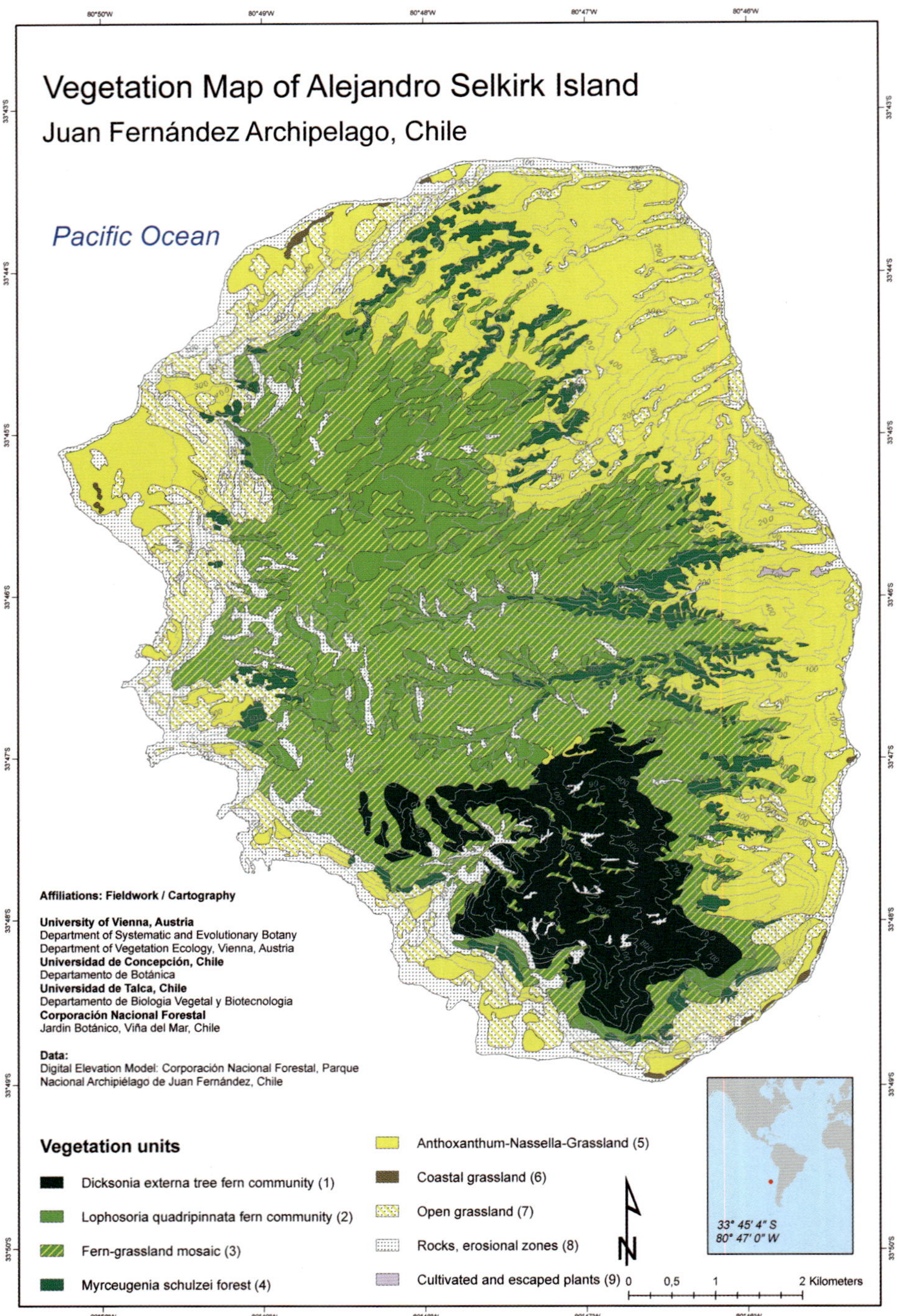

Figure C104. Vegetation map of Alejandro Selkirk Island. From Greimler et al. (2013).

Figure C105. Open grassland, Playa Ancha, Alejandro Selkirk Island.

Figure C106. Southeastern coastal deep valleys (quebradas) on Alejandro Selkirk Island.

Figure C107. Quebrada Óvalo, Alejandro Selkirk Island.

Figure C108. Grassland on side of Quebrada de Las Vacas, Alejandro Selkirk Island.

Figure C109. Quebrada Mono, grassland, Alejandro Selkirk Island.

Figure C110. Ridge up from Quebrada de Las Vacas with grasses and *Myrceugenia schulzei*, Alejandro Selkirk Island.

Figure C111. Toward Tres Torres, *Blechnum*, *Dicksonia*, *Lophosoria*, Alejandro Selkirk Island.

Figure C112. Quebrada Las Casas, with *Gunnera masafuerae*, Alejandro Selkirk Island.

Figure C113. Quebrada Sándalo, with ferns and other cryptogams, Alejandro Selkirk Island.

Figure C114. Looking down toward Quebrada Sándalo, Alejandro Selkirk Island.

Figure C115. Looking upward toward Los Inocentes, Alejandro Selkirk Island.

Figure C116. Quebrada dominated by *Gunnera masafuerae*, *Dicksonia externa* and other ferns, Alejandro Selkirk Island.

Figure C117. Quebradas near Tres Torres, Alejandro Selkirk Island, with *Lophosoria quadripinnata* and grasses.

Figure C118. Collecting by rope on side of cliff at upper reaches of Quebrada Las Casas, Alejandro Selkirk Island.

Figure C119. Cordon Inocentes and Tres Torres, Alejandro Selkirk Island.

Figure C120. North end of Alejandro Selkirk Island, *Lophosoria quadripinnata* with grasses.

Figure C121. Quebrada Guatón, on western side of Alejandro Selkirk Island.

Figure C122. Cordón Atravesado, Alejandro Selkirk Island, with typical fog cover.

Figure C123. Region of Los Inocentes, Alejandro Selkirk Island.

Figure C124. Cordón Inocentes, mat of *Marchantia berteroana*, Alejandro Selkirk Island.

This unit covers most of what Skottsberg (1953a) termed the "rock face carpets" that include some *Gunnera* and forest patches on the southern and western declivities. On our map, however, this unit covers a much higher proportion of the island because we include all "open" vegetation, in which *Lophosoria* (and/or *Dicksonia* in the south) and grassland are the main components. In the highest elevations (ca. 1,300 m) of the summit area of Cerro de Los Inocentes, *Abrotanella linearifolia* adds to many of the subantarctic elements mentioned earlier.

4 *Myrceugenia schulzei* Forest

Myrceugenia schulzei, often covering 100%, is the dominant tree, occasionally accompanied by *Fagara externa* in this lower montane forest of canyons/ravines (quebradas) at mostly between 200 and 700 m of elevation (Fig. C110), reaching up to the lower tableland in the very northern part of the island. Soil conditions are often dry with high variation of the organic layer and the sandy-silty texture. On the slopes and bottoms of the inner parts of the ravines (Quebrada Casas, Q. Vacas, and adjacent ravines), *Aristotelia chilensis* is occasionally abundant. The proportion of alien plants is often high, and the higher number of those plants found in some relevés divides the forest roughly into two subunits. Subunit (a) is characterized by high abundance of *Peperomia fernandeziana*, the tall *Pteris chilensis*, and small ferns (*Blechnum mochaenum*, *Asplenium dareoides*, and *Hymenophyllum* spp.), whereas in subunit (b) aliens such as *Anthoxanthum odoratum*, *Geranium core-core*, and *Oxalis* spp. are more often found. In addition to dense stands with tall *Myrceugenia* trees, these forests are frequently open, consisting of several small groves with many saplings and small trees around them. However, we do not know whether those saplings have emerged from seedlings or clonal ramets.

From remote observation, it seems that more diverse *Myrceugenia* forest patches including more *Drimys* and *Coprosma* may be present on the nearly inaccessible western cliffs, which agrees with Danton (2006), who distinguishes a "Myrtisylve de zone basse" from a more rich "Myrtisylve d´altitude" found in those zones and in higher elevations of the southern part of the island. Regarding the problem of delimiting the upper from the lower montane forest, we agree with Skottsberg (1953a) that *Dicksonia* is a good indicator of a more humid upper montane "forest" on Alejandro Selkirk, although he argues that it is impossible to "fix a line between lower and an upper montane belt." Our data on this island indicate a distinct lower montane *Myrceugenia* forest. However, due to constraints of time and logistics, we were not able to access the *Myrceugenia* forests on the higher elevations of the southern part of the island and the western cliffs.

5 *Anthoxanthum-Nassella* Grassland

Nassella laevissima is the dominant grass from sea level up to about 600 m, whereas *Anthoxanthum odoratum* is present nearly everywhere (Figs. C108 and C109). *Anthoxanthum* is the only grass species we found on the highest elevations around 1,300 m, whereas the two *Nassella* species are less frequent and as a rule of low abundance in higher elevations. The aliens *Aira praecox* and *A. caryophyllea* are often found at elevations above 1,000 m. Accordingly, this unit can be divided into two communities.

5.1 ***Aira praecox–Anthoxanthum odoratum* community.** This is the typical grassland of the higher elevations, where it is often limited to small areas. This community characterizes not only the upper end of the grassland unit, again with the subantarctic elements and endemics mentioned earlier, but also occurs as the major grassland component of the fern-grassland mosaic. Bryophytes are occasionally of high abundance in positions that are protected from the strong winds by the local landscape configuration. Soil conditions change roughly along an altitudinal gradient from an often-marked organic layer on solid rock or not at higher elevations to sandy silty at medium elevations.

5.2 ***Nassella neesiana–Nassella laevissima* community.** A number of weeds unsuited for higher elevations (e.g., *Anagallis arvensis*, *Briza minor*, *Sonchus asper*, and *Vulpia bromoides*) are found in this grassland. The number of weeds increases toward the lowest elevations and near the settlement. Bryophytes are missing or of low abundance on the dry sandy-silty or sandy soil that is often covered by a dense layer of dry litter. In coastal areas, this grassland is often found on naked rock.

The entire Unit 5 is congruent with Skottsberg's (1953a) "native & secondary grassland," in which *Anthoxanthum* is by far the most abundant "foreign grass." We add to this, however, that *Anthoxanthum* is also the most abundant of all taxa in our grassland relevés, including the fern-grassland mosaic. This taxon also occurs with the highest overall frequency (69%) in our relevés, slightly ahead of *Rumex acetosella* (68%), and also was found in all eight relevés of Cuevas (2002). From observations about 100 years ago, Skottsberg (1953a) reasoned that "it is hard to imagine what the upland country looked like before the arrival of these intruders." Several episodes of burning (Skottsberg 1953a; Cuevas 2002; Danton 2006) and the huge goat population are certainly among the agencies favoring these aliens. Space preemption by those species in arising gaps between the ferns may have added to their success.

6 Coastal Grassland with *Juncus procerus*

The tall rush *Juncus procerus* often forms large clusters on coastal rocks, gravel, and sand (Fig. C33). The two *Nassella* species are still of high abundance, but some typical plants of the coastal cliffs (*Erigeron rupicola*, *Asplenium obtusatum* var. *sphenoides*, and *Spergularia masafuerana*) and ruderals (*Spergularia confertifolia* and *Polycarpon tetraphyllum*) indicate a stronger impact from the salt spray. This community often forms the lower fringe of the grassland against the sea.

7 Open Grassland

This unit comprises mainly all the open grassland along the ridges, on the cliffs around the island, and on many slopes of the canyons and also includes some special communities at the base of the coastal cliffs (Fig. C105).

7.1 **Open *Anthoxanthum-Nassella* grassland.** This is the open variant of Unit 5 covering less than 40% of the ground. The open grassland is found on the coastal cliffs and the steep slopes of the valleys, especially in the northern, drier part of the island.

7.2 **Liverwort crusts** are found typically along the open ridges at around 1,000 m in the southern *Dicksonia*-dominated part. Several layers of thallose liverworts have produced a thick rubber-like crust, saturated with moisture and resembling a raised bog (Fig. C124). *Marchantia berteroana* is the dominant species, although there may be other taxa too. Within and around these liverwort crusts we found evidence of fire damage, due probably to the fire of February 1996 (Barría 1996).

7.3 ***Spergularia masafuerana–Erigeron rupicola* community.** The heterogeneous plant assemblages on the coastal cliffs are probably best characterized by these two endemics of Alejandro Selkirk. *Erigeron rupicola* occasionally enters the outer parts of the canyons, whereas *Spergularia masafuerana* keeps to the coastal habitat. *Sarcocornia fruticosa* and *Asplenium obtusatum* are mainly found in this context. Locally, the alien *Bahia ambrosioides* becomes dominant, and another alien, *Chaetotropis imberbis*, was also found mostly on the coastal rocks.

8 Rocks, Erosional Zones

Vegetation is entirely missing or only single plants or small groups of plants are found covering notably less than 10% of the ground surface. Those zones are found especially on the cliffs, on coastal gravel and sand, and on eroded summits, ridges, and slopes of the canyons. As on Robinson Crusoe Island, our data were recorded in late summer, so some overestimation of the proportion of these areas is possible. This unit includes the single tiny settlement with about 20 small buildings (fishermen's homes and the National Park Administration).

9 Cultivated and Escaped Plants Near the Settlement

There are a few trees of *Eucalyptus globulus*, *Ailanthus altissima*, *Pinus radiata*, and *Acacia melanoxylon* near the settlement. The bark and cortex of tall *Eucalyptus* trees have been cut off by the park rangers, so those trees have died. At present, one can see the tall skeletons of those dead trunks.

Plant Assemblages that Are Not Shown on the Map of Alejandro Selkirk

There are some plant assemblages that occur either in small clusters (i, iv) or in narrow linear arrangement in and along the valleys and canyons (ii, iii) that cannot be displayed reasonably on the map at the given resolution. Because they do occur in various vegetational contexts, they are included in the respective surrounding vegetation units.

(i) ***Drimys confertifolia*** often forms small clusters or larger groves within the tall ferns in the shallow valleys of the highest elevations, especially in the southern part of the island and there on the slopes. Such forest patches of *Drimys* within the fern matrix are either mixed with those tall ferns or occasionally very dense with nearly no other plants between or below the trees. The larger groves are found within the *Dicksonia* assemblage (Unit 1), while the smaller clusters occur in the *Lophosoria* assemblage (Unit 2). Regeneration of *Drimys* seems to be poor below the dense canopy of old trees, forming small groves surrounded by a dense *Dicksonia* and *Lophosoria* matrix. However, when the tall fern cover becomes more open, sometimes small trees of *Drimys* occur between them in the southern part of the island. Single small trees were also found rarely within the dense *Lophosoria* community of the central western part of the island.

(ii) The ***Gunnera masafuerae* community** is found typically in shallow depressions, in moderately deep canyons of higher elevations (Fig. C116), and in the deep and dark canyons in lower elevations, where the plants grow tall (up to 5 m) as nowhere else (Fig. C75). The *Gunnera masafuerae* community does occur in the Units 1, 2, and 3 in higher elevations, whereas in lower elevations it is found within Units 4 and 5. A similar community in the canyons of Robinson Crusoe is formed by *G. peltata* (Greimler et al. 2002a). Furthermore, *G. masafuerae* is occasionally highly abundant in the mixture of *Lophosoria* with everything on the cliffs termed "rock face carpets" by Skottsberg (1953a), which is mostly covered by our Units 2 and 3.

(iii) The small-scale variation of the **plant assemblages found on the nearly vertical walls of the canyons** does not allow inclusion of this vegetation into a single one of the above-presented vegetation units because floristic elements of all of them are found there. Similar to the groups of *Gunnera masafuerae*, larger or smaller clusters of trees or scrub of *Myrceugenia* and *Aristotelia* can be found in the two deepest and biggest canyons (Quebrada Casas and Q. Vacas) that could be entered up to an elevation of approximately 200 m. The narrow, shady, and moist inner parts of those canyons harbor many bryophytes and ferns, whereas toward the dryer outer parts with higher insolation there are fragments of *Nassella-Anthoxanthum* grassland. Carpets of *Megalachne masafuerana* and/or *M. berteroana* were found on steep walls of the narrower parts of the canyons under moderately moist conditions with little insolation. These assemblages can occur in the context of Units 2 through 5.

(iv) ***Histiopteris incisa*** often fills the gaps between the tree ferns and tall ferns at higher elevations of the southern part, where it also occurs in the grassland of lower elevations.

Common Features and Differences between the Islands

The two islands are very different in their physical setting essentially due to their different erosional patterns and different ages (Stuessy et al. 2005a; see also Chapter 3). Different age and geomorphology have a strong impact on assembly and spatial distribution of the vegetation that is more homogeneous on Alejandro Selkirk Island with its huge tableland dissected by deep canyons. In contrast, the highly eroded Robinson Crusoe Island has many narrow ridges connecting the more or less distinct summits and the single high mountain, El Yunque, which exceeds the other elevations by nearly 300 m.

Among the tall ferns and tree ferns, *Thyrsopteris elegans*, which is very rare on Alejandro Selkirk Island, forms a tree fern community together with *Dicksonia berteroana* on Robinson Crusoe Island. *Blechnum cycadifolium*, so common on Robinson Crusoe, occurs occasionally on Alejandro Selkirk but is never dominant, whereas *Lophosoria quadripinnata*, a moderately common tall fern in the upper montane forest of Robinson Crusoe Island, is the omnipresent and dominant fern on Alejandro Selkirk Island. Regarding the single island endemic tree ferns, *Dicksonia externa* tends to form closed, dense, and often monodominant stands on Alejandro Selkirk Island, while *D. berteroana* on Robinson Crusoe Island occurs in mixed assemblages together with other tall ferns and the trees *Drimys confertifolia* and often also *Nothomyrcia fernandeziana*. Another conspicuous fern, *Histiopteris incisa*, which is widespread in the southern hemisphere (Wace 1960; Pickard 1983), forms large clonal clusters on both islands.

The most important trees, *Drimys confertifolia*, *Myrceugenia schulzei*, and *Nothomyrcia fernandeziana*, play very different roles on the two islands. On Robinson Crusoe Island, *Drimys* is often found in the forests at lower elevations (below 350 m) together with *Nothomyrcia fernandeziana*, whereas on Alejandro Selkirk Island, we never observed *Drimys* together with *M. schulzei* and never below 700 m. Although occurring mostly in a mixed forest with *Drimys* on Robinson Crusoe Island, *N. fernandeziana* becomes the dominant tree toward the lowest elevations (Fig. C100). The other endemic on Alejandro Selkirk Island, *M. schulzei*, forms a forest at between 200 and 700 m on the slopes of the deep canyons that are separated by ridges with grassland and fern heath. In addition to dense stands, these forests on Alejandro Selkirk Island are frequently open, consisting of several small groves with many saplings and small trees around the tall ones. However, we do not know whether those saplings have emerged from seedlings or clonal ramets. The general impression is that *M. schulzei* on Alejandro Selkirk Island regenerates better than *N. fernandeziana* on Robinson Crusoe Island, although Vargas et al. (2010) also found ample below-canopy regeneration in the latter species. The other trees (*Rhaphithamnus venustus* and *Coprosma pyrifolia*) are rare in the lower montane forest of both islands, whereas another taxon occurring in the lower montane forest of Robinson Crusoe Island, *Boehmeria excelsa*, is missing on Alejandro Selkirk Island.

The two endemic species of *Gunnera, G. peltata* on Robinson Crusoe Island and *G. masafuerae* on Alejandro Selkirk Island, form typical communities in shallow depressions and canyons. On the latter island, however, *G. masafuerae* is also often highly abundant on the wet cliffs together with tall ferns and grasses (Fig. C112). The third endemic species of *Gunnera* in the archipelago, *G. bracteata*, is usually found in the higher elevations of Robinson Crusoe Island together with the tree ferns

Dicksonia and *Thyrsopteris* as well as on wind-exposed ridges together with *Blechnum cycadifolium*.

In the grasslands, *Nassella laevissima* is the dominant native taxon on both islands. *Nassella neesiana* is more abundant and *Piptochaetium bicolor* is very rare on Alejandro Selkirk Island. Recently, Baeza et al. (2007) considered the two *Nassella* species (*N. laevissima* and *N. neesiana*) as introduced (so treated in Table 5.2). There may be an argument for indigenous status for these species, however, based on the sediment core analysis of Haberle (2003) on Alejandro Selkirk Island. Grass pollen dominance, as well as a high percentage of pollen of the scrub *Pernettya*, was found at approximately 8,000 BP, indicating drier conditions. The only candidates among the recent grasses are the *Nassella* species because *Anthoxanthum* and other aliens were certainly introduced only since the sixteenth century (with a corresponding peak in the pollen data).

In contrast to the many exotic trees that are planted around the village on the eroded slopes of Robinson Crusoe Island, there are only a few alien trees on Alejandro Selkirk Island. The only probably proliferating scrub of *Cestrum parqui*, a taxon observed as early as in 1917 by Skottsberg (1953b), is found a few hundred meters inside Q. Casas on Alejandro Selkirk Island.

Outlook

From the sediment core analysis on Alejandro Selkirk Island (Haberle 2003), one can conclude that the vegetation of the islands has experienced dramatic changes due to climatic shifts through the Holocene. Also, disturbance by fire is documented from this sediment core analysis through the late Holocene before and after the first Europeans entered the islands. Since the arrival of the first humans, however, a dramatic shift in the vegetation occurred (Skottsberg 1953a; Greimler et al. 2002b; Dirnböck et al. 2003) due to the impact of alien biota. Among the most dangerous alien plants are the woody species *Aristotelia chilensis*, *Rubus ulmifolius*, and *Ugni molinae* (Saunders et al. 2011; Arellano 2012), although their role in the two islands is different (see Chapter 8). Introduced animals such as goats, rabbits, and rats have a negative impact on the native flora of the islands. Programs have been developed to control the alien biota (Cuevas and van Leersum 2001), but these programs require long-term commitments by village residents, park administrators, and scientists, as recommended by the management plan of the National Park Archipelago Juan Fernández (CONAF 2009). Smith-Ramírez and Arellano-Cataldo (2013) even proposed employment of the Chilean army to control alien plant invasion.

Part IV

Plant Conservation

Throughout our expeditions and investigations on the flora of the Juan Fernández Archipelago, we have been keenly aware of the need for conservation of the endemic plants. It is a remarkable sensation to be eating lunch under some large tree, leaning back against its trunk, and then realizing that this species occurs nowhere else on Earth. If we do not take proper care of the populations of this tree, it will no longer be a part of our biological heritage. A very striking moment occurred when we first visited Morro Spartan, which is a very small, flat islet separated by a water channel from Santa Clara Island (southwest off the coast of Robinson Crusoe Island). This location is the only known locality of *Chenopodium sanctae-clarae*. It would be so easy for this population of several dozen individual plants to be struck by some disease or even to be accidentally cut by people. It was at this moment that the conservation imperative struck home in full force. Because of the fragility of existence of so many rare plants in this archipelago, it is essential that resources be dedicated to their future preservation.

People have not been particularly kind to the vegetation of this archipelago. In Chapter 7, impacts resulting from more than 400 years of human activity are chronicled. Much of the forest along the eastern side of Robinson Crusoe Island has been severely cut, leaving eroded hillsides in some areas. Historical documentation allows a summary of human impacts such that the archipelago serves as another instructive case study of the negative effects that can occur in isolated oceanic archipelagos. Along with deliberate alteration of the landscape, people have brought cultivated plants for food and enjoyment, some of which have escaped and become serious pests. Other introductions of aggressive plants have occurred accidentally, riding along with foodstuffs or other supplies to the islands. Chapter 8 discusses the worst of the invasive plants in the archipelago, indicating which ones are particularly harmful to the native vegetation. Finally, in Chapter 9 we discuss the general conservation picture, with summaries of the flora in IUCN categories. It comes as no surprise that more than three-fourths of the endemic flora is endangered at some level. There is no escaping the conclusion that conservation must have a very high priority for the plants of this archipelago.

7 Impacts on the Vegetation

Tod F. Stuessy, Clodomiro Marticorena, Ulf Swenson, Josef Greimler, and Patricio López-Sepúlveda

Every current visitor to the Juan Fernández Islands quickly becomes aware that extensive parts of the landscape are devoid of vegetation. This raises the question of whether these areas are naturally barren or have resulted from more recent human (anthropogenic) influence. In this chapter we attempt to answer this question by surveying natural and human impacts on the vegetation and, in particular, by use of early historical documentation.

The geological changes that have occurred in the past 4 million years, which constitute the ontogeny of the islands, were chronicled in Chapter 3. There is no doubt that the combination of island subsidence and erosion has had a huge impact on the landscape and most likely resulted in loss of specific and genetic diversity (Stuessy et al. 2005a). This would have surely been more significant on Robinson Crusoe Island, which is two to four times older than Alejandro Selkirk Island (Stuessy et al. 1984).

The human impact on the Juan Fernández Archipelago during the past 400 years has been substantial. Because no indigenous peoples lived in the islands prior to their discovery in 1574 (Anderson et al. 2002; Haberle 2003), the impact must have been less than has occurred in some Polynesian islands (such as Hawaii) that have much longer colonization histories (Cuddihy and Stone 1990; Hobdy 1993). Furthermore, because of the ruggedness of the terrain and diverse types of forest trees, there has never been large-scale agriculture or commercial forestry. A very important point is that historical records regarding the Juan Fernández Archipelago can provide documentation of the original state of the vegetation and the human influences since discovery.

Natural Factors

The natural factors impacting the native vegetation in the Juan Fernández Archipelago are due to both physical and biological influences. The changing geological ontogenies of the two major islands in the Juan Fernández Archipelago have had a profound impact. After formation and through time, the islands have subsided and eroded, as already documented in detail in Chapter 3 (cf. Figs. 3.5 and 3.6). As the Nazca Plate moves slowly eastward toward the South American coast, it inclines downward, eventually to be subducted under the South American Plate. As it moves, the islands also drop until eventually they become seamounts under the ocean. Accompanying subsidence is the constant erosion of the lava and ash from wind, rain, and wave action. The only really solid portions of oceanic islands are basalts, which often form ridges that endure much longer than the surrounding volcanic deposits.

The combination of these natural physical factors results in less surface area and also reduced ecological habitats (Stuessy et al. 2005a). As the elevation of the islands lowered, the climate also became modified, bringing about changes in precipitation. The soil surely changed, too, due to patterns of weathering over time. Occasional natural fires occurred (Haberle 2003). Earthquakes and tsunamis have also impacted the islands, as in 1835 (Sutcliffe 1839) and more recently in 2010 (see Chapter 1), but it is doubtful that these had a very permanent impact on the native vegetation.

In addition to the many changes brought about by physical forces in the Juan Fernández Archipelago, biotic changes also occurred. These are nearly impossible to document clearly, but loss of ecological and specific diversity would certainly change the dynamics of the ecosystem. The interactions resulting from competition, predation, and pollination would constantly be changing due to the physical alterations of the islands. Again, all these shifts would have been more marked on the older island (Robinson Crusoe). Some species would be keystones, which, when reduced in number of populations or individuals, could bring about a cascade of other shifts (and losses). As the land surface reduced, previously separated populations of related species would have been brought into contact, and hybridization surely resulted. At the present time, there is little interspecific hybridization in the islands, with the only known examples being in *Gunnera* (Pacheco et al. 1991a), the unusual intergeneric hybrid ×*Margyracaena* (Crawford et al. 1993a), and between two native species of ferns, *Pleopeltis macrocarpa* and *P. masafuerae* (Danton et al. 2015; see further comments in Chapters 13 and 15).

Human Factors

Although impacts from natural factors on the native vegetation of the Juan Fernández Archipelago must have been substantial, there is no *direct evidence* for specific modifications. One can only infer changes based on a comparison between the two major islands as well as comparisons with other oceanic archipelagos. With human factors, however, the case is quite different. Historical records exist that help document what humans have done and when. Because the archipelago was an important stopping place for ships sailing around the world and for pirates from various European countries for attacking the Spanish colonies on the South American continent, many accounts of these trips and adventures exist. It is possible, therefore, to present historical evidence for the impact that humans have had on the native vegetation during this time. Table 7.1 lists major sources that cover the period of discovery and early exploration and trading in seventeenth and eighteenth centuries.

Brief Review of Human Activities in the Archipelago

It is convenient to divide the impact of human-associated activities into four phases: (1) discovery and early exploration (1574–1742), (2) the trading period (1750–95), (3) the

Table 7.1 Published Historical Accounts of Voyagers to the Juan Fernández Islands during the Seventeenth and Eighteenth Centuries

Voyager	Year visited	Type of documentation available
Schouten	1616	Drawing
Le Hermite	1624	Drawing, text
Ovalle	1649	Text
Rosales	1665	Text
Vea	1675	Text
Ringrose	1680	Drawing, map, text
Cowley	1684	Text
Dampier	1684	Text
Wafer	1686	Text
Funnell	1704	Drawing, text
Cooke	1709	Drawing, text
Rogers	1709	Drawing, text
Clipperton	1719	Text
Shelvocke	1720	Drawing, text
Roggeveen	1722	Text
Anson	1741	Drawing, map, text
Ulloa	1742	Map, text
Sobrecasas	1751	Text
Byron	1765	Text
Carteret	1767	Drawing, text
Magee	1792	Text
Vancouver	1795	Text

sealing and whaling period (1798–1808), and (4) Chilean independence (1811–present). The human impact on the Juan Fernández Archipelago began with its discovery in 1574 by Juan Fernández (Medina 1974). This was occasioned by the desire to find a faster route from Lima (Callao) south to Valparaíso, Chile, which often took several months. The fast-flowing Humboldt Current along the western South American coast impedes the progress of sailing ships southward. By first sailing further west, Juan Fernández was able to progress quickly southward, arriving in Valparaíso in only several weeks. As part of this trip, he passed by Robinson Crusoe Island, which was the beginning of its exploration by European visitors. The voyage around Cape Horn was the preferred route to the Pacific Ocean, but this passage was often very difficult, resulting in damage to sailors and vessels alike. The Juan Fernández Islands were well situated to allow a chance for repairs and recuperation. Robinson Crusoe Island, with its broad natural bay, fresh water, and large trees for lumber for repairs to damaged ships, provided a salvation for voyagers either continuing westward into the Pacific Ocean or for those intent on attacking the Spanish colonies.

The second period of impact covers the trading period 1750–95. During the exploration period, many raiders and pirates used the Juan Fernández Islands to raid the Spanish colonial coast and to intercept the galleons that sailed from Callao with silver from the mines at San Luis Potosí, Bolivia. As a result of these attacks, Spain decided to establish

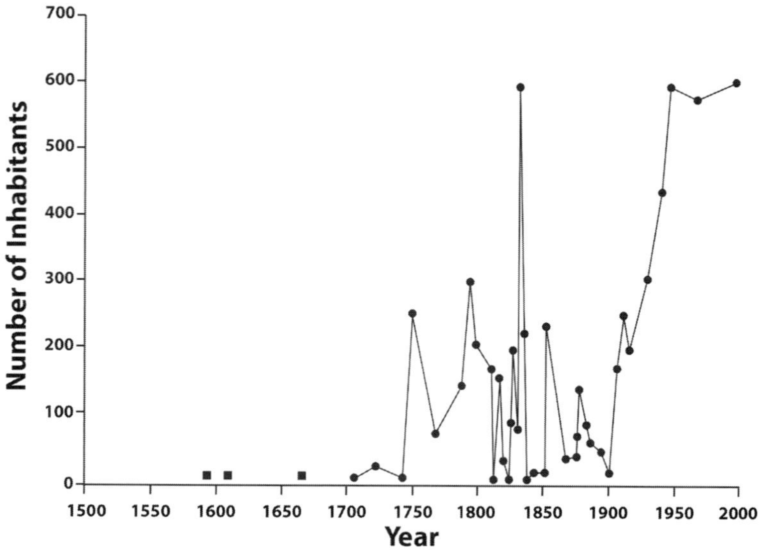

Figure 7.1 Size of the human population during the last 400 years in the village of San Juan Bautista, Robinson Crusoe Island. Squares = island inhabited but exact number of people unknown. (Data from Woodward 1969.)

a permanent colony in the islands in 1759 with approximately 250 persons (Woodward 1969). That number has since fluctuated between almost none and about 600 inhabitants twice during the past 200 years (Fig. 7.1). A fort, Santa Bárbara, was also built on the hill overlooking the bay, which with strategically placed canons effectively gained control of the area. With a permanent population also came the need for cultivated plants and domesticated animals for food. Trading with passing ships was also encouraged as a means of obtaining much-needed supplies.

The third period was short (1790–1810) and focused on Alejandro Selkirk Island. Due to the strong new Spanish defenses of Robinson Crusoe Island, sailing ships now preferred to anchor at the younger western island. Although no protected bay is available, anchorage is possible near shore due to the steep drop-off. Fresh water was abundant. The item of commerce located here was principally the endemic fur seal, which abounded (Hubbs and Norris 1971; Goldsworthy et al. 2000). During these two decades, as well as earlier and later, millions of animals were killed, skinned, and their hides sent to Europe, the United States, and the Orient for trade (Torres 1987). This activity was so excessive, in fact, that the species was driven to near extinction. It was regarded as completely extinct until some animals were found in 1965 (Bahamonde 1966), which, when protected, have rebounded to currently yield a surviving population. Because human activity was largely confined to harvesting fur seals along the coast, the impact on the natural vegetation was minimal.

The fourth phase of human impact starts with Chilean independence from Spain in 1811 and continues to the present time. Chile was later reconquered by Spanish loyalists, and some of the Republicans were sent to Robinson Crusoe Island as punishment. There they carved out an existence and lived in caves in the side of the hill adjacent to Fort Santa Bárbara, and survived for several years until Chile was again in the hands of the patriots, this time permanently. Search for food and water was obviously now a priority, as was the sale of wood from native trees for use by passing ships. A particularly large influx of visitors occurred during the years of the California gold rush (1848–55), whereby prospectors took ships from the eastern United States around Cape Horn and then up the coast to California (Hale 1923; Lewis 1949; Monaghan 1973). Stopping at Robinson Crusoe Island provided a useful rest.

Human impact in the Juan Fernández Archipelago from 1574 to the time of establishment of the island as a national park in 1935 can be categorized into (1) cutting and harvesting of forest trees, (2) introduction of animals, (3) introduction of plants, and (4) fire.

Direct Cutting and Harvesting of Forest Trees

This human activity derives from the obvious need to cut trees for lumber to build houses and repair passing ships, and most occurred on Robinson Crusoe Island. Considerable documentation exists on this point, beginning in 1599 (Table 7.2). Much later, but particularly noteworthy, is the report in a letter from Pascual Jaque

Table 7.2 Historical Documentations for Cutting of *Juania australis*, *Santalum fernandezianum*, and Forest Trees in General on Robinson Crusoe Island

Date	*Juania*	*Santalum*	Forest trees
1599			X
1624			X
1680	X		X
1684	X		
1704	X		
1709	X		X
1719	X		X
1720	X		X
1722			X
1741	X		X
1751	X		X
1767			X
1800		X	
1822–9		X	
1835 (Jun)			X
1835 (Nov)		X	
1856	X	X	
1869			X (1,200 tons)
1877			X

Sources: Data based on compilations in Woodward (1969) plus other original sources. From Stuessy et al. (1998b, p. 249).

to José Correa of May 14, 1869 (cited by Woodward 1969, p. 206) of 1,200 tons of wood from forest trees having been accumulated on the shore for sale to passing ships, presumably to fire steam boilers. Another shocking report is that during the early exploration period, sailors sought vegetables to help alleviate the effects of scurvy, and one important source was the large apex of the endemic palm, *Juania australis* (Fig. C30), called the "cabbage tree." To obtain the edible apical meristem, felling the entire tree was required, which led to eventual absence of the species in the lowlands and confinement to the inaccessible ridges and secluded valleys of the island. It now numbers approximately 1,000 plants (Moore 1969; Stuessy et al. 1983; Gunther and Mahalik 1990; Zizka 1991). Another sad case of overexploitation was the harvesting to complete extinction of the endemic sandalwood *Santalum fernandezianum*. First described as new to science in 1892 by F. Philippi, it was already being harvested for sale to Oriental markets because of its highly scented wood, and the species was mostly gone in 1875 (e.g., Moseley 1892, p. 467). The harvesting was sufficiently severe that the last tree was seen by Carl Skottsberg in 1908 (Skottsberg 1910), and it perished soon thereafter.

Introduced Animals

As with many oceanic islands, animals have been introduced to the Juan Fernández Archipelago. Historical records demonstrate that goats, cattle, sheep, donkeys, pigs, rabbits, rats, dogs, cats, and coati mundis have been brought to the islands and later escaped and become feral (Woodward 1969; Wester 1991) (Fig. 7.2). A traditional approach was to let some animals loose on the island in hopes of their reproducing and hence providing fresh meat on subsequent voyages to the archipelago. Most of these animals had an impact on Robinson Crusoe Island, particularly because this was the site of most traffic to the archipelago and also where the permanent Spanish settlement was established.

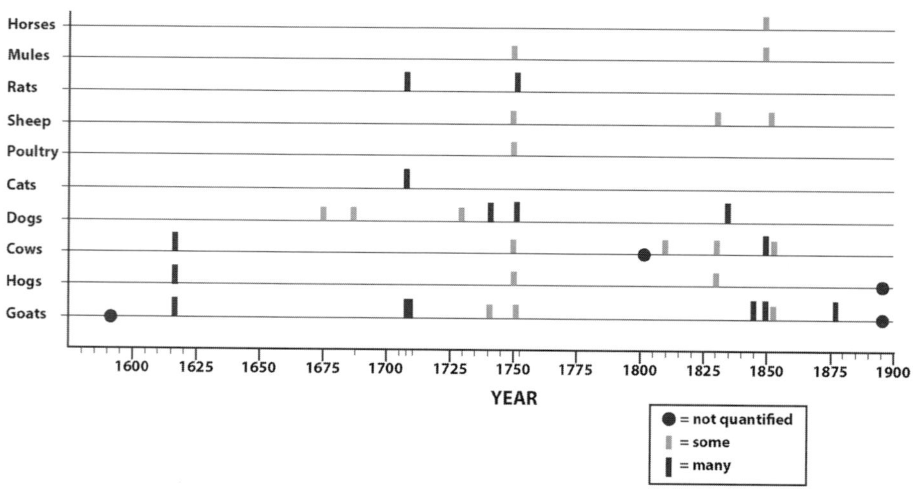

Figure 7.2 Historical mention of domesticated and/or feral animals on Robinson Crusoe and/or Alejandro Selkirk Islands. (From data in Woodward 1969.)

Figure 7.3 The introduced coati (*Nasua nasua*; juveniles) on Robinson Crusoe Island.

Most of the introduced animals have had only a minor negative impact on the vegetation of the Juan Fernández Islands. Hogs were introduced early to Robinson Crusoe Island (Wester 1991), but they never developed large populations and were controlled by the early nineteenth century (Fig. 7.2). Cats have been infrequently reported, although Shelvocke (1726; he visited the island in 1720) noted their abundance. Rats have surely always been on Robinson Crusoe Island, arriving with the early ships, but they have never been a major pest in the archipelago. With the establishment of the permanent Spanish village around Cumberland Bay, mules, sheep, poultry, and later horses were also introduced along with the colonists. Dogs were brought by the Spanish to control the goats. The hope was that if the goats could be exterminated, the pirates who kept attacking the coast would have less to eat when hiding out in the islands and might, therefore, diminish in number. The coati (*Nasua nasua*) (Fig. 7.3) was introduced as a domesticated pet in the twentieth century, and it has escaped into the vegetation and become feral. The population is small, but the animals eat eggs of endemic birds and burrow, which loosens the soil and hastens erosion.

Most injurious to the vegetation of the islands have been cattle, rabbits, and goats. Cattle were first cited, with some uncertainty, as having been present in 1616 (Schouten

Figure 7.4 Harvesting by CONAF of introduced rabbits (*Oryctolagus cuniculus*) on Robinson Crusoe Island.

1619). In our early expeditions to Robinson Crusoe Island in the 1980s, cattle were still feral in the forests, but these animals were shortly thereafter captured and restricted to pasture (of introduced plants) in Puerto Inglés. In a 27-month study of cattle exclusion in this same valley (Cuevas and Le Quesne 2006), the vegetation did not change in species diversity during this time. The species involved, however, were mostly introduced, and therefore, the results do not bear on recuperation of native vegetation.

At the present time, the two most injurious feral animals are the rabbits (*Oryctolagus cuniculus*) (Fig. 7.4) on Robinson Crusoe Island and the goats (*Capra aegagrus* f. *hircus*) (Fig. 7.5) on Alejandro Selkirk Island. Rabbits were apparently introduced in the 1930s (Saiz and Ojeda 1988) and are routinely hunted by the

Figure 7.5 The goat (*Capra aegagrus* f. *hircus*), introduced in the late sixteenth century in the Juan Fernández Archipelago. (A) Group on the upper ridges of Alejandro Selkirk Island. (B) Captive individual.

CONAF guides, but these animals reproduce more rapidly than they can be killed. Studies on biological control using the myxoma virus (causing fatal myxomatosis) have been explored (Saiz et al. 1982), but it might require the simultaneous introduction of fleas to serve as a vehicle of transmission within the rabbit populations, with unknown ecological consequences. At this point, nothing has yet been found that completely controls the rabbit population. Goats have been hunted relatively successfully on Robinson Crusoe Island due to the many colonists there in the village. On Alejandro Selkirk Island, however, which has more difficult terrain and no permanent settlement, goats have survived and proliferated. It is estimated that approximately 4,000 goats exist at the present time on this island (Biodiversa 2009a). Elimination of these animals by hunting is perhaps possible, but it would take a long time and be very costly. During the conservation project financed by the Dutch government from 1997 to 2003, some success was obtained in reducing the numbers of goats and rabbits in the archipelago. Without persistent follow-up efforts, due to limited resources, the populations have now rebounded.

Introduced Plants

Plants were introduced to the Juan Fernández Archipelago both deliberately and accidentally. Because of the known food and health value of fresh fruits and vegetables, plants were introduced on purpose in the seventeenth century (Johow 1896; Matthei et al. 1993). After the curative potentials of fruits against scurvy were realized in the eighteenth century (Carpenter 1986), it became traditional to grow fruit trees for use on return voyages. With the arrival of permanent colonists in the mid-eighteenth century, including women and children, many other plants also were imported for decorative as well as nutritive and medicinal purposes (Swenson et al. 1997) (Table 7.3).

Among the introductions, which later escaped, the most aggressive and damaging to the native vegetation have been the zarzamora (*Rubus ulmifolius*) and maqui (*Aristotelia chilensis*). Both species were brought to the islands for their edible fruits. *Rubus* appeared relatively early in the twentieth century (Looser 1927) and *Aristotelia* even earlier (Johow 1896, p. 108). These species have now formed extensive patches that are so dense that almost no other species can grow underneath. The impact that these two species have had on the native vegetation has been enormous, and eradicating them will

Table 7.3 Plants Introduced to Robinson Crusoe Island, Compiled from Historical Accounts

Voyager	Year visited	Plants mentioned
Le Hermite (in Callander 1768)	1624	Clover, quince
Cooke (1712)	1709	Turnip
Rogers (1712)	1709	Fever-few, parsley, purslein, sithes, turnip, watercress
Shelvocke (1726)	1720	Pumpkin, turnip, watercress, wild sorrel
Anson (in Walter and Robins 1748)	1741	Clover, oats, Sicilian radish, turnip
Sobrecasas (in Medina 1923)	1751	*Grasses*: cortadera, enea, theatina *Medicinal herbs*: anthemisa, berros, cardo santo, chicoria silvestre, culantrillo, malva, mastuerzo, moreta, mostaza, nabo silvestre, romasa, trifolio, vinagrillo
Magee (1795)	1792	Pumpkin, radish, small potato

be a major challenge. They both sprout from the base, which makes cutting them back to the ground a futile exercise. Only poisoning after cutting might yield success, but the job is immense and hence very expensive. Biological control of *Rubus* using the fungus *Phragmidium violaceum* has been suggested as another possible avenue (Oehrens and Garrido 1986). If left unchecked, both species will become greater threats to the native vegetation in the near future (Dirnböck et al. 2003).

To provide wood for making the lobster fishing boats, for building houses, and for other construction needs, plus to help retain the volcanic soil, cedars, pines, and eucalypts have been planted around the village of San Juan Bautista. Although not native species, they have provided a "green" area around the village, which is desirable for the population. CONAF monitors these species carefully to keep them from spreading into the native vegetation, and so far, the plan seems to be working reasonably well.

Along with plants that were introduced purposefully, some species were introduced inadvertently, some becoming noxious weeds. Chapter 8 deals in more detail with these species, but *Acaena argentea* ("trun"; Rosaceae) is one notable example. This species covers many hectares, especially on Robinson Crusoe Island. On the positive side, it can colonize eroded areas and is useful for holding the soil, but with its aggressive stolon production, it crowds out and hinders native plants from establishing.

Fire

It is somewhat sad that most cases of fire in the Juan Fernández Archipelago have been due to human activities. Woodward (1969) chronicles many of these events, as does Wester (1991). Table 7.4 provides additional details for the major fire incidents. Most fires have occurred during historical times, but a recent fire took place on Alejandro Selkirk Island in February of 1996 (Barría 1996). This was a fisherman's campfire that jumped into the vegetation and burned more than 72 hectares before it went out. Fortunately, due to the frequent fog and rain on Alejandro Selkirk Island, this did not

Table 7.4 Historical Records of Fire on Robinson Crusoe Island

Date	Location	Circumstance
November 26, 1795	Summit of El Yunque; burned for 8 days	Ramón Negrete, Francisco Clavel, Pedro José Gutiérrez, and Marcelo Boza set fire to prove that they had climbed the peak
January 5, 1816	Began in chaplain's huts and spread over "the entire island"	Apparently an accidental fire
November 18, 1837	San Juan Bautista was set on fire	Peruvians
May 19, 1849	Whole valley behind Puerto Inglés	Twenty Californians (49ers) set fires
March 25, 1862	Huge forest fire in Puerto Francés area	Accidental fire set by Rengifo party
May 18, 1869	"Frequent" forest fires in various locations	Lumberjacks of Fernández López
February 1872	One square mile of good timber burned down	Sailor carelessly set fire

Sources: Based on Woodward (1969); from Stuessy et al. (1998b, p. 251).

destroy too much of the vegetation. Effects of the burning, however, were still evident during our expedition in 2011 (J. Greimler, personal observations).

Despite these damages to the native vegetation, there is no evidence that any extinction has occurred from burning of the native forest. Furthermore, the forest has proven remarkably resilient and has always rebounded. Perhaps the greatest danger to the vegetation after a burn comes from introduced plant species, which are particularly aggressive and able to establish preferentially in these newly opened areas. Refer to Chapter 8 for more details on threats from these introduced taxa.

Historical Records of Present-Day Native Vegetation

The extensive historical records of human visits to the Juan Fernández Archipelago, and documentation of activities, allows some interpretations of what the original vegetation might have been. Most reports have dealt with Robinson Crusoe Island because this was where the natural bay offered protection and where most ships dropped anchor. There are, in fact, very few early commentaries on the vegetation of Alejandro Selkirk Island because most human activity there centered on the harvesting of the fur seals, which occurred on the shores around the island.

Robinson Crusoe Island

Among the various early reports on the vegetation of this island, three general points can be profitably discussed: (1) savannah-like vegetation, especially in some of the valleys, (2) vegetation of the dry western portions, and (3) the absence of trees and the erosion on the eastern side of the island. Grasslands were apparently common in the native vegetation, as can be seen more clearly now on Alejandro Selkirk Island. Cowley (visited 1684; report seen in Callander 1768) commented that "... the sides of the hills [are] partly covered with wood, and partly savannas, or places naturally clear of wood, bearing fine grass." Dampier (1697), a keen observer, pointed out that "The sides of the mountains are part savannah, part woodland. Savannas are clear pieces of land without woods ...." It is not clear, however, which native species would have made up these extensive grasslands, but *Danthonia malacantha* and *Piptochaetium bicolor* are possibilities. At the present time, these areas are covered mainly by introduced species (Skottsberg 1921, 1953a; Baeza et al. 2007), especially *Anthoxanthum odoratum*. It seems probable, therefore, that with the introduction of exotic grasses, a gradual turnover took place where the invasive species have outcompeted and replaced the natives.

The low western part of Robinson Crusoe Island (such as near the airstrip) is very dry and nearly devoid of native vegetation (Fig. 7.6). Some weeds survive, such as *Papaver somniferum*, germinating and turning green in the austral spring with seasonal rain and then drying brown later in the year. The earliest report for this part of the island was Ringrose (1680; in Esquemeling 1685), who commented "... whereas in the place where we had first anchored, not one stick of wood nor tuft of grass was to be found." Anson (1741; in Walter and Robins 1748) echoed with his report of this area "... being dry,

Figure 7.6 Permanently very dry area at the western end of Robinson Crusoe Island and adjacent Santa Clara Island.

stony and destitute of trees, but very flat and low." Cowley (visited 1684; seen in Callander 1768), Dampier (1697), and Juan and Ulloa (1748) made the same basic observations, but they also mentioned the presence of some grasses. It is very likely, therefore, that this western part of Robinson Crusoe Island was originally low and dry by the time that Europeans starting arriving. Desiccation of the area presumably occurred in conjunction with island subsidence and erosion that led to a reduction of water and the disappearance of much of the flora from this part of the island.

The eastern part of Robinson Crusoe Island, however, is another story. Many older textual descriptions and maps show trees covering this part of the island. This is where the ships nearly always dropped anchor, especially in Cumberland Bay and just outside Puerto Inglés and Puerto Francés. Schouten (1619), Ovalle (1649), Dampier (1697), Funnell (1707), Shelvocke (1726), Roggeveen (visited 1721–2; published 1838), and Anson (1741; in Walter and Robins 1748) all comment on the extensive forests along the eastern side of this island (Table 7.5). The drawings and maps of Ringrose (1680; in Esquemeling 1685) (Fig. 7.7), Funnell, and Anson (Fig. 7.8) also show trees in this region. Therefore, the present barren areas on the eastern side of Robinson Crusoe Island were caused by felling of trees, which led to loss of vegetation and serious erosion.

Santa Clara Island

This small island, barely more than 2 km^2, is now mainly covered by introduced grasses and is without any trees. Historical records are somewhat inconsistent. Schouten (1619)

Table 7.5 Historical Evidence for Extensive Forests on the Eastern Part of Robinson Crusoe Island

Voyager (reference)	Year visited	Comments
Schouten (1619)	1616	"... hath many trees."
Ovalle (1649)	1646	"... has a great variety of trees, and much grass."
Ringrose (in Esquemeling 1685)	1680	"... many trees ...."
Dampier (1697)	1684	"The sides of the mountains, are part savannah, part woodland."
Funnell (1707)	1704	"The woods afford several sorts of trees ...."
Shelvocke (1726)	1720	"... the woods which cover the island ...."
Roggeveen (1838)	1722	"The hills are covered with tall trees of various kinds ...."
Anson (in Walter and Robins 1748)	1741	"The northern part of this island is ... generally covered with trees." "For the woods which covered most of the steepest hills .... "

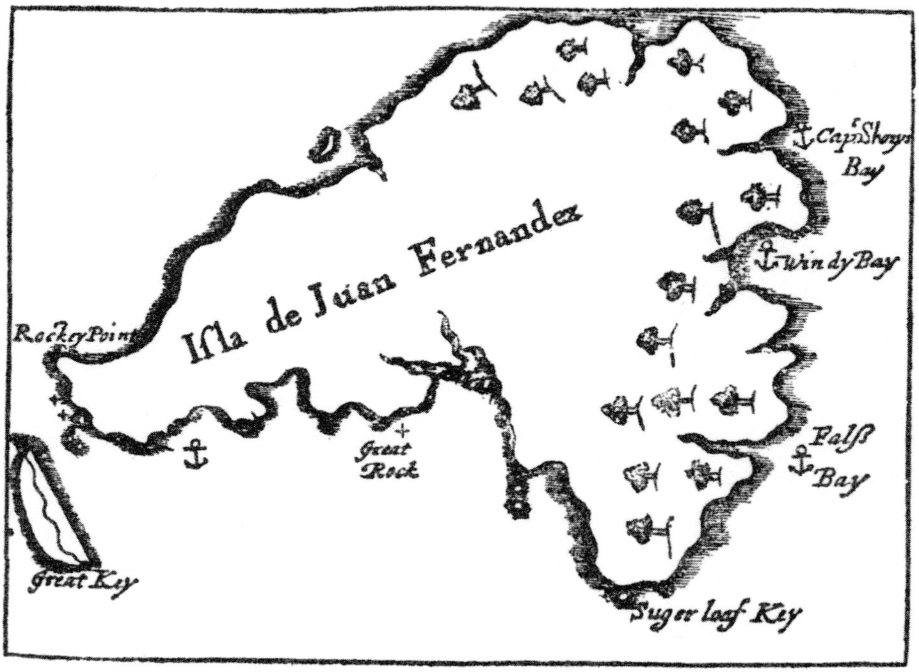

Figure 7.7 Rough map of Robinson Crusoe and Santa Clara ("Great Key") Islands in the Juan Fernández Archipelago. (From Ringrose; in Esquemeling 1685.)

remarked that this is "... a very dry bare island, with nothing on it but bare hills and cliffs." Ringrose (1680; in Esquemeling 1685) included a rough map (Fig. 7.7) that showed this island without trees, as did Juan and Ulloa (1748). However, Anson (1741; in Walter and Robins 1748), normally a very careful observer, provided a map that showed "Goat Island," an earlier name for this island, as all forested. Because the western part of Robinson Crusoe Island is also dry and nearly barren, as pointed out by Anson

Figure 7.8 View of landscape on Robinson Crusoe Island adjacent Bahía Cumberland, recorded by Commodore George Anson in 1740. (From Walter and Robins 1748.)

(1741; his map copied by Bellin 1764) (Fig. 7.9), we suspect that this was also the original condition of Santa Clara Island when it was first seen by Europeans and that the Anson map was erroneous. Certainly having goats on this small island could not have helped preservation of the herbaceous flora, but trees were likely absent prior to the arrival of humans.

Alejandro Selkirk Island

No detailed early comments exist with reference to the vegetation of this younger island. Because of the ruggedness of the terrain, most visitors have kept close to the shore, especially in what is now called Las Casas. This is the small fishing village that functions during eight months of the year. It is also the site of a former penal colony for Chile, but it has always been small. Most of the human activity occurred during the sealing and whaling period (1798–1808), when the native sea lions were extirpated from this island and the endemic fur seal was hunted nearly to extinction (Hubbs and Norris 1971). This species was only rediscovered in 1965 (Bahamonde 1966), and successful population recovery has now taken place under strict protection from CONAF.

Estimates of Loss of Vegetation and Specific Diversity

Taking all natural and human factors into consideration allows us to estimate the loss of vegetation and specific diversity in the Juan Fernández Archipelago. First, with the

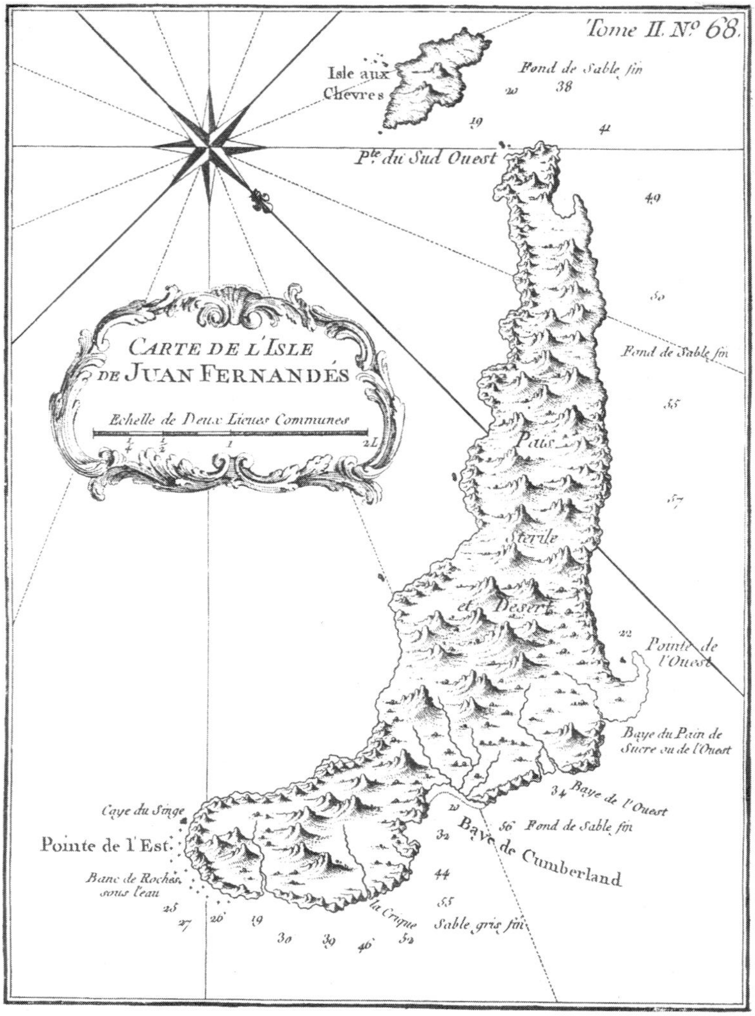

Figure 7.9 Map of Robinson Crusoe and Santa Clara Islands. (From Bellin 1764, but redrawn from Anson; in Walter and Robins, 1748.)

hypothesized natural loss of 95% of surface area on Robinson Crusoe Island (Stuessy et al. 1998a, 1998b) due to subsidence and erosion and doubtless also loss of ecological zones and the sustained cutting of the forests, we suggest that 25% of the specific diversity on this island might have been lost. Clearly, much of the present flora has been compacted into a refugium. The native forest only occurs in and around the highest ridges (Greimler et al. 2002a), thus representing only a fraction of its original extent. This loss is not seen in Alejandro Selkirk Island (Greimler et al. 2013; see also Chapter 6), which is much younger geologically and has suffered much less from human disturbance.

There is no way to know with certainty how many individual species may have gone extinct during the past 4 million years of existence of the archipelago. All we can say is that during attempts to document the flora, beginning with the reasonably comprehensive treatment by Hemsley (1884), some species are now regarded as extinct. The most dramatic example is the sandalwood, which was endemic to Robinson Crusoe Island. The last plant was seen and photographed in 1908 by Skottsberg (1910). The grass *Podophorus bromoides* was described as new to science by Philippi (1856b), but it has not been collected since and is now regarded as extinct. Two species of the genus *Robinsonia* are also now regarded as extinct: *R. macrocephala* and *R. berteroi*. The former has not been seen for many decades (Danton et al. 2006), and the last plant of the latter died recently (Danton 2005). ×*Margyracaena skottsbergii* survives in the CONAF garden on Robinson Crusoe Island, but it is gone from the wild. *Eryngium sarcophyllum* is presumed extinct, and *Empetrum rubrum* and *Notanthera heterophylla* have been extirpated from the archipelago. See Chapter 9 for more discussion of the conservation status of species in the islands.

8 Invasive Species

Josef Greimler, Tod F. Stuessy, Ulf Swenson, Patricio López-Sepúlveda, and Carlos M. Baeza

One of the very positive achievements over the past half century has been the acceptance in human society of the need to preserve global biodiversity. While arguments may still rage over the need to save a particular species of plant or animal in a particular part of the world, especially in consideration of other human needs such as economic development, there will be little dispute that the survival of the human species depends on preservation of global biodiversity. This need has many dimensions, including providing foods, clothing, and shelter; offering opportunities for new pharmaceuticals; and maintaining a stable climate (e.g., Malhi et al. 2008). Biodiversity also plays a significant role in helping maintain a positive viewpoint on life in general, as we have all experienced with a refreshing walk in the woods on a fine summer afternoon (so-called forest bathing) (Li 2010; Lee et al. 2011).

Despite the acknowledged positive value of biodiversity, many pressures combine to cause concern regarding the obvious need for its preservation. The growing human population demands increasing resources to meet even the simplest daily needs, and this is exacerbated because of widespread poverty in many countries of the world, such as in Africa. Land is cleared for the growing of crops and domesticated animals and for the building of roads, houses, schools, hospitals, and factories. Armed conflicts among peoples of different countries, ethnic backgrounds, and religions also combine to have a negative impact on biodiversity protection. Corruption is also a factor that interferes with good intentions for establishing nature parks and reserves (Smith et al. 2003; Laurance 2004).

More recently, awareness has been heightened on the significance of introduced plants and animals into natural areas and the often negative impacts that this can have on the autochthonous biota. Although the impact that exotic organisms can have on an ecosystem has long been known (e.g., Fosberg 1967), it has only been in recent decades that focus has been placed on this problem (e.g., Williamson 1996; Cronk and Fuller 2001; Kueffer et al. 2010). In effect, it makes little sense to worry only about preservation of native or endemic plants when huge numbers of aggressive exotics are being introduced at the same time. Invasive species, therefore, now have become another important part of the challenge of protecting global biodiversity; a scholarly journal, *Biological Invasions*, publishes pertinent articles.

Definition of Invasives

Numerous definitions of invasive species exist, and hence a few comments are in order. Lambelet-Haueter (1990, 1991) provides an exhaustive list of different terms that have been applied to introduced species of plants, and these include exotics, invasives, weeds, undesirable elements, unwanted plants, introduced species, pests, adventitious plants, anthropogenic flora (anthropophile, anthropophyte), colonizers, pioneers, ubiquists, opportunists, neophytes, aliens, synanthropes, nuisance species, and ruderals. Use of the term "alien" has been especially popular (e.g., Ryves et al. 1996; Reynolds 2002). Each of these terms has a slightly different meaning, and to some extent they group under popular, economic, or ecological perspectives. For our purposes in the Juan Fernández Archipelago, the term "invasive species" refers to one that has already arrived in the archipelago and has caused significant alteration of the habitats. This follows the definition suggested by Hunter and Gibbs (2007), in which invasive species (they used the term "exotic") are those that have already invaded an ecosystem (or have the potential to invade) and that have caused significant problems of an ecological, human health, or economic nature.

Dealing with invasive species in any part of the world is not a simple matter and provides challenges on several fronts. The greatest difficulty is that for control of any weedy species, there must be some sort of funding, and for this to occur, the political will must be present. In other words, the initial challenge is to convince people in the area that there is actually a problem with a particular invasive species (Simberloff et al. 2013). Some of them may be visually very attractive, and others may be useful for edible berries, fencing, lumber, and so on. Plants are, after all, extremely important in the daily lives and in life events of people everywhere, and this is particularly the case with persons in isolated oceanic islands. The intent of introduction may have been completely innocent, for helpful purposes, and with the plan to contain them as garden plants. No one, however, is anxious to be found culpable of having introduced something that has now escaped and is creating a huge problem for the community. These issues relate directly for the need of conservation education so that the problem with invasive species can be better understood, which might lead eventually to concerted action. Another challenge is to be certain that the taxonomic understanding of the invasive species is correct (Pyšek et al. 2004, 2013). Some taxa are well known as dangerous weeds, and close relatives may be benign. Having the proper determination ties the invasive species to possible known biological attributes already recorded in the literature, such as the ability to reproduce asexually, which can provide a more realistic picture of the ecological dangers involved. Having the proper identification is essential before addressing the political issues for possible control.

Invasive Species in Oceanic Islands

Perhaps in no other place on Earth has the problem of invasive species been greater than in oceanic islands. Humans love islands because of the often moderate or tropical climates, the nice beaches, and the holiday feeling that comes with the hotels, restaurants, and shops that inevitably develop as people are attracted to live and visit. As happens nearly everywhere in the world, with humans comes human activity, and this usually causes perturbation of the ecosystem, often for the worse (Walker and Bellingham 2011). All these factors are conducive to the arrival, establishment, and invasion of alien species in the island setting. Disruption of the native environment, in which the autochthonous species are well adapted, opens up opportunities for introduced species to become invasive. If the environment becomes substantially degraded, some weedy species are more efficient in use of the lessened resources and can outcompete the native vegetation (Funk and Vitousek 2007). Particularly dangerous is when an invasive species, such as the brown tree snake in Guam introduced from Australasia, begins eating indiscriminately at different levels of the trophic chain and causes havoc within the entire ecosystem (Fritts and Rodda 1998). There are numerous examples of the general problem of impact on ecosystems from alien plant species in islands, and the Hawaiian Archipelago provides an illustrative case. No one needs to be reminded of the high attraction of Honolulu or other cities in the Hawaiian Islands for vacationing tourists. Along with people have come many invasive species, as documented clearly by many studies (e.g., Smith 1989; Cuddihy and Stone, 1990; Stone et al. 1992; Staples and Cowie 2001; Kueffer et al. 2010). Ferns can also become problematical in Hawaii (Wilson 1996) in part because the environment invites growing exotic ferns in gardens, which then have opportunities for escaping into the wild. Another example is provided by the Bonin (Ogasawara) Islands, which are located 1,000 km south of Japan, in which a number of alien plant and animal species have caused substantial damage to the ecosystem (Kawakami and Okochi 2010).

Introduced and Invasive Plant Species in the Juan Fernández Archipelago

Alien plants certainly arrived in the archipelago with the first sailing ships landing there. *Rumex* pollen appears first approximately 500 years ago, with a peak approximately 300 years ago, whereas grass pollen shows a notable increase since about 200 years on the more remote Alejandro Selkirk Island (Haberle 2003). This is most likely due to the introduced Euro-Mediterranean grasses. Precise information on introduced plants is available since 1823 with Mary Graham (Johow 1896; see Chapter 2). A short history on the inventory of the introduced and most aggressive invasive plants on Robinson Crusoe Island is given in Greimler et al. (2002b), together with a diagram showing a steady increase in the number of introduced species over 150 years. By the end of the twentieth century, it became evident that the number of introduced taxa had exceeded the number of native taxa (Swenson et al. 1997, Marticorena et al. 1998). In the twenty-first century, Cuevas et al. (2004) identified additional introduced species, providing a total number of 260 alien compared with 211 native taxa. This high proportion (55%) of introduced plants on the Juan Fernández Archipelago is only surpassed by Easter Island (75%) among the Chilean oceanic islands and is fairly high

Table 8.1 The Most Frequent Invasive Species Shown in Percentage of Our 196 Vegetation Plots (Relevés) in the Juan Fernández Archipelago, and Probable Dispersal Agents

Taxon	Alexander Selkirk Island	Robinson Crusoe Island	Dispersal agents
Shrubs			
Aristotelia chilensis[a]	7	28	Birds, endozoochory
Rubus ulmifolius[b]	0[c]	20	Birds, endozoochory
Ugni molinae[a]	0[c]	32	Birds, endozoochory
Herbs			
Acaena argentea[b]	6	36	Mammals, epizoochory
Hypochaeris radicata	60	8	Wind
Plantago lanceolata	23	7	Endo- and epizoochory
Rumex acetosella[b]	67	30	Wind; endozoochory
Sonchus asper/oleraceus	36	10	Wind
Grasses			
Aira caryophyllea	27	31	Wind; mammals, epizoochory
Aira praecox	23	0[c]	Wind; mammals, epizoochory
Anthoxanthum odoratum	69	27	Wind; mammals, epizoochory
Briza minor	20	20	Wind; mammals, epizoochory
Holcus lanatus	22	0	Wind; mammals, epizoochory

[a] Not clonal in the proper sense but vigorously branching from the base.
[b] Clonal.
[c] Present on the island but not observed when collecting data in the plots.
Source: Greimler et al. (2002a, 2013).

compared with other oceanic archipelagos such as the Hawaiian Islands (44%) and the Galápagos Islands (29%) based on numbers published recently (Cuevas et al. 2004). A few years later, however, Danton et al. (2006) provided a total number of 503 introduced plants for the archipelago. Their statistics contain many plants of gardens or plants that have escaped within the village but have not yet become invasive. These plants do not satisfy the criteria applied by Cuevas et al. (2004). Considering all sources (Matthei et al. 1993; Swenson et al. 1997; Marticorena et al. 1998; Cuevas et al. 2004; Danton et al. 2006), we recognize 267 taxa as having been introduced to the archipelago, which includes invasive and indifferent/harmless plants that are naturalized and found outside garden fences (see Table 5.2). In this chapter we focus on those that are frequent, some of which have already caused great damage (Table 8.1). We also add comments on a few species that have not yet become problematical but that could potentially become threats to the native and endemic plants (Tables 8.2 and 8.3).

The Most Harmful Invasive Plant Species in the Juan Fernández Islands

Table 8.1 provides an overview of the most frequent invasive plants in the archipelago based on our 196 vegetation relevés (Greimler et al. 2002a, 2013). The most harmful

Table 8.2 Introduced Plants and Their Current Status in the Juan Fernández Archipelago that Are Found in at Least Five Island Archipelagos and Have Become Invasive in Some of Them

Species	Family	Status
Ailanthus altissima	Simaroubaceae	Naturalized
Carpobrotus edulis	Aizoaceae	Naturalized
Cytissus scoparius	Fabaceae	Cultivated
Delairea odorata	Asteraceae	Escaped from garden(s)
Erigeron karvinskianus	Asteraceae	Cultivated
Fuchsia magellanica	Onagraceae	Cultivated
Holcus lanatus	Poaceae	Naturalized
Lantana camara	Verbenaceae	Cultivated
Oxalis pes-caprae	Oxalidaceae	Escaped from garden(s)
Paraserianthes (= Albizia) lophantha	Fabaceae	Escaped from garden(s)
Phormium tenax	Agavaceae	Escaped from garden(s)
Psidium cattleianum	Myrtaceae	Cultivated

Sources: From Danton et al. (2006) and Kueffer et al. (2010).

invasives among the woody species (including subshrubs) are *Aristotelia chilensis* (Figs. 8.1 and 8.2), *Rubus ulmifolius* (Figs. 8.3 and 8.4), and *Ugni molinae* (Figs. 8.5 and 8.6), all with a high impact on the vegetation of Robinson Crusoe Island (e.g., Swenson et al. 1997; Greimler et al. 2002b; Dirnböck et al. 2003). They are, however, less frequent on Alejandro Selkirk Island. *Aristotelia* was observed in 1917 by Skottsberg (1953a), and this shrub does occur in a few of our relevés of this island in contrast to *Rubus* and *Ugni*, which are both still rare on this more remote island with its different geomorphological setting. A detailed summary of the introduced species found on only one hectare around the fishermen´s huts on Alejandro Selkirk is given by López et al. (2013a).

Among the herbs and grasses, *Rumex acetosella* and *Anthoxanthum odoratum* occur with very high or the highest frequencies on both islands. This was observed years ago by Skottsberg (1953a, p. 932), who commented: "It is hard to imagine what the upland country [of Alejandro Selkirk] looked like before the arrival of these intruders." *Hypochaeris radicata*, however, which is so common now on Alejandro Selkirk Island, was not observed at all by Skottsberg. This could, therefore, be a more recent introduction with subsequent rapid spread across the island. There are two peculiarities regarding the introduced grasses: *Holcus lanatus* found in 20 of our 90 plots (relevés) on Alejandro Selkirk Island is not listed in the recent floristic summaries (Marticorena et al. 1998; Danton et al. 2006). However, the small annual *Aira praecox* occurring at high frequency on Alejandro Selkirk Island was not recorded in any of the 106 relevés on Robinson Crusoe Island. This taxon is listed in the above-mentioned summaries. It is apparently less common on this older island than its congener *A. caryophyllea*.

Table 8.3 Invasive or Potentially Invasive Plants Found in the Juan Fernández Archipelago on Alexander Selkirk (AS), Robinson Crusoe (RC), and Santa Clara (SC) Islands that Are Naturalized, Have Escaped from Gardens, but Are Stil within the Limits of the Village or Still Residing in Gardens

Taxon	Islands	Life form	Status
Acacia (Racosperma) dealbata	AS, RC	Herb	Naturalized
Acacia (Racosperma) melanoxylon	AS, RC	Herb	Naturalized
Acaena argentea	AS, RC, SC	Herb	Naturalized
Ailanthus altissima	AS	Tree	Naturalized
Albizia lophantha	RC	Tree	Naturalized
Anthoxanthum odoratum	AS, RC	Herb	Naturalized
Aristotelia chilensis	AS, RC	Tree	Naturalized
Arundo donax	RC	Herb (cane)	Cultivated
Avena barbata	AS, RC, SC	Herb	Naturalized
Canna indica	RC	Herb	Naturalized
Carduus pycnocephalus	RC	Herb	Naturalized
Carpobrotus aequilaterus/edulis	AS, RC	Herb	Naturalized
Centaurea melitensis	AS, RC, SC	Herb	Naturalized
Chasmanthe aethiopica	RC	Herb	Within the village
Chusquea culeou	RC	Herb	Cultivated
Cortaderia selloana	RC	Herb	Within the village
Cupressus goveniana	RC	Tree	Naturalized
Cupressus macnabiana	RC	Tree	Naturalized
Cupressus macrocarpa	AS, RC	Tree	Probably naturalized
Delairea odorata	RC	Herb	Within the village
Eucalyptus globulus	AS, RC	Tree	Naturalized
Ficus carica	AS, RC	Tree	Naturalized
Galium aparine	AS, RC	Herb	Naturalized
Hedychium flavescens	RC	Herb	Within the village
Lantana camara	RC	Shrub	Cultivated
Lardizabala biternata	RC	Twining vine	Naturalized
Lobelia tupa	RC	Herb	Naturalized
Lonicera japonica	RC	Shrub	Within the village
Papaver somniferum	RC, SC	Herb	Naturalized
Phormium tenax	RC	Herb	Within the village
Pinus pinaster	RC	Tree	Within the village
Pinus radiata	AS, RC	Tree	Naturalized
Pittosporum crassifolium	RC	Shrub	Within the village
Pittosporum eugenioides	RC	Shrub	Naturalized
Pittosporum tenuifolium	RC	Shrub	Cultivated
Robinia pseudoacacia	RC	Tree	Within the village
Rubus ulmifolius	AS, RC	Shrub	Naturalized
Santalum album	RC	Tree	Cultivated
Sorghum halepense	RC	Herb	Within the village
Sporobolus indicus	RC	Herb	Cultivated
Teline monspessulana	RC	Shrub	Naturalized
Ugni molinae	AS, RC	Shrub	Naturalized
Vinca major	RC	Herb	Within the village
Watsonia borbonica	RC	Herb	Within the village
Watsonia meriana	RC	Herb	Within the village
Zantedeschia aethiopica	AS, RC	Herb	Naturalized

Source: Data from Danton et al. (2006).

Figure 8.1 *Aristotelia chilensis* (Elaeocarpaceae), with edible fruits.

Acaena argentea (Figs. 8.7 and 8.8) is a special case among the invasives (see also Chapters 6 and 7). It is the most frequent invasive herb on Robinson Crusoe Island, where its fruits are obviously dispersed successfully by the abundant rabbits. Moreover, it produces long stolons for clonal spread and forms very dense carpet-like populations. Cuevas and Le Quesne (2006) consider this species to be appropriate for plantations in eroded sites; they argue that because of its shade intolerance, it is not threatening for native forest vegetation. While this may be the case, we do not favor further encouragement of *A. argentea* because germination of any native species under it is impeded.

Introduced Animal Species

Livestock (cattle) entering the native forests and forest-grassland ecotone and foraging on tree seedlings have been a severe problem on Robinson Crusoe Island. Rabbits, introduced on this island in 1935, have increased in number to an immense quantity and were estimated by Cuevas and van Leersum (2001) to number about 50,000 individuals, which are extremely harmful to the native vegetation (Stuessy et al. 1998d; Camus et al. 2008). The same is true for the goats that were probably introduced to the archipelago on its discovery (Cuevas and van Leersum 2001). These authors give estimates of 200 to 500 goats on Robinson Crusoe Island, whereas their number has been estimated to be about 2,000 (Cuevas 2002) on Alejandro Selkirk Island; estimates by local people in 2011 suggest up to 4,000. The presence of goats is especially critical for very rare species that grow in steep, rocky places that are accessible to these sure-footed herbivores (Cuevas and van Leersum 2001). Introduced rats and coatis are dangerous to the native bird fauna.

Figure 8.2 Thick grove of *Aristotelia chilensis* on Robinson Crusoe Island.

There are numerous interactions between introduced animals and native plants but also with introduced plants. For example, the large size of the rabbit population and abundance of livestock on Robinson Crusoe Island have together helped spread fruits of *Acaena argentea* over the entire island. On Alejandro Selkirk Island, however, *A. argentea* is less abundant, although it was observed there in 1917 (Skottsberg 1953a). This is due in part to the different geomorphology with lower erosion on Alejandro Selkirk Island but also likely due to the lower numbers of both rabbits (Camus et al. 2008) and livestock. The impact of livestock on

Figure 8.3 *Rubus ulmifolius* (Rosaceae), illustrating attractive edible fruits.

Alejandro Selkirk Island is limited to small parts of the coastal area and adjacent canyon bottoms. Another interaction between a native bird, the austral thrush (*Turdus falcklandii*), and the invasive shrubs *Aristotelia chilensis*, *Rubus ulmifolius*, and *Ugni molinae* produces a dilemma for conservation management. All these introduced plant species have fleshy fruits that are dispersed primarily by this native bird. Smith-Ramírez et al. (2013) argue that control or even eradication of the bird would significantly limit the spread of these invasive plants.

Predicting Future Threats to the Native and Endemic Flora in the Islands

The detrimental impact of alien species on island ecosystems is generally recognized and has been on the agenda of the Convention on Biological Diversity (UNEP 2003). The note released by the Subsidiary Body on Scientific, Technical and Technological Advice states that "Invasive alien species are believed to be the most significant driver of the decline of plant and animal populations and species extinctions in island ecosystems." As a rule, introduced animals are disastrous for native island plants to a much greater extent than introduced plants, at least in the short run. While direct competition from alien plants so far has been a minor cause of native plant extinctions, the steady increase in alien plants on oceanic islands may have a delayed effect via increased propagule pressure and together with the often long life span of plants may produce an "extinction debt" in native island plants (Sax and Gaines 2008).

Figure 8.4 Dense patch of *Rubus ulmifolius* on Robinson Crusoe Island on the path from San Juan Bautista to Selkirk's lookout.

At present, there are only a few documented extinctions in the archipelago (Danton et al. 2006; see Chapters 5 and 9). Direct competition from invasive plants is certainly insignificant in those cases. Some populations of the rare endemics, however, have become so small that the above-mentioned extinction debt may come due in the near future. The flora of the Juan Fernández Archipelago is highly

Figure 8.5 *Ugni molinae* (Ericaceae).

vulnerable due to habitat loss and the impact of feral and invasive animals and invasive plants. In a comparison of Skottsberg's (1953a) data collected in 1916–17 with our data collected in 1999–2000 (Greimler et al. 2002a) on Robinson Crusoe Island, we concluded that the native forest on this island has decreased significantly due to the impact of the most aggressive invasive plants (Dirnböck et al. 2003). Extrapolating the speed of invasion of *Ugni molinae* and *Aristotelia chilensis* during the last 80 years in a linear way results in the prediction that approximately 50% of the native forest could be invaded or replaced by one or both of these species in another 80 years. In contrast to *Ugni* and *Aristotelia*, the younger invasive subshrub *Rubus ulmifolius* was not significantly constrained by the habitat conditions used in this modeling approach (Dirnböck et al. 2003). Predictions on future performance of *Rubus* are therefore difficult. The plant was not observed on the single highest summit of the island by Kunkel (1957) and is still seen less frequently in remote

Figure 8.6 Salsipuedes ridge on Robinson Crusoe Island covered with *Ugni molinae*.

areas. But a more recent expedition to the summit of El Yunque recorded single individuals of *Rubus* (Danton, personal communication, 1999). *Rubus* is also present at low frequency on Alejandro Selkirk Island, and its performance there is now being carefully monitored by CONAF.

The increasing impact from invasive shrubs, herbs, and grasses, together with pressure from introduced animals and human activity, promotes dominance of introduced plants and homogenization of the flora on the islands. In a list of problematical species of plants in five oceanic archipelagos provided by Kueffer et al. (2010) (Table 8.2), twelve of these species are among the introduced plants in the Juan Fernández Archipelago. So far only three of these plants (*Ailanthus altissima*,

Figure 8.7 *Acaena argentea* (Rosaceae) showing fruits with hooks that aid in dispersal.

Carpobrotus edulis, and *Holcus lanatus*) are naturalized in the Juan Fernández Islands, whereas four of them have escaped from the gardens but still reside within the village, and five are found only within garden fences. There may in fact be many more potentially harmful taxa among the garden and village plants, with many of them recorded for the first time by Danton et al. (2006). We have selected 46 species that were judged by these authors to be most dangerous for the native flora and ecosystems (Table 8.3). Of these, 27 have been naturalized, 13 have escaped from gardens (but still exist within the village), and six are still contained behind garden fences (see also Fig. 8.9).

Evaluation of future threats to the native vegetation of the islands depends to some extent on pessimistic or optimistic judgments of researchers, their experience with

Figure 8.8 Area between Pangal and Centinela ridge on Robinson Crusoe Island covered by *Acaena argentea*.

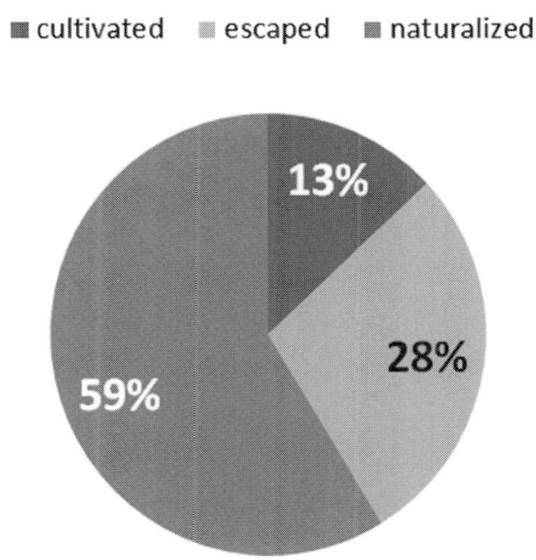

Figure 8.9 Percentages of invasive or potentially invasive plants found in the Juan Fernández Archipelago that are naturalized, escaped from gardens but within the village, or still cultivated behind garden fences. (After Danton and Perrier 2006.)

other islands, the local ecological setting, sanitary measures by the National Park Administration, and (the often lacking) information on basic biology of invasive taxa. The shrub *Lantana camara*, for instance, was recorded in a few gardens of the village San Juan Bautista in 1996 (Swenson et al. 1997) and remained there ten years later

(Danton et al. 2006). This species is not listed among the introduced and naturalized plants in Table 5.2, but it is a potentially aggressive invader that has caused much damage in other oceanic islands, for example, in Hawaii (Wagner et al. 1999; August-Schmidt et al. 2015). Various modeling techniques, such as those applied by Taylor and Kumar (2014), may help in improving predictions, but these approaches are often limited due to the coarse grid of available environmental data.

Efforts for Control of Invasive Species in the Juan Fernández Archipelago

It is more difficult to eradicate an invasive plant than a vertebrate from an island, although success is possible (Simberloff 2001). Because of this difficulty, some workers (e.g., Davis et al. 2011) are now advocating abandoning such eradication efforts and learning to live with the aliens as best as possible. One must also be careful that the elimination of a well-established invasive does not create additional problems, such as endangering an endemic that is now using the invasive for some purpose (e.g., food, home) because its original endemic species host is now extinct (Lampert et al. 2014). Physical removal of *Aristotelia* and *Rubus*, due to their broad distributions, would be nearly impossible. *Ugni molinae* is so established in the upper parts of the islands that it would be impossible to eradicate it, and total removal would expose the wind-whipped ridges and slopes to erosion. Similarly, removal of the herb *Acaena argentea*, which holds the soil in many eroded areas, would be problematic. The preferable approach at this time may be to leave these alone until native species can be planted as replacements.

9 Conservation of Native and Endemic Species

Tod F. Stuessy, Carlos M. Baeza, Patricio López-Sepúlveda, Roberto Rodríguez, and Ulf Swenson

It is well known that the floras of oceanic archipelagos have been long under pressure from human activities (Caujapé-Castells et al. 2010; Trusty et al. 2011; Gillespie et al. 2012; Moreira-Muñoz et al. 2014). Oceanic islands, with their often mild climates, attractive beaches, and fascinating forms of wildlife, have served as magnets to draw people to visit as well as to settle permanently. With increased numbers of inhabitants have come more roads, hotels, plantations of cultivated plants, pastures for cattle and sheep, condominiums, restaurants, and shops (Heywood 2011). All of these developments have led to the clearing and burning of forests, planting of exotics, construction of highways and buildings, and the bulldozing of dunes and vegetation for smoother beaches. One only needs to recall the impact that the city of Honolulu has had on the island of Oahu in the Hawaiian Archipelago.

The Juan Fernández Archipelago is another oceanic island system that has suffered from human impact. On the positive side, because the islands were uninhabited until discovered by the seafarer Juan Fernández in 1574 (Medina 1974; Woodward 1969), human impact has been restricted to the past 400 years. During this period, much alteration of the natural environment has occurred (Vargas et al. 2011; see Chapter 7). Most conspicuous has been the cutting of the forests on Robinson Crusoe Island, especially along the eastern coast around Bahía Cumberland, Puerto Francés, and Puerto Inglés, as well as along the northern coast in La Vaquería. These areas provided convenient anchorage for ships so that crews could obtain wood to rebuild boats and procure fresh meat (from feral goats) and vegetables (from introduced and native plants) for food. The combined result of these activities has led to a loss of forest along the eastern coast such that few of the native or endemic plants remain below 350 m elevation. In fact, as was well documented in Chapter 6, the original vegetation on this island now consists of only about 20% of its original extent (Greimler et al. 2002a). Documentations of human impacts over time have been chronicled by Johow (1896), Skottsberg (1953a), Woodward (1969), Muñoz Pizarro (1974), Wester (1991), Stuessy et al. (1998b), Danton (2006a), Danton et al. (2013), and many others.

A very positive development is that the conservation context in the Juan Fernández Archipelago has much improved in recent decades. Most important, the islands were declared a Chilean national park in 1935 and a UNESCO biosphere reserve in 1977. These categories have provided a legal framework of protection (Valenzuela 1978), along with other Chilean legislation, within which conservation initiatives can be undertaken. The Corporación Nacional Forestal (CONAF) is the agency of the Chilean government responsible for protecting the natural resources of the archipelago.

Because the village of San Juan Bautista has a long history, having been established as a permanent Spanish colony in 1750 (Woodward 1969; Orellana 1975), it continues to exist within the national park. Many of the families have lived there for generations (Hernández and Monleon 1975; Ortiz R. 1982), and there have been no attempts to downsize the village and relocate the people, presumably due to the needs of national security (Agurto 1943) and hence to maintain a strong permanent presence in the archipelago. The successful politic, therefore, must combine conservation of the natural areas of the park while also attending to the needs of the villagers. Although numerous conflicts between villagers and CONAF existed in the early 1980s, good cooperation now prevails. This improved situation has resulted from several converging factors: a loss of revenue from a drop in the lobster harvest due to overfishing (Arana E. 1985) (other sources of marine life, mainly fish, have partially compensated for this loss), an augmentation of international interest in biodiversity within the archipelago, and development of ecotourism.

IUCN Categories

Evaluation of conservation for the native and endemic species of the flora of the Juan Fernández Archipelago requires adoption of a set of categories and criteria for their definition. As used in most areas of the world, we here also apply the International Union for the Conservation of Nature (IUCN) categories (IUCN 2001), which are data deficient (DD), least concern (LC), vulnerable (VU), endangered (EN), critically endangered (CR), and extinct (EX). *Data deficient* (DD) means simply that not enough data are in hand for setting the status for a particular species. *Least concern* (LC) means that the species is thriving and is out of danger at the present time. Table 5.1 lists all the native and endemic vascular taxa of the archipelago and provides IUCN conservation status for all of them. These categorizations have come from our own research experience and those assigned by Danton et al. (2006) and Ricci (2006) and observations on the status of critical species by Leiva et al. (2013).

A summary of the data on conservation status from Table 5.1 is given In Table 9.1, in which taxa have been organized into the broad taxonomic units of ferns and fern allies and angiosperms, the latter of which have been divided into basal angiosperms (Archaeangiospermae), monocots, and dicots. The ferns and fern allies contain fewer endemic species and fewer endemics restricted to single islands (see Chapter 13), which means that in general they are less threatened than angiosperms. Ferns tend to disperse successfully to both islands of the archipelago, no doubt due to the ease of dispersal of the lightweight spores by strong winds (Tryon 1971). Nonetheless, 45 fern taxa (79%) are still regarded as vulnerable, endangered, or critically endangered. The basal angiosperms are a small group with only six species, but all are either vulnerable or endangered. The monocots contain 38 taxa, of which 28 (74%) are vulnerable, endangered, or extinct. The dicots, which contain the largest number of endemic and native taxa, also

Table 9.1 Conservation Status of the Native and Endemic Taxa of the Juan Fernández Archipelago

Taxon	Total number of species	Data deficient	Least concern	Vulnerable	Endangered	Critically endangered	Extinct[a]
Ferns and fern allies	57	2	10	23	18	4	
Archaeangiospermae	6			3	3		
Monocotyledonae	38	2	8	15	12		1
Dicotyledonae	107	2	5	27	50	16	7
Totals	208	6 (3%)	23 (11%)	68 (33%)	83 (40%)	20 (10%)	8 (4%)

[a] Including two extirpated and one extinct in the wild.
Source: Based on data from Table 5.1.

harbor the largest number (100) of taxa with a conservation threat assessment of vulnerable or higher. Five taxa are extinct: *Eryngium sarcophyllum* (Apiaceae), *Podophorus bromoides* (Poaceae; Baeza et al. 2002), *Robinsonia berteroi* (Asteraceae; Danton and Perrier 2005), *Robinsonia macrocephala* (Asteraceae), and *Santalum fernandezianum* (Santalaceae). ×*Margyricaena skottsbergii* is apparently extinct in the wild, but it survives in the CONAF garden on Robinson Crusoe Island. *Empetrum rubrum* (Ericaceae) and *Notanthera heterophylla* (Loranthaceae) have been extirpated from the archipelago. Summing the data, we arrive at the totals for the entire vascular flora of 6 (3%) taxa that are data deficient, 23 (11%) that are of least concern, 68 (33%) that are vulnerable, 83 (40%) that are endangered, 20 (10%) that are critically endangered, and 8 (4%) that are extinct (including one extinct in the wild and two extirpated). Putting it another way, 134 taxa (89%) of the native and endemic angiosperms are vulnerable or in worse condition. It is obvious, therefore, that the flora of the archipelago is in a fragile condition and needs to be monitored carefully now and into the future. This conclusion has also been reached with similar data by Ricci (2006).

Within the categories of critically endangered, endangered, and extinct taxa are several species that are worth highlighting. One success story is *Dendroseris litoralis* (Fig. 9.1). This species has only six small populations left in the wild on Robinson Crusoe Island (Leiva et al. 2013), but the species is self-compatible (Bernardello et al. 2001), yields good seed, and the seeds germinate well. Plants seem well adapted to cultivation in gardens in the islands as well as on the continent. Even though critically endangered in the wild, hundreds, if not thousands, of individuals of this species exist throughout the world, including in botanical gardens in Europe. Another interesting case is *Chenopodium sanctae-clarae* (Fig. 9.2). In 1980, this species consisted of one population of only 58 individuals (Stuessy et al. 1984) restricted in the wild to one single, small rock, Morro Spartan (Fig. 9.3). This is located near Santa Clara Island, separated only by a small water channel. This short distance has kept it isolated from feral animals, especially goats, and saved it from extinction. The species is doing well on this rock, but obviously the wild population is endangered. Genetic variation in *C. sanctae-clarae* is known to be low (Crawford et al. 1988). Years ago, an effort was made to cultivate it in the village, and it resulted in successful hedges

Figure 9.1 *Dendroseris litoralis* (Asteraceae) growing in the plaza of San Juan Bautista on Robinson Crusoe Island. These plants were destroyed in the tsunami of February 2010.

Figure 9.2 *Chenopodium sanctae-clarae* (Chenopodiaceae), a shrub endemic to Morro Spartan near Santa Clara Island.

Figure 9.3 Morro Spartan, a small, isolated rock just off the northern coast of Santa Clara Island (in background), which is the only known locality for *Chenopodium sanctae-clarae*.

(Fig. 9.4). This positive development, however, did not persist long because introduced microbes apparently caused much of it to die back.

Two species are known to have gone extinct in historical time, *Santalum fernandezianum* and *Robinsonia berteroi*. The last individual of *Santalum fernandezianum* was seen on August 28, 1908 (and photographed by Skottsberg 1910; Fig. 9.5). Material of this last tree is today stored in the Lund Botanical Museum (LD) and the Herbarium (S) of the Swedish Museum of Natural History, Stockholm. The fragrant wood of this species, suitable for lacquered boxes, was harvested for sale to Oriental markets (for comments on the importance of the Pacific sandalwood market, see Fanning 1833, p. 322). The discovery of semifossil wood of a species of sandalwood (reported by Skottsberg 1921) on Alejandro Selkirk Island suggests that some taxon of the genus grew there, but in the absence of specimens with leaves and flowers, it is unclear whether this material represented the same or a different, species. Another sad, and quite recent, loss is that of *Robinsonia berteroi* (Fig. C58). Only one male plant was known in the native forest from Villagra valley on Robinson Crusoe Island. Because the genus is dioecious, this alone represented an insurmountable survival condition for the species, but before any focused rescue efforts (such as tissue culture) could be mounted, this plant succumbed from damage apparently caused by rats (Danton 2005). There is a recent report from CONAF on Robinson Crusoe Island that another individual (male) has been located (Silva 2016), but this has yet to be confirmed.

Figure 9.4 Hedges of *Chenopodium sanctae-clarae* in the village of San Juan Bautista. These cultivated efforts eventually failed because the species succumbed to pests, apparently from the mainland.

Phylogenetic Perspectives in Conservation

Conservation of biodiversity is not an easy matter, and priorities can be placed on species, the level at which most legislation is written, but also on families, genera, subspecific taxa, populations, and genes. One way to decide on which level to focus on is to emphasize phylogenetic position or evolutionary information content. The idea is that a taxon (often family, genus, or species) that is divergent from other groups should receive high conservation priority (Purvis et al. 2005). To lose these unique lineages is to lose more phylogenetic information than for other lineages (Faith 1994), and therefore, the former merit more efforts toward their conservation.

Without doubt, the two most significant evolutionary lineages in the Juan Fernández Archipelago are the monospecific families Thyrsopteridaceae, with the tree fern *Thysopteris elegans* on both Robinson Crusoe and Alejandro Selkirk Islands, and Lactoridaceae, containing the fragile shrub *Lactoris fernandeziana* on Robinson Crusoe Island. These are, in fact, two of only three families of vascular plants that are restricted to an oceanic archipelago (the other is Degeneriaceae, restricted to Fiji) (Bailey and Smith 1942; Miller 1988, 1989). Both these genera have long evolutionary histories. *Thyrsopteris* (Fig. 9.6) can be related anatomically to *Thyrsopterorachis*, known from fossils from the Cretaceous of Japan (Nishida and Nishida 1979). *Lactoris* (Fig. 9.7) has unique tetrad pollen (Zavada and Taylor 1986), which has been found (as *Lactoripollenites*) in fossil borehole sediments off the southwestern coast of southern Africa (Late Cretaceous) (Zavada and Benson 1987), off the coast of Australia

(A) (B) (C)

Figure 9.5 *Santalum fernandezianum* (Santalaceae). (A) The last known individual, photographed by C. Skottsberg in 1908. (B) Herbarium specimen. (C) Wood sample, presumably from the same tree, collected by Percy Quensel in September, 1908 (now in the Stockholm Herbarium [S]). (A and B from Skottsberg 1910.)

Figure 9.6 *Thyrsopteris elegans*, the only species of Thyrsopteridaceae, endemic to the Juan Fernández Archipelago.

Figure 9.7 *Lactoris fernandeziana*, the sole representative of Lactoridaceae, endemic to Robinson Crusoe Island.

and also on the mainland (Campanian to Early Oligocene) (MacPhail et al. 1999), and now from eastern Patagonia in Argentina (Early Miocene) (Gamerro and Barreda 2008). Furthermore, molecular phylogenetic analyses of early angiosperms consistently reveal *Lactoris* as among the basal angiosperms (Qiu et al. 1999; Wanke 2007). These two families therefore represent very ancient lineages that used to have much broader distribution and that are now restricted to the Juan Fernández Archipelago. Fortunately, neither of the species is critically endangered at this time. *Thyrsopteris* is reasonably common on both islands and in no obvious peril. Although it occurs along the path down from Mirador Selkirk into Villagra valley, most individuals are in much more secluded forest locations.

Decades ago it was believed that *Lactoris* was nearly extinct, known from only a few plants (Lammers et al. 1986), but due to outstanding reconnaissance efforts on Robinson Crusoe Island by the CONAF guides (and with the help of D. Wiens in 1990 and 1991), many more plants and populations have been located, and all are in very secluded, moist valleys. It is estimated that about 1,000 plants survive (Ricci 2001). Using isozyme markers, little genetic variation exists within the species (Crawford et al. 1994), which places a greater susceptibility on this species to resist disease or changes in the environment. Low genetic variation within populations is known to increase probability toward extinction (Newman & Pilson 1997). With more sensitive RAPD markers (Brauner et al. 1992), a bit more genetic variation can be documented, and with even more sensitive ISSR markers, genetic variation can be seen more clearly (Crawford et al. 2001a). Nonetheless, these levels of variation are at the lower end in comparison with other endemic species in the archipelago (Crawford et al. 2001b).

Endemic Genera

Endemic genera also have a high priority for conservation. These are (in addition to *Lactoris* and *Thyrsopteris* described earlier) *Centaurodendron*, *Dendroseris*, *Greigia*, *Juania*, *Megalachne*, *Robinsonia*, *Selkirkia*, and *Yunquea*. *Selkirkia* and *Greigia* are extremely rare, although Leiva et al. (2013) have recently reported 29 individuals in three populations of *Greigia berteroi*. *Juania* is the endemic palm (genus and species) that used to be plentiful, but the historical records show it to have been harvested for its edible apex, making a nice boiled vegetable. This, of course, caused the death of the entire tree. The common name of "palm-cabbage" (Shelvocke 1726, p. 245) or "cabbage tree" (Funnel 1707, p. 18; Anson 1741, in Walter and Robins, 1974, p. 118; Russ 1923, p. 51) therefore was early applied to this species. This latter name, however, is now sometimes also applied to *Dendroseris litoralis* (Hind and Johnson 2006), perhaps due to its large cabbage-like leaves. The good news with *Juania australis* is that it occurs only in deep, secluded forests or high ridges far from most human activities (Moore 1969; Stuessy et al. 1983).

The four endemic genera of Asteraceae are definitely worth conserving. *Yunquea*, with its sole species *Y. tenzii*, is confined to the highest valley on Robinson Crusoe

Island, at the top of El Yunque (Skottsberg 1958), and consists of about only forty plants (Ricci 2006). Because this secluded area is nearly inaccessible to people (Danton 2000), the taxon is not under threat from normal human activities. The most recent investigation of this region in 2015 (P. López-Sepúlveda, personal communication), however, has revealed many plants of *Rubus ulmifolius* and *Aristotelia chilensis*, which are both aggressive invasive species (see Chapter 8). Nothing is still known about reproduction, pollination, breeding systems, genetic variation, or phylogenetic relationships in *Yunquea tenzii*. *Centaurodendron* includes two species, *C. palmiforme* and *C. dracaenoides*, both endangered. We have seen very few individuals of the former species and none of the latter on our research excursions to Robinson Crusoe Island. Brooke (1987a) has recorded hummingbirds visiting *C. palmiforme*, but whether this is the typical mode of pollination needs to be determined.

Robinsonia and *Dendroseris* (Asteraceae) are the largest endemic genera in the Juan Fernández Archipelago with 8 and 11 species, respectively. Together they make up 15% of the endemic vascular plant species in the islands. Both have species that range from vulnerable to critically endangered (and even extinct). The species of *Dendroseris* are never common. *Dendroseris litoralis* is a conservation success story because although very few individuals exist in the wild, the species has been cultivated successfully in the village as well as on the continent and hence has been rescued from extinction. *Dendroseris macrantha* may be extinct. We saw a plant of this species in a garden in 1980 (owned by the parents of Alvis González), but it disappeared long ago. Furthermore, this particular individual may actually have been a hybrid between *D. macrantha* and *D. litoralis* (personal observation). Ricci (2001) supports the idea that the species may no longer occur in the islands. *Dendroseris neriifolia* exists with only a few trees in eroded valleys near Piedra Agujereada, and its fate is uncertain. *Dendroseris pinnata*, *D. berteroana*, and *D. regia* are doing better, but they exist in very small populations in the forest. *Dendroseris micrantha* is the least endangered, but *D. gigantea* on Alejandro Selkirk Island is known from only one or two plants (Ricci 2006).

Robinsonia is somewhat better off than *Dendroseris*, but not by much. *Robinsonia berteroi*, previously known from only one male plant, is now presumed extinct (Danton and Perrier 2005). *Robinsonia macrocephala* was recorded once in 1989 in Puerto Frances (Ricci 2006) but has not been seen since. The newly described *R. saxatilis* (Danton 2006b) occurs only in one small population in Corrales de Molina. *Robinsonia evenia* is now occasionally seen in the forest, but its close relative, *R. thurifera*, is quite rare. Gunckel (1968, p. 24) wrote that this latter species was harvested for treating headaches on the continent, which led to its scarcity on the island. In our recent expedition to Robinson Crusoe Island in 2010, we saw only one plant. The most common species of this endemic genus, although still regarded as vulnerable, is *R. gayana*, which grows on exposed basaltic ridges. *Robinsonia gracilis*, with its attractive small leaves and heads, survives well on the highest edges of forest at about 700 m.

Recommendations for Conservation of the Native and Endemic Flora

The increased awareness of the need for conservation of the native and endemic flora of the islands has led to many ecological studies and programs of conservation. Most important and the basis for further developments has been that presented in 1976 by CONAF, the Plan de Manejo. Another in-depth series of studies was that of Ortiz (1982), which assessed the biotic and abiotic dimensions of the two major islands and made recommendations for effective conservation. The most recent comprehensive analysis and recommendations were prepared in 2009 (Biodiversa 2009a, 2009b), in preparation for an international workshop in Viña del Mar, Chile, in that same year. Soon thereafter, a detailed program for conservation was published (Cavieres et al. 2011).

If one looks at the archipelago, several points become immediately obvious. First, the human impact on Robinson Crusoe Island has been much more severe than on Alejandro Selkirk Island due to the higher number of airline visits and ship traffic to Bahía Cumberland. Approximately 20% of the native forest remains on Robinson Crusoe Island (Greimler et al. 2002a; Stuessy et al. 2005a), whereas little such reduction can be documented on Alejandro Selkirk Island (Greimler et al. 2013). Second, many parts of Robinson Crusoe Island are not only devoid of native vegetation but also are suffering from serious erosion (Fig. 9.8). Such erosional extremes are not known on Alejandro Selkirk Island. Third, the native and endemic plant species of the archipelago are in a fragile state, and this is true for both islands. Fourth, introduced animals, especially goats (*Capra aegagrus* f. *hircus*) on Alejandro Selkirk Island and rabbits (*Oryctolagus cuniculus*) on Robinson Crusoe Island, have taken a high toll on the vegetation. Fifth,

Figure 9.8 Seriously eroded area in the vicinity of Puerto Francés, Robinson Crusoe Island.

a thriving village community exists on Robinson Crusoe Island that is not going to go away, and in fact, it is being rebuilt after the destructive tsunami of February 2010 (see Chapter 1). This means that measures to maintain the economic viability of the community need to be ensured. This realization points to obvious recommendations for conservation in the archipelago.

Because the village of San Juan Bautista is an integral part of the physical nature of the national park, conservation education is paramount for present and future generations. Ecotourism has appeared as a solution that coincides well with the objectives of the national park. There is now a new secondary school being built on Robinson Crusoe Island, and in this, as well as the primary school, teaching of the value of the native and endemic plants will take place. The dangers from invasive species also need to be understood. The series of publications by Philippe Danton and collaborators (Danton et al. 1999; Danton 2004) have helped considerably to draw attention to the natural richness of the islands and the need for their conservation. Lectures, workshops, and training sessions will all be needed to educate the villagers about successful ecotourism.

Because the village of San Juan Bautista needs food and materials from the continent to survive, many packages arrive each month by plane and boat. So far there has been no rigorous inspection of these arrivals nor lists of prohibited items, especially plants. A system of phytosanitation for inspection of boats, cargo, and passengers has just been inaugurated through CONAF, and one hopes that the outcome will be positive. For conservation for the longer term, importation of problematical invasive species needs to be prohibited.

Elimination of the population of goats on Alejandro Selkirk Island and rabbits on Robinson Crusoe Island is also of high priority. It is estimated that more than 4,000 goats live on Alejandro Selkirk Island (Biodiversa 2009a), and perhaps upward of 50,000 rabbits on Robinson Crusoe Island (Saiz et al. 1982; Bourne et al. 1992). These originally domesticated animals were brought to the archipelago centuries ago to provide food for sailors and residents, but they became feral and now place pressure on the vegetation. Getting rid of the goats on Alejandro Selkirk Island is feasible but difficult due to the terrain and agility of the goats and hence very costly. As a comparison, Isabela Island in the Galápagos Archipelago was cleared of its feral goats and other introduced mammals, but it took nearly ten years and considerable expense (Galapagos Conservancy 2015). The rabbits on Robinson Crusoe Island are easier to confront because they burrow in accessible, often partially eroded areas, but there are more of them. Direct capturing and biological control seem to produce the best solutions to this challenge. Rabbits have also created huge problems on the Chilean mainland (Jaksic and Fuentes 1991).

Eliminating the two worst invasive plant species, the maqui (*Aristotelia chilensis*) and the zarzamora (*Rubus ulmifolius*), is an even greater challenge. Both species are now on both islands, the latter having been commented on by Looser (1927) more than 90 years ago. Montenegro et al. (1991) mention that the zarzamora was introduced to mainland Chile by German colonists in 1860. Fortunately, at this time, few plants exist on Alejandro Selkirk Island, and it would be propitious to cut and poison these for complete eradication. On Robinson Crusoe Island, the challenge is immense, with both species consisting literally of hectares of dense populations. They grow so dense, in fact, that very little else can grow underneath due to lack of light. The extent of both species is so vast and the vegetative ability of both for regeneration from underground rhizomes is so strong that extirpation seems almost impossible. These sorts of challenges have led some workers (Davis et al. 2011) to suggest that it may be more realistic to contain and live with these aliens rather than attempting total removal. The problem with both these species is that they produce large quantities of fruits, edible for birds (e.g., *Turdus falcklandii*) that eat them and drop seeds elsewhere on the island (Smith-Ramírez et al. 2013). Tight confinement, therefore, is nearly impossible.

Additional studies on genetic variation within and among endemic species in the archipelago is an important way to determine the resilience of these species to changing environmental parameters. As shown in Chapter 14, much has already been done in the species of *Chenopodium*, *Dendroseris*, *Drimys*, *Erigeron*, *Lactoris*, *Myrceugenia*, *Nothomyrcia*, *Rhaphithamnus*, *Robinsonia*, and *Wahlenbergia*. Many more species, however, might profitably be analyzed. The Next Generation Sequencing (NGS) nucleotide techniques are so sophisticated now (Soltis et al. 2013) that very little leaf material is needed to obtain many molecular markers inexpensively. These avenues should be encouraged because they not only aid evolutionary understanding but also help in establishing conservation priorities.

Further development of the garden at CONAF headquarters in San Juan Bautista is also to be encouraged. Such a facility in an oceanic island serves many purposes (Oldfield 2011). First, it allows rare plants to be maintained and studied so that their attributes can be better understood. This allows better protection of them when they are reintroduced to the natural environment. Second, the garden serves as an educational tool for villagers and visitors. By seeing the rare plants, people learn of their value, and their appreciation grows proportionately. This is also the place where the new private ecoguides can train for their new careers. The only caution involved with such a common garden is that many congeners on the islands can hybridize, thus leaving hybrid seed to be harvested. Ideally, these should not be reintroduced to the wild unless there is no reasonable alternative. Great care therefore must be taken in the common garden to discourage hybridization between native species. Focus should be on collecting seeds that only contain the germ lines of each endemic species. It should be mentioned that CONAF also maintains a successful garden of Juan Fernández plants in the National Botanical Garden at Viña del Mar, Chile.

What is lacking to achieve these many conservation goals in the Juan Fernández Archipelago are financial resources. It is simply that and nothing more. The Chilean government through CONAF has made many positive efforts toward these goals, but much more is needed. The infusion of US$2 million from the Dutch government in 2004–8 was of huge assistance in constructing buildings for CONAF for both administration and research. Steps to control erosion in selected locations on Robinson Crusoe Island were also implemented, and an improved garden was developed. Through it all, however, these important steps are just a beginning for successful conservation efforts for the flora of the archipelago. It has been estimated (Cavieres et al. 2011) that a minimum of US$6 million might be needed to really make an impact, but for such a large sum, international financing would have to be forthcoming.

Part V

Patterns of Character Diversity

Our understanding of the evolution and biogeography of the plants of the Juan Fernández Islands is based on the gathering of comparative data and their evaluation. Over the past decades, we have investigated the secondary metabolites, cytology (chromosome numbers), and reproductive biology of many endemic species. In this section of this book, we summarize the important results from these different studies.

Secondary plant products, especially flavonoids, frequently used in the 1960s and 1970s to determine relationships, have been employed within many groups of the Juan Fernández Archipelago. By comparison with progenitor taxa on the Chilean continent, we have in some cases been able to test the hypothesis of loss of secondary metabolites in oceanic islands in response to reduction of predator pressure. Flavonoid data from *Dendroseris*, *Erigeron*, *Gunnera*, *Peperomia*, and *Robinsonia* are presented in Chapter 10.

Chromosome numbers, reflecting the condensation of DNA that allows for successful transfer of genetic material during mitosis and meiosis, often give clues to evolutionary relationships as well as processes of evolution. A strong survey of the endemic and native species therefore has been done, and the results are given in Chapter 11. Chromosomal changes during speciation have not occurred in the archipelago, which is a pattern typical of most oceanic island floras.

To understand the meaning of patterns of genetic variation requires also having a grasp of the reproductive biology of the endemic flora. Species that are inbreeding, for example, will obviously be expected to harbor less genetic variation than those that are outcrossing. Perhaps even more significantly, knowledge of type of pollination has much to do with conservation because as specific pollination vectors disappear, so do the plants in the mutualistic associations. Because of the low natural diversity of insects in the archipelago, some surprises have been found in the flora, as chronicled in Chapter 12.

10 Flavonoid Compounds

Daniel J. Crawford, Mario Silva O., Patricia Pacheco, and Hugo Valdebenito

Vascular plants produce a vast array of constituents often designated as "secondary" compounds (Harborne and Turner 1984; Seigler 1998). While no hard and fast criteria distinguish primary and secondary metabolites, in general, secondary compounds do not function directly in fundamental activities such as growth and development. The original designation "secondary" was applied because the compounds were thought to be "waste" products from primary metabolic pathways such as photosynthesis. It is now clear that these compounds have many functions, and thus the original name was a gross misnomer (Bohm 1998a, chap. 7). These compounds are typically of low molecular weight and are often also referred to collectively as "micromolecules" (Giannasi and Crawford 1986; Seigler 1998).

Rationale for the Use of Flavonoids and Other Micromolecules in Systematic Studies

Beginning in the late 1950s and early 1960s, micromolecules were employed for taxonomic-systematic purposes and for studying processes such as hybridization and the origin of polyploid species (Bate-Smith 1958, 1962; Alston and Turner 1963; Swain 1963; Harborne and Turner 1984). This new hybrid discipline of natural products chemistry and plant systematics was dubbed "chemical taxonomy," "chemosystematics," "biochemical systematics," or other similar names (Alston and Turner 1963; Giannasi and Crawford 1986).

For any new kind of comparative data introduced into plant systematics, perspectives on the value and utility of the data change through time. There is often an initial euphoria about the potential contributions of new information, and this is tempered through time until eventually the data are woven into the fabric of plant systematics as part of Constance's (1964) unending synthesis. Harborne and Turner (1984, pp. 42, 44, tab. 4.2) compared and contrasted the advantages and disadvantages of micromolecules and morphology. Some of the perceived strengths of chemical characters are worthy of mention. The structures of micromolecules can be elucidated, and thus they represent distinct, contrasting characters as compared with morphological characters such as leaf shape, where many loci contribute to the phenotype of the leaf. Further, molecular structures are the products of biosynthetic pathways and usually more closely reflect, or may be equated to, genetic differences

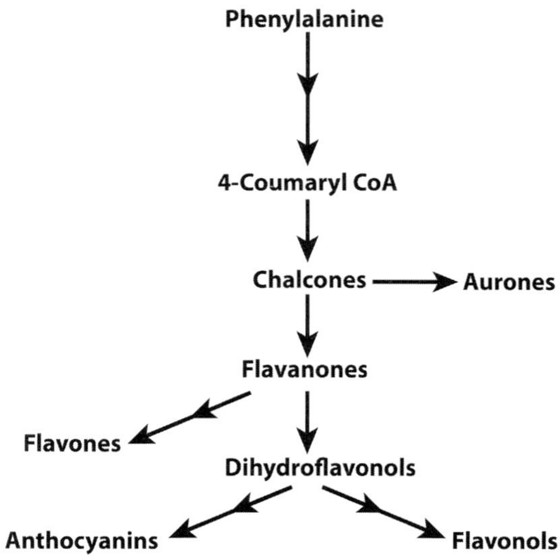

Figure 10.1 Simplified diagram of biosynthetic pathway to different classes of flavonoid compounds.

than is possible with morphology. With compounds such as flavonoids, it was initially thought that particular structures could be scored as either present or absent (detected or not), making them easier than scoring quantitative characters. The presence/absence was also seen as an advantage for examining potential cases of hybridization and introgression because compounds typical of each putative parent are either detected or not detected in their presumed hybrid. By contrast, quantitative features may be difficult to evaluate in hybrids. As discussed below, not all the alleged advantages proved to hold in all cases.

Despite the many classes of compounds produced by plants (e.g., terpenoids, alkaloids, betalains, alkanes and related hydrocarbons, glucosinolates, and polyacetylenes), flavonoids emerged as the most popular constituents for systematic studies (Harborne 1967; Bohm 1998a, chap. 3; Bohm and Stuessy 2001). They occur universally in vascular plants, so comparisons may be made at different taxonomic levels. Flavonoids exhibit tremendous structural diversity, so there is variation for comparative purposes (Bohm 1998a, pp. 120, 121). Thus any group of plants will have detectable quantities of flavonoids and likely exhibit variation among the taxa being studied (Harborne and Turner 1984; Giannasi and Crawford 1986). Practical considerations include the structural stability of flavonoids so that herbarium material or material collected and dried in the field may often be used, and it makes extraction of compounds relatively easy (Mabry et al. 1970; Markham 1982; Bohm 1998a, pp. 120, 121). Lastly, small amounts of plant material are needed for at least partial characterization of compounds.

A discussion of the biosynthesis of flavonoids is beyond the scope of this chapter. However, a highly simplified diagram of the pathways to the compounds discussed in this chapter is given in Fig. 10.1. Flavonoid biosynthesis is now the focus of studies at the level of the gene, and a general appreciation of the enzymes involved in the pathways and the regulation of flavonoid biosynthesis may be obtained from Ferreyra et al. (2012).

Analysis of Flavonoid Data

Any discussion of flavonoid data and plant systematics/phylogeny must consider how the data are to be analyzed and used. Bohm (1998a, pp. 121–131) provides a thoughtful overview of the different philosophies or approaches for using flavonoids as taxonomic, systematic, or phylogenetic data. In the most general sense, the systematic application of chemical components can be broken down into two general approaches or methods One is to use the compounds as taxonomic characters or markers. The simplest method is to compare, in narrative form, the presence/absence of different structural types among taxa. A second method is to use a coefficient to calculate an overall similarity of compounds in pairwise comparisons of taxa (Giannasi and Crawford 1986; Stuessy 2009, chap. 19). Both methods are basically phenetic approaches.

Secondary compounds have rarely been used for constructing phylogenies (but see Humphries and Richardson 1980; Seaman and Funk 1983; Stuessy and Crawford 1983), and over the past 25 years there has been even less interest in using them phylogenetically because of the explosion in the application of DNA, initially restriction-site mutations and subsequently sequence data, for phylogenetic reconstruction (Soltis et al. 2005; Judd et al. 2007). However, the availability of molecular phylogenies facilitates the mapping of secondary compounds onto a phylogeny, providing insights into the pattern of structural evolution within a plant lineage.

Practical Issues in Using Flavonoids for the Study of Island Plants

Flavonoids have several advantages for studying island plants. The aforementioned ability to use dried material is especially valuable in the island setting because it is simpler than freezing fresh material in the field or doing extractions and then freezing the extracts in the field. The logistics of getting to populations in isolated, remote areas of distant islands can be difficult, but secondary metabolites, particularly flavonoids, make the enterprise easier because plant material can be collected and processed later with minimal effort.

As indicated earlier, small amounts of plant material are sufficient for analysis and at least partial characterization of flavonoids. This is an especially important advantage when sampling rare species, and in some cases it permits the sampling of individual plants from populations (Ganders et al. 1990). Because dried material can be used as a source of flavonoids, the weight of the accrued material is much less than for fresh plants or extracts. This is a considerable advantage in transporting collections by air from island to continent, especially when using small chartered airplanes for transport to a relatively isolated archipelago such as Juan Fernández.

Flavonoid Studies of Juan Fernández Plants: General Purposes

Our flavonoid studies of Juan Fernández plants were initiated with several purposes in mind. One was to determine the utility of flavonoids for distinguishing endemic and native taxa and to infer relationships among the taxa within lineages. As mentioned earlier, secondary compounds have rarely been used to construct phylogenies. However, the availability of morphological and particularly molecular phylogenies facilitates the mapping of secondary compounds onto a phylogeny and visualizing the changes (or evolution) of chemical profiles during the radiation and diversification of lineages in oceanic archipelagos such as Juan Fernández. A last purpose was to compare the flavonoids of Juan Fernández plant lineages with their presumed closest continental relatives (ancestors). These comparisons could be used to see whether flavonoids support a hypothesis (or hypotheses) on the progenitor of island plants and to assess flavonoid changes between continental and island plants. Of special interest in this regard would be the general loss of compounds or the gain of novel compounds.

Case Studies of Juan Fernández Plants

Bohm (1998b) provided an excellent comprehensive discussion of the use of secondary compounds in systematic or evolutionary studies of island plants. The following discussion will incorporate and evaluate points raised by him. Most important, there have been advances in understanding the relationships of Juan Fernández plants, especially insights from molecular phylogenetic studies, since Bohm's (1998b) review.

Lactoris fernandeziana (Lactoridaceae or Aristolochiaceae)

This species is arguably the most fascinating endemic plant of the archipelago. Its interest stems from having long been a species of uncertain affinities within the flowering plants and its rarity on and restriction to Robinson Crusoe Island. Crawford et al. (1986) examined flavonoids from one individual from each of two populations of *L. fernandeziana* (at that time, the species was thought to consist of fewer than 10 plants, but it is now known to be more abundant, albeit still rare) and detected six glycosides of the two flavonols kaempferol and isorhamnetin (Fig. 10.2B). The presence of flavonols was viewed as evidence that the affinities of the species rest within the "archaic" or "primitive" flowering plants (now generally referred to as the "basal angiosperms," Archaeangiospermae) because flavonols are generally prevalent throughout these basal lineages (Gornall and Bohm 1979). Although flavonoids were not very informative in suggesting the closest affinities of the species within these large basal clades, the presence of isorhamnetin in *Lactoris* was used to suggest that it may be closely allied with Laurales, particularly the monospecific Gomortegaceae and Monimiaceae (Crawford et al. 1986).

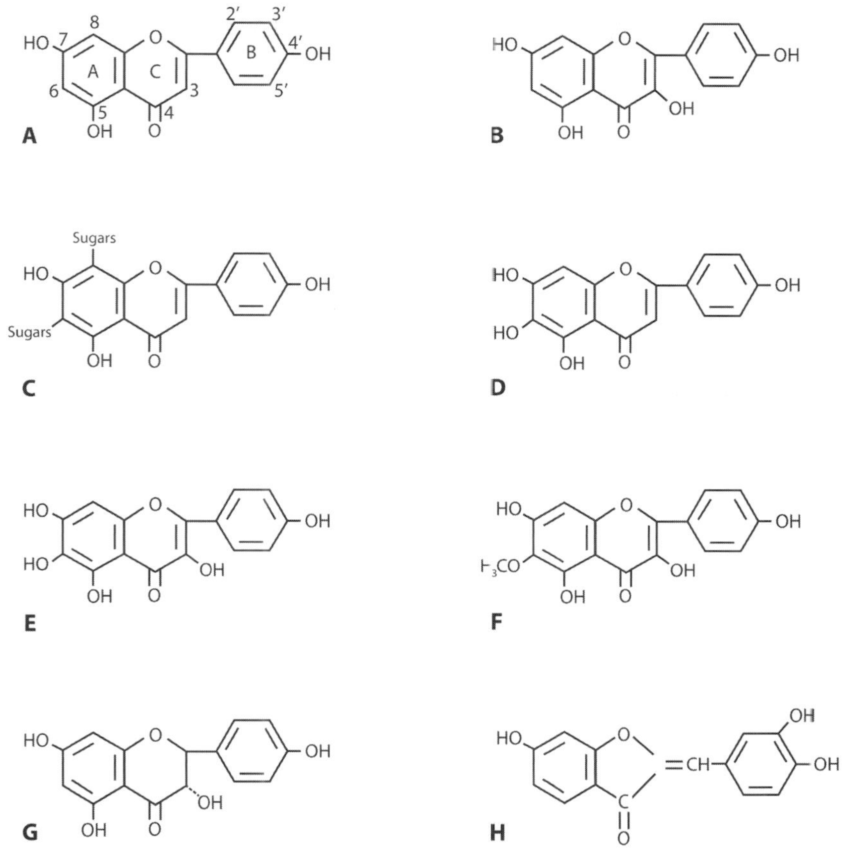

Figure 10.2 Structures of several classes of flavonoid compounds. (A) Flavones, with the standard numbering of positions on the molecules shown; –OH (hydroxyl) at 4′ = apigenin, at 3′ and 4′ = luteolin. (B) Flavonols; –OH at 4′ = kaempferol, at 3′ and 4′ = quercetin, $-OCH_3$ (methyl) on 3′ and –OH at 4′ = isorhamnetin. (C) C-Glycosylflavone sugars may be at either the 6 or 8 position or at both. (D) 6-Hydroxyflavone. (E) 6-Hydroxyflavonol. (F) 6-Methoxyflavone. (G) Flavanone (no –OH at 3 position) and dyhydroflavonol (–OH at 3 position). (H) Aurone.

It is now known that *Lactoris* is closely related to and is nested within or closely related to the Aristolochiaceae in molecular phylogenies (Soltis et al. 2005; see also Chapter 13), so the conversation must turn to a comparison of flavonoids of *Lactoris* with Aristolochiaceae. Aristolochiaceae have a very diverse flavonoid chemistry with various glycosides of flavonols and flavones, including flavone C-glycosides (Fig. 10.2C), flavanones (Fig. 10.2G), and even an aurone (Fig. 10.2H) (e.g., Iwashina et al. 2005). The flavonoids of *Lactoris* represent a small subset of the array of compounds sequestered by Aristolochiaceae. It must be emphasized, however, that the fossil record shows an ancient divergence (93 to 76 Mya) of Aristolochiaceae and Lactoridaceae, which long predates dispersal of *Lactoris* (or an ancestor thereof) to Robinson Crusoe Island (Gamerro and Barreda 2008). This ostensibly provided ample time for divergence of flavonoids between *Lactoris* and its closest exant relative, as well as for diversification of

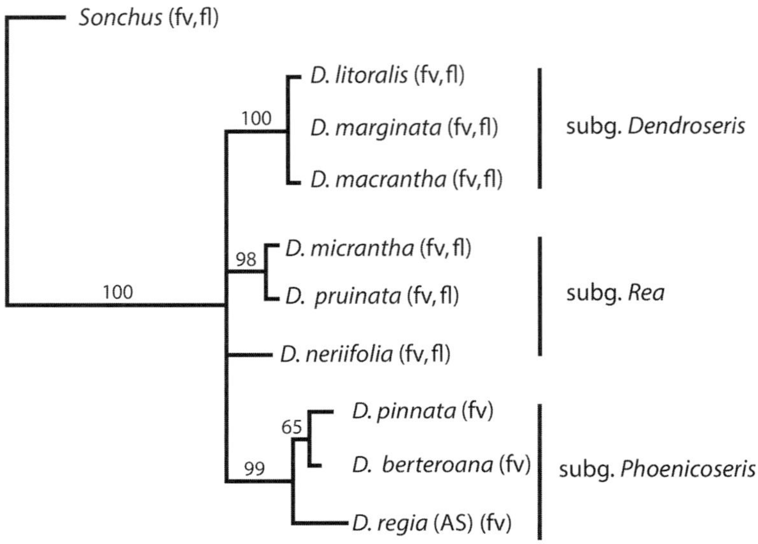

Figure 10.3 Phylogeny of *Dendroseris*. Strict consensus tree of the four most parsimonious trees resolved by ITS sequences with classes of flavonoids detected in each species shown in parentheses: fv, flavones; fl, flavonols. *D. regia* is endemic to Alejandro Selkirk Island (AS); all other taxa are on Robinson Crusoe Island. Bootstrap percentages (1,000 replicates) given above branches. (Modified from Sang et al. 1994.)

flavonoids during the evolution of Aristolochiaceae. In the end, flavonoids proved to be of little value for identifying the closest relative of the enigmatic *Lactoris*.

Dendroseris (Asteraceae: Cichorieae)

The relationships of this endemic genus (the largest in the archipelago) have been the subject of considerable debate for several decades (Pacheco et al. 1991b and references therein). Pacheco et al. (1991b) considered both *Hypochaeris* and *Hieracium* as possible ancestors of *Dendroseris* based on the views of earlier workers (Sanders et al. 1987) and concluded that the latter genus was the more likely colonizing ancestor. However, it is now known from molecular phylogenetic studies that neither is ancestral to the island endemics (Kilian et al. 2009). Kim et al. (2007) showed *Dendroseris* nested within the genus *Sonchus* and suggested that it be treated as a member of that genus. All studies indicate that *Dendroseris* originated from a single colonization; that is, it is monophyletic (see Chapter 13). Within *Dendroseris*, phylogenies generated from morphology (Sanders et al. 1987), restriction sites of ribosomal nuclear and plastid DNA (Crawford et al. 1992a), and ITS sequences (Sang et al. 1994) are all basically the same, with the two subgenera *Dendroseris* and *Phoenicoseris* holophyletic and subg. *Rea* paraphyletic (Fig. 10.3). While morphological and ecological data suggest that members assigned to subg. *Rea* are most primitive (basal) and that subg. *Phoenicoseris* is most highly derived (Sanders et al. 1987), molecular data fail to resolve relationships among the subgenera (Fig. 10.3).

Pacheco et al. (1991b) conducted an extensive survey of flavonoids (44 populations representing all 11 species) in *Dendroseris*; two of the species, *D. macrophylla* (subg. *Dendroseris*) and *D. gigantea* (subg. *Rea*), both rare endemics on Alejandro Selkirk Island, were not included in the ITS phylogeny (Fig. 10.3) for lack of material. The predominant compounds detected were glycosides of the common flavones apigenin and luteolin (Fig. 10.2A), which were present in all species. Either one or two glycosides of the flavonol quercetin (Fig. 10.2B) were detected in seven species, but they were of questionable occurrence (low concentrations) in some populations of the species. A summary of flavonoids in *Sonchus*, the closest continental relative of *Dendroseris*, was presented by Bohm and Stuessy (2001, p. 337, table 20.12); there has been rather extensive sampling of species compared with many genera of comparable size (over 70 species). As in *Dendroseris*, simple flavones such as apigenin and luteolin are prevalent in *Sonchus*, with the sporadic occurrence of flavonols and other flavonoids reported. Thus it appears that the flavonoid chemistry of *Dendroseris* reflects the types of compounds present in its closest relatives, and there has been minimal divergence from its colonizing ancestors.

Flavonoids do not distinguish all species in the genus *Dendroseris*. Three of the four species of subg. *Dendroseris* and two of the three species of subg. *Phoenicoseris* are not separable, but all four species of the paraphyletic subg. *Rea* have unique arrays of flavonoids. At the subgeneric level, species of subg. *Phoenicoseris* have the most distinctive flavonoid profiles because they lack flavonols (Pacheco et al. 1991b), suggesting that there was a loss of flavonols in the common ancestor of the subgenus (Fig. 10.3). The two remaining subgenera, *Dendroseris* and *Rea*, have a similar array of compounds, and it should also be noted that the two clades comprising the paraphyletic subg. *Rea* are not distinguishable by flavonoid chemistry (Fig. 10.3). One species from each subgenus of *Dendroseris* occurs on Alejandro Selkirk, each the result of a separate dispersal to the younger island. *Dendroseris regia* of subg. *Phoenicoseris* is endemic to Alejandro Selkirk and does not differ from the two species of the subgenus on Robinson Crusoe in flavonoid components. *Dendroseris macrophylla* (subg. *Dendroseris*) from Alejandro Selkirk lacks one of the luteolin 7-*O*-glucosides, and the same compound is also missing from the Alejandro Selkirk species of subg. *Rea*, that is, *D. gigantea*. Bohm (1998b) discussed the possible structural differences between the four luteolin 7-*O*-glucosides detected in *Dendroseris* and emphasized the need for additional data to elucidate distinctions among the four.

In summary, the array of flavonoid compounds produced by *Dendroseris* matches closely the classes of flavonoids present in the genus *Sonchus*, its closest continental relative. The only major evolutionary change in flavonoids during the radiation of *Dendroseris* in Juan Fernández was the loss of flavonols in subg. *Phoenicoseris*.

Robinsonia (Asteraceae: Senecioneae)

Robinsonia is the second largest genus endemic to Juan Fernández, with eight species, two of which appear to be extinct and a new one recently described (Danton 2006a; Danton and Perrier 2005), making six extant species. While *Robinsonia* is very distinct

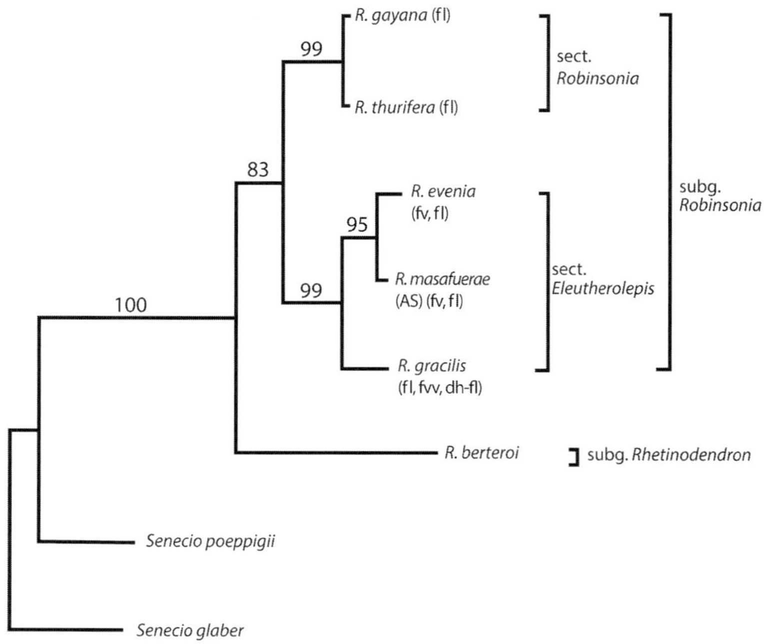

Figure 10.4 The most parsimonious tree for *Robinsonia* from analyses of ITS sequences, with classes of flavonoids detected in each species shown in parentheses. fv, flavones; fl, flavonols; fvv, flavanones; dh-fl, dihydroflavonols. *R. masafuerae* is endemic to Alejandro Selkirk Island (AS); other species occur only on Robinson Crusoe Island. Numbers above branches are bootstrap support percentages (1,000 replicates). (Modified from Sang et al. 1995.)

morphologically, consisting of rosette dioecious trees, there has been little question that its affinities lie with the very large genus *Senecio* (Sanders et al. 1987). A broadly based molecular phylogenetic study by Pelser et al. (2007) shows that *Robinsonia* is nested within *Senecio s. str.*, and Pelser et al. (2007) transferred the species to *Senecio*. However, we prefer to recognize it as a distinct genus. Because of the large number of species in *Senecio*, the closest continental relative of *Robinsonia* remains unknown.

Sanders et al. (1987) produced a hypothesis of relationships in *Robinsonia* based on morphology, and Sang et al. (1995) used ITS sequences to produce a phylogeny for the genus (Fig. 10.4). The morphological phylogeny contained all species excluding the recent one described by Danton (2006a), whereas the ITS phylogeny did not include the presumably extinct *R. macrocephala* nor the new *R. saxatilis*. The ITS phylogeny indicates that the genus is holophyletic. However, Pelser et al. (2007) show the now-extinct *R. berteroi* as far removed from the other species in their large phylogeny, but given the morphological characters uniting the genus, this seems a curious result and would necessitate the parallel origin of several characters otherwise very rare in extant *Senecio*, including dioecy (see Chapter 13). The ITS phylogeny is highly resolved for the six species, with all clades receiving strong support (Fig. 10.4). This phylogeny will be used for tracing the evolution of flavonoids in *Robinsonia*.

Pacheco et al. (1985) surveyed flavonoids in all seven species of *Robinsonia* known at that time, including the already then-extinct *R. macrocephala* and the now-extinct *R. berteroi*. The genus produces a wide array of flavonoid classes, including common flavone and flavonol glycosides, two flavanones, and a dihydroflavonol (Fig. 10.2), and the distribution of the compounds within the genus is of systematic or evolutionary interest (Fig. 10.4).

Three species of *Robinsonia* (*R. evenia*, *R. gracilis*, and *R. masafuerae*) are distinguishable by flavonoid components. *Robinsonia gayana* and *R. thurifera*, collectively, have the same array of flavonoids, which is concordant with morphology (Sanders et al. 1987), ITS sequences (Sang et al. 1995), and microsatellite and AFLP markers (Takayama et al. 2015a) in suggesting a close relationship between these two species. With the exception of the absence of the flavonol aglycone quercetin (Fig. 10.2B) in *R. berteroi* (subg. *Rhetinodendron*), the profiles of it and *R. macrophylla* (subg. *Symphyochaeta*) are identical and consist of the same two glycosides of quercetin. The presence of quercetin aglycone may be an artifact of extracting flavonoids from a herbarium specimen several decades old because, with time, there may be some loss of sugars from flavonoid glycosides in dried plants.

In the presumed ancestor of *Robinsonia*, some still unknown element of *Senecio*, flavonols alone, or flavones and flavonols together (Fig. 10.2A, B) are quite common (summarized by Bohm and Stuessy 2001, pp. 351–3, table 15.1). Thus the known occurrence of flavonoids in *Senecio* indicates that any combination of these arrays of flavonoids could have been ancestral in *Robinsonia*. If the ancestral state in *Robinsonia* is assumed to be the presence of flavonols, which seems reasonable because this is the common condition in the genus *Senecio*, and the Chilean species of *Senecio* surveyed by Pacheco et al. (1985) had only flavonols, then the most parsimonious explanation for the present distribution of compounds is the single gain of flavones in the ancestor of *R. evenia* and *R. masafuerae* (Fig. 10.4). However, in the seemingly less likely scenario that the ancestral state is the presence of both classes of compounds, then one would hypothesize the loss of flavones in the common ancestor of *Robinsonia* followed by their aforementioned gain in the ancestor of *R. evenia* and *R. masafuerae* (Figs. 10.2 and 10.4). In three of the 13 populations of *R. evenia* surveyed by Pacheco et al. (1985), no flavones were detected. Whether the difference is qualitative with a genetic/biosynthetic basis or quantitative with the inability to detect smaller amounts of flavones in three populations is unknown. Quantitative differences could be an artifact of the amount of material available for extraction, or they could have a genetic/environmental component.

The two extinct species are *R. berteroi*, the only known representative of subg. *Rhetinodendron*, and *R. macrocephala*, the sole member of sect. *Symphyochaeta*. Both species have reduced flavonoid profiles consisting of only the flavonol quercetin (Fig. 10.2B) and several of its glycosides. *Robinsonia berteroi* is sister to all other species in the ITS phylogeny (Fig. 10.4) and is very distinct morphologically (Sanders et al. 1987; Sang et al. 1995). However, only one plant was examined (the only one known at the time), and given the variation detected between plants of some species, the paucity of compounds could be a reflection of the existence of only one plant of the species and not an accurate picture of the flavonoids once present when the species consisted of more plants and populations. Likewise, only one herbarium specimen of *R. macrocephala* was examined. The flavonoid components present in these two species when they consisted of multiple populations will never be known, but lacking evidence to the contrary, it seems safe to say that they probably produced flavonols exclusively.

The remaining five species of *Robinsonia* comprise subg. *Robinsonia* sects. *Eleutherolepis* and *Robinsonia*, which occur as sister groups in a strongly supported clade (Fig. 10.4). *Robinsonia gayana* and *R. thurifera* (sect. *Eleutherolepis*) produce an identical array of quercetin glycosides, and no variation was detected in the glycosides of 28 populations of the two species (Pacheco et al. 1985). Flavonoid chemistry is condordant with allozymes (Crawford et al. 1992b), ITS sequences (Sang et al. 1995), and AFLP and microsatellite markers (Takayama et al. 2015a) in showing a very close relationship between these two species. The greatest diversity of flavonoids in *Robinsonia* occurs in sect. *Eleutherolepis*, where *R. gracilis* is sister to a strongly supported clade consisting of *R. evenia* and *R. masafuerae* (the only species on the island of Alejandro Selkirk). As indicated earlier, the latter two species are the only ones producing flavone glycosides, which serve to unite these two sister species (Pacheco et al. 1985).

The flavonoid chemistry of *Robinsonia gracilis* is very distinct from all other species in the genus. The flavonols kaempferol and isorhamnetin occur in addition to quercetin. In addition, both flavanones and a dihydoflavonol (Fig. 10.2) are sequestered by all populations of this species. Pacheco et al. (1985) provided possible biosynthetic explanations for the occurrence of these compounds, and Bohm and Stuessy (2001, p. 500) supplied additional hypotheses for the presence of these compounds. Flavanones are precursors to flavones and also to dihydoflavonols (Fig. 10.1), which, in turn, may be converted to flavonols (Bohm and Stuessy 2001, chap. 5). One possibility is a gain mutation that would cause a decrease in enzymes converting flavanones to dihydroflavonols and then subsequently to flavonols (Fig. 10.1). Alternative explanations include an increase in carbon flowing through the system due to control changes for enzymes such as phenylalanine ammonia lyase or chalcone synthase, and this could cause the accumulation of the intermediates (i.e., flavanones and dihydroflavonols) because the capacity of the two enzymes to convert all the intermediates had been exceeded (Bohm and Stuessy 2001, p. 500). The occurrence of glycosides of the flavonol isorhamnetin (Figs. 10.1 and 10.2B) in *R. gracilis* suggests a gain mutation for *O*-methylation.

In summary, there has been considerable flavonoid diversification in *Robinsonia* during its radiation in the Juan Fernández Islands. Assuming that the colonizing ancestor produced flavonol glycosides, there has been gain in classes of compounds as well as a unique substitution in flavonols (*O*-methylation). It is remarkable that one species, *R. gracilis*, exhibits more gains in compounds than all other species in the genus combined.

Erigeron (Asteraceae: Astereae)

Although *Erigeron* is a relatively large genus with some 390 species worldwide, it has been the subject of very little flavonoid research (Bohm and Stuessy 2001, pp. 193, 194). The most extensive taxon sampling was done by Valdebenito et al. (1992a) in a study of *Erigeron* in Juan Fernández. In addition, several species from continental South America were surveyed. *Erigeron* is the third largest genus in Juan Fernández, with six endemic species. Although there has not been a molecular phylogenetic study of the group, morphological data suggest that it is holophyletic (Valdebenito et al. 1992a). In addition, Noyes (2000) conducted a molecular phylogenetic study of *Erigeron* that included two species (*E. fernandezia* from Alejandro Selkirk and *E. rupicola*, which was mistakenly identified as *E. rosulatus*) from the Juan Fernández Islands, and the two species grouped together with relatively strong support. Thus available data suggest that *Erigeron* in Juan Fernández originated from a single introduction. All six species occur on Alejandro Selkirk, with one species, *E. fernandezia*, also present on Robinson Crusoe Island. This is an unusual distribution of endemics in the archipelago because the older island near mainland South America, Robinson Crusoe, typically contains more species of a lineage than the more distant and younger Alejandro Selkirk Island (Stuessy et al. 1990).

Flavonoids of the Juan Fernández *Erigeron* endemics consist of a quercetin 7-*O*-glycoside and a 6-hydroxyquercetin 7-*O*-glycoside, two 6-methoxy luteolin 7-*O*-glycosides, two flavone 7-*O*-glycosides, and a C-glycosyl 7-*O*-glycoside (Fig. 10.2) (Valdebenito et al. 1992a). Two of these compounds, luteolin 7-*O*-diglucoside and 6-C-glucosylacacetin, are of particular interest because they were used by Bohm (1998b, pp. 249–51) to hypothesize relationships in *Erigeron*. The most striking feature of the flavonoid components is the extensive interpopulational variation (Valdebenito et al. 1992a). In fact, half or fewer of the compounds detected in any species occur in every population of the species. Every species in which two or more populations were examined (only one population of the rare *E. stuessyi* was examined) exhibited variation among populations. In several instances, certain populations of one species were identical to populations of another species but differed from conspecific populations by several flavonoid components. While the extensive variation seriously compromises the value of flavonoids for distinguishing *Erigeron* species in the Juan Fernández Islands, the variation is of interest when compared with other data suggesting that the species are of very recent divergence or perhaps in the process of diverging.

Eleven populations representing five species of mainland South America *Erigeron* contained an array of flavonoids similar to those found in the island endemics, including the 6-substituted compounds (Fig. 10.2D–F) (Valdebenito et al. 1992a). The rather limited sampling of continental *Erigeron* species precludes making strong statements about the chemistry of the colonizing ancestors, but available data suggest that they contained flavonoids with A-ring substitutions at the 6-position in both flavones and flavonols (Fig. 10.2D, E). The three mainland South American species, in which two or three populations were examined, revealed interpopulation variation.

Valdebenito et al. (1992a) suggested that interpopulation variation in Juan Fernández *Erigeron* limits the taxonomic and phylogenetic utility of the flavonoid compounds. Bohm (1998b) and Bohm and Stuessy (2001, pp. 483, 484) argued that the flavonoids, instead of being of limited phylogenetic utility, suggest a scenario for the radiation of *Erigeron* in the Juan Fernández Archipelago. The central question is whether the

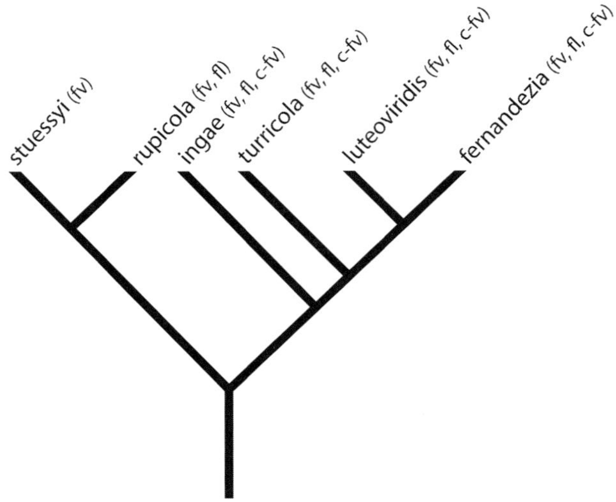

Figure 10.5 Phylogenetic hypothesis of relationships among *Erigeron* species endemic to the Juan Fernández Archipelago based on morphological characters. Six species from continental South America were used as outgroups. *E. fernandezia* occurs on both islands, whereas other species are restricted to Alejandro Selkirk Island. Flavonoid classes are shown in parentheses. fv, flavones; fl, flavonols; c-fv, C-glycosylflavones. (Modified from Valdebenito et al. 1992a.)

ancestor of *Erigeron* colonized the older, closer Robinson Crusoe Island followed by dispersal to Alejandro Selkirk or vice versa (see Chapter 13). If the former were true, then *E. fernandezia* would be the basal element in the island clade because it is common to both islands, whereas all other species are endemic to Alejandro Selkirk, and presumably they originated from an *E. fernandezia* colonizer from Robinson Crusoe. However, the phylogeny based on morphology does not place *E. fernandezia* basal, or even close to basal, to the other insular species (Fig. 10.5), which is concordant with an initial dispersal to Alejandro Selkirk and not Robinson Crusoe. Despite the morphological/phylogenetic evidence implicating Alejandro Selkirk as the site of the initial colonizing ancestor for *Erigeron*, Bohm (1998b) and Bohm and Stuessy (2001, pp. 483, 484) argued that the distribution of flavonoids may be interpreted in a phylogenetic context and taken alone suggest that the island lineage originated on Robinson Crusoe. First, four of the five species of *Erigeron* from mainland South America contain the two aforementioned compounds luteolin 7-*O*-diglucoside and 6-C-glucosylacacetin (Fig. 10.1B, C), but it should be emphasized that three of the nine populations of these four continental species lack one or both of the compounds. These two compounds occur together in 17 of the 26 populations of *E. fernandezia*, meaning that nine populations lack one or both of them. The two compounds were not found in any of the 12 populations from Alejandro Selkirk and are of sporadic occurrence in the Alejandro Selkirk endemics (Fig. 10.5). They were not detected in any of the ten populations of *E. rupicola* or the single population of *E. stuessyi* but were of variable occurrence in the remaining three species. One population of *E. ingae* had both and another lacked both, and the same situation was found for the two populations of

E. turricola. In *E. luteoviridis*, one population lacked one compound and a second population lacked the other.

Bohm (1998b) and Bohm and Stuessy (2001, pp. 483, 484) suggested that at least three genes are probably responsible for the presence of the two compounds, one to add the second glucose to produce the luteolin 7-*O*-diglucoside, one for C-glucosylation, and one for 4'-*O*-methylation to produce the 6-C-glucosylacacetin (Fig. 10.2B, C). They argued that this represents a considerable genetic/biosynthetic difference between plants producing and lacking the compounds. Since most of the mainland species have the compounds, Bohm (1998b) and Bohm and Stuessy (2001, pp. 483, 484) hypothesized that the colonizing ancestor of the Juan Fernández lineage sequestered the two components, and since Robinson Crusoe *E. fernandezia* has the compounds, it is reasonable to assume that colonization occurred first on that island, with subsequent loss of the compounds on Alejandro Selkirk. The sporadic occurrence of luteolin 7-*O*-diglucoside and 6-C-glucosylacacetin in the species endemic to Alejandro Selkirk, including absence from two species, would be interpreted as evidence for their derivation from *E. fernandezia* lacking the compounds. Bohm (1998b) and Bohm and Stuessy (2001, pp. 483, 484) argued that this scenario is more likely than loss of compounds during the radiation on Alejandro Selkirk (assuming that the ancestor had them) and then gain of the compounds in *E. fernandezia* on Robinson Crusoe. Given the assumed genetic complexity of the different chemical states, Bohm (1998b) and Bohm and Stuessy (2001, pp. 483, 484) suggested that it would be difficult to have repeated losses and gains under this scenario.

The interpretation of flavonoids by Bohm (1998b) and Bohm and Stuessy (2001, pp. 483, 484) for the radiation of *Erigeron* on the Juan Fernández Archipelago is worthy of consideration. Unfortunately, data are lacking to make a strong inference about the island of origin. First, there is no rigorous nucleotide phylogeny for the island clade, and this would be highly desired for inferring relationships, especially the position of *E. fernandezia*. This is not meant to disparage the morphological phylogeny of Valdebenito et al. (1992a), but an independent assessment of relationships would be most valuable for inferring the biogeographical pattern of the radiation of the genus in the archipelago. However, recent data from molecular markers (microsatellites and AFLPs) (López-Sepúlveda et al. 2015a) and several other factors suggest an origin on Alejandro Selkirk. A potential weakness of the Bohm and Stuessy hypothesis is the assumption of the considerable genetic difference underpinning the presence/absence of the two compounds in plants because there is interpopulational variation for the two compounds in several of the species, including *E. fernandezia* on Robinson Crusoe and three of the five species on Alejandro Selkirk for which more than one population was sampled. Whatever the genetic/biosynthetic basis of the different chemotypes, it seems clear that they can vary with impunity among populations, which, in turn, suggests that little evolutionary/phylogenetic weight can be placed on their presence/absence. Clearly, more data are needed to interpret the radiation of *Erigeron* in the Juan Fernández Archipelago.

The reason(s) for the extensive intraspecific variation in *Erigeron* is unknown, and it should be pointed out that the distribution of compounds indicates that all species have the capacity to produce many of the same flavonoids. The lack of consistent differences among species may be the result of recent divergence from a colonizing ancestor, and ancestral flavonoid polymorphisms have not yet sorted out in the species. Morphological similarity among the species supports the interpretation of recent divergence (Valdebenito et al. 1992a). As discussed below, a recent study by López-Sepúlveda et al. (2015a) using two molecular markers indicates low divergence among species. Because plants were bulked from populations for flavonoid analysis, chemical variation at the individual level is not known. The structurally identified components were either present in high concentrations or not detected at all in populations, suggesting that most individuals in a given population either do or do not produce a given compound. However, it is also possible that within a population the observed presence/absence pattern results not from an absolute difference for all plants but from the vast majority of plants having one condition or the other. Regardless of the situation, the overall pattern that emerges is that the species of Juan Fernández *Erigeron* are very similar in their biosynthetic capacities to synthesize flavonoids. Also, it does not appear that the flavonoids of the endemics have diverged appreciably from those of their ancestors. Lastly, all the species of Juan Fernández *Erigeron* are high polyploids ($n = 27$) (Sanders et al. 1983; Spooner et al. 1987; Valdebenito et al. 1992a) and thus could harbor considerable genetic diversity for flavonoid biosynthesis, and the sorting of the diversity could be slowed by the higher ploidal levels.

A recent population genetics study by López-Sepúlveda et al. (2015a), mentioned earlier, is concordant in several respects with the morphology and flavonoids of Juan Fernández *Erigeron*. Microsatellites and AFLP markers grouped the three species *Erigeron ingae, E. luteoviridis*, and *E. turricola*, but with no clear distinctions among them. These species are quite similar morphologically, and importantly, all produce the same array of flavonoids (Valdebenito et al. 1992a). In contrast, the morphologically similar *E. rupicola* and *E. stuessyi* differ consistently from other species by invariably lacking C-glycosyl-flavones (Fig. 10.1C). However, it is important to note that in other species there is interpopulational variation for the presence of C-glycosyl-flavones, so the distinction is not absolute over all populations of all species. Although *E. fernandezia* is morphologically distinct from other species in the Juan Fernández Archipelago, populations on the two islands are not separable morphologically, yet individuals from the two islands do not group together with either of the molecular markers. As indicated earlier, two flavonoids, particularly one flavone and C-glycosylflavones (Fig. 10.2B, C), are useful for distinguishing plants from the two islands because the compounds were not detected in any populations from Alejandro Selkirk, but one or both were found in nearly 90% of the 26 populations examined.

In summary, flavonoid data are concordant with morphology and population genetic studies in suggesting that overall, the species of *Erigeron* on the Juan Fernández Archipelago are of recent divergence and are likely still in the process of the sorting of ancestral diversity. However, there do appear to be three distinct groups: *E. rupicola* and *E. stuessyi; E. ingae, E. luteoviridis*, and *E. turricola*; and *E. fernandezia*.

Peperomia (Piperaceae)

The genus *Peperomia* is one of the largest in the flowering plants, with some 1,600 species, and has long been recognized as a complex and difficult genus (Wanke et al. 2006; Samain et al. 2009). Four species from this large genus occur in the Juan Fernández Archipelago (Valdebenito et al. 1992b), but unfortunately, none of them has been included in a published molecular phylogenetic study of the genus (Wanke et al. 2006; Samain et al. 2009; Symmank et al. 2011; see also Chapter 13). Valdebenito et al. (1992b) suggested on the basis of morphology that *P. fernandeziana*, which occurs on both islands of the Juan Fernández Archipelago, as well as in continental Chile, is a member of subg. *Sphaerocarpidium* (= subg. *Micropiper*) (Frenzke et al. 2015). The remaining three species include *P. berteroana* (on both islands), *P. margaritifera* (on Robinson Crusoe Island), and *P. skottsbergii* (on Alejandro Selkirk Island), and Valdebenito et al. (1992b) considered these species as members of subg. *Tildenidium*. Frenzke et al. (2015), however, now place these species also in subg. *Micropiper*. Because of divergent morphologies, it is probable that the species of *Peperomia* in the Juan Fernández Archipelago resulted from two dispersal events, one for *P. fernandeziana* and another for the ancestor of the other three species. Molecular data would be particularly desirable to test this hypothesis given their demonstrated value for assessing relationships in *Peperomia* (Wanke et al. 2006; Samain et al. 2009; Symmank et al. 2011).

The flavonoid chemistry of insular *Peperomia* is dominated by flavones, with a rich array of flavone *O*-glycosides, including those with *O*-methylation in the B-ring (Fig. 10.2), and there is an even larger number of C-glycosylflavones, again with B-ring methylation (Fig. 10.2A, C). Three flavonol glycosides were detected, two of them with *O*-methylation (Fig. 10.2B). Lastly, several sulfated flavones, which have not been commonly reported from plants, were detected, with all of them based on the same basic structures as the *O*-glycosides (Valdebenito et al. 1992b). These compounds have a sulfate group ($SO_3{}^-$), often in the form of esters, in positions on the molecules where hydroxyl or methoxyl groups are otherwise located.

Extensive interpopulational variation in flavonoids was detected in all species of *Peperomia* examined except for *P. margaritifera* (two populations sampled) and the two populations of continental *P. fernandeziana*. In addition, only one population each of five mainland South American species was studied. Despite the interpopulational variation, there were consistent differences among the three insular taxa that ostensibly evolved from a common ancestor. In addition, populations of *P. berteroana* from the two islands are distinguishable on the basis of flavonoid chemistry, and *P. fernandeziana* populations on the two islands differ consistently by one compound, whereas other compounds may occur in all populations of one island but vary among populations of the other island (Valdebenito et al. 1992b). Flavonoids are not useful for showing a closer relationship between continental *P. fernandeziana* and populations of the species on either of the two islands. Thus flavonoids provide no clues as to which of the islands was colonized first with subsequent dispersal to the other. The possibility of independent dispersal events to each of the islands, while not likely, cannot be dismissed with

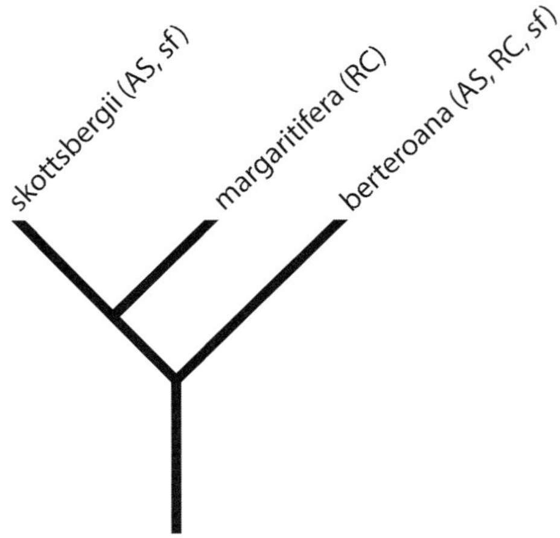

Figure 10.6 Hypothesis of relationships in species of *Peperomia* endemic to the Juan Fernández Islands based on morphological characters. Taxa of subg. *Sphaerocarpidium* (= subg. *Micropiper*) (Frenzke et al. 2015) were used as a single combined outgroup. AS, occurrence on Alejandro Selkirk Island; RC, on Robinson Crusoe Island; sf, presence of sulfated flavonoids. (Modified from Valdebenito et al. 1992b.)

available data. Despite extensive interpopulational variation, flavonoid chemistry is useful for delimiting the three endemic species of *Peperomia* in the Juan Fernández Islands. The most obvious distinction is the presence of sulfated compounds in *P. berteroana* and *P. skottsbergii* and their absence in *P. margaritifera* (Fig. 10.6). The latter species is distinguished from the former two species by the exclusive possession of two C-glycosylflavones, the absence of both flavone glycosides, and a large array of C-glycosylflavones of variable occurrence in the other two species (Valdebenito et al. 1992b). The three endemic species are, in turn, separable from the native, nonendemic *P. fernandeziana* by three C-glycosylflavones and one flavonol.

Inferring the evolution of flavonoids for the three island endemics is limited by several factors, including the lack of a robust phylogeny for *Peperomia* that includes both the endemics and a wide taxomonic sampling from throughout the large genus and the paucity of data on the distribution of flavonoids in *Peperomia*. Valdebenito et al. (1992b) hypothesized relationships among species of insular *Peperomia* based on morphology (Fig. 10.6), and flavonoid evolution in *Peperomia* endemic to the Juan Fernández Archipelago will be discussed using this hypothesis. The most noteworthy aspect of flavonoids is the presence of the rare and ostensibly advanced or derived sulfated compounds in *P. berteroana* and *P. skottsbergii* but not in *P. margaritifera*. Given the phylogeny, one explanation is the presence of sulfated compounds in the common ancestor of the three species, followed by loss in *P. margaritifera* (Fig. 10.6). A less parsimonious explanation would be to assume an ancestral condition of absence of sulfated flavonoids, with two independent gains, one in *P. berteroana* and another in *P. skottsbergii* (Fig. 10.6). While sulfated flavonoids are apparently otherwise unknown in *Peperomia*, the aforementioned paucity of flavonoid reports in the large genus and the lack of a robust phylogeny suggest caution in inferring whether presence or absence of sulfated compounds is the ancestral condition for the insular taxa. The occurrence of both flavone and flavonol *O*-glycosides in three of the four continental species suggests that their presence was the ancestral condition in the insular taxa. If this were the case, then there has been loss of flavone *O*-glycosides in *P. margaritifera* (Fig. 10.6).

A particularly interesting example of the systematic utility of flavonoid chemistry in *Peperomia* is the evidence it provides for a close relationship between *P. berteroana* and *P. tristanensis* on Tristan de Cunha Island in the South Atlantic Ocean, some 5,000 km from the Juan Fernández Islands (Valdebenito et al. 1990a, 1990b). Most interestingly, the flavonoid profiles of plants from Alejandro Selkirk and Tristan de Cunha are identical and differ from the profile of plants from Robinson Crusoe. Further, morphometric studies (Valdebenito et al. 1990b) show that plants from Alejandro Selkirk and Tristan de Cunha are more similar to each other than either is to populations from Robinson Crusoe. The data prompted a taxonomic change, with the populations from the South Atlantic relegated to a subspecies of *P. berteroana* (Valdebenito et al. 1990a).

Gunnera (Gunneraceae)

The genus *Gunnera* consists of some 20 to 30 species distributed in the southern hemisphere. It is the only genus of Gunneraceae, which has an isolated position phylogenetically, being in a clade (with another family) that is sister to all other eudicots (Soltis et al. 2005, chap. 6). In parallel with the uncertain affinities of the genus is the difficulty of inferring relationships within it (Wanntorp et al. 2001, 2002). The three species endemic to the Juan Fernández Archipelago were studied by Pacheco et al. (1991a, 1993). The species are *G. bracteata* and *G. peltata* from Robinson Crusoe and *G. masafuerae* from Alejandro Selkirk. Pacheco et al. (1993) hypothesized that a single introduction was responsible for the endemic species, with *G. tinctoria* as the closest continental species. Pacheco et al. (1993) did consider the possibility of two introductions because of the distinctive morphology of *G. bracteata* but ultimately rejected the idea.

A molecular phylogenetic study by Wanntorp et al. (2002) using nuclear ITS sequences and sequences from two plastid regions resolved *G. masafuerae* sister to *G. tinctoria* in a strongly supported clade and *G. bracteata* and *G. peltata* sister in a moderately supported clade, with both of these clades, in turn, forming a clade with very weak support. These four species occur together with other Central and South American taxa in a very strongly supported clade but with little internal resolution among species (Fig. 10.7). Ruiz et al. (2004) and Stuessy et al. (2005b), in addition to confirming the results of Wanntorp et al. (2002) of low ITS sequence divergence among Juan Fernández species, also documented low allozyme divergence (high identity) among the taxa. The results of Wanntorp et al. (2002) provide strong evidence for *G. tinctoria* as the closest continental relative and presumed ancestor (or sister) of *G. masafuerae* but leaves open the question of the ancestor of *G. bracteata* and *G. peltata*. The data suggest that the endemic species of the Juan Fernández Archipelago originated from two introductions. Allozyme similarity (genetic identity) values were discussed by Ruiz et al. (2004) in the context of phylogenetic relationships

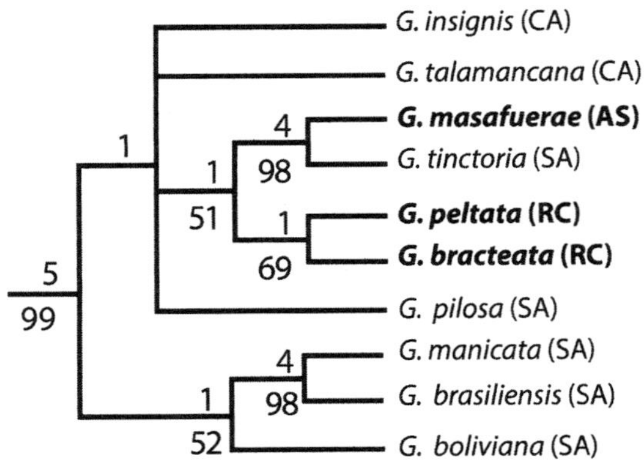

Figure 10.7 Strict consensus tree of 30 most parsimonious trees generated from ITS sequences showing relationships in a clade containing species of *Gunnera* subg. *Panke* from South (SA) and Central America (CA). Juan Fernández endemics (**bold**) designated by island (AS, Alejandro Selkirk; RC, Robinson Crusoe). Decay values are given above branches and bootstrap values below. (Modified from Wanntorp et al. 2002.)

in *Gunnera*. However, the values should be used with some caution because they are all quite high (genetic identities of 0.89–0.99, mean of 0.94 for all pairwise comparisons among species) and the differences between most species are quite small (Ruiz et al. 2004). In addition, only five loci were used for comparison of the species. The very low ITS sequence divergence between *G. tinctoria* and *G. masafuerae* is concordant with a recent divergence of the latter from the former. The six times higher divergence between *G. tinctoria* and *G. peltata* suggests an earlier dispersal to Robinson Crusoe, and the identical sequences for *G. bracteata* and *G. peltata* support the recent divergence of these two species on Robinson Crusoe. The high divergence between endemics on the two islands is additional support for the hypothesis of separate introductions (see Chapter 13).

The flavonoid data will be discussed within the context of the presumed origin of the Juan Fernández endemics from two introductions, one to each island, with divergence and speciation on Robinson Crusoe. Based on the molecular phylogeny, *Gunnera tinctoria*, or a species similar to it, is the presumed ancestor of *G. masafuerae*. While the ancestor of the two endemics on Robinson Crusoe Island has not been elucidated, the totality of data suggests that they could have originated earlier than *G. masafuerae* from a *G. tinctoria*–like common ancestor, though this remains an open question because of the very low support for the clade containing this species and the insular endemics (Fig. 10.7). The most prominent feature of the flavonoid chemistry of the three island endemics is its uniformity and the minimal flavonoid divergence between them and the continental South American species examined (Pacheco et al. 1993). All species produce glycosides of common flavones and flavonols (Fig. 10.2A, B). The most obvious and consistent difference between island and continental taxa is the presence of kaempferol in all continental species except *G. tinctoria*. The lack of kaempferol in this continental species and the endemics is concordant with an earlier suggestion

(Ruiz et al. 2004) that *Gunnera* in Juan Fernández resulted from two dispersal events, both from *G. tinctoria* or perhaps from a *tinctoria*-like ancestor. Under this scenario, one colonization resulted in *G. bracteata* and *G. peltata* on Robinson Crusoe Island, and the second dispersal produced *G. masafuerae* on Alejandro Selkirk Island.

Consider first the flavonoids of the two Robinson Crusoe endemics *G. bracteata* and *G. peltata*. One change relative to *G. tinctoria* is the loss of one compound (quercetin 3-*O*-digalactoside) during evolution on the island because both endemics lack the component (Fig. 10.2B). There has also been the gain of quercetin 3-*O*-xyloglucoside in *G. bracteata*. It is generally assumed that *G. peltata* is ancestral to *G. bracteata* (Pacheco et al. 1993), and if this is the case, then there was the gain of the compound (since it is absent from *G. tinctoria*) followed by its loss in *G. bracteata*. The flavonoid chemistry of *G. masafuerae* shows the gain of quercetin 3-*O*-xyloglucoside relative to *G. tinctoria*, which parallels the gain in *G. bracteata*. Unlike the Robinson Crusoe endemics, *G. masafuerae* has retained rather than lost quercetin 3-*O*-digalactoside (Pacheco et al. 1993). When flavonoids are viewed within the best estimate of the phylogeny of *Gunnera*, it appears that there have been gains (including parallel gains in the two islands) and losses of compounds during the origin and evolution in the Juan Fernández Archipelago.

Myrceugenia (Myrtaceae)

Myrceugenia consists of about 40 species of trees or shrubs occurring in temperate to subtropical South America, with two species endemic to the Juan Fernández Archipelago (Landrum 1981a, 1981b). *Myrceugenia fernandeziana* is endemic to Robinson Crusoe Island, and *M. schulzei* is found only on Alejandro Selkirk Island (Landrum 1981a). A recent molecular phylogenetic study with comprehensive taxon sampling within the genus and the inclusion of related genera has provided definitive evidence for the relationships of the two species endemic to the Juan Fernández Archipelago (Murillo-A. et al. 2012). As shown in Fig. 10.8, *M. schulzei* from Alejandro Selkirk is closest to the Chilean species *M. colchaguensis*. However, the Robinson Crusoe endemic, heretofore referred to as *M. fernandeziana*, is distantly related to *M. schulzei*, and it is best treated as a member of the genus *Nothomyrcia* with its closest relatives in the genus *Blepharocalyx* from Argentina (Fig. 10.8). Two earlier studies (Ruiz et al. 2004; Stuessy et al. 2005) had also suggested that the two endemic species were not closely related, although the data were not as conclusive as those presented by Murillo-A. et al. (2012).

Ruiz et al. (1994) surveyed flavonoids in the Juan Fernández Archipelago, selected continental species of *Myrceugenia*, and found an array of flavones and flavonols as well as two flavanones and one dihydroflavonol (Fig. 10.2A, B, G); the flavonoid results may be viewed within the context of the molecular phylogeny of Murillo-A. et al. (2012). A phenogram based on flavonoid similarity shows that the two Juan Fernández endemics cluster at a much higher similarity with Chilean species than with the two species from Brazil (Fig. 10.9). This is concordant with the molecular phylogeny, where the Brazilian species occur in a strongly supported clade (Fig. 10.8). The second major conclusion reached by Ruiz et al. (1994) from flavonoid data is that the two endemics are the result

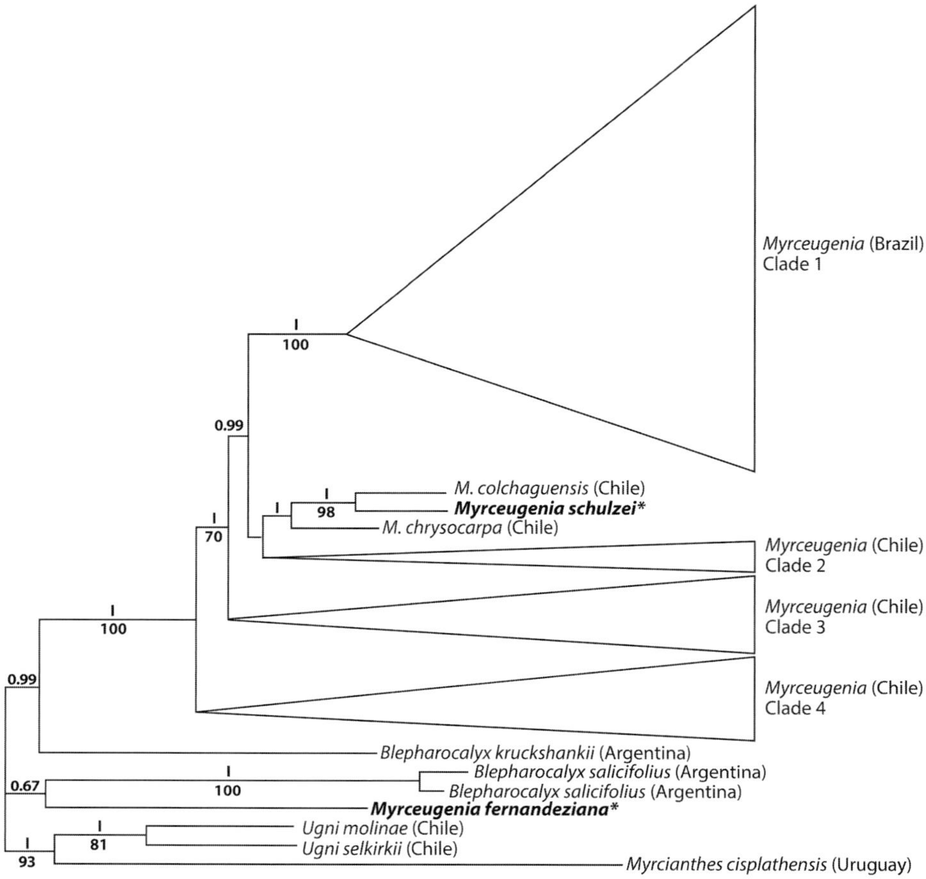

Figure 10.8 Phylogeny (Bayesian tree) for *Myrceugenia* and related genera based on nuclear and plastid DNA seqeuences; numbers above branches are Bayesian posterior probabilities; numbers below branches are bootstrap percentages. As discussed in the text, *Myrceugenia fernandeziana* is now treated as the sole species of *Nothomyrcia* (Murillo-Aldana and Ruiz 2011). Juan Fernández species are shown in bold with asterisks. (Modified and condensed from Murillo-A. et al. 2012.)

of independent dispersals to the archipelago, and the results of Murillo-A. et al. (2012) strongly support this conclusion (Figs. 10.8 and 10.9). Much earlier, Landrum (1981b) had also suggested that the two Juan Fernández endemics did not originate from a single introduction.

However, at a finer scale, flavonoid similarity is not concordant with relationships of the endemics resolved with molecular data. For example, *M. schulzei* is strongly supported as sister to *M. colchaguensis* in the molecular phylogeny (Fig. 10.8), whereas it clusters at low flavonoid similarity with this species (Fig. 10.9). The low similarity between the two sister species is largely the result of having a different array of glycosides of the flavonols kaempferol and quercetin (Fig. 10.2B). In addition, one flavone was detected in the continental species but was not found in the three populations of *M. schulzei* examined. The high flavonoid similarity between *Nothomyrcia*

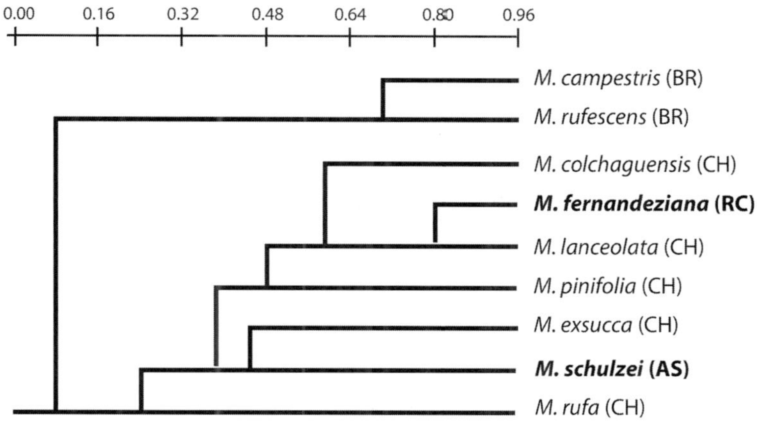

Figure 10.9 Phenogram of relationships among selected species of *Myrceugenia* generated from flavonoid compounds. The Jaccard coefficient was used to calculate flavonoid similarities, and UPGMA was used to construct the phenogram. The two Juan Fernández endemics are indicated by island (AS, Alejandro Selkirk; RC, Robinson Crusoe). Continental species are designated by BR (Brazil) and CH (Chile). (Redrawn from Ruiz et al. 1994.)

fernandeziana from Robinson Crusoe Island and *M. lanceolata* is not a reflection of a close relationship between the two species and shows the potential limitations of flavonoids alone for assessing relationships, especially for more distant relationships.

Sophora (Fabaceae)

This genus has historically been viewed as consisting of 40 to 50 species, but it is now known not to be monophyletic (Crisp et al. 2000). The two species endemic to the Juan Fernández Archipelago are *S. fernandeziana* on Robinson Crusoe and *S. masafuerana* from Alejandro Selkirk; both are members of sect. *Edwardsia*. This section includes species widely distributed throughout the southern hemisphere, many of which occur on islands (Mitchell and Heenan 2002). With the exception of New Zealand, all islands have a single endemic species of *Sophora*. The section appears to be monophyletic, though not all species were included in a molecular phylogenetic study (Mitchell and Heenan 2002), and there is very little ITS divergence among any of the taxa sampled to date.

The two species endemic to the Juan Fernández Archipelago have not been included in a comprehensive phylogenetic study. Ruiz et al. (2004) and Stuessy et al. (2005b) reported high genetic identities between the two endemic species at allozyme loci and identical ITS sequences. However, the generally high similarity in allozymes and ITS sequences combined with the few continental species of *Sophora* that have been examined precludes drawing strong inferences about whether the two endemics are the result of a single introduction and their relationships to continental species (Ruiz et al. 2004).

Ruiz et al. (1999) examined flavonoids in the endemic *Sophora* and two other species, one from Chile and another from Chile and New Zealand. A number of flavones, flavanones, flavonols, and dyhydroflavonols were detected in the species (Fig. 10.2). The phenetic analysis of flavonoids showed that the two Juan Fernández species did not cluster together, leading Ruiz et al. (1999) to suggest that the endemics did not originate from a single dispersal event to the islands. Peña et al. (1996, 2000) used cladistic analysis of morphology and alkaloid chemistry to infer relationships in *Sophora*, and the two island endemics were not resolved as sister species. However, because of limited taxon sampling from sect. *Edwardsia*, these results must be viewed with caution. The widespread distribution of *Sophora* sect. *Edwardsia* throughout islands of the southern hemisphere (Hurr et al. 1999; Mitchell and Heenan 2002) indicates the dispersal ability of these plants and offers the possibility of independent dispersals to the two islands of the Juan Fernández Archipelago.

Discussion and Overview of Flavonoids of Juan Fernández Plants

Flavonoids have been examined in a variety of plant taxa native and endemic to the Juan Fernández Islands. The purposes of the studies have been to use the compounds taxonomically to distinguish species in the archipelago and to compare the compounds of insular plants with continental taxa to infer the closest relatives of the insular plants. A second purpose was to infer the gain/loss of compounds between the island endemics and their continental relatives and the evolution of flavonoids during the diversification of insular lineages. Although it has long been recognized that flavonoids per se are of limited value for inferring phylogenetic relationships among plants (Crawford 1978), it is of interest to map the distribution of flavonoids onto available phylogenies to infer the evolution of compounds during speciation and diversification in a lineage. As documented in this chapter, these studies have been done for a number of Juan Fernández lineages, with the elucidation of several different interesting patterns of variation and evolution.

Flavonoid chemistry is generally useful for distinguishing congeneric species in Juan Fernández lineages. As discussed earlier, interpopulational variation is common in endemic species, but the causes of the observed differences are not known. Detailed studies of plants endemic to the Hawaiian Islands by Bohm and Fong (1990), Ganders et al. (1990), and Bohm and Yang (2003) documented variation among individuals in populations, and the extensive variation in some cases precluded use of the compounds for delimiting species (Ganders et al. 1990). In the two largest genera of Juan Fernández, *Dendroseris* and *Robinsonia*, each species has a distinctive array of flavonoids, the only exception being two very closely related species (based on morphology, allozymes, and ITS sequences) of *Robinsonia*. Despite extensive interpopulational variation, flavonoids are useful in distinguishing the three species of *Peperomia* that originated from a single colonizing event (Valdebenito et al. 1992b). In contrast, in *Erigeron*, flavonoids are of limited value for distinguishing species because of extensive interpopulational variation that overlaps species boundaries (Valdebenito et al. 1992a).

Flavonoids by themselves have been of limited value in elucidating the closest continental relatives of Juan Fernández endemics. Rather, it has been useful to compare flavonoids of island and continental taxa once explicit methods, usually molecular phylogenetics, have been employed to suggest the closest continental taxa. For example, in the case of the enigmatic *Lactoris*, flavonoid compounds are of no value in choosing among the various continental taxa proposed as its closest relative (Crawford et al. 1986), which has now been shown by molecular phylogenetic studies to be Aristolochiaceae (Soltis et al. 2005). Comparison of flavonoids of *Lactoris* and Aristolochiaceae show that the former has a profile that is a small subset of the classes of flavonoids found in the latter. *Lactoris* and Aristolochiaceae represent an ancient divergence as judged by the fossil record and the high sequence divergence between them, and this could account for the differences in flavonoid profiles. There are few other examples of significant flavonoid divergence between continental and insular taxa, and this could be because of the relatively recent divergence compared with *Lactoris*. The two best examples of divergence are the occurrence of apparently novel classes of flavonoids in insular taxa of *Robinsonia* and *Peperomia*, but the assertion of novelty in the insular plants must be tempered by the caveat that additional sampling in continental relatives could reveal the presence of the ostensibly novel compounds.

Flavonoid chemistry for several lineages suggests that the endemics are not the result of a single dispersal event. The flavonoids of the two endemic species of *Myrceugenia* (Ruiz et al. 1994) support the hypothesis of Landrum (1981b), subsequently documented by Murillo-A. et al. (2012) with DNA sequence data, that the species are the result of separate introductions and indeed argue that the Robinson Crusoe endemic should be recognized as a distinct genus. While flavonoids are relatively uniform in *Gunnera* (Pacheco et al. 1993), they do suggest that the endemics are the result of more than one dispersal event. Flavonoid data in Juan Fernández *Sophora* suggest that species endemic to each of the islands are the result of independent ancestral colonizations (Ruiz et al. 1999), although the results should be viewed with some caution because of the small taxonomic sampling from continental and other insular areas.

Perhaps the most striking example of the value of flavonoids per se for inferring evolutionary-biogeographical relationships in the Juan Fernández flora comes from *Peperomia*. Flavonoids show a close relationship between a species of *Peperomia* endemic to the Juan Fernández Islands and one occurring on Tristan de Cunha Island in the South Atlantic Ocean. In addition, flavonoids (and morphology) indicate a closer relationship between *Peperomia* on Alejandro Selkirk and Tristan de Cunha than between populations on Robinson Crusoe and Alejandro Selkirk (Valdebenito et al. 1990a, 1990b).

11 Chromosome Numbers

Tod F. Stuessy and Carlos M. Baeza

To reveal evolutionary relationships and to understand modes of speciation in the flora of any part of the world, having chromosome numbers available is essential (Stebbins 1971; Levin 2002; Stuessy et al. 2014a). Strict genetic and epigenetic forces that control heredity and hence affect evolution are most important to investigate, and these have formed a focus of many of our investigations (see Chapters 14 and 15). The external expression of these genetic contents and regulators is seen in the morphology, which has always been another important dimension for assessing relationships and understanding evolutionary processes (e.g., speciation and hybridization). In between are chromosome numbers, which are the packages of DNA and histones that form during mitosis and meiosis to allow reproduction of all types to occur. This level of the organizational hierarchy can be expected to reveal much about evolutionary events that have taken place in the flora of any region and, in particular, the Juan Fernández Archipelago. Although having chromosome counts for any plant in the archipelago is valuable, the most important for evolutionary purposes come from the endemic taxa.

To utilize the value of chromosome numbers first requires having counts from the species. During our research trips to the archipelago, a focused effort has always been placed on collecting materials for chromosome counting both from young flower buds (for meiotic configurations) and from germinated seeds (for mitotic data). Our first set of reports was published by Sanders et al. (1983), followed by Spooner et al. (1987) and Sun et al. (1990), and a few additional counts have been added since, such as Baeza et al. (2004) and Kiehn et al. (2005). Only angiosperms have been counted; ferns provide added challenges and were therefore not a part of our investigations.

All known reports for endemic, native, and introduced species from collections in the Juan Fernández Archipelago are provided in Table 11.1. The results come mostly from the dicots s.l., with the only monocots being the endemic *Ochagavia elegans* (Bromeliaceae) and the native *Libertia chilensis* (Iridaceae). The data show that 100% (six of six species) of endemic basal angiosperms (Archaeangiospermae) and 38% (41/107 species) of endemic dicot taxa have been recorded. Five native and eight introduced dicot taxa have also been counted. Although these investigations do not provide data for all endemic taxa of the archipelago, they do give good insights into many genera, especially *Dendroseris*, *Erigeron*, *Peperomia*, *Robinsonia*, and *Wahlenbergia*.

Sanders et al. (1983) analyzed levels of polyploidy among the endemic species of the archipelago. Two general types of polyploids can be recognized with reference to evolution (not to mode of formation): neopolyploids and palaeopolyploids. In the Juan

Table 11.1 Published Chromosome Counts from Collections of the Introduced, Native, and Endemic Angiosperm Flora of the Juan Fernández Archipelago

Taxon	Chromosome count *n*	Chromosome count 2*n*	Reference
ARCHAEANGIOSPERMAE			
Lactoridaceae			
Lactoris fernandeziana		40	Raven et al. 1971 (ca. 40); Tobe et al. 1993
Piperaceae			
Peperomia berteroana	22 II + 2 I		Sanders et al. 1983 (22–24); Spooner et al. 1987 (*2*, both ca. 22); Sun et al. 1990; Valdebenito et al. 1992b (*8*, one 22–24, four ca. 22, two ca. 23, one ca. 22 II + 2 I)
Peperomia fernandeziana	22 ± 2		Sanders et al. 1983; Spooner et al. 1987 (*2*, 23 ± 2 and ca. 22); Valdebenito et al. 1992b (12; one 23 ± 2, three ca. 22)
Peperomia margaritifera	ca. 24		Sanders et al. 1983; Valdebenito et al. 1992b
Peperomia skottsbergii	22–24		Sanders et al. 1983; Spooner et al. 1987 (ca. 23); Valdebenito et al. 1992b (*4*, two ca. 23)
Winteraceae			
Drimys confertifolia	ca. 43		Sun et al. 1990 (*3*)
MONOCOTYLEDONAE			
Bromeliaceae			
Ochagavia elegans	50		Gitai et al. 2005, 2014
Iridaceae			
Libertia chilensis	57		Sanders et al. 1983
DICOTYLEDONAE			
Aizoaceae			
Carpobrotus aequilaterus	9		Sun et al. 1990
Apiaceae (Umbelliferae)			
Conium maculatum	11		Spooner et al. 1987
Eryngium bupleuroides	16		Sanders et al. 1983 (*2*)
Eryngium bupleuroides	(22)-24		Kiehn et al. 2005 (many meiotic irregularities)
Asteraceae (Compositae)			
Abrotanella crassipes		18	Sun et al. 1990
Dendroseris litoralis	18		Tomb et al. 1978; Sanders et al. 1983 (*2*)
Dendroseris litoralis		36	Carlquist 1967; Strother 1976; Tomb et al. 1978; Lack et al. 1978 (ca. 36); Baeza et al. 2004
Dendroseris macrantha	18		Sanders et al. 1983
Dendroseris macrophylla		36	Stebbins et al. 1953
Dendroseris micrantha	18		Sanders et al. 1983
Dendroseris micrantha		ca. 36	Lack et al. 1978
Dendroseris neriifolia	18		Sanders et al. 1983; Spooner et al. 1987 (*3*)
Dendroseris pinnata		36	Stebbins et al. 1953
Dendroseris pruinata		36	Carlquist 1967

Table 11.1 (cont.)

	Chromosome count		
Taxon	*n*	2*n*	Reference
Erigeron fernandezia	27		Solbrig et al. 1969 (*2*); Sanders et al. 1983 (*2*); Spooner et al. 1987 (*2*); Valdebenito et al. 1992a (*5*)
Erigeron ingae	27		Valdebenito et al. 1992a (*4*)
Erigeron luteoviridis	27		Valdebenito et al. 1992a (*4*)
Erigeron rupicola	27		Sanders et al. 1983; Sun et al. 1990; Valdebenito et al. 1992a (*7*)
Erigeron stuessyi	27		Spooner et al. 1987 (as *E.* cf. *rupicola*)
Erigeron turricola	27		Valdebenito et al. 1992a
Hypochaeris radicata	4		Sun et al. 1990 (*3*)
Lagenophora hariotii	7 II + 4 I		Sun et al. 1990
Robinsonia gayana	20		Sanders et al. 1983
Robinsonia gracilis	20		Sanders et al. 1983 (*5*)
Robinsonia thurifera	20		Kiehn et al. 2005
Taraxacum officinale		24	Baeza et al. 2013
Brassicaceae (Cruciferae)			
Nasturtium officinale	16		Sun et al. 1990
Campanulaceae			
Lobelia tupa	21		Spooner et al. 1987 (*2*, one ca. 21)
Wahlenbergia fernandeziana	11		Sanders et al. 1983; Spooner et al. 1987
Wahlenbergia fernandeziana		22	Kiehn et al. 2005 (*5*; one also *n* = 11; one as *W.* cf. *fernandeziana*; another as *W. larrainii*)
Wahlenbergia masafuerae	11		Sun et al. 1990 (*6*); Spooner et al. 1987 (ca. 11)
Caryophyllaceae			
Spergularia confertiflora var. *confertiflora*	36		Sanders et al. 1983 (*2*); Sun et al. 1990 (*3*, as *S. confertiflora*)
var. *polyphylla*	36		Sanders et al. 1983
Ericaceae			
Pernettya rigida	33		Sanders et al. 1983 (*2*); Sun et al. 1990
Escalloniaceae			
Escallonia callcottiae	12		Sanders et al. 1983
Escallonia callcottiae		24	Kiehn et al. 2005 (*3*, two 22–24)
Euphorbiaceae			
Dysopsis hirsuta	13		Sanders et al. 1983
Gunneraceae			
Gunnera bracteata		34	Pacheco et al. 1993
Gunnera masafuerae		34	Pacheco et al. 1993
Gunnera peltata		34	Pacheco et al. 1993
Haloragaceae			
Haloragis masatierrana	7		Sanders et al. 1983
Haloragis masatierrana		14	Kiehn et al. 2005
Fabaceae (Leguminosae)			
Sophora fernandeziana		18	Stiefkens et al. 2001

Table 11.1 (cont.)

	Chromosome count		
Taxon	*n*	2*n*	Reference
Lamiaceae (Labiatae)			
Cuminia eriantha	22		Sanders et al. 1983 (*2*); Kiehn et al. 2005 (22–24)
Cuminia fernandezia	22		Sanders et al. 1983
Cuminia fernandezia		ca. 44	Kiehn et al. 2005
Myrtaceae			
Myrceugenia schulzei	11		Sanders et al. 1983; Sun et al. 1990.
Myrteola nummularia	22		Sun et al. 1990
Nothomyrcia fernandeziana	11		Sanders et al. 1983 (as *Myrceugenia fernandeziana*)
Ugni molinae	11		Sanders et al. 1983
Ugni selkirkii	22		Sanders et al. 1983 (*2*)
Orobanchaceae			
Euphrasia formosissima	44		Sun et al. 1990 (*3*)
Rubiaceae			
Coprosma pyrifolia	22		Sun et al. 1990
Galium aparine	32		Spooner et al. 1987
Nertera granadensis	22		Sun et al. 1990
Salicaceae			
Azara serrata var. *fernandeziana*	9		Sanders et al. 1983; Spooner et al. 1987
Solanaceae			
Nicotiana cordifolia	12		Narayan 1987
Nicotiana cordifolia		24	Tatemichi 1990
Solanum robinsonianum	36		Sanders et al. 1983
Urticaceae			
Parietaria debilis	8		Sun et al. 1990 (*2*)
Urtica glomeruliflora	ca. 11		Sun et al. 1990 (as *U. fernandeziana*)
Verbenaceae			
Rhaphithamnus venustus	18		Sanders et al. 1983 (ca. 18); Sun et al. 1990

Note: Numbers within parentheses are numbers of populations sampled (in *italics*; if more than one) or minor variations from the listed number.

Fernández Islands, taxa that come from clearly polyploid continental progenitors would be regarded as palaeopolyploids. Those that apparently arose after colonization in the islands would be neopolyploids. Providing such estimates is now fraught with more difficulty due to the mounting molecular evidence that most, if not all, angiosperms have had at least one polyploid event at some time in their evolutionary history (Soltis et al. 2009). Nonetheless, if one takes a general view of polyploidy, that is, a haploid number more than $n = 12$, then 69% of the species analyzed by Sanders et al. (1983) show polyploidy.

The data show clearly that only 2% (*Spergularia confertiflora*, *Ugni selkirkii*) of species having neopolyploidy exist in the endemic angiosperm flora. As for ancient polyploidy, however, 66% of the species can be recorded as already being at the polyploid level on arrival in the islands. That so many colonizing lineages exist at higher ploidy levels suggests that selective forces have acted on genomes in regions of the continental relatives, and this means largely southern South America (see Chapter 16). The great diversity of climates in this region and the strong climatic impacts in the Andean region during the Pleistocene (Vuilleumier 1971; Simpson 1975; Seltzer et al. 2003) may have been conducive to chromosomal change leading to such high polyploid levels.

More conspicuous in the data is that there has been little chromosomal change during the evolution of species *within* the archipelago. Sanders et al. (1983) concluded that almost no change in chromosome number has occurred between continental progenitors and island lineages. The only exceptions are in *Dysopsis hirsuta* and *Wahlenbergia fernandeziana*. Uniformity of chromosome number is seen clearly in *Dendroseris*, *Peperomia*, *Robinsonia*, *Wahlenbergia*, and *Erigeron*. The last genus is perhaps particularly diagnostic because the complex has evolved on Alejandro Selkirk Island, the younger island, during the past 1 to 2 million years. During this rapid speciation, via adaptive radiation (Takayama et al. 2015b), no change of chromosome number has occurred.

The lack of change in chromosome number in floras of oceanic islands has been commented on by several studies. Carr (1998) showed that chromosome stability is the rule in the endemic flora of the Hawaiian Islands. Ono and Kobayashi (1985) presented similar data for the Bonin (Ogasawara) Islands, as did Castro and Rosselló (2007) for the Balearic Islands. Mandáková et al. (2010) stressed this point for *Pachycladon* (Brassicaceae) of Tasmania and the South Island of New Zealand. The same picture prevails here in the Juan Fernández flora. In these four island systems, some chromosomal variation exists within and among the endemic species (of the same genus), but this is uncommon. At a broader scale, Stuessy and Crawford (1998) compared chromosome counts in five oceanic archipelagos (Bonin Islands, Canary Islands, Galápagos Islands, Hawaiian Islands, and the Juan Fernández Archipelago) with those of the Queen Charlotte Islands off the coast of western Canada (Taylor and Mulligan 1968), islands of continental origin and geographically very close to the continent. The oceanic islands reveal very few changes during evolution of the island lineages, and the latter show much more variation due to dysploidy and polyploidy, as much as typically found in any continental region.

The low level of observed chromosomal change during speciation of plants in oceanic islands, including the Juan Fernández Islands, appears to be a real phenomenon and requires explanation. Although many factors may be involved, we offer the following hypothesis: When a colonist arrives in a new island setting and a successful immigrant population is developed, the next step is often dispersal into different habitats on the same island or among islands. If these new secondary colonists survive, then the different environmental factors surrounding each colonist begin to shape their adaptations, leading after many generations to distinct species with dramatically different morphologies. A change in chromosome number during the process of adaptive radiation would likely be so disruptive as to lead to lineage extinction. Chromosomal stasis, therefore, appears to be the general rule during plant speciation in oceanic islands.

12 Reproductive Biology

Gregory J. Anderson and Gabriel Bernardello

We begin with what often concluded our papers on reproductive biology over the past decades, that is, the essential nature of data on pollination and reproductive biology for successful conservation and preservation efforts. There are relatively few islands in the world with much native vegetation left. The US Virgin Islands, apart from St. John, provide a good example of extirpation of the native flora. Fortunately, there are conservation programs for many archipelagos, even when the extent of natural vegetation is limited (Caujapé-Castells et al. 2010). We are fortunate that the Juan Fernández Archipelago has fairly extensive tracts of land left undeveloped. However, the impact of introduced and invasive species even in this archipelago is huge (see Chapters 7 through 9) in part because island biota are especially susceptible to invasive plants (e.g., *Rubus*; see more on invasives in Chapter 8) and introduced animals (e.g., the impact of the introduced coati mundi on the key native and the endemic animal pollinators – hummingbirds) (see Colwell 1989). Evolution in isolation, including on islands, has yielded some of the great wonders of the biological world, but it has also produced some species that do not compete well with hardened "world travelers," that is, invasives or broadly distributed species. Thus, conservation plans even for islands with more extensive native tracts, such as the Juan Fernández Islands, are important to establish immediately and with comprehensive knowledge regarding the flora. Successful conservation plans require deep knowledge of the pollination and reproductive biology. The usual data sets such as edaphic factors, climate, and habitat preferences are important, but if conservationists are not aware of the required pollinators and/or they do not understand the sexual system, the compatibility, or the mating system, island species will only be represented in and by the present generation. That is, without providing settings appropriate for sexual reproduction – facilitated by maintenance of required pollinators and promoted by population genetics structures appropriate for successful breeding – species will disappear because plants will not thrive beyond the lifespan of current ramets. Thus knowledge of reproductive biology is not just an interesting addendum to distribution and so on. Instead, it is information that is essential for ensuring species continuance in successful conservation and/or restoration programs (e.g., Ellstrand and Elam 1993; Crawford et al. 2011, 2013; Anderson et al. 2015).

Studies of the reproductive biology on the Juan Fernández Archipelago were not initiated with our work. Skottsberg (particularly 1928) and some earlier explorers (e.g., Hemsley 1884; Johow 1896) made an impressive beginning with observations in the field of many plants in flower, including some comments on insect visits – but not many insect visits. They combined those few observations with a good deal of speculation regarding pollination. Thus most suggestions for pollination syndromes of Juan Fernández plants were based on projections of floral visitors *likely* to visit plants with a certain floral morphology and flower color. These early researchers essentially used given knowledge of floral pollination syndromes associated with pollinator classes elsewhere (as represented in the classical pollination syndromes summarized, for example, in Willmer 2011). There was virtually no work published on reproductive biology following the Skottsberg publications until we began our studies (first publication by Sun et al. in 1996).

Over the past three decades, there have been many studies providing intensive data on individual species that have led to an emerging picture of the Juan Fernández flora as a whole. These studies are valuable not only for understanding the establishment and evolution of the Juan Fernández flora but also to generate actual data from observations and experiments that provide the foundation for generalizations on island floras (e.g., Carlquist 1974, 2009). While there are a good number of pollination studies of individual species or even genera, there have been few *comprehensive* studies of whole-island floras (but cf. Abe 2006 for a comprehensive study of the pollination of Bonin Island plants). Similarly, there are many studies of breeding systems and reproductive systems of species and groups of species, but there are almost no comprehensive analyses of these phenomena. Our work on these topics for the Juan Fernández Archipelago has been presented in a series of papers (e.g., Bernardello et al. 1999, 2000, 2004; Anderson et al. 2000a, 2000b) and largely summarized in two papers, one from the *Botanical Review* (Bernardello et al. 2001) and the other from the *American Journal of Botany* (Anderson et al. 2001b). Here we review these studies and provide a comparative context. Our studies of reproductive biology have largely been conducted on Robinson Crusoe Island, and analyses presented in the text following for the Juan Fernández or the Fernandezian flora are focused on the species found on Robinson Crusoe Island. The numerical calculations are based on the 151 total taxa (i.e., 109 endemic + 42 native) recorded for Robinson Crusoe Island.

We begin by reference to the alternate name for the archipelago – the Robinson Crusoe Islands, derived from Daniel Defoe's (1719) inveterate and persevering hero. Defoe's Robinson Crusoe and his compatriot, Friday, became legendary literary characters in large part because they so cleverly "made do" with what they had at hand. They salvaged some materials from the wreckage of their ship but then had to improvise and make the best of what they had left. That seems perhaps a useful metaphor for thinking about the reproductive biology of this special flora. The pollination, if not the reproductive biology overall, of the native flora seems to have followed a "make do" paradigm. We suggest that this "Robinson Crusoe phenomenon" is an appropriate paradigm for many elements of the biota of islands in general and, specifically herein, island plants that succeed reproductively do so with the narrow genetic base the colonists brought with them (their biology), facilitated or impeded by the often strongly limited options for pollination (the ecology) present on new(er) islands (Crawford et al. 2011, 2013; Anderson and Bernardello 2012).

As indicated earlier, Skottsberg (1928) made many useful direct observations. In addition, the floral morphology of a number of species apparently suggested to him the usual (i.e., the continental) array of pollination systems. Despite Skottsberg's excellent work in so many contexts, we have not found support for the pollination diversity he suggested for the flora. With the exception of two species of hummingbirds, the archipelago is devoid of the usual pollinating agents, that is, bees, pollinating flies, butterflies, and moths. As you will read below, we concluded that the Robinson Crusoe phenomenon is manifest in anemophily (wind pollination), which serves for a large proportion of the flora. We argue that anemophily applies even to a number of species that, based on floral features, were considered "bee" or "moth" or "butterfly" or "fly" flowers based on the character syndromes from standard (and very useful first approximations) pollination biology (e.g., as listed in Willmer 2011). Generally, island faunas are smaller – including the array of animals that constitutes the pollinator fauna (e.g., Rick 1966). This conclusion is particularly appropriate for islands that are characterized by one or more of such features as relative geological youth, small size, and considerable distance from continental or other island source floras. As we pointed out in 2001 (Bernardello et al. 2001), and as you will read below, the Juan Fernández Archipelago can now be considered a, or perhaps even *the*, type exemplar of a highly restricted pollinator fauna.

We now consider three topics related to overall reproductive biology: pollination, breeding system, and the sexual systems, in particular, the extent of manifestation of dioecy. In each section we make comparisons with work done with other island systems. We use the circumscriptions of these plant reproductive biology categories (e.g., breeding system, sexual system, and a topic we do not review in depth, mating system) as proposed by Neal and Anderson (2005).

Pollination Biology

First, we consider the overall picture of pollination per se, followed by more general considerations of the floral biology. The flora is relatively small but, as pointed out, very rich in endemics, particularly in endemics per unit area, and it is a botanical opportunity and pleasure of the first order for botanists to visit and work with this flora. We have analyzed the floral morphology in the context of pollination (Bernardello et al. 2001). The flowers show the usual range of sizes (though, as for most islands, most flowers are small to very small), shapes (most are dish/bowl shapes, after Willmer 2011 and other pollination biology compendia), and colors (green is the most frequent flower color). The only other island flora where the pollination biology has been studied comprehensively (i.e., not the excellent studies of individual species or groups or useful literature compilations [e.g., Chamorro et al. 2012] but of the flora as a whole) is that of the Bonin Islands (also known as the Ogasawara Islands) (Abe 2006), with which we will make comparisons where appropriate. The fauna associated with pollination on the Juan Fernández Archipelago is even smaller than the flora; the pollinating fauna is tiny at best. In fact, save for the new bee species we discovered and believe to be very recently introduced (Engel 2000; Anderson et al. 2001a), the only animals dedicated to pollination are two species of hummingbirds. There are

Figure 12.1 Male of *Sephanoides fernandensis* (Juan Fernández firecrown) taking nectar on *Cuminia eriantha* flowers. (Photograph by Héctor Gutiérrez Guzmán.)

incidental insect *visits* to flowers recorded for some flies and moths, but these insects are of no reproductive consequence (here we make a strong and important distinction between flower "visitors" versus "pollinators," as does Willmer 2011). This contrasts with the estimated 74% of angiosperms being animal pollinated on the Bonin Islands (Abe 2006). One species, the Fernandezian endemic "picaflor" (*Sephanoides fernandensis*) (Fig. 12.1) is particularly notable as the only hummingbird species native (and endemic in this case) to an oceanic island (Colwell 1989). Both it and its sister species (*Sephanoides sephaniodes*), native to the islands and to continental South America, are important as pollinators for what we estimate is about 10% of the flora. But beyond that, the Juan Fernández angiosperms, as with Crusoe and Friday, have had to improvise – to "make do." They reproduce sexually, but without the usual aid of biotic pollinators. These conclusions are based on both our own 300+ hours of observations over several field seasons where we recorded fewer than 25 visits by native insects to flowers and Skottsberg's extensive time in the field in the early 1900s. In over six months, Skottsberg (1928) recorded insect visitors to flowers of only seven plant species (Bernardello et al. 2001).

To be sure, in the context of reproductive biology, a significant proportion of the flora is capable of self-pollination or is at least self-compatible (~80% of the flora we studied is likely self compatible), similar in this way to most other island species studied. However, most of the species that we studied are not autogamous. Instead, it seems that selfing mediated by geitonogamy is the most frequent method of pollen transfer (Anderson et al. 2001b; Bernardello et al. 2001). Thus, in terms of pollination, like most island floras, a number of the species do not necessarily need the services of animals to move the pollen among flowers or between plants. However, for species that depend on, or require, out-crossing (plants for which seed set derives only from transfer of pollen between genetically distinct individuals) or geitonogamous selfing (because herkogamy or dichogamy prevents autogamy, and thus transfer of pollen among flowers on the same individual is required), the biotic options are severely limited, basically to the two species of hummingbirds. Our team has made two detailed studies of the interdependence of endemic plants with their hummingbird pollinators (Sun et al. 1996; Bernardello et al. 2004). For one of these, that is, *Sophora fernandeziana* (Fig. C74), we found highly concentrated nectar serving as a reward for the hummingbird pollinators to this mostly self-compatible species (Bernardello et al. 2004). When *Sophora* blooms, it is the major food source for the hummingbirds (Skottsberg 1928). Thus there is a clear interdependence of the pollinator and the plant pollinated. Unfortunately, the number of *Sophora* on the islands has, and is diminishing rapidly. This is, obviously, an issue for plant conservation, but given the interdependence of plant and pollinator, it is an issue for long-term survival of the iconic hummingbirds as well (Bernardello et al. 2004; see also Hahn et al. 2009).

In terms of pollination, it is useful to record all visitors to flowers (sometimes these observations lead to breakthrough studies), but it is equally, or perhaps more, important to study the behavior of the *visitors* at least to a level sufficient to be able to predict whether floral visitors could be actual/effective floral *pollinators* (e.g., Willmer 2011; Anderson et al. 2013). Consider moths. From time to time, we observed moths on flowers (during the day – none are observed at night, though we made observations late at night and in predawn hours). But study of those uncommon moth-flower interactions showed that the moths were as likely to visit a leaf, the soil, or detritus as they were to visit another flower of the same species (e.g., Anderson et al. 2000). Thus moths *could* transmit pollen, as any insect might do that visits a flower if the pollen is presented in a way that "mess and soil" (random walk) pollination can occur (particularly in the classical "dish/ bowl" open-type of flower, combined with abundant pollen on anthers). But such transmission is incidental at best, and equating it to pollination often misrepresents and confounds conclusions about the functional pollinator fauna for a given species or in a particular area and confuses the identification of the important pollinator syndromes on an island (or anywhere for that matter) (e.g., Willmer 2011). For instance, Skottsberg (1928) promoted the cryptically dioecious *Pernettya rigida* (Fig. C71) as "entomophilous" but recorded no visitors to any flowers (Anderson et al. 2000b). In dozens of hours working with this and related species in the same area, we never observed any biotic pollinators.

Based on our observations of floral visitors, combined with the direct observations by Skottsberg (and not including the speculations based on flower morphology), we have direct, not inferential, observations on about 35% of the species. Contrary to speculations based on flower features, we have concluded that for the Juan Fernández Archipelago, there is virtually *no* effective and dependable insect pollination. The newly discovered bee (*Lasioglossum fernandenzis*) could *become* important, but this is not clear at this point (Engel 2000; Anderson et al. 2001a), and that newly discovered and newly described bee was not part of the biota of the prehistorical Juan Fernández Archipelago. Of particular interest is that our conclusion is true despite our estimates that over 50% the flora offers some sort of nectar reward (Bernardello et al. 2000, 2001; this is similar to the Bonin Islands [Abe 2006], where about 64% of the angiosperms offer nectar) and about 2% of the flora offers a pollen reward. Thus island species have the evolutionary opportunity to express a range of forms (Carlquist 1974, 2009) unfettered by the narrower niches imposed by competition on continents, and that variation is certainly manifest in some groups on the Juan Fernández Archipelago (e.g., the spectacular variety in the Asteraceae genera *Dendroseris* [Figs. C36 through C52] and *Robinsonia* [Figs. C58 through C67]) (cf. Sang et al. 1994, Crawford et al. 1998). However, insular species at the same time may well suffer the limitation of access to dedicated and diverse pollinators, species essential to pollination, but that arrive independently of the plant diaspores. However, many archipelagos, those large enough, and old enough, have at least a limited bee fauna associated with them. For example, Hawaii has a number of *Hylaeus* (Daly and Magnacca 2003); the Galápagos (Linsley 1966; McMullen 1989, 1993) and Bonin Islands (Sugiura 2008) possess a large *Xylocopa* (different endemic species for each archipelago), and the Canary Islands have a large *Bombus* (Prohens et al. 2007; Anderson et al. 2014) among many bee genera and species. With the limited exception of the new bee species described earlier, the Juan Fernández Archipelago has no bees. We suspect that the new species is a recent introduction from South America, a conclusion based on the lack of any other sightings by any of our team in countless hours in the field over many seasons or by Skottsberg in months on the islands. There are ants, and they can (infrequently) be functional in pollen movement (e.g., in *Wahlenbergia*, Figs. C68 and C69) (Anderson et al. 2000a), but they are not broadly effective pollinators on the Juan Fernández Archipelago or anywhere else. Unfortunately, the Argentine ant has been introduced to the Juan Fernández Archipelago (Ingram et al. 2006), as it has been to so much of the rest of the world. In Hawaii, this highly invasive ant has had negative effects on the populations of many arthropods, including the native *Hylaeus* bees (Cole et al. 1992). It obviously may have multiple effects on the balance of biota on the Juan Fernández Archipelago, but given the absence of insect pollinators, and the predominantly urban focus of the ants, it will not have much effect in that context.

Pollen-to-ovule (P/O) ratios can be an informative and useful indicator of the efficiency of pollen transport, with implications of biotic or abiotic pollination (e.g., Philbrick and Anderson 1987; Mione and Anderson 1992). It is thus interesting that for a number of Fernandezian species, the P/O ratios we found imply biotic pollination, or at least they are not high enough to be in the range of most known anemophilous species (Anderson et al. 2001b). Thus P/O ratios, like flower color and morphology from the whole flora, and the floral nectar measured from some species (Bernardello et al. 2000) make the picture of the reproductive biology of this flora quite intriguing. The P/O ratios, the presence of floral nectar, and a flower size and color for a number of species would seem to provide further support for the improvisational, "make do" nature of Fernandezian reproductive biology. That is, the P/O ratios and these other features are all more indicative of biotic pollination, like the proposed continental colonizing progenitors (Bernardello et al. 2006) than they are of species with a reproductive system adapted to an archipelago that lacks most biotic pollinators.

To reiterate, biotic pollination is limited to birds, and ornithophily accounts for only 10% of the diversity of the flora. Thus we have concluded that, as for many islands (Carlquist 1965, 1974), anemophily is a significant element for the native Juan

Fernandez flora as well, here responsible for pollen transfer in over 46% of the species we have studied in detail. For some species, for example, *Lactoris fernandeziana* (Bernardello et al. 1999; Fig. C26), *Wahlenbergia berteroi* (Fig. C68), and *W. fernandeziana* (Anderson et al. 2000a; Fig. C69), we have been able to demonstrate a direct effect of wind-in-pollen transmission and in others, for example, *Pernettya rigida* (Fig. C71), the highly likely role for anemophily (Anderson et al. 2000b). Abiotic pollen transfer is obviously imprecise, but, it is dependable, especially in a situation where the wind is omnipresent, as it is for many oceanic islands, including the Juan Fernández Archipelago. Comparatively, wind does not play a very prominent role in the pollination of Galápagos natives (Rick 1966; McMullen and Close 1993); there is a much more diverse biotic pollinator fauna there (McMullen 1993, 2009a, personal communication). On the subtropical Bonin Islands of Japan (~25°N, 140°E), the estimates are that about 26% of the species are anemophilous (Abe 2006). We do not yet have firm estimates for the Canary Islands (~28°N, 15°W), but that flora is much older than the Juan Fernández flora, much closer to source areas, and has abundant and diverse biotic pollinators (e.g., Garcia 2000; Olesen et al. 2002; Dupont and Skov. 2004; Anderson et al. 2006, 2014; Prohens et al. 2007; Ollerton et al. 2009; González and Fuertes 2011). Thus we have predicted that the Canary Islands will not have the same high percentage of anemophily as the Juan Fernández Islands (Anderson et al. 2013; Crawford et al. 2013).

Anemophily is abiotic and thus an inexact method of pollen distribution among plants, much less precise than insect- or bird-mediated pollen delivery. But it may have an advantage, a kind of preadaptation, to move pollen within flowers and among plants in the absence of pollinators in an environment with no or few pollinators. That is, there is considerable concern over the impact of invasives on the pollinator fauna of island plants (e.g., Moragues and Traveset 2005; Traveset and Richardson 2006; Dohzono and Yokoyama 2010). There is evidence in various studies that invasive plants may outcompete the endemics in many ways, including for the service of native pollinators. Obviously, endemics that have the capacity to reproduce via wind-distributed pollen will not suffer pollinator loss (and reduced seed set) to invasives and thus will have a much better chance of surviving the inevitable competition with invasives. Perhaps many elements of the spectacular Juan Fernández endemic flora apparently disadvantaged by a lack of biotic pollinators may actually benefit from the otherwise apparent drawbacks of the much more imprecise but omnipresent anemophily.

Over several field seasons we have had the opportunity to observe floral visitors (which we have concluded are incidental and not pollinators in virtually all cases) for about a quarter of the endemic species and to experimentally study the pollination of about a fifth of the flora (Anderson et al. 2001b, Bernardello et al. 2001, 2004). However, we have analyzed the morphology, directly or from the literature, of virtually all the native and endemic species. The full data tables are included in Anderson et al (2001) and Bernardello et al. (2001). Thus, from the literature and our own field work, we have conducted pairwise comparisons for a number of features important to reproductive biology for the flora as a whole (Bernardello et al. 2001). We do not infer from these data what biotic pollinators are likely. Instead, we used these data to look at associations

Table 12.1 Statistical Tests (Chi-Square and General Association Coefficient) of Features of the Juan Fernández Flora that Were Positively Associated

Feature	Chi-square	General association	*p*-Value
Inconspicuous and green flowers	48.50	48.17	<0.0001
Very small and green flowers	50.09	49.77	<0.0001
Inconspicuous and very small flowers	118.47	117.69	<0.0001
Very small flowers and perennial herbs	31.65	31.44	<0.0001
Very small flowers and current wind pollination	24.49	24.21	<0.0001
Very small/small flowers and current wind pollination	40.27	39.81	<0.0001
Inconspicuous flowers and current wind pollination	27.41	27.10	<0.0001
Green flowers and current wind pollination	6.22	6.14	0.01
Green/brown flowers and current wind pollination	47.03	46.53	<0.0001
Perennial herbs and current wind pollination	13.96	13.80	0.0002
Trees and wind pollination of colonizers	4.76	4.72	0.03
Very small flowers and wind pollination of colonizers	36.31	35.88	<0.0001
Green flowers and wind pollination of colonizers	31.02	30.65	<0.0001
Inconspicuous flowers and wind pollination of colonizers	48.43	47.86	<0.0001
Large-/medium-sized flowers and current bird pollination	25.50	25.20	<0.0001
Bright colored flowers and current bird pollination	59.16	58.44	<0.0001
Bright colored flowers and bird pollination of colonizers	44.82	44.27	<0.0001
Bell-shaped and white flowers	12.07	11.95	0.0005
Dish-shaped flowers and insect pollination of colonizers	26.51	26.20	<0.0001
White flowers and insect pollination of colonizers	24.77	24.46	<0.0001
Hermaphroditic and white flowers	13.17	13.07	0.0003
Monoecious and green flowers	17.21	17.09	<0.0001

Note: In all cases, degrees of freedom = 1. p = probability for both chi-square and general association coefficient.

between reproductive characters. We include here a summary table (Table 12.1) of some of the highly significant associations between features important in pollination. For context, we consider similarities/differences in the analyses to the comprehensive study of pollination on the Bonin Islands (Abe 2006). The Bonin Islands are similar in being oceanic, of volcanic origin, and a great distance from continental areas. They are, however, 10 times older than the Juan Fernández Islands, there are 10 times more islands, and they are subtropical.

Most of the Fernandezian flora is composed of perennials. There are more small and inconspicuous flowers (41%) than any other size category, there are more green flowers than any other color (41%), and most flowers are hermaphroditic (70%). Though the flowers of a significant percentage of the Bonin flora are also small (about 39%) with green or white flowers (58%), it is clear that small green/brown/white and nondescript flowers are perhaps more emblematical of the Juan Fernández flora. *Lactoris fernandeziana* (Fig. C26), the endemic monotype of the family Lactoridaceae, constitutes a notable exemplar of many of these features (Stuessy et al. 1998c). Some of the associations between characters on the Juan Fernández Islands are similar to those from continental floras, where small flowers are often green colored, inconspicuous, and wind pollinated. In addition, a number of perennial herbaceous plants are wind pollinated. We speculate that the bulk of the species with the small green/white flowers are wind pollinated. Obviously, those that are bird pollinated are characterized by more brightly colored flowers. We also found a strong association between green flowers and monoecy; the meaning of this association is not clear; perhaps it is a spurious correlation derived from the relatively small sample size the Fernandezian flora constitutes.

Our studies on the Juan Fernández Archipelago are not the first to confirm the association among very small, inconspicuous, and green flowers on islands (e.g., Carlquist 1974; Ehrendorfer 1979). However, a comparison with purported ancestral colonizers implies that there has not been selection, or at least not strong selection, for change in flower size or color among some of the species with larger or brightly colored flowers. The latter persist in the flora, some of them with nectar and many with typically biotic-syndrome P/O ratios. So what is it that leads the species with small, green, and inconspicuous flowers to predominate in the flora (and not just of the Juan Fernández Archipelago but, to the extent we know other insular floras, for them as well (e.g., Carlquist 1974)? This is a matter of considerable interest for further study, but obviously there are likely associations among dispersal and establishment capabilities and some of these features. This small-green-flowered group of species does not show a higher percentage of self-compatibility than other species (Anderson et al. 2001b; Bernardello et al. 2001). And given the lack of all but bird biotic pollinators, it also cannot be argued that the small, simple, nonrestrictive flower shapes and muted colors facilitate pollinator service by a larger spectrum of pollinators (i.e., nondescript flowers with an open-morphology could be pollinated by any group). In fact, Rick (1966) suggested the opposite for the Galápagos, where he thought that the lack of entomophilous-mediated selection may have led to the preponderance of small, dull-colored flowers. These floral features are associated with wind pollination in continental situations. Thus colonization by such species to the Juan Fernández Archipelago may at least have preadapted them to wind pollination, in essence to the "make do" pollination setting in which they persist. And wholly speculatively, perhaps some of the earlier colonists with brighter-colored, larger, biotically pollinated flowers failed to establish on an archipelago without insect pollinators. Thus the current flora has a larger representation of small, dull-colored, wind-pollinated survivors.

Breeding System: Compatibility

Although there is a clear perception that self-compatibility is a key element of island breeding systems (e.g., Baker 1955, 1967; Carlquist 1974; Pannell 2015) and, furthermore, that it is a fundamental component for plant species establishment, there have been very few studies of island plant breeding systems – at least in a comprehensive way. A number, really many, of species have been studied from several archipelagos, but for almost no insular floras has a substantial portion of species been studied. And there is a good reason for this: it is much easier to document the pollinators of the flora than it is to document the breeding system (see definitions and distinctions of breeding and mating systems in Neal and Anderson 2005). Pollinators can be proposed by careful observation of flowers and visitor behavior and well done with observations and analyses of pollinator pollen loads and depositions on stigmata. Documentation of breeding systems requires multiple (at least three) visits to the same plants to (1) remove open flowers and bag flowers, (2) conduct controlled crosses, and (3) harvest and fix flowers for analysis of pollen tube growth (e.g., Anderson et al. 2001b) or, even better, to return much later to score fruit and seed set (as in Rick 1966; McMullen 2007). We have studied all the species accessible to us, and in flower, over several field seasons on Robinson Crusoe Island. Thus we have been able to report on the breeding system for about a third of the 71 endemic species of the Robinson Crusoe Island (Anderson et al. 2001b; Bernardello et al. 2001). From these studies we have documented that over 80% of the 25 species we tested are self-compatible. This is not surprising, but it is important. As noted earlier, although high self-compatibility is expected for islands, the breeding system for significant portions of island floras is not widely documented. For the theory associated with the establishment, evolution, and radiation of insular floras to be most accurate and applicable, it needs support from data from many archipelagoes. Data regarding physical factors, for example, island size, age, distance from sources, and position relative to wind and water currents, need to be combined with dependable data regarding pollinators (versus just floral visitors) as well as breeding systems, compatibility, and ultimately, mating systems (Crawford et al. 2015). Understanding "island biology" becomes more important virtually every day not only for oceanic islands but also for the urgency of understanding continental floras as well as more and more of the world's native biota becomes effectively island-like as a result of habitat loss and disruption of contiguous distributions; this a concept dating back at least to Wallace in 1880. The only other insular flora for which there is a broad survey at this point is for the Galápagos. There, McMullen (1987, 1990, 2009a, 2009b, personal communication) and Rick (1966) before him estimated that a similar percentage of species (~80%) are self-compatible (from an analysis of nearly 70 species; however, the surveyed group included introduced species along with endemics and natives). In a more recent compilation (that includes Rick's and McMullen's work), Chamorro et al. (2012) estimate that 80% of the 55 native species they studied (10% of the Galápagos flora) are self-compatible (and that ~94% of those selfers are autogamous as well). As noted earlier, the pollination of a large number of the species on Bonin has been carefully recorded (99 of 269 spp., 37%), but the breeding system for the species has not been studied as exhaustively. Similarly, a number of species have been carefully studied for Hawaii (e.g., Wagner et al. 2005), but there are no comprehensive estimates of the breeding system for the flora as a whole.

As has been pointed out elsewhere (Crawford et al. 2013; Pannell 2015), it is important to keep in mind that the breeding systems at this point, when we are assessing such systems on various archipelagoes, are likely, but not necessarily, the breeding systems that plants manifested as initial colonists. This is, of course, most important in old archipelagoes (e.g., such as the Canaries) (Crawford et al. 2013) versus younger archipelagoes (such as the Juan Fernández). In the former, the breeding system (perhaps self-compatibility and autogamy) that helped one to a few colonizing plants to become established could be very different from the breeding system manifest several hundred thousand (or even a million or more) years later in the extant flora where competition within a more complex flora and the consequences of inbreeding depression are of significant evolutionary importance. Furthermore, more detailed experimental studies need to be conducted on many more island species to determine how even the systems where outbreeding, induced by self-incompatibility or dioecy, is seemingly required. As detailed studies are performed, there is increasing evidence of "leakiness" in both self-incompatibility (Crawford et al. 2015) and dioecy (Anderson et al. 2015). Such a leaky system may constitute evolution in process (from systems with self-crossing) or a system that is very well suited to both (1) colonization and (2) subsequent expansion into multiple habitats and radiation.

Dioecy

Most of the Juan Fernández angiosperm species are hermaphroditic flowered (~70%, which compares with ~75% for the Bonin Islands) (Abe 2006). But it is the percentage of dioecy on island systems that has and continues to be a topic of considerable interest in large part because island floras tend to have or are thought to have (there are more comprehensive studies of this) a larger percentage of dioecious members than continental floras. The occurrence of dioecy worldwide is estimated at about 6% (Renner and Ricklefs 1995; Renner 2014). Almost all island systems have figures significantly larger than this (estimates up to 15% to 20% for the outliers) (Abe 2006 presents a compilation of these data). We know the flora of the Juan Fernández Archipelago well and estimate that 9% of the species are dioecious (and another 9% are monoecious and 7% gynomonoecious, with a single gynodioecious species and a few andromonoecious species among the sedges and composites) (Bernardello et al. 2001). This is greater than the Canary Islands (estimated at 3%) but less than the 16% estimated for the Galápagos (Chamorro et al. 2012), 15% for Hawaii (Sakai et al. 1995), and 13% for the Bonin Islands (Abe 2006). Clearly, these percentages are important to know because much of island theory regarding breeding/mating systems relates to speculation on establishment of an outcrossing program that will compensate for the genetic uniformity proscribed by limited propagule establishment, accompanied by self-compatibility. And, of course, conservation programs and restoration efforts need to be aware of a reproductive feature that requires two or more plants.

The figures for dioecy both for the Juan Fernández flora and for many of the other island floras studied/to be studied around the world are likely underestimates in another context. That is, dioecy is not always obvious and is often discovered only following very detailed, manipulative experiments (often involving garden or greenhouse-grown plants). Thus, for instance, for the Juan Fernández Archipelago, we reported that previously unsuspected dioecy characterizes *Pernettya rigida* (Anderson et al. 2000b) (Fig. C71). This matches reports of previously undocumented dioecy from the Canary

Islands (*Withania*, Solanaceae) (Anderson et al. 2006) and the Virgin Islands (*Solanum*, Solanaceae) (Anderson et al. 2015).

Finally, it is worth reinforcing the brief discussion earlier that initial colonists *can be dioecious*, in particular if the dioecy is "leaky" (first described by Baker and Cox 1984). Increasingly, as insular taxa are studied carefully, not only are new dioecious species discovered, but "leakiness" in expression of the dioecious system is documented (e.g., Anderson et al. 2015). The occasional hermaphroditic flower (often produced by the otherwise all-male plants in our experience), coupled with self-compatibility, allows a very small population of invaders to constitute successful colonists, and colonists with a reproductive system that de facto promotes outbreeding (to avoid some elements of inbreeding depression from obligate selfing that Carlquist 1966 was concerned with). As noted earlier, "leakiness" in the compatibility system is being documented, increasingly, in self-incompatibility systems as well (e.g., Crawford et al. 2009a, 2015). Such leakiness means that species that are self-incompatible may also qualify as good colonists, with a system that both allows effective colonization (leakiness) and promotes outcrossing (to minimize inbreeding depression).

Conclusions

The Juan Fernández flora is not large. It is notable in terms of the proportion of endemics per unit area (the greatest for any island archipelago) (Lowry 2009) and for unusual species (e.g., *Lactoris* and the many unusual composites in *Dendroseris* and *Robinsonia*) (see Chapters 5 and 13). In terms of reproductive biology, it is notable for several reasons. The level of dioecy is not particularly high, higher than that on the Canaries and significantly less than in Hawaii. As studies move beyond the necessary alpha-systematic and alpha-ecological first steps to careful in-depth analyses of various taxa, our understanding of breeding systems and of the incidence of dioecy will change. The pollination is notable for two reasons. First, it is notable because bird pollination, hummingbirds in this case, is prominent, serving some 10% of the flora. As noted earlier, one of the hummingbirds is the only hummingbird endemic to an oceanic island (Fig. 12.1). Second, it is notable because other than the hummingbirds, *there are virtually no animal pollinators*. The impressive pioneering work by Skottsberg cited a number of likely zoophilous flowers in the flora, but given that there are virtually no bees (though the newly discovered and newly described species *might* become more important) or flower-visiting Lepidoptera or Diptera, animal pollination is restricted to the ornithophilous species. Thus we have concluded that the island-ubiquitous autogamy or geitonogamy, and anemophily, are the main sexual reproductive mechanisms. That lack of potential animal pollinators, coupled with a relatively young flora, has led to another particularly notable feature: many more of the island endemics and natives have retained features that are similar to their presumed ancestors than might be expected. To be sure, a significant portion of the flora fits the usual island model that Carlquist (1974) recognized, where many of the flowers are small, green, and, at least in a pollination sense, not showy. However, a number of the features would seem to manifest retention of features of the colonizing forms (such things as larger brightly colored corollas, the expected pollen-to-ovule ratios for biotically pollinated taxa, and the presence of nectar – all characteristic of zoophilous pollination) presumably as a result of the lack of selection among species that are autogamous or geitonogamous and/or are wind pollinated. The flora of these little islands, the Juan Fernández or Robinson Crusoe Islands, is indeed interesting in so many ways. And it seems, at least for the reproductive biology, to manifest an ability to "make do," to improvise success and sustainability, characteristics shared with its alternative namesake from the literature.

Part VI

Evolutionary Processes

The focus of our investigations in the Juan Fernández Islands over the past 35 years has been, and continues to be, the evolution of the endemic angiosperms. Evolutionary aspects include numerous dimensions, but they all begin with phylogeny. To be able to understand any aspect of the evolutionary process, an appreciation of relatedness among taxa is paramount. It is obviously impossible to unravel modes of speciation if we are not comparing species that have a joint evolutionary history. For this reason, we have earlier placed emphasis on phylogenetic reconstruction based on morphology but more recently using molecular markers, especially DNA sequences. These data are presented in Chapter 13 as a series of models to allow comparisons among genera.

Evolution begins with differentiation within populations, and different markers, such as isozymes, AFLPs, and microsatellites (SSRs), are needed to reveal genetic processes. Because endemic plants of oceanic islands tend to be quite similar in their genetic profiles, population genetics markers have also helped to clarify phylogenetic relationships in many cases. These population data, given in detail in Chapter 14, are also essential for testing hypotheses regarding modes of speciation.

Speciation is the end result of evolutionary processes acting at the population level that conclude with reproductive isolation. Modes of speciation in oceanic islands tend to be less diverse than in continental regions, and evolution via chromosomal change is very rare. During the course of investigations on patterns of phylogeny in the archipelago, however, it became clear that a number of species in the islands had evolved by simple transformation, or anagenesis. This has led to a comparison of genera with molecular markers that have species originating by transformation in contrast with those evolving via splitting (cladogenesis). The latter process is typical of many island groups and has often led to adaptively radiated complexes. These results, only recently published, are reviewed in detail in Chapter 15.

13 Patterns of Phylogeny

Tod F. Stuessy, Daniel J. Crawford, and Eduardo A. Ruiz

The reconstruction of phylogeny is one of the first steps to understanding patterns and processes of evolution in any group of plants in any part of the world. Without a framework of relationships that show relative affinities of organisms, it is impossible to infer modes of speciation and completely inefficient to attempt to test hypotheses. Long recognized as a significant part of evolutionary biology since Darwin (1859), only in the past 30 or more years have explicit methods of phylogeny reconstruction become available. This has now led to numerous approaches for phylogenetic inference (Felsenstein 2004), so much so that the challenges have become more statistical and mathematical (e.g., Paradis 2006) instead of strictly organismal. Fueling many of these advances in phylogeny reconstruction has been the outpouring of huge quantities of nucleotide data now more easily available from next-generation sequencing (NGS) techniques (Soltis et al. 2013; Hörandl et al. 2015). The challenge has become computational, to assess relationships from among the millions of available base pairs and to develop new avenues of phylogeny reconstruction. These approaches establish precise informational patterns that can be used to infer biogeographical and evolutionary processes.

The application of phylogeny reconstruction to understanding relationships in oceanic islands has many obvious benefits to interpreting speciation and biogeography. The great advantage is that in archipelagos that are geologically dated, which is the case now for most of the islands on Earth, a time frame or sequence sets up a hypothesis of biogeographical probabilities such that the older island in an archipelago should be colonized first, before the younger islands were formed. As the archipelago develops geologically, the flora continues to migrate to the new islands, setting up a biogeographical progression. This is the "progression rule" that has been so well elaborated in the Hawaiian Archipelago (Funk and Wagner 1995). This biogeographical directionality also suggests hypotheses of evolutionary directionality such that the most ancient species in the chain of islands would be expected to occur on the oldest island. Many exceptions to this rule exist, of course, but it serves as an extremely powerful backdrop in which to offer biogeographical and evolutionary hypotheses for testing. The Juan Fernández Archipelago is particularly well positioned in this regard because the island (Robinson Crusoe) closest to the major source area (South America) is also the oldest, making it the site of initial colonization for most groups.

The greatest concentration of endemic taxa occurs on the older island, Robinson Crusoe (Table 13.1). A total of 90 endemic taxa occurs on this island, 71 angiosperms and 19 ferns. In contrast, 66 taxa occur on Alejandro Selkirk Island, 47 angiosperms and 19 ferns; of the latter, only seven taxa occur *only* on Alejandro Selkirk Island. Explanation of the greater diversity on the older island probably relates to its geological age and its proximity to the South American continent. Although the major islands are presently about the same area and even nearly the same elevation (Alejandro Selkirk reaches 1,319 m and Robinson Crusoe 915 m), it is believed that Robinson Crusoe Island was previously much larger, perhaps 95% more (Sanders et al. 1987; see also Chapter 3). Hence Robinson Crusoe Island has been available for colonization longer, it has been closer to the primary source region (South America), and it was undoubtedly much more ecologically diverse when it originated, perhaps even to 3,000 m elevation. Habitat diversity is a major factor for species diversity in any ecosystem (Hortal et al. 2009).

Phylogeny reconstruction has been central to all biogeographical and evolutionary studies in the Juan Fernández Archipelago. Our earliest explicit phylogenies in the 1980s were based on parsimony analysis with morphological data, followed by nrITS sequences and chloroplast markers in the 1990s. Because many species within genera of oceanic islands are genetically quite similar, population genetic markers of isozymes, AFLPs, and SSRs have also been emphasized, particularly beginning in the late 1980s. Explicit phylogenetic information in our investigations has now been obtained for the following genera: *Berberis*, *Coprosma*, *Cuminia*, *Dendroseris*, *Drimys*, *Erigeron*, *Gunnera*, *Myrceugenia*, *Nothomyrcia*, *Peperomia*, *Rhaphithamnus*, *Robinsonia*, *Sophora*, and *Wahlenbergia*. Together these involve 48 species, or 46% of the endemic angiosperms of the archipelago. In addition, many other laboratories using nucleotide data have been investigating genera that contain one or more of the island endemics, which adds considerably to the phylogenetic information.

The aim of this chapter is to summarize all available data relating to phylogenetic relationships among endemic species of vascular plants plus continental relatives. We seek to sketch all patterns of distribution that exist in the archipelago, to determine geographical sources for origin of the endemics (as a correlate of these data also presented in Chapter 16), to document phylogenetic patterns and processes for all endemic taxa, and to synthesize information regarding speciation (complementary to Chapter 15) and biogeographical events in the entire flora. Although not enough data are in hand to explain definitively all these aspects for all endemic species, the available information does allow an overview of these patterns and processes that is the most complete for the vascular flora of any oceanic archipelago.

Patterns of Distribution

To understand phylogeny in the Juan Fernández Archipelago requires seeing clearly the patterns of distribution of the endemic taxa of ferns and angiosperms in the two islands. This involves listing all the endemic taxa in the flora at the specific and infraspecific levels (Table 13.1) so that patterns can be seen. These data also allow statistics of endemism to be summarized (Table 13.2). For the endemic ferns, there is one family (Thyrsopteridaceae), one genus, 25 species, two subspecies, and one variety. For the

Table 13.1 Endemic Species, Subspecies, and Varieties of Vascular Plants in the Juan Fernández Archipelago Showing Island Distributions, Geographical Origin, and Modes of Speciation

Endemic species	Distribution	Geographic origin	Mode of speciation
ANGIOSPERMS			
Endemic genera			
Centaurodendron dracaenoides	RC	SA	C
C. palmiforme	RC	SA	C
Cuminia eriantha	RC	SA	C
C. fernandezia	RC	SA	C
Dendroseris berteroana	RC	WP	C
D. gigantea	AS	RC	A
D. litoralis	RC	WP	C
D. macrantha	RC	WP	C
D. macrophylla	AS	RC	A
D. marginata	RC	WP	C
D. micrantha	RC	WP	C
D. neriifolia	RC	WP	C
D. pinnata	RC	WP	C
D. pruinata	RC	WP	C
D. regia	AS	RC	A
Juania australis	RC	SA	NA
Lactoris fernandeziana	RC	SA	NA
×*Margyracaena skottsbergii*	RC	SA	H
Megalachne berteroana	RC	SA	C
M. masafuerana	AS	RC	A
M. robinsoniana	RC	SA	C
Nothomyrcia fernandeziana	RC	SA	A
Podophorus bromoides	RC	SA	A
Robinsonia berteroi	RC	SA	C
R. evenia	RC	SA	C
R. gayana	RC	SA	C
R. gracilis	RC	SA	C
R. macrocephala	RC	SA	C
R. masafuerae	AS	RC	A
R. saxatilis	RC	SA	C
R. thurifera	RC	SA	C
Selkirkia berteroi	RC	Unclear	A
Yunquea tenzii	RC	SA	C
Endemic species			
Acaena masafuerana	AS	SA	A
Agrostis masafuerana	AS	SA	A
Apium fernandezianum	RC	WP	A
Azara serrata var. *fernandeziana*	RC	SA	A
Berberis corymbosa	RC	SA	A
B. masafuerana	AS	RC	A
Boehmeria excelsa	RC	SA	A
Cardamine kruesselii	AS	SA	A
Carex berteroniana	AS, RC	SA	C
C. fernandezensis	RC	SA	C

Table 13.1 (cont.)

Endemic species	Distribution	Geographic origin	Mode of speciation
C. stuessyi	AS	SA	A
Chenopodium crusoeanum	RC	SA	C
C. nesodendron	AS	RC	A
C. sanctae-clarae	SC	SA	C
Chusquea fernandeziana	RC	SA	A
Colletia spartioides	RC	SA	A
Coprosma oliveri	RC	WP	C
C. pyrifolia	AS, RC	WP	C
Drimys confertifolia	AS, RC	SA	A
Dysopsis hirsuta	RC	SA	A
Erigeron fernandezia	AS, RC	SA	C
E. ingae	AS	SA	C
E. luteoviridis	AS	SA	C
E. rupicola	AS	SA	C
E. stuessyi	AS	SA	C
Eryngium bupleuroides	RC	Unclear	C
E. ×fernandezianum	RC	RC	H
E. inaccessum	RC	Unclear	C
E. sarcophyllum	AS	Unclear	A
Escallonia callcottiae	RC	SA	A
Euphrasia formosissima	AS	SA	A
Galium masafueranum	AS	SA	A
Gavilea insularis	AS	SA	A
Greigia berteroi	RC	SA	A
Gunnera bracteata	RC	SA	C
G. masafuerae	AS	SA	A
G. peltata	RC	SA	C
Haloragis masafuerana			
var. *asperrima*	AS	RC	C
var. *masafuerana*	AS	RC	C
H. masatierrana	RC	WP	A
Luzula masafuerana	AS	SA	A
Machaerina scirpoidea	RC	SA	A
Margyricarpus digynus	RC	SA	A
Myrceugenia schulzei	AS	SA	A
Nicotiana cordifolia			
subsp. *cordifolia*	AS	SA	A
subsp. *sanctaclarae*	SC	AS	A
Ochagavia elegans	RC	SA	A
Peperomia berteroana subsp. *berteroana*	AS, RC	RC	C
P. margaritifera	RC	SA	A
P. skottsbergii	AS	RC	C
Pernettya rigida	AS, RC	SA	A
Plantago fernandezia	RC	Unclear	A
Ranunculus caprarum	AS	SA	A
Rhaphithamnus venustus	AS, RC	SA	A
Santalum fernandezianum	RC	WP	A

Table 13.1 (cont.)

Endemic species	Distribution	Geographic origin	Mode of speciation
Solanum fernandezianum	RC	SA	A
Sophora fernandeziana			
var. *fernandeziana*	RC	SA	C
var. *reedeana*	RC	SA	C
S. masafuerana	AS	RC	A
Spergularia confertiflora			
var. *confertiflora*	AS, RC, SC	SA	C
var. *polyphylla*	AS, RC	SA	C
S. masafuerana	AS	RC	A
Ugni selkirkii	RC	SA	A
Uncinia aspericaulis	AS	SA	A
U. costata	AS	SA	Unclear
U. douglasii	AS, RC	SA	Unclear
U. macloviformis	AS	SA	A
Urtica glomeruliflora	AS, RC	SA	C
U. masafuerae	AS	SA	C
Wahlenbergia berteroi	RC, SC	AS	A
W. fernandeziana	RC	Unclear	C
W. grahamiae	RC	Unclear	C
W. masafuerae	AS	RC	C
W. tuberosa	AS	RC	C
Zanthoxylum externum	AS	RC	A
Z. mayu	RC	SA	A
FERNS			
Endemic genus			
Thyrsopteris elegans	AS, RC	SA?	NA
Endemic species			
Argyrochosma chilensis	AS, RC	SA	A
Arthropteris altescandens	AS, RC	WP	A
Asplenium macrosorum	AS, RC	SA?	A
A. stellatum	AS, RC	SA?	A
Blechnum cycadifolium	AS, RC	SA	A
B. longicauda	AS	SA	A
B. mochaenum var. *fernandezianum*	AS, RC	SA	A
B. schottii	AS, RC	SA	A
Dicksonia berteroana	RC	WP	A
D. externa	AS	RC	A
Hymenophyllum rugosum	AS, RC	SA	A
Megalastrum glabrius	AS	RC	C
M. inaequalifolium	RC, SC	SA	A
M. masafuerae	AS	RC	C
Ophioglossum fernandezianum	RC	SA	A
Pleopeltis ×cerro-altoensis	RC	RC	H
Polyphlebium ingae	RC	SA?	A
P. philippianum	RC	Unclear	A
Polypodium intermedium			

Table 13.1 (cont.)

Endemic species	Distribution	Geographic origin	Mode of speciation
subsp. *intermedium*	RC	SA	A
subsp. *masafueranum*	AS	RC	A
P. masafuerae	AS	SA?	A
Polystichum tetragonum	AS, RC	WP	A
Pteris berteroana	AS, RC	SA	A
Rumohra berteroana	AS, RC	SA	A
Sticherus lepidotus	AS	SA	A

Note: Distribution: AS, Alejandro Selkirk Island; RC, Robinson Crusoe Island; SC, Santa Clara Island. Sources of origin: WP, western Pacific; SA, South America; AS and RC, here referring to interisland origins. Mode of speciation: A, anagenesis; C, cladogenesis; H, hybridization; NA, not applicable because the taxon is ancient and did not speciate in the archipelago.

Table 13.2 Numbers of Endemic Taxa of Ferns and Angiosperms in the Flora of the Juan Fernández Archipelago

	Family	Genera	Species	Subspecies	Varieties
Ferns	1	1	25	2	1
Angiosperms	1	12	104	2	7
Totals	2	13	129	4	8

endemic angiosperms, there is one family (Lactoridaceae), 12 genera, 104 species, two subspecies, and seven varieties. The distribution of these endemic taxa on the major island is also given in Table 13.1, as well as their origins and modes of speciation, which will be discussed in the next sections of this chapter.

The endemic taxa in the archipelago are distributed between the islands in 19 different patterns (Table 13.3), and the genera within which these taxa occur are given in Table 13.4. With 134 total endemic specific and infraspecific taxa, it is of interest to learn which patterns are prevalent, which islands have the greatest diversification, and what differences can be seen between ferns and angiosperms.

From an overall assessment of the patterns of distribution, it can be seen that most genera harbor only one or two endemic species either in one or both of the islands (Table 13.5). In the ferns, 63% of the genera have only one endemic species and 25% have two, making up 88% of the ferns in the archipelago. For angiosperms, the pattern of only one species occurs at nearly the same frequency, with 62%. With two species, it drops to 13%. This means that speciation in the islands, in general, has not led to accumulations of large adaptively radiated species complexes, which are so common in other archipelagos such as Hawaii. Most speciation is transformational, yielding one endemic species, or perhaps via different modes of transformation or splitting, resulting in two island endemics. A few exceptions can be noted in the angiosperms, however, with *Dendroseris* (Asteraceae) with 11 endemic species and *Robinsonia* (Asteraceae) with eight.

Table 13.3 Patterns of Distribution of Endemic Taxa within Genera in the Vascular Flora of the Juan Fernández Islands

	Pattern		Number of genera with pattern		Percent genera with pattern	
	AS	RC	Ferns	Angiosperms	Ferns	Angiosperms
1		A	2	22	13	40
2	A	A	7	3	44	5
3	A		1	9	6	16
4	B	A	2	3	13	5
5	A B	A	0	1	0	2
6		A B	1	2	6	4
7	A	A B	0	1	0	2
8	A B	A B	1	0	6	0
9	A B	A C	0	2	0	4
10	C	A B	0	4	0	7
11	A B	C	1	1	6	2
12	A B C	A B	0	1	0	2
13	A	B C D	0	1	0	2
14	A B C D	A	0	1	0	2
15	A B C D	A B C	1	0	6	0
16	A B	C D E	0	1	0	2
17	A D B E C	A	0	1	0	2
18	A	B E H C F D G	0	1	0	2
19	A B C	D G J E H K F I	0	1	0	2
	Totals		16	55	100	101

Note: A, B, C, etc. refer to endemic taxa within a single genus.
Source: Modified from Stuessy et al. (1990).

Table 13.4 Patterns of Distribution of Endemic Taxa within Genera of Vascular Plants of the Juan Fernández Archipelago

Pattern of distribution	Genera	
	Ferns	Angiosperms
1	*Ophioglossum*, *Pleopeltis*	*Apium*, *Azara*,[a] *Boehmeria*, *Chusquea*, *Colletia*, *Dysopsis*, *Escallonia*, *Greigia*, *Juania*, *Lactoris*, *Macherina*, ×*Margyracaena*, *Margyricarpus*, *Nothomyrcia*, *Ochagavia*, *Plantago*, *Podophorus*, *Santalum*,[c] *Selkirkia*, *Solanum*, *Ugni*, *Yunquea*
2	*Argyrochosma*, *Arthropteris*, *Hymenophyllum*, *Polystichum*, *Pteris*, *Ruhmora*, *Thysopteris*	*Drimys*, *Pernettya*, *Rhaphithamnus*
3	*Sticherus*	*Acaena*, *Agrostis*, *Cardamine*, *Euphrasia*, *Galium*, *Gavilea*, *Luzula*, *Myrceugenia*, *Ranunculus*
4	*Dicksonia*, *Polypodium*[b]	*Berberis*, *Nicotiana*,[b] *Zanthoxylum*
5		*Urtica*
6	*Polyphlebium*	*Centaurodendron*, *Cuminia*
7		*Coprosma*
8	*Asplenium*	
9		*Carex*, *Peperomia*[b] (in part)
10		*Chenopodium*, *Gunnera*, *Megalachne*, *Sophora*[a] (in part)
11	*Megalastrum*	*Haloragis*[a] (in part)
12		*Spergularia*[a] (in part)
13		*Eryngium*
14		*Uncinia*
15	*Blechnum*[b] (in part)	
16		*Wahlenbergia*
17		*Erigeron*
18		*Robinsonia*
19		*Dendroseris*

Note: Refer to Table 13.3 for the patterns.
[a] Pattern seen at the varietal level.
[b] Subspecific level.
[c] Extinct.
Source: Modified from Stuessy et al. (1990).

Another obvious point that arises from the patterns of distribution is the striking difference between ferns and angiosperms. The number of endemic angiosperm taxa (108) is more than four times that of the endemic ferns (26) (Table 13.2). More diverse patterns exist among the angiosperm genera (Table 13.3), with 11 unique patterns in angiosperms and only two unique ones among ferns. In fact, the data in Table 13.3 show that the patterns found among the fern genera are restricted to pattern numbers 1, 2, 3, 4, 6, 8, 11, and 15. In general, ferns show lower levels of endemism than do seed plant taxa

Table 13.5 Numbers and Percentages of Genera Containing Different Numbers of Endemic Taxa in the Vascular Flora of the Juan Fernández Islands

		Genera			
		Ferns		Angiosperms	
Numbers of endemic species	Pattern (from Table 13.3)	Number	%	Number	%
1	1, 2, 3	10	63	34	62
2	4, 5, 6, 7, 8	4	25	7	13
3	9, 10, 11, 12	1	6	8	15
4	13, 14, 15	1	6	2	4
5	16, 17	0	0	2	4
8	18	0	0	1	2
11	19	0	0	1	2

Source: Modified from Stuessy et al. (1990).

(Smith 1993). One would expect, therefore, fewer different patterns of distribution and phylogeny among the fern representatives. The main message from these data is that once ferns arrive in the archipelago, which is relatively easy with air-dispersed spores (Tryon 1971; Kramer 1993; Driscoll and Barrington 2007), they tend to also disperse to the other island, assuming that it has already been formed and is available for colonization. The majority pattern of distribution of ferns in the archipelago therefore is for endemic species and infraspecific taxa to occur on both islands (56%) rather than having one on each island (44%). For angiosperms, only 35% of genera have taxa on both islands (Table 13.3). These data suggest that most ferns have had colonists that arrive in the archipelago and transform into a new species, either remaining on the older island or usually dispersing to the younger island, but without additional divergent speciation. The more powerful dispersal capability of ferns is most likely the explanation for these differences in patterns of distribution, not only for the ability to move across the landscape and from island to island, but also as a genetic unifier that interferes with the formation of geographically isolated populations that might diverge further leading to speciation. Continuing spore dispersal within and among islands helps to maintain gene flow among populations and can place a brake on geographical speciation, even when populations are isolated on different islands. This same point has also been suggested for spore-producing bryophytes (Vanderpoorten et al. 2008).

Modes of Speciation

The patterns of distribution and phylogeny presented in this chapter provide a spatial and temporal frame in which to infer modes of speciation. Details of the processes of speciation, especially at the genetic level, are given in Chapter 15, but the phylogenetic patterns presented in this chapter allow distinctions into two major mechanisms:

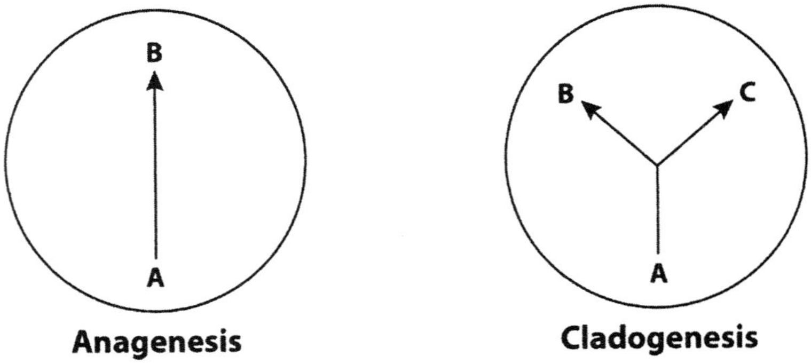

Figure 13.1 Diagrams of patterns of phylogeny (and inferred modes of speciation) in the endemic flora of the Juan Fernández Archipelago: (A) ancestral immigrant population; (B, C) derivative species. The models show single islands, but the processes apply to the entire archipelago (i.e., both major islands).

cladogenesis and anagenesis (Fig. 13.1). Cladogenesis involves the origin of new species by splitting events, that is, by geographical separation that leads eventually to morphological and genetic differences among populations at the specific level. In oceanic islands, with their often diverse habitats, this process is usually accompanied by adaptive radiation (see Chapter 15). Anagenesis is a form of transformational speciation or phyletic evolution (Simpson 1953) whereby a colonist arrives and successfully establishes in an island but does not disperse widely into different ecological zones and therefore does not adaptively radiate. Genetic change occurs via mutation, recombination, and drift, and gene flow over the landscape keeps the population unified. This is particularly the mode of choice for colonists that arrive on an island that is ecologically homogeneous, such as in Ullung Island, Korea (Sun and Stuessy 1998). A third mechanism, anacladogenesis, has also been recognized as an intermediate type of pattern (Stuessy et al. 1990) whereby divergence leads to a splitting of a derivative species, but the initial lineage survives (i.e., does not go extinct, as in strict cladogenesis). This might also be viewed as the beginning of cladogenesis.

These terms refer to both patterns of phylogeny and inferred modes of speciation. It is useful to describe these patterns and summarize their import for the archipelago as a whole. This was done previously in Stuessy et al. (1990), but more data are now available to allow more precise assessments. Vaux et al. (2016) have opined that the terms cladogenesis and anagenesis are unnecessary to describe evolutionary phenomena largely because other ways of describing lineages, splits, or divergence already exist. In some ways, this is quite true. Another viewpoint stresses that all speciation is anagenetic (Victor Rodych, personal communication). Despite these alternative perspectives, we believe that the distinction between these two contrasting modes of speciation in oceanic islands is very helpful as a way to organize information about evolution and phylogeny in endemic species, and therefore, we continue to use the terms here. Patiño et al. (2014) also found the terms helpful.

Genera that contain two or more closely related species on one island are assumed to have undergone cladogenesis on that island. If there are two closely related endemic species in the Juan Fernández Archipelago but distributed one on each island, they are assumed to represent two cases of anagenesis. This is so because the older island became available for colonization at least 2 million years prior to the younger one, and it is assumed that the species on the younger island came from an immigrant from the older island. This represents a case of two anagenetic speciations resulting from only one introduction to the archipelago. If the species are not closely related, then this represents two independent introductions and hence also two independent anagenetic speciation events.

Phylogenetic Reconstructions

Considerable information is now available regarding phylogeny of the endemic species of the Juan Fernández Islands. Some of the new data have come from morphology, but most of the new insights have derived from molecular phylogenetic studies. In fact, it is remarkable how many of these genera have been investigated in recent years, both from our own research and from other laboratories. It is appropriate, therefore, to discuss what is now known about the continental origins of these groups as well as relationships among taxa within the archipelago.

Several points need to be mentioned for the discussion that follows. In Chapter 5, Table 5.1 lists all the endemic species of the Juan Fernández Archipelago. Different biogeographical and evolutionary categories are represented by these taxa, involving generic and specific endemism and cladogenetic and anagenetic modes of speciation. Because the reproductive systems are so different between angiosperms and ferns, these groups are treated separately in the discussions that follow.

As an aid to presenting the different conditions, models of phylogeny are used (Fig. 13.2) involving the two major islands and the mainland, represented by South America, which is the source for three-fourths of the endemic and native flora (see Chapter 16). Nine percent of the specific origins can be traced more specifically to Chile. In some cases close continental relatives are known and in some cases not. Examples of phylogenetic reconstructions for genera investigated in the Juan Fernández Archipelago, which contain two or more species, subspecies, or varieties, are summarized diagrammatically in the models of Fig. 13.2A–X. The discussions that follow present the genera in alphabetical order within each of the four categories: (1) endemic angiosperm genera, (2) endemic angiosperm species, (3) endemic fern genus, and (4) endemic fern species.

Angiosperms: Endemic Genera

Twelve genera of angiosperms are known to be endemic to the Juan Fernández Archipelago. In general, due to the striking degree of morphological divergence from continental progenitors, and hence the reason for their being regarded taxonomically as distinct genera, it is more challenging to determine origins.

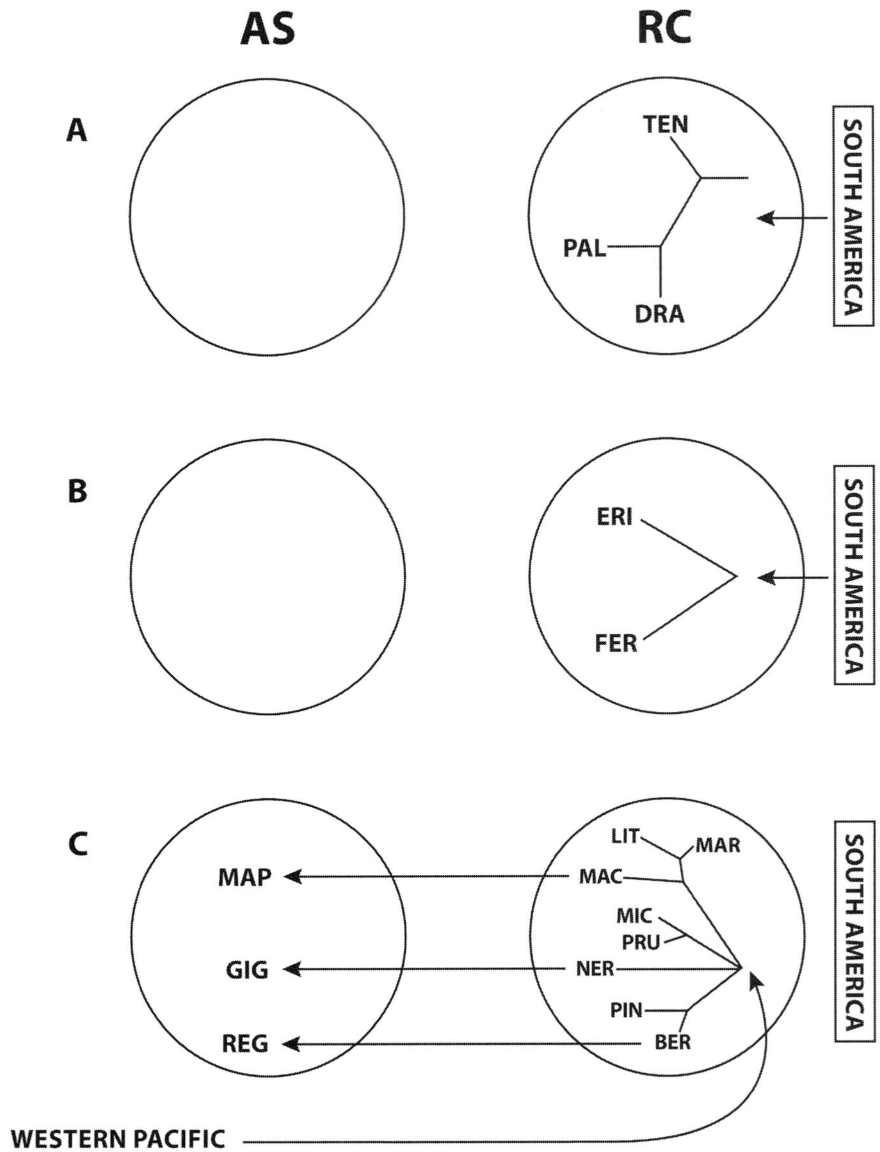

Figure 13.2 Models of phylogeny for genera with two or more endemic species, subspecies, or varieties superimposed on the two major islands of the Juan Fernández Archipelago. Island on the left, Alejandro Selkirk (1–2 Ma); on the right, Robinson Crusoe (4 Ma). (A) *Centaurodendron* and *Yunquea* (DRA, *C. dracaenoides*; PAL, *C. palmiforme*; TEN, *Y. tenzii*). (B) *Cuminia* (ERI, *C. eriantha*; FER, *C. fernandezia*). (C) *Dendroseris* (BER, *D. berteroana*; GIG, *D. gigantea*; LIT, *D. litoralis*; MAC, *D. macrantha*; MAP, *D. macrophylla*; MAR, *D. marginata*; MIC, *D. micrantha*; NER, *D. neriifolia*; PIN, *D. pinnata*; PRU, *D. pruinata*; REG, *D. regia*. (D) *Megalachne* (BER, *M. berteroi*; MAS, *M. masafuerana*; ROB, *M. robinsoniana*. (E) *Robinsonia* (BER, *R. berteroi*; EVE, *R. evenia*; GAY, *R. gayana*; GRA, *R. gracilis*; MAC, *R. macrocephala*; MAS, *R. masafuerae*; SAX, *R. saxatilis*; THU, *R. thurifera*. (F) *Berberis* (COR, *B. corymbosa*; MAS, *B. masafuerae*). (G) *Carex* (BER, *C. berteroniana*; FER, *C. fernandezensis*; STU, *C. stuessyi*). (H) *Chenopodium* (CRU, *C. crusoeanum*; NES, *C. nesodendron*; SAN, *C. sanctae-clarae*). (I) *Coprosma* (OLI, *C.*

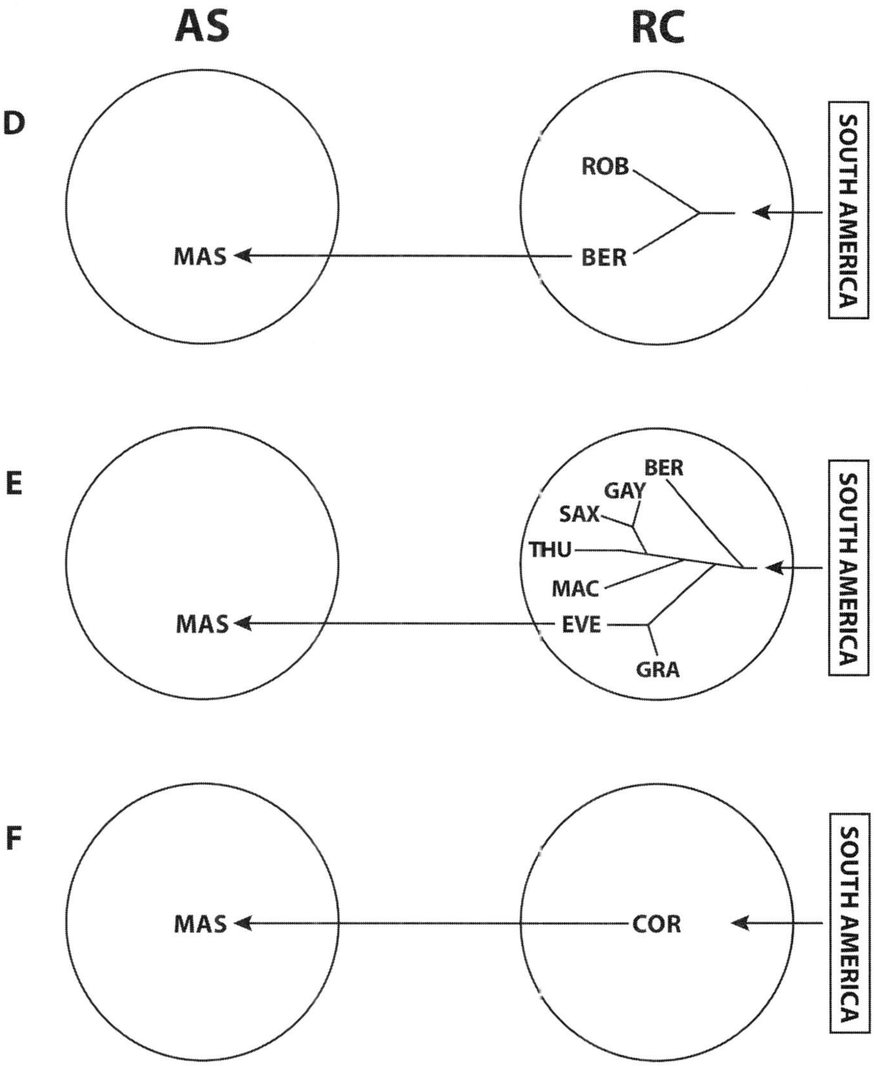

Caption for Figure 13.2 (cont.)

oliveri; PYR, *C. pyrifolia*). (J) *Erigeron* (FER, *E. fernandezia*; ING, *E. ingae*; LUT, *E. lutea*; RUP, *E. rupicloa*; STU, *E. stuessyi*). (K) *Eryngium* (BUP, *E. bupleuroides*; FER, *E. ×fernandezianum*; INA, *E. inaccessum*; SAR, *E. sarcophyllum*). (L) *Gunnera* (BRA, *G. bracteata*; MAS, *G. masafuerae*; PEL, *G. peltata*). (M) *Haloragis* (MAF, *H. masafuerana* vars. *masafuerana* and *asperrima*; MAT, *H. masatierrana*). (N) *Peperomia* (BER, *P. beteroana* subsp. *berteroana*; MAR, *P. margaritifera*; SKO, *P. skottsbergii*). (O) *Sophora* (FER, *S. fernandeziana* vars. *fernandeziana* and *reedeana*; MAS, *S. masafuerana*). (P) *Spergularia* (CON, *S. confertiflora* vars. *confertiflora* and *polyphylla*; MAS, *S. masafuerae*). (Q) *Urtica* (GLC, *U. glomeruliflora*; MAS, *U. masafuerae*). (R) *Wahlenbergia* (BER, *W. berteroi*; FER, *W. fernandeziana*; GRA, *W. grahamiae*; MAS, *W. masafuerae*; TUB, *W. tuberosa*). (S) *Zanthoxylum* (EXT, *Z. externa*; MAY, *Z. mayu*). (T)

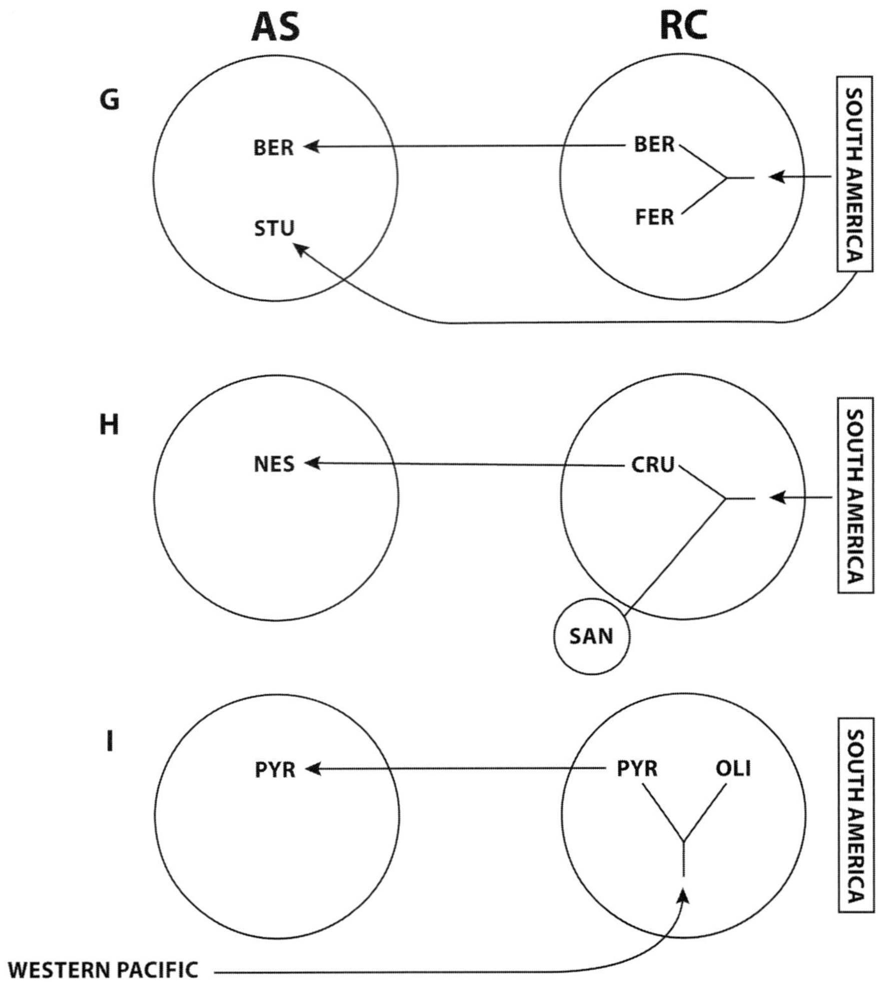

Caption for Figure 13.2 (cont.)

Asplenium (MAC, *A. macrosorum*; STE, *A. stellatum*). (U) *Blechnum* (CYC, *B. cycadifolium*; LON, *B. longicauda*; MOC var. FER, *B. mochaenum* var. *fernandezianum*). (V) *Dicksonia* (BER, *D. berteroana*; EXT, *D. externa*). (W) *Megalastrum* (GLA, *M. glabrius*; INA, *inaequalifolium*; MAS, *M. masafuerae*). (X) *Polypodium* (INT subsp. INT, *P. intermedium* subsp. *intermedium*; INT subsp. MAS, *P. intermedium* subsp. *masafueranum*).

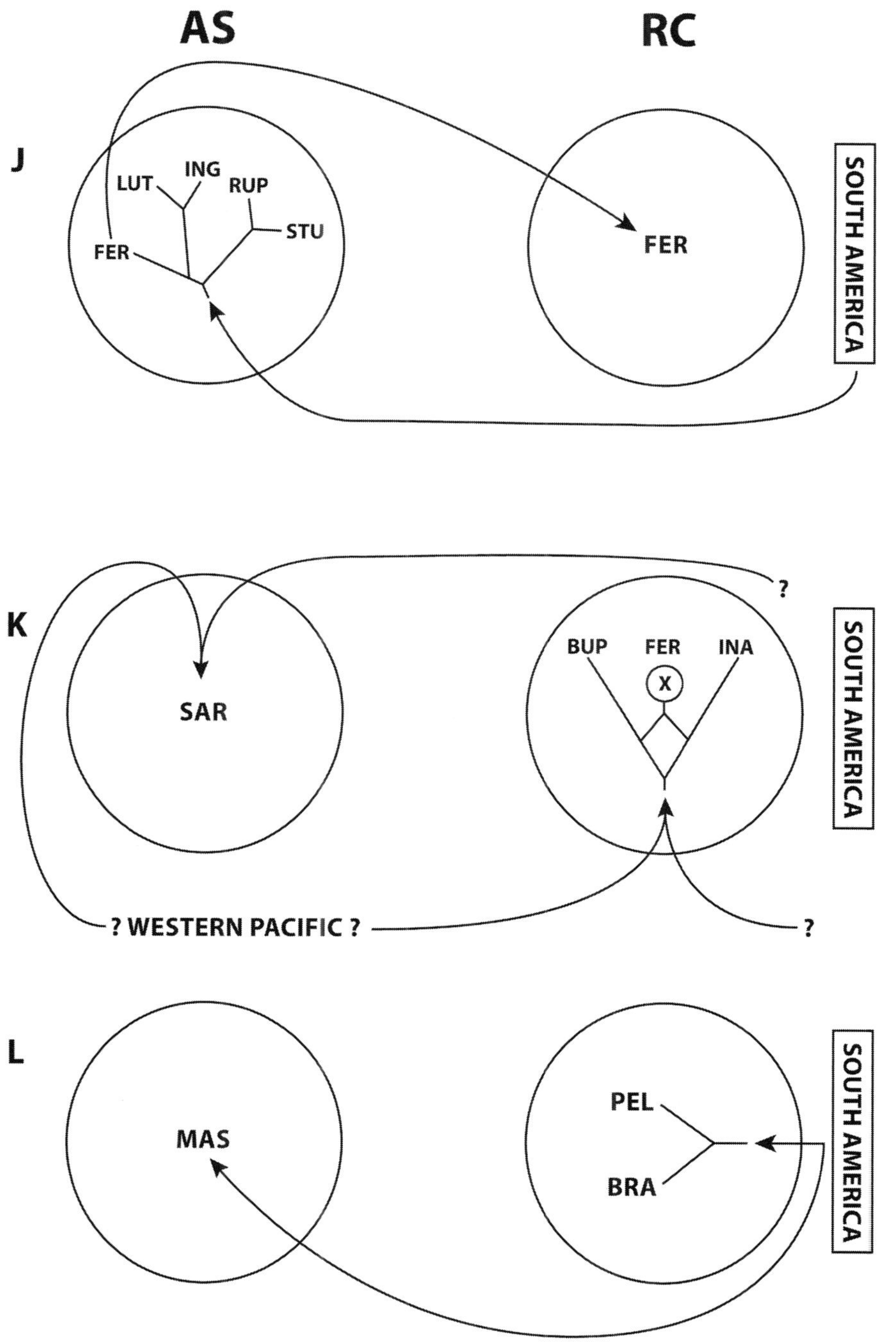

Figure 13.2 (cont.)

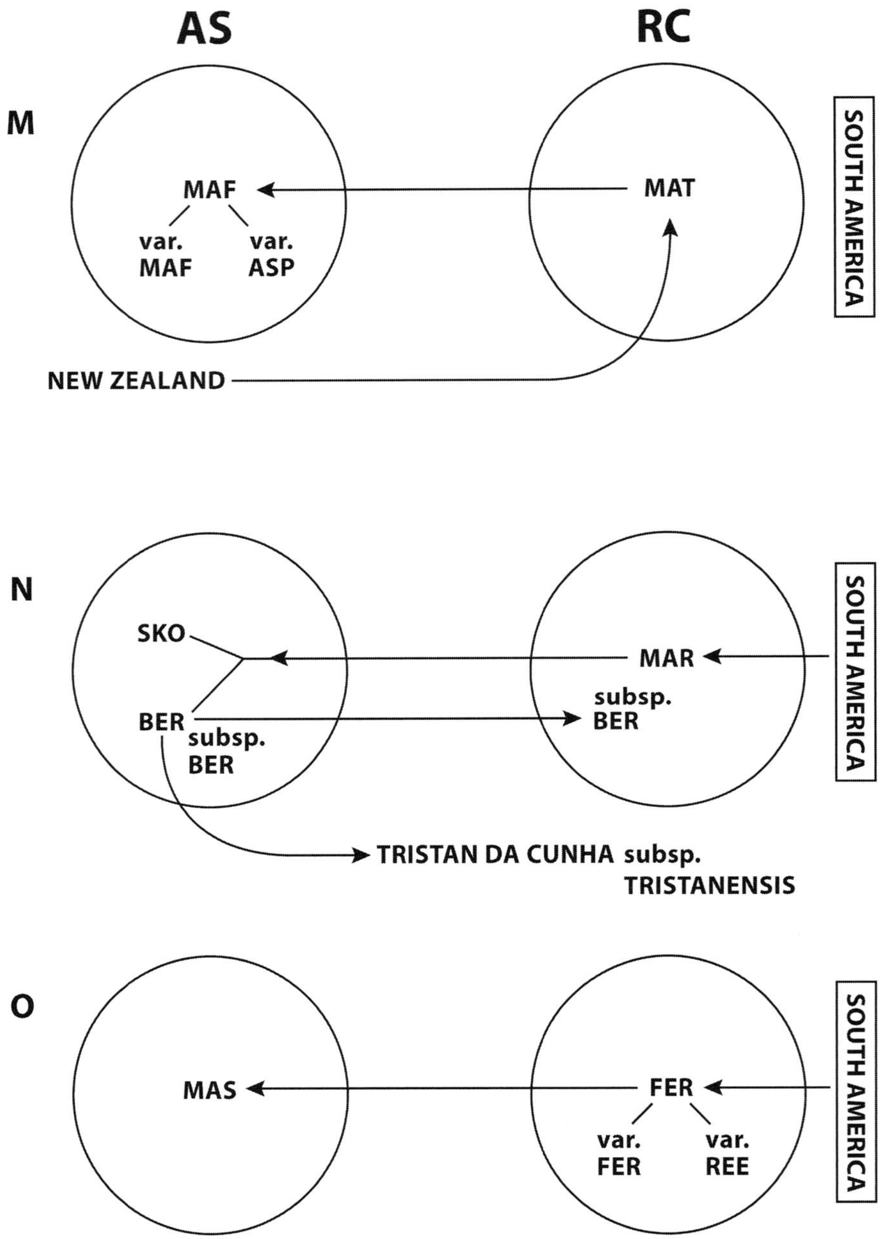

Figure 13.2 (cont.)

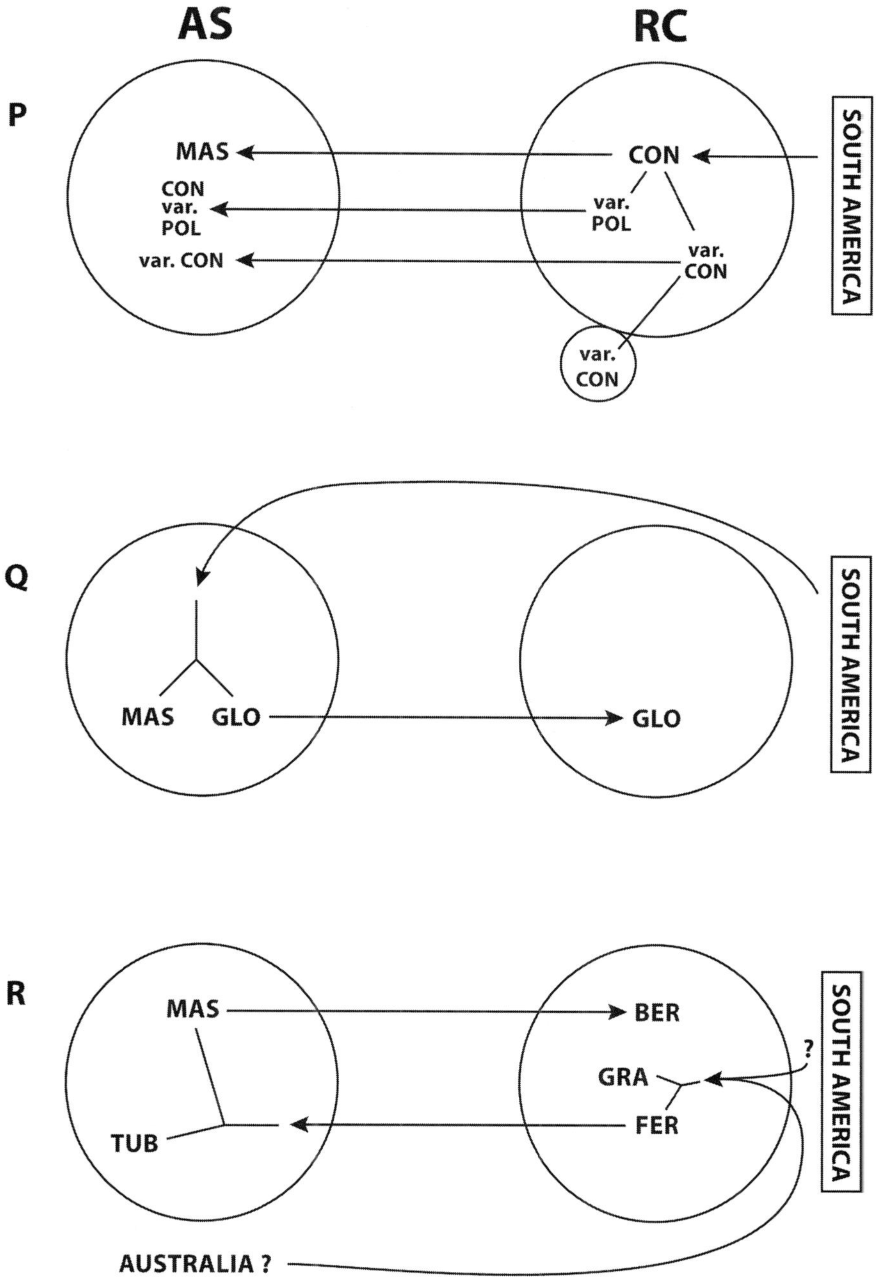

Figure 13.2 (cont.)

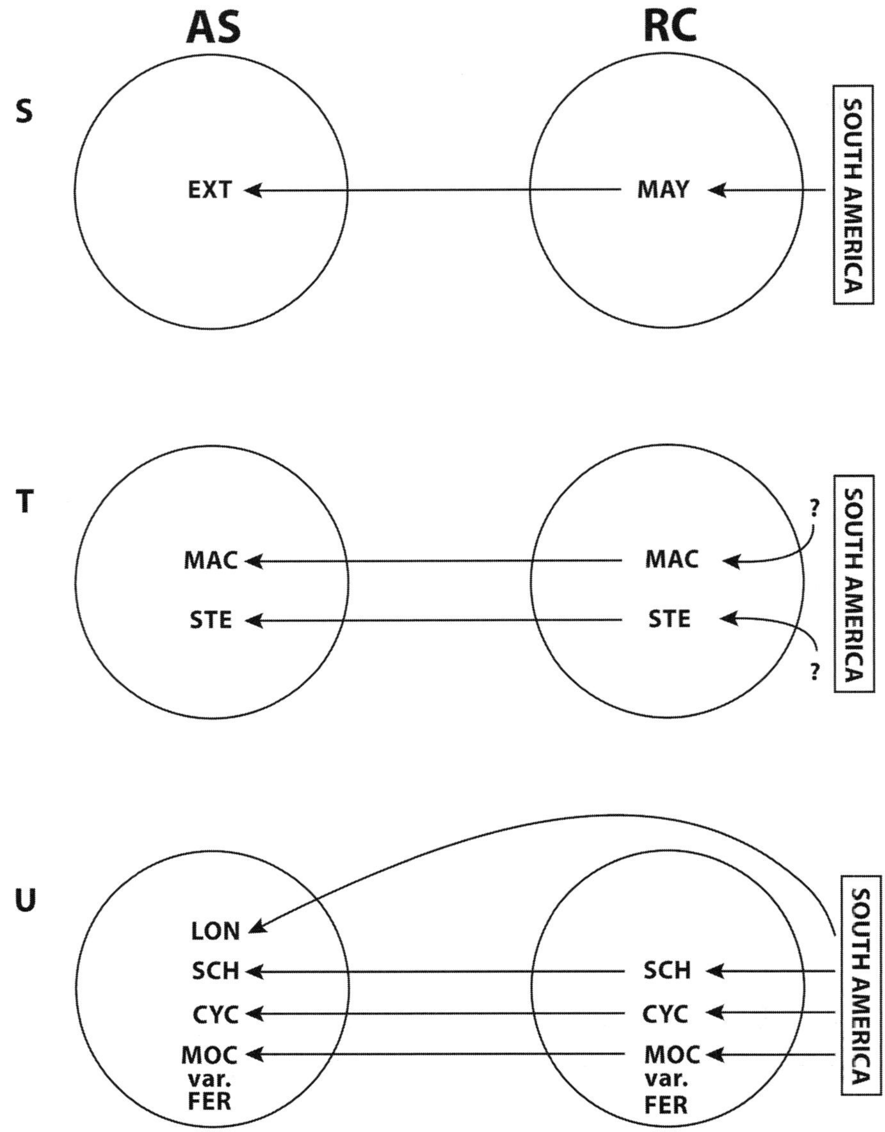

Figure 13.2 (cont.)

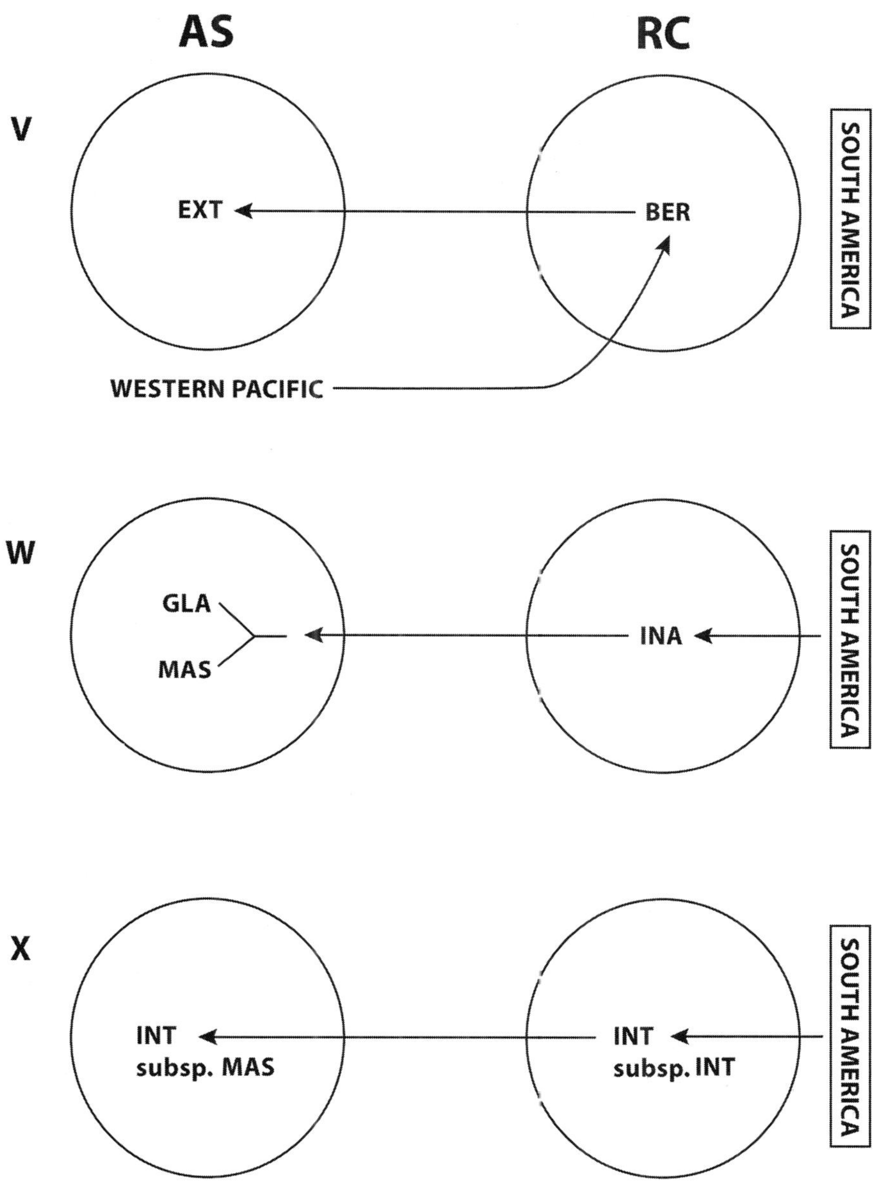

Figure 13.2 (cont.)

The first endemic genus is *Centaurodendron*, with two species (*C. palmiforme* and *C. dracaenoides*; Fig. C35) of Asteraceae (tribe Cardueae), and both are restricted to Robinson Crusoe Island. Susanna et al. (2011), using ITS-ETS plus three chloroplast markers, showed *C. palmiforme* to be nested within *Plectocephalus*, which has a disjunct distribution in Ethiopia, North America, and South America, as well as with the South American *Centaurea cochinalensis* and *C. floccosa*. The genus has been maintained by Susanna et al. (2011) based on the morphological divergence of *Centaurodendron* from these relatives. Both of these species are rare in the island, and hence no detailed study has been done on their evolution within the archipelago. Presumably, the two species diverged cladogenetically when the island was younger (Fig. 13.2A), and perhaps even into different habitats, but it is difficult to know what happened because these species are now so scarce and isolated. *Centaurodendron* probably descended from the same introduction as did *Yunquea* (see below).

Cuminia (Lamiaceae) is another endemic genus with two species on Robinson Crusoe Island (Fig. 13.2B), *C. eriantha* (Fig. C77) and *C. fernandezia*. Molecular phylogenetic studies have revealed a connection to *Monardella* and *Minthostachys* (Trusty et al. 2004) or even more broadly to the New World *Clinopodium* group (Bräuchler et al. 2010), but nothing more specific about origins can be said at this time. Studies by Ruiz et al. (2000) have shown identical ITS-1 sequences but 1.3% total sequence divergence in ITS-2. This molecular difference coincides with a consistent morphological distinctiveness in leaf size and shape plus pubescence on leaves and floral parts, supporting recognition of two distinct species. Earlier, however, Harley (1986) treated them as varieties. The present distribution of the two species is not habitat-differentiated, at least not as seen from our field observations. As Ruiz et al. (2000) suggested, it appears that these two species diverged cladogenetically from an original lineage that descended from the founder population. Over time as the island aged and reduced in size and ecological diversity, these two species were restricted to higher-elevation forests along the main ridges of the island and now grow in close proximity to each other (for distributions, see Ruiz et al. 2000, fig. 1).

Dendroseris is the largest endemic genus in the archipelago (Figs. C36–C52) with 11 species, eight restricted to Robinson Crusoe Island and three occurring on Alejandro Selkirk Island (Fig. 13.2C). The genus was first recognized as distinct by Don (1832), and Skottsberg (1953b) even divided it into four genera, which emphasizes the differences among the species as well as the divergence from continental relatives. More recently, based on morphological cladistic approaches, Sanders et al. (1987) elected to recombine these segregates back into a single genus. Molecular phylogenetic studies using nrITS (Sang et al. 1994) confirmed the monophyly of the group, indicating that it originated from a single introduction to the islands. This was followed by an ITS analysis among genera of subtribe Sonchinae of Asteraceae (tribe Cichorieae) (Kim et al. 1996), another using intergenic spacers *psbA* and *trnH* in chloroplast DNA, and most recently an analysis of a larger set of data also including *matK* sequences (Kim et al. 2007). From these analyses, it appears that *Dendroseris* has evolved from out of the large and diverse continental genus *Sonchus*. Based on these results, and in order to avoid leaving *Sonchus* paraphyletic in a cladistic sense of classification, Mejías and Kim

(2012) submerged all species of *Dendroseris* into *Sonchus*, thus eliminating the former as an endemic genus in the archipelago. We reject this treatment because it ignores the high degree of divergence between the progenitor and the island-derivative lineage (see Chapter 5 for more perspectives on classification and its implications for recognizing genera and species in the archipelago). Although the molecular data clearly show *Dendroseris* originating from out of *Sonchus*, the sample does not conclusively reveal from which specific progenitor the island lineage evolved. The most recent phylogeny based on combined data (Kim et al. 2007) shows *Dendroseris* to be sister to a clade of species of *Sonchus* subg. *Sonchus* (sects. *Maritimi* and *Arvenses*, both with a broad distribution) and the genera *Actites* (endemic to Australia), *Embergeria* (found only in the Chatham Islands, New Zealand), and *Kirkianella* (endemic to New Zealand) (Kim et al. 1999). These phylogenetic results therefore offer no conclusive evidence to the geographical direction of *Dendroseris* in the Juan Fernández Archipelago. Based on all available evidence, however, it would seem most likely that the original propagules to Robinson Crusoe Island, when it first originated, came from the western Pacific (Fig. 13.2C). Certainly there are no close relatives in South America, other than *Thamnoseris* in the Desventuradas Islands, which is a close relative of *Dendroseris* (Kim et al. 2007), and might, in fact, have derived from it (or from a common early colonizer). After initial colonization, *Dendroseris* speciated cladogenetically within Robinson Crusoe Island to produce the three distinct subgenera now recognized taxonomically (Sanders et al. 1987). From within each of these subgenera, dispersal occurred to the young island when it emerged, yielding the anagenetic species *D. macrophylla* (from subg. *Dendroseris*), *D. gigantea* (from subg. *Rea*), and *D. regia* (from subg. *Phoenicoseris*).

Juania (Fig. C30) is an endemic genus of palms (tribe Ceroxyleae) with the single dioecious species *J. australis*. This is an isolated genus with origins probably on the South American continent. Several premolecular studies have been done on the morphology and anatomy of this species (Moore 1969; Tomlinson 1969; Uhl 1969), but other than tribal disposition, little is known about its evolutionary origins. We do know, however, that approximately 1,000 to 2,000 individuals exist on the highest ridges and secluded high valleys of Robinson Crusoe Island (Stuessy et al. 1983; Ricci 2006). Molecular phylogenetic investigations by Trénel et al. (2007) using three plastid and two nuclear DNA sequences show that *Juania* lies within tribe Ceroxyleae and is sister to *Ceroxylon*. A New World tropical origin is obviously suggested by these data. It is of interest that a molecular clock applied to this analysis (Trénel et al. 2007) gives an age of divergence of *Juania* at about 25 million years, which greatly predates the known age of Robinson Crusoe Island. This suggests that the genus speciated on the continent prior to arrival in the archipelago.

Lactoris is one of the most interesting genera in the archipelago and also surely one of the most ancient. The morphology of *L. fernandeziana* (Figs. C26 and 9.7) is so unusual that understanding relationships with other families has been problematical (Lammers et al. 1986; Stuessy et al. 1998). This isolation is what has led to the placement of this species in its own family, Lactoridaceae, the only family of flowering plants endemic to an oceanic island. Carlquist (1964, p. 434) suggested that the relationships of this family were "unquestionably annonalian." Lammers et al. (1986), based on phenetic and cladistic morphological analyses, opined that the family belonged somewhere in Magnoliales, perhaps near Anonnaceae. Kelly and González (2003) also concluded that Lactoridaceae belonged with Magnolialean genera such as *Drimys* and *Calycanthus*. Recent molecular analyses based on *rbcL* (Qiu et al. 1993), 18S rDNA (Soltis et al. 1997), and a 17-gene analysis (Soltis et al. 2011) suggest that *Lactoris* nests within Aristolochiaceae near *Aristolochia*. Another analysis (Wanke et al. 2007), using the chloroplast markers *matK* and *trnK*, show *Lactoris* as sister to the rest of Aristolochiaceae. Even more recently, Naumann et al. (2013) with a 19-gene analysis of chloroplast and nuclear sequences show Lactoridaceae as sister to both Aristolochioideae and Hydnoraceae. From a morphological perspective, the family is distinct with trimerous floral symmetry, zigzag branching with fragile stems, and tetrad pollen (Zavada and Taylor 1986; Sampson 1995). An emphasis on these features suggests that Lactoridaceae might belong in Aristolochiales or perhaps more broadly in Piperales (Carlquist 1990, 1992; Tobe et al. 1993; Tucker and Douglas 1996; González and Rudall 2001; Wagner et al. 2014). Because Lactoridaceae tie to ancient angiosperm families, it must have a long evolutionary history, and this makes assessing geographical origins difficult. Lammers et al. (1986), based on microfossil evidence of related families and a 3D morphological phyletic analysis, suggested that the family should be at least 69 million years old. More recent reports have shown fossil pollen of *Lactoris* (as *Lactoripollenites*) to be present in cores off the western coast of South Africa to an age of 70 to 90 million years (Zavada and Benson 1987), in Australia in sediments approximately 50 to 70 million years old (also as *Lactoripollenites*) (Macphail et al. 1999), in southern South America in deposits approximately 19 million years old (Gamerro and Barreda 2008), and possibly also in Canada, the United States, India, and the Antarctic Peninsula (see Gamerro and Barreda 2008 for a summary). If these fossil reports are indeed Lactoridaceae, and the evidence does seem reasonable, then the family was widely distributed, especially in the southern hemisphere, long before the Juan Fernández Islands existed. Recent dispersal to the islands, and extinction from mainland sites, has resulted in the pattern seen today. Speciation in the archipelago, therefore, probably would not have occurred again, and thus this species, as with *Juania australis* and *Thyrsopteris elegans*, cannot be used realistically as an example of any particular type of speciation within the islands.

×*Margyracaena skottsbergii* (Rosaceae) was originally described as an intergeneric hybrid on Robinson Crusoe Island by Bitter (1921; republished in Skottsberg 1921, pp. 134–135). It was hypothesized to have resulted from a cross between one endemic parental species, *Margyricarpus digynus*, and one introduced species, *Acaena argentea*. Crawford et al. (1993a) tested the parentage of this hybrid with RAPD markers and convincingly confirmed its origin. It is unknown whether this hybrid still exists on the island. It was seen only twice by us near the Centinela Ridge, up from Quebrada Pangal, and these collections were sterile. This is the second case of an intergeneric hybrid in the Juan Fernández Archipelago, the other being the fern *Pleopeltis* ×*cerro-altoensis* (Danton et al. 2015). In other oceanic archipelagos, such as Hawaii, intergeneric hybridization is more frequent (Carr 2003).

Megalachne (Poaceae) is an endemic genus with three species, *M. berteroana*, *M. masafuerana*, and *M. robinsoniana* (Peña et al. 2017). The first two had been cited as occurring on both the major islands (Baeza et al. 2002), but the recent revision (Peña et al. 2017) places *M. berteroana* and *M. robinsoniana* as only occurring on Robinson Crusoe Island, with *M. masafuerana* restricted to Alejandro Selkirk Island. The genus is morphologically close to *Bromus*, and at times it has been included within this genus (Pilger 1920; Skottsberg 1921, 1953). More recently, *Megalachne* has been regarded as close to *Bromus* and *Festuca* (Matthei 1974, 1986; Watson and Dallwitz 1992) or *Festuca, Podophorus*, and six other genera in subtribe Loliinae of tribe Poeae (Soreng et al. 2000). A molecular phylogenetic study of *Bromus* has been done (Saarela et al. 2007), but none of the species of *Megalachne* was included in the analysis. Schneider et al. (2011) successfully obtained ITS data from *M. berteroniana* (*Wiens et al. 11203*) and *M. masafuerana* (*Skottsberg 183*) and showed the closest affinities with *Podophorus*. The topological relationship is curious in that *M. berteroniana* ties preferentially to *P. bromoides* rather than to *M. masafuerana*, although the entire complex of three species has 100% bootstrap support. Peña et al. (2017) conducted morphometrical analyses of the three species that showed *M. masafuerana* as being more closely related to *M. berteroniana* than to *M. robinsoniana*. The data suggest (Fig. 13.2D), therefore, that an immigrant arrived on Robinson Crusoe Island from out of *Bromus* (or possibly *Festuca*) in South America, followed by splitting (cladogenesis) into two lineages (species). From out of *M. berteroniana* came dispersal to the younger island when it became available for colonization, followed by anagenetic speciation leading to *M. masafuerana*.

Nothomyrcia (Myrtaceae), with its single species *N. fernandeziana* (Fig. C80), was previously regarded as a species of *Myrceugenia* (Skottsberg 1921; Landrum 1981a, 1981b), but molecular phylogenetic studies (Murillo-A. et al. 2012) have shown that it does not belong in this genus. It appears to be isolated (Murillo-A. et al. 2012), with affinity near *Blepharocalyx* from South America (Murillo-Aldana and Ruiz 2011; Retamales and Scharaschkin 2015). This is a clear case of anagenesis, but the specific continental progenitor is unknown, which precludes comparisons regarding origins.

Podophorus (Poaceae) is a monospecific genus with the sole species *P. bromoides*. It has not been collected again since the original few samples on Robinson Crusoe Island by Germain in 1854 (Philippi 1856). Skottsberg (1921) reported that he searched diligently for this species but was unsuccessful. We have also been unable to locate this species, and therefore, we have sadly concluded that it is most probably extinct (see also Baeza et al. 2002; Danton et al. 2006; Ricci 2006). Despite this categorization, nucleotide data of ITS1 and 5.8S rRNA (Schneider et al. 2011) have been successfully extracted from original specimens of Philippi at HAL and SGO, and comparisons have been made with other genera. The closest genus is clearly *Megalachne* (72% bootstrap support), another endemic genus of Poaceae in the archipelago, and both are nested within the Aveneae/Poeae complex (Schneider et al. 2011). It is possible that *Podophorus* evolved from *Megalachne* and acquired very distinctive morphological features within only 4 million years, but it is also possible that these genera derived from similar ancestors on the continent.

The second largest genus in the archipelago is *Robinsonia* (Figs. C58–C67) of Asteraceae (tribe Senecioneae) with eight species. Because of the nomenclatural complexity in this group, a few comments on the history of naming are in order because they help us better understand the taxonomic structure used in this book. The genus was first described by Candolle (1833), but without a listed species. At the same time (1833), he described a related genus from the same islands, *Balbisia*, with the single species *B. berteroi*. Decaisne (1834) also recognized *Balbisia* DC. and *Robinsonia* DC., and in the latter he included four species (*R. macrocephala*, *R. thurifera*, *R. gayana*, and *R. gracilis*). In 1838 Candolle recognized both genera, and for *Robinsonia* he described three sections containing the four species treated by Decaisne: sect. *Symphyochaeta* (*R. macrocephala*); sect. *Eleutherochaeta* (*R. gayana*, *R. thurifera*); and sect. *Eleutherolepis* (*R. gracilis*). Meisner (1839) changed *Balbisia* to *Rhetinodendron* because two earlier homonyms were revealed: *Balbisia* Willd. (Compositae, 1803; now a *nom. rej.*) and *Balbisia* Cav. (Geraniaceae; now a *nom. conserv.*). Hemsley (1884) and Johow (1896) both recognized *Rhetinodendron* and *Robinsonia*. Skottsberg (1921) also treated the monotypic *Rhetinodendron* as distinct, and he divided *Robinsonia* into subgenera and sections: subg. *Symphyochaeta* (with R. *macrocephala*); subg. *Eleutherochaeta*, with sect. *Symphyolepis* (*R. gayana, R. thurifera*); and sect. *Eleutherolepis* (*R. evenia*, *R. masafuerae*, *R. gracilis*). In 1951, however, Skottsberg elevated the monotypic subg. *Symphyochaeta* to generic status. Pacheco et al. (1985) submerged *Rhetinodendron* into *Robinsonia* as a subgenus, a treatment also followed by Sanders et al. (1987), based on assumptions that the group evolved from a single introduction to the archipelago. They also recognized three sections in subg. *Robinsonia* (sect. *Eleutherolepis* with *R. evenia*, *R. gracilis*, and *R. masafuerae*; sect. *Robinsonia* with *R. gayana* and *R. thurifera*; and sect. *Symphyochaeta* with *R. macrocephala*). Recent ITS molecular phylogenetic studies (Sang et al. 1995) confirmed the monophyly of *Robinsonia* and supported the sectional classification. Even more recently, Danton (2013) described a new species of *Robinsonia*, *R. saxatilis*, which proves closely related to *R. gayana* (Takayama et al. 2015a), bringing the total to eight endemic species. Additional ITS investigations using a much larger sample of species in *Senecio* (Pelser et al. 2007, 2010a) suggest that *R. berteroi* (Fig. C58) might have resulted from a separate introduction to the archipelago. With a much expanded analysis of cpDNA markers (*trnL-trnL-F*, *psbA-trnH*, 3′*trnK*, 5′*trnK*, *ndhF*, and *psbJ-petA*), however, these same authors show the genus to be holophyletic as it is also with combined ITS-ETS and plastid data (Pelser et al. 2010b). In our opinion, the bulk of evidence suggests that the genus did result from a single introduction, and this is what is presented in the model in Fig. 13.2E. From initial arrival, the lineage divided early into what eventually became subg. *Rhetinodendron* and subg. *Robinsonia*, and in the latter, cladogenesis continued to eventually yield three lineages (sections) following the taxonomic structure discussed earlier. After Alejandro Selkirk Island appeared 2 to 3 million years later, a dispersal event occurred to the younger island from out of sect. *Eleutherolepis* (perhaps from *R. evenia* or its immediate ancestor), where it diverged anagenetically into *R. masafuerae*. The principal difficulty of understanding the continental relatives of *Robinsonia* is that it has evolved from out of *Senecio* s.l., which is one of the largest genera of flowering plants, with more than 1,000 species (Mabberley 2008). Pelser et al. (2010a, 2010b), with an admirable effort, show *Robinsonia* as sister to the *Senecio acanthifolius–S. fistulosus* clade from South America, but only 13 species of this group were sampled. The problem again is that *Senecio* is so huge, with over 200 species just in Chile alone (Cabrera 1949), that it will take a large amount of additional sampling and analyses to pinpoint more precisely the closest relatives of *Robinsonia*.

Selkirkia (Boraginaceae) contains the single endemic species *S. berteroi*. Molecular phylogenetic analyses have been done within the family (Långström and Chase 2002; Nazaire and Hufford 2012), but *Selkirkia* was not included. Generic affinities, therefore, have not yet been precisely determined either morphologically or at the molecular level.

Yunquea (Asteraceae, Cardueae) is an endemic genus with the single species *Y. tenzii* that is restricted to the top of the highest peak on Robinson Crusoe Island, El Yunque. This genus has never been collected well, and even morphological details are still lacking for the floral features. It has recently been placed in synonymy with *Centaurodendron* (Susanna and Garcia-Jacas 2001) based on limited morphological data, but no molecular information is available despite surveys of the tribe (Garcia-Jacas et al. 2001; Susana et al. 2011). It may be that *Yunquea* is best synonymized with *Centaurodendron*, but in our opinion, further data are needed before this decision can be taken with confidence. The biogeographical origins of these two endemic genera are likely from South America (Fig. 13.2A), but due to a lack of evidence of close continental relatives, it is impossible to be more specific at this time.

Angiosperms: Endemic Species

For purposes of revealing evolutionary origins, measuring degrees of divergence, and assessing modes of speciation, analysis of endemic *species* in the archipelago offers much more information than genera. In some cases ties to continental relatives are very clear and in other cases not so obvious. We tend to talk about close relatives as progenitors of island derivatives rather than using the general term "ancestor." This is so because, due to the youth of the archipelago, little time has transpired since the islands were formed, and the island environment has been changing much more rapidly than most continental areas. We assume, therefore, that when a clear and close tie exists between an island and continental species, the former was derived from out of the latter. This is the well-documented phenomenon of progenitor-derivative speciation (Crawford 2010). It may be that the continental taxon has also diverged somewhat from the original progenitor population from which successful dispersal to the islands occurred, but this cannot be viewed as cladogenesis in the normal sense of a splitting event that divides the population into two major segments. Cladogenesis has occurred in the Juan Fernández Archipelago, but only in genera after arrival in the islands (Fig. 13.1). As mentioned earlier, the discussions list the genera in alphabetical order for ease of reference.

Acaena masafuerana (Rosaceae) is endemic to the younger Alejandro Selkirk Island. The genus was recognized to contain 110 species and numerous infraspecific taxa in the comprehensive monograph by Bitter (1910), but more recent estimates suggest approximately 43 species (Marticorena 2006). The group is especially concentrated in the southern hemisphere (Marticorena 2006), with South America containing about 22 species (Grondona 1964), New Zealand 17 (Lee et al. 2001), and additional taxa in Australia, Hawaii, and South Africa. The island endemic was placed in sect. *Acrobyssinae* by Skottsberg (1921) and Marticorena (2006) along with the Chilean relatives *A. antarctica* and *A. tenera*. The closest relative of the Juan Fernández endemic seems to be *A. antarctica*, which is distributed in Tierra del Fuego (Moore 1983). In her Masters thesis dealing with the Chilean species of the genus, presented in 1996 prior to her publication of the same treatment in 2006, Marticorena commented (p. 61) that "Es probable que corresponda a una variedad de aquella [*A. antarctica*]." This comment does not appear in her 2006 publication (Marticorena 2006), but the thesis at minimum suggests that there is a close morphological resemblance between *A. masafuerana* and *A. antarctica*. Dispersal from the continent to the islands would certainly be no challenge because the terminal barbs on the fruits are admirably suited for dispersal by attachment to bird feathers (e.g., Lee et al. 2001). The only molecular phylogenetic study involving *A. masafuerana* has been done by Kerr (2004) and presented in her unpublished Ph.D. thesis. She successfully extracted and analyzed a *trnL/F* sequence from herbarium material (*Kuschel 215*, US). The analysis revealed that *A. masafuerana* is quite different from other species of the genus, and these data therefore provide no help in suggesting ties to continental relatives. Kerr (2004) also applied a molecular clock to the *trnL/F* data among genera and species of tribe Sanguisorbeae, and this suggested paradoxically that *A. masafuerana* would have split from other species approximately 8.8 million years ago, which is much older than the 1 to 2 million years for Alejandro Selkirk Island to which it is presently confined. Haberle (2003) found fossil pollen of *A. masafuerana* in a core from the "alpine" region of the island with inferred dating of approximately 10,000 years or so, but speciation would doubtless have taken place much earlier. In any event, data in hand recommend a hypothesis that a propagule of *A. antarctica* (or similar taxon) arrived on the younger island after it formed and there speciated anagenetically.

Agrostis masafuerana is another endemic grass species in the archipelago, this one restricted to the younger island. Little is known about affinities with other species. Pilger (1920, p. 388), when describing this species as new, remarked that it had the appearance ("... in der Tracht ...") of *A. magellanica* Lam. The genus has about 175 species (Mabberley 2008) worldwide, and in the Southern Cone there are approximately 130 that have been mentioned (Rúgolo de Agrasar and De Paula 1978). In Chile, 31 species have been listed (Rúgolo de Agrasar and Molina 1997). A partial revisionary treatment of the genus was done by Romero García et al. (1988), but this only dealt with species in the Iberian Penninsula. A revision of the Chilean species has been completed by Rúgolo de Agrasar and Molina (1997), but no indication of the affinities of *A. masafuerana* was provided. The molecular data are of no particular help with origin of the island endemic because in the only study that focused on the genus (Amundsen and Warnke 2012), only 15 species were sampled, and unsurprisingly, the authors did not include the island taxon.

Broader investigations in the Poaceae have used a similar number of taxa of *Agrostis* (e.g., 17 species) (Saarela et al. 2010) but again without *A. masafuerana*. We can only speculate, therefore, that *A. masafuerana* arrived to the younger island from an ancestor in southern South America and there in the island diverged anagenetically.

Apium fernandezianum (Apiaceae) is endemic to Robinson Crusoe Island. In addition to the island endemic, 19 species of the genus occur as natives in Chile (Marticorena and Quezada 1985). Skottsberg (1921) suggested that *A. fernandezianum* was related to *A. prostratum*, which is a common species native to Australia and New Zealand but also found in mainland Chile. At first glance, it may seem plausible that the island endemic originated from South America. Recent molecular phylogenetic studies by Spalik et al. (2010), however, shed a different light on the problem. They included *A. fernandezianum* plus many accessions of the genus, including seven samples of *A. prostratum*. The tie of the island endemic is clearly with samples from New Zealand, which strongly suggests that dispersal came from the western Pacific to the older island. A molecular clock estimate on this divergence shows it to be very recent, within the past 1 million years (cf. Spalik et al. 2010, fig. 2).

The genus *Azara* (Salicaceae, formerly in Flacourtiaceae) has ten species in the tropical, subtropical, and temperate zones of America (Sleumer 1977). *Azara serrata* var. *serrata* is found from central to southern Chile, and var. *fernandeziana* is found on Robinson Crusoe Island (Fig. C85). Sleumer (1977) points out that these two varieties are only minimally different ("... nur wenig verschieden," p. 175). One might prefer not to recognize these varieties, but there are consistent leaf features (i.e., texture, shape, and size) that separate them. Skottsberg (1921) mentioned similarity between the two taxa but treated them as distinct species ("... upheld with hesitation," p. 146) More study is obviously needed, but in the meantime we maintain var. *fernandeziana* as an endemic taxon. Leskinen and Alström-Rapaport (1999) examined several genera of Flacourtiaceae for ITS variation, and they included two species of *Azara, A. integrifolia* and *A. serrata*. Voucher data were not given for these samples because they were apparently obtained from the Melbourne Botanical Garden, but presumably the sampled plant of *A. serrata* came originally from the Chilean mainland. A molecular phylogenetic study of *Azara* has been done and an abstract presented (Gabel et al. 2000), but full details of the study apparently have never been published. Obviously, dispersal to Robinson Crusoe Island must have come from var. *serrata* or its direct ancestor, and the immigrant population then diverged anagenetically at the varietal level. That only varietal divergence has occurred suggests that the dispersal and establishment may have taken place relatively recently (i.e., 1 million years or less), although morphological stasis is an alternative possibility.

Berberis corymbosa (Berberidaceae) is endemic to Robinson Crusoe Island, and *B. masafuerana* occurs only on Alejandro Selkirk Island. The exact progenitors of these island endemics have not been completely resolved. Landrum (1999) suggested, based on leaf shape, that *B. corymbosa* is similar to *B. microphylla* and *B. rotundifolia* from continental Chile. Based on style morphology, however, he regarded *B. corymbosa* as appearing more similar to *B. jujuyensis* and *B. jobii* from Argentina. These hypotheses partially coincide with the classification of Ahrendt (1961), who placed *B. corymbosa,*

B. masafuerana, *B. rotundifolia*, and *B. setigrifolia* within sect. *Corymbosae* and *B. microphylla* in sect. *Microphyllae*. Isozyme studies by Ruiz et al. (2004) have revealed a very close tie between *B. corymbosa* and *B. microphylla* with no ITS divergence and a genetic identity of 0.999. The ITS sequence divergence between *B. microphylla* and *B. masafuerana*, however, is also 0.0, but the genetic identity from isozymes is less with 0.843. The genetic identity between the two endemic species is higher, at 0.936. Complicating the situation is that there is also no ITS divergence between *B. rotundifolia* on the continent and *B. masafuerae*, whereas there is an ITS divergence of 1.8 between the former and *B. corymbosa*. At the nucleotide level, two attempts at sampling the genus have been done by Kim et al. (2004) from 79 taxa and Adhikari et al. (2015) from 68 accessions, but none of the island endemics was examined. Likewise, ITS and AFLP surveys have been completed on species of *Berberis* from Patagonia (Bottini et al. 2002, 2007), but again, none of the island species was included. Taking all evidence into consideration, the most robust hypothesis at the moment (Fig. 13.2F) is that the colonist to the archipelago came from *B. microphylla* or its immediate ancestor and arrived first on Robinson Crusoe Island, where it diverged anagenetically. When the younger island appeared, dispersal occurred to that island, and another anagenetic speciation eventually resulted in *B. masafuerana*.

Boehmeria excelsa (Urticaceae) (Fig. C86) is closely related morphologically to *B. pavonii* found in moist tropical to montane forests from Guatemala south to Paraguay (Wilmot-Dear and Friis 1996). These authors suggest specifically that *B. excelsa* originated from a dispersal event from South America to Robinson Crusoe Island, which would then have been followed by anagenetic speciation. Few molecular studies have been done so far in the genus (e.g., Kang et al. 2008; Yu et al. 2015), and none has included *B. excelsa*.

Cardamine kruesselii (Brassicaceae) is the only member of a large genus of approximately 150 to 200 species (Franzke et al. 1998; Carlsen et al. 2009) that is endemic in the archipelago and confined to Alejandro Selkirk Island. It is also one of the few annual endemic species in the islands. Skottsberg (1921) remarked that O. E. Schulz related it to *C. vulgaris* Phil. from Chile, but there are no comments to this effect in his monograph (Schulz 1903) under either species, and *C. kruesselii* was placed in "Species incertae" at the end of the treatment. Marticorena and Quezada (1985) do not list this species for the flora, presumably because Sjöstedt (1975) placed *C. vulgaris* in synonymy with *C. glacialis* (G. Forster) DC, which *is* listed for Chile. Despite the intense interest in *Arabidopsis* and related genera of Brassicaceae (e.g., Bailey et al. 2006; Warwick et al. 2010), the taxonomy of *Cardamine*, especially in South America, appears to need more attention. Schulz (1903) recognized 27 species from Central and South America. Sjöstedt (1975), however, departed from this estimate and treated just five species with many synonyms, but *C. kruesselii* was not included in his treatment. Al-Shehbaz has suggested 20 species in South America (in Carlsen et al. 2009, p. 215). Despite the many molecular investigations among genera of this family (for a bibliography, go to brassibase.cos.uni-heidelberg.de), such as Carlsen et al. (2009), who included 110 species, none has included *C. krusselii*. The seeds of *Cardamine* have good dispersal ability, as evidenced by the many disjunct patterns of close relatives (Carlsen et al.

2009). The seeds can become mucilaginous when moist and may adhere to birds (Al-Shebaz 1988). *Cardamine glacialis*, for example, is most closely related to species from Australia, Tasmania, and New Zealand and apparently originated from somewhere in those regions by long-distance dispersal (Carlsen et al. 2009). It should prove no obstacle, therefore, for a propagule of *C. glacialis* to have dispersed from the Chilean mainland and arrived on Alejandro Selkirk Island, followed by anagenetic speciation. An alternative explanation would have been a completely independent dispersal from the western Pacific from the common ancestor of *C. vulgaris*, but this would have been a much longer distance for dispersal and hence less probable.

Carex (Cyperaceae) is a very large genus of some 2,000 species (Reznicek 1990) with a very broad distribution in temperate areas of the world (Ball 1990). Four species of the genus occur in the Juan Fernández Archipelago, one being native (*C. phalaroides* on Robinson Crusoe Island) and the other three endemic. *Carex stuessyi* occurs only on Alejandro Selkirk Island, and *C. fernandezensis* is only found on Robinson Crusoe Island. *Carex berteroniana*, also endemic, occurs on both islands. Wheeler (2007) places *C. berteroniana* and *C. fernandezensis* into sect. Echinochlaenae, along with *C. lamprocarpa* and *C. poeppigii* from the Chilean mainland (Wheeler 1988), which suggests that these two endemic Juan Fernández species most probably originated in Chile. Wheeler (2007) offered a hypothesis for the evolution of these endemic species, suggesting that *C. fernandezensis* on Robinson Crusoe Island evolved from *C. berteroniana* by anacladogenesis. *C. stuessyi* on Alejandro Selkirk Island, however, ties to *C. banksii* in southern South America, which ranges from Tierra del Fuego up to 39°S. (Moore 1983). This would represent a case of anagenesis, but from a separate introduction directly to the younger island (Fig. 13.2G).

Chenopodium (Chenopodiaceae) is a genus of more than 150 species with worldwide distribution (Sukhorukov and Zhang 2013). In the Juan Fernández Archipelago, there are three endemic species, *C. crusoeanum* restricted to Robinson Crusoe Island, *C. nesodendron* confined to Alejandro Selkirk Island, and *C. sanctae-clarae* (Figs. C70, 9.2, and 9.4) found only on a small rock, Morro Spartan (Fig. 9.3), just separated by a small water channel on the northern side of Santa Clara Island. Skottsberg (1921) suggested that these three endemic species of *Chenopodium* were related to *C. paniculatum*, which ranges from Chile and Perú northward into North America. He also suggested a possible connection to *C. sandwicheum* from the Hawaiian Archipelago. In the most recent comprehensive taxonomic treatment of the genus, Scott (1978) placed these endemics, along with *C. oahuense* from Hawaii, together in *Chenopodium* subg. *Chenopodium* sect. *Skottsbergia*. Molecular work has been done in the genus (Fuentes-Bazan et al. 2012a, 2012b), but these studies have not included any of the endemic island taxa. The only molecular study that involved one of these endemic island species was that of Kadereit et al. (2010), who showed, using *rbcL* data and with a sample of only 12 species in the genus, a connection of *C. sanctae-clarae* with *C. acuminatum* from Eastern Asia and *C. frutescens* from Central Asia. Kadereit et al. (2010) did include *C. oahuense* in their analysis with *atpB-rbcL*, and it falls close to *C. acuminatum*, but these data were not available for *C. sanctae-clarae*. We regard the suggested ties to taxa of Asia or the Hawaiian Islands as possibilities that need to be investigated further with a much larger sample of species, but we suspect that the closest ties may, in fact, lie with taxa of South America. Much more work is obviously needed to determine progenitors and source areas for the island endemic species. If the endemic species of *Chenopodium* in the Juan Fernández Islands are monophyletic, then one might hypothesize (Fig. 13.2H) that a single introduction came from South America first to Robinson Crusoe Island, followed by anagenetic divergence into the lineage that gave rise to *C. crusoeanum*. When erosion and subsidence to this island resulted in the formation and isolation of the remnant small island, Santa Clara, this is when *C. sanctae-clarae* speciated. Dispersal of *C. crusoeanum* to the newly formed Alejandro Selkirk Island led to the anagenetic origin of *C. nesodendron*.

Chusquea is a genus of bamboos (Poaceae, tribe Bambuseae, subtribe Chusqueinae) with 169 described species (Fisher et al. 2014) found in tropical and semitropical America and from Mexico to southern South America (Clark 1989). In southern Chile alone there are 15 species (Judziewicz et al. 1999). *Chusquea fernandeziana* is the Juan Fernández endemic, and it is restricted to Robinson Crusoe Island. Determination of precise affinities is not yet available, but among the continental Chilean species, *C. macrostachya* and *C. uliginosa* have been placed with the island species in the same part of a morphologically based key (Parodi 1945). Considerable effort has been expended on molecular phylogenetic analyses in *Chusquea*, but the focus has been mainly on determining the subtribal and generic relationships, which have involved using only one to four species of *Chusquea* as generic placeholders (Sungkaew et al. 2009; Kelchner et al. 2013; Wysocki et al. 2015). Kelchner and Cark (1997) examined relationships within *Chusquea* using chloroplast *rpl16* intron with 23 species, but *C. fernandeziana* was not included. Likewise, Fisher et al. (2014) examined a much broader sample of about 40% of the species in the genus, but again, *C. fernandeziana* was omitted. One might expect that the continental relatives might reside in the complex of *C. cumingii*, *C. quila*, *C. uliginosa* of subg. *Chusquea* and *C. gigantea*, *C. montana*, *C. culeou* of subg. *Swallenochloa*, all of which are distributed in southern Chile and adjacent Argentina, but further work will have to include *C. fernandeziana* for comparison. We assume, however, that a dispersal took place from somewhere in southern South America to Robinson Crusoe Island followed by anagenetic speciation that yielded *C. fernandeziana*.

Colletia (Rhamnaceae) consists of five species distributed in South America from Ecuador to Argentina (Mantese and Medan 1993), with *C. spartioides* being endemic to Robinson Crusoe Island. In the most recent revision of the genus, Tortosa (1989) placed the island endemic very close to *C. hystrix* from Chile. In a morphological cladistic analysis of *Colletia* and other genera in tribe Colletieae (Aagesen 1999), *C. spartioides* joined with the other species of the genus in a monophyletic unit, but the specific affinities among the species varied with the particular methods used. Some character information in *C. spartioides* was also lacking. Molecular phylogenetic studies have been done on the genus using the *trnL* intron and *trnL-F* spacer (Aagesen et al. 2005), but *C. spartioides* was not included. Relationships have been examined with *rbcL* and *trnL-F* among many genera of Rhamnaceae (Richardson et al. 2000), but only *C. ulicina* was used for these comparisons. Taking the information that is known at present, it seems highly likely that long-distance dispersal to the older island occurred from a progenitor similar to *C. hystrix* on the Chilean mainland, followed by anagenetic speciation that yielded *C. spartioides*.

Coprosma (Rubiaceae) is a Pacific genus of approximately 110 species (Cantley et al. 2014). *Coprosma oliveri* is endemic to Robinson Crusoe Island, and *C. pyrifolia* is endemic to both islands. Skottsberg (1921) suggested that *C. oliveri* (as his *C. triflora*) might be related to *C. foliosa* from Hawaii and that *C. pyrifolia* was closer to *C. laevigata* from Rarotonga and other species in this Pacific region. Oliver (1935) mentioned that *C. oliveri* (as *C. hookeri*; for details of this nomenclature, consult Fosberg 1968) was not closely related to any other species but that it might be related to *C. acutifolia*, *C. lucida*, *C. macrocarpa*, or *C. robusta* from New Zealand. He regarded *C. pyrifolia* as approaching *C. taitensis* from Tahiti in the shape of leaves and stipules and presence of domatia but also differing in a number of other features. Fosberg (1968), in addition to straightening out the nomenclature of *C. oliveri*, opined that the two endemic island species are close to each other and then possibly to *C. cookie* from Rapa. Heads (1996) placed the two Juan Fernández species into his group "Pyrifoliae" along with nine other species grouped from the Society Islands (five species) and the island complex of Rarotonga, Rapa, and Pitcairn Island (four species). At the morphological level, therefore, the affinities of the Juan Fernández endemics are apparently from the western Pacific, but the specific ties have been unclear. Charlotte Taylor (unpublished manuscript) pointed out that there may be hybridization between the two species on Robinson Crusoe Island (e.g., *Crawford et al. 11889*, *Stuessy & Soto 11907*), which suggests that they are genetically similar. Hybridization is not rare in the genus, with several hybrids having been detected between species in New Zealand (Wichman et al. 2002). Eduardo Ruiz et al. (unpublished manuscript) have analyzed ITS sequences from the two Juan Fernández endemics, and they are identical, which also suggests that these species are very closely related and have descended from a single introduction to the archipelago. The most comprehensive molecular study in the genus has been done by Cantley et al. (2014) using ITS, ETS, and *rps16*. Although the focus was on Hawaiian endemic species, *Coprosma pyrifolia* was included for comparison. In a combined data analysis, this Juan Fernandez endemic falls between a group of species from Hawaii, the Marquesas, and Rapa Iti Island and on the other side with species from New Zealand, the Kermadec Islands, and the Chatham Islands. All that can be said at this time, therefore, is that the Juan Fernández endemic has arrived from the western Pacific and probably dispersed by birds (Fig. 13.2I). It is likely that the first colonist arrived on the older island and speciated cladogenetically because both endemic species occur there. After the younger island was formed, another dispersal event occurred with *C. pyrifolia*, perhaps very recently, because this lineage has not yet diverged at the specific level.

Drimys confertifolia (Winteraceae) (Fig. C29) is endemic to the archipelago and is one of the conspicuous forest trees, especially on Robinson Crusoe Island. Earlier ITS studies (Ruiz et al. 2008) revealed insignificant sequence variation among *D. confertifolia* and the continental congeners *D. winteri* and *D. andina*. With indel analysis, however, the island species appeared somewhat distant from the continental relatives. Recent molecular genetics studies using AFLP and nuclear microsatellites

(López-Sepúlveda et al. 2015b) have shown that *D. confertifolia* does relate to *D. winteri* (and *D. andina*) from mainland Chile. Although the populations on the two islands are placed in the same species based on morphological similarity, genetic differentiation has occurred between the islands. It seems most likely that dispersal took place from *D. winteri* to the first formed and nearest island, followed by anagenetic speciation. More recently, dispersal has also taken place to the younger and further island but not yet accompanied by marked differentiation.

Dysopsis hirsuta (Euphorbiaceae) (Fig. C73) is an endemic species of the archipelago restricted to Robinson Crusoe Island that ties to the only other two species of the genus, both of which occur in South America. *Dysopsis paucidentata* occurs in western Venezuela, Colombia, and Ecuador with isolated disjunct populations in Costa Rica, Perú, and Bolivia (Lozano and Murillo-A. 2001). *Dysopsis glechomoides* is distributed in southern Chile (with a few localities over the Andes into Argentina). It is presumed that the Juan Fernández endemic was derived from *D. glechomoides* on the continent, but more study is needed to confirm this hypothesis. In any event, *D. hirsuta* is derived from South America, and it evolved anagenetically after arrival on Robinson Crusoe Island.

One of the most biogeographically and evolutionarily interesting genera in the Juan Fernández Archipelago is *Erigeron* (Asteraceae, Astereae) (Figs. C53–C57). This is a large genus of nearly 400 species with upwards of 23 species in South America (Solbrig 1962). In the archipelago, it occurs primarily on Alejandro Selkirk Island, where the following five endemic species are found: *E. fernandezia*, *E. ingae*, *E. luteoviridis*, *E. rupicola*, and *E. stuessyi*. *Erigeron fernandezia* is also present on Robinson Crusoe Island, but this is presumably a more recent introduction, perhaps even during historical time (López-Sepúlveda et al. 2015). *Erigeron corrales-molinensis* has been added recently for Robinson Crusoe Island (Danton 2014), but this appears to be only a diminutive morphological variant of *E. fernandezia*. Molecular phylogenetic analyses using ITS have been done on *Erigeron* by Noyes (2000), and *E. fernandezia* and *E. rupicola* (erroneously cited as *E. rosulatus*, which does not occur in the islands) tie to *Conyza floribunda* and *C. bonariensis* in the analysis, both from Latin America. The relationships between these two (and other) genera of Astereae are complex (Cronquist 1943, 1947; Harling 1962; Nesom 2008), and it may be that the island species came from out of *Conyza* or what is now so regarded. Andrus et al. (2009), also using ITS markers, showed that the two island species (the same ones) tied most closely to *C. bonariensis* but also to *Erigeron luxurians*, *E. fasciculatus*, and *Darwiniothamnus alternifolius* (endemic to the Galápagos Islands). Based on morphological phenetic and cladistic analyses of 22 species from South America, Valdebenito et al. (1992a) suggested that the island endemics might have derived from something similar to *E. leptorhizon* from coastal Perú, a species not included in the above-cited molecular surveys. At minimum, the ITS data reveal the two island species to be closely related, which strengthens the hypothesis that the group in the islands is monophyletic, resulting from a single introduction. Most interesting, however, is that *Erigeron* apparently did not establish successfully on the older island, Robinson Crusoe, but only on the younger island. After arrival on Alejandro Selkirk Island, cladogenesis ensued to produce the complex of five species now seen, all taking place within a maximum of 1 to 2 million years (Fig. 13.2J). There is a clear ecological differentiation among some of the species such that *E. rupicola* is confined to coastal rocks; *E. stuessyi* occurs in the moist, cool parts of the quebradas; *E. fernandezia* is more adapted to open areas, especially on the middle slopes and ravine margins; and *E. ingae* and *E. luteoviridis* grow mixed together in the "alpine" portion of the highest ridges of the island, where high humidity results from frequent fog cover. Because of the loss of surface area and habitat on the older island, as well as substantial human inpact, the present distributions of taxa there, for example, in the larger genera *Dendroseris* and *Robinsonia*, do not reflect well the original ecological conditions that may have encouraged adaptive radiation on that island. *Erigeron*, therefore, represents the best example of radiation, presumably adaptive, within the archipelago.

Eryngium (Apiaceae) consists of four endemic species in the Juan Fernández Archipelago. *Eryngium bupleuroides* (Fig. C34), *E. inaccessum*, and *E. ×fernandezianum* occur on Robinson Crusoe Island, and *E. sarcophyllum* used to occur only on Alejandro Selkirk Island, but it may be extinct (Danton et al. 2006), having last been recorded by Johow (1896); Ricci (2006), however, has reported one visual sighting. Skottsberg (1914, p. 17) originally described *E. fernandezianum* as a good species, although with some uncertainty whether it might be a new genus. Later (1921), after describing another endemic species, *E. inaccessum*, he saw the possibility, based on intermediate morphological features, that *E. fernandezianum* might be an interspecific hybrid between *E. inaccessum* and *E. bupleuroides*. Skottsberg (1921, p. 161) also presented a table of quantitative morphological features among the three taxa that further showed intermediacy for *E. fernandezianum*. Results by Gerhard Jakubowsky at the University of Vienna (abstract, Jakubowsky and Stuessy 1999), using 18 floral and vegetative features with PCA morphometric analyses, have confirmed the intermediate nature of *E. ×fernandezianum*, and we treat it as having a hybrid origin. It is not known whether the plants seen and named by Skottsberg represent a stabilized species of hybrid origin or are only occasional F_1 hybrids. Skottsberg commented (1921, p. 160) that "No ripe seeds were found," but this may not necessarily mean that the individuals were sterile. All these taxa are rare and grow in secluded localities.

Determining origins of *Eryngium* is also difficult. It is a very large genus, with 220 to 250 species distributed in Eurasia, North Africa, Australia, and North and South America (Wörz 2005, 2011). Calviño et al. (2008a) presented the most comprehensive molecular phylogenetic results within the genus using the nrITS region and cp *trnQ-trnK* 5′ exon and with a sampling of 118 species. Included from the Juan Fernández Archipelago was *E. bupleuroides* (*E. inaccessum* is also placed in the trees, e.g., in Calviño et al. 2008a, fig. 5, showing analysis of combined data, but their table 1 seems to indicate that only *E. bupleuroides* was sampled). The combined analysis reveals these two species as falling within the "Pacific" clade close to *E. articulatum*, *E. vaseyi*, *E. ovinum*, *E. rostratum*, and *E. vesiculosum*. These taxa are found in Chile, California, and Australia, which leaves doubt regarding the direction of origin of the Juan Fernández complex, perhaps most likely from either the western Pacific or southern

South America. Calviño et al. (2010) again presented these same results and this time stated (in their fig. 1) that this Pacific clade involves "Several dispersals probably from Chile to California, Australia, and the Juan Fernandez Islands." The molecular results in Calviño et al. (2008a, 2008b), however, clearly show that *E. bupleuroides* is most closely related to *E. vesiculosum* from Australia/New Zealand, a more ancient land region more suggestive of being a source rather than recipient area. Kadereit et al. (2008) also examined relationships among 52 species of *Eryngium* and other genera in a broader survey of ITS and cp *rps16* intron variability in subfamily Saniculoideae tribe Saniculeae. They also found, based on the ITS data, that *E. bupleuroides* from Juan Fernández nested within a clade containing species mostly from Australia and the Americas. The best that can be said based on these data, therefore, is that *Eryngium* might have arrived in the archipelago from either southern South America or from the western Pacific. Wörz (2011), in the first part of his new monograph on the genus, has suggested that *Eryngium* in South America derived from Mediterranean progenitors that eventually arrived to the Juan Fernández Archipelago. From whichever direction, the genus probably arrived first on the older Robinson Crusoe Island. Because *E. sarcophyllum* on Alejandro Selkirk Island may be extinct, it is more difficult to know its relationships with the other three species. Herbarium material of *E. sarcophyllum* indicates that this species is distinct from the others on the older island, particularly in its narrow leaves (Jakubowsky and Stuessy 1999). Our best guess at this point is that there were two introductions to the archipelago (Fig. 13.2K). It has been suggested previously that *E. sarcophyllum* ties more closely to relatives in South America, such as with *E. ciliatum* from Brazil (Jakubowsky and Stuessy 1999). In any event, the colonist to Robinson Crusoe Island diverged cladogenetically into two lineages, *E. bupleuroides* and *E. inaccessum*, which subsequently hybridized to produce *E.* ×*fernandezianum*, perhaps even in recent times. This is compatible with the demonstration by Calviño et al. (2008) that numerous hybridizations have taken place during evolution of the genus.

Escallonia callcottiae (Escalloniaceae) (Fig. C72) is a shrub endemic to Robinson Crusoe Island. Recent molecular phylogenetic studies (Sede et al. 2013) on the genus *Escallonia* using the chloroplast markers *trnS-trnG* and 3'*trnV-ndhC* intergenic spacers, plus the *ndhF* gene, have shown *E. callcottiae* to be closely related to the southern Andean species *E. virgata*. In another molecular phylogenetic study, however, Zapata (2013), using *trnH-psbA*, the third intron from nitrate reductase (NIA), and the first intron of a MYC-like anthocyanin regulatory gene, obtained slightly different results. He showed a tie with *E. rubra* using NIA and with *E. florida* using MYC. The results with *trnH-psbA*, however, showed the closest ties with *E. rosea* and *E. virgata*. All these associated species have distributions in Chile and therefore strongly underline the idea that no matter which particular progenitor was involved, *E. callcottiae* resulted from a dispersal from the Chilean mainland to Robinson Crusoe Island followed by anagenetic speciation.

Euphrasia formosissima (Orobanchaceae) is an extremely distinct endemic species within this genus of approximately 170 species (Wu et al. 2009), so much so that it has been placed in its own section, *Paradoxae*, by Pugsley (1936). Skottsberg (1921) pointed out that this species has nothing in common with the other species from Chile (now listed at 16 species by Marticorena and Quezada 1985) nor does it fit well with the species from Australia and New Zealand. Wettstein (in Skottsberg 1921), the only comprehensive monographer of the genus (1896), provided a long commentary on these relationships and concluded that the Juan Fernández endemic was not close to species in South America but rather closer to species of Japan and New Zealand. Barker (1982), in a comprehensive treatment of the genus in Australia, who also provided an overview of all sections plus a cladistic analysis at the sectional level, indicated that *E. formosissima* (sect. *Paradoxae*) is closest to species in sect. *Novaezeelandiae*. The molecular phylogenetic study by Gussarova et al. (2008) analyzed relationships among 51 species of the genus, and they did include *E. formossisima*. The results are quite surprising, in view of the affinities discussed earlier based on morphological data. With three cp regions (*trnL* intron, *trnL-trnF*, and *atpB-rbcL* intergenic spacers), the Juan Fernández endemic ties using parsimony (bootstrap 51% support) and Bayesian inference (0.88 posterior probability) with *E. antarctica* and *E. chrysantha* in sect. *Trifidae*, both known from continental Chile (Marticorena and Quezada 1985). The support values, however, are not strong in support of this node. With ITS sequences, *E. formossima* joins with *E. chrysantha* from Chile and *E. hookeri* from Tasmania, and this time with more substantial 86% bootstrap support. The preponderance of molecular evidence, therefore, contradicts morphology and suggests a long-distance dispersal event from a mainland Chilean progenitor to Alejandro Selkirk Island followed by anagenetic speciation. Barker (1982) concluded that seeds of *Euphrasia* are not well adapted for long-distance dispersal, but in view of the many disjuncts in the genus, this would seem the only plausible explanation. Recently, Danton (2014) described a new subspecies within *E. formossima* (subsp. *cucharensis*), but population studies need to be completed before this new taxon can be accepted.

Skottsberg (1921) described *Galium masafueranum* (Rubiaceae) as new, from the "alpine" region of Alejandro Selkirk Island, found ". . . trailing in the moss mats . . .." (p. 174). He associated this new taxon with other Latin American species, such as *G. uncinulatum*, *G. canescens*, *G. andicola*, and *G. fuegianum*, the latter from Chile. At the present time, 16 other species are also known from continental Chile (Marticorena and Quezada 1985). Dempster (1980) placed the island endemic in sect. *Lophogalium*, which contains species that have fruits with straight hairs. The distribution of this section is given by her from southern Ecuador to Bolivia, to near Concepción, Chile, and to Neuquén in southern Argentina. She offers no specific comments about relationships of the island endemic, but in her key, the species falls next to *G. kilippii* (from Perú) and *G. plumosum* (with two subspecies, one from Bolivia and the other from Argentina). As for the species to which Skottsberg referred, *G. uncinulatum* is from Central America, *G. canescens* from Venezuela, and *G. fuegianum* from Tierra del Fuego. These are in a group of 16 species that have uncinate hairs on their fruits (Dempster 1981). Once again, a number of molecular phylogenetic studies have been done on *Galium* (e.g., Manen et al. 1994; Manen and Natali 1995; Natali and Manen 1995; Manen 2000; Soza and Olmstead 2010) but with approximately 400 species, the sampling challenge is immense, and so far no one has analyzed *G. masafuerana*. It is apparent that at present there is no clarity on the progenitor of the Juan Fernández endemic. Taking all facts in hand, however, dispersal presumably took place from some progenitor in South America that arrived at Alejandro Selkirk Island during the past 1 to 2 million years, followed by anagenetic speciation.

Gavilea insularis (Orchidaceae) is the only endemic orchid in the Juan Fernández Archipelago, and it is restricted to Alejandro Selkirk Island. The species had not been collected since the original specimen by G. Kuschel in 1956 from Cerro Correspondencia at 1,200 m on Alejandro Selkirk Island and hence was considered extremely rare, if not extinct. Danton (1998), however, was able to relocate it again in 1997, and we found it during our expedition to the island in 2011. The genus consists of 16 species and is distributed in the southern humid regions of Argentina and Chile (Correa 1956; Chemisquy 2009). Twelve species are found in Chile, with five of them endemic to the country (Lehnebach 2003). *Gavilea insularis* was described rather recently by Correa (1968) as a relative of *G. supralabellata*, which has a distribution in southern Argentina in Santa Cruz province (Correa 1966). Correa, however, pointed out that even though these species are morphologically similar, they belong to different sections of the genus (sect. *Gavilea* and sect. *Anadenia*, respectively). There is also some evidence that this latter species might even be an intergeneric hybrid between *Gavilea* and *Chloraea* (Chemisquy and Morrone 2012). Several molecular phylogenetic studies have been done on some of the species of the genus. Cisternas et al. (2012) focused on intergeneric relationships in Chloraeinae and examined seven species of *Gavilea*. Chemisquy and Morrone analyzed six species, but a much broader sampling was completed by Chemisquy and Morrone (2012) with 11 species, but again, *G. insularis* was not included. Although a particular continental relative of the island endemic has not been identified, dispersal must have taken place from southern South America to the younger island followed by anagenetic speciation. That orchid seeds have good dispersal ability over large distances has been summarized well by Arditti and Ghani (2000), hence presenting no obstacle to reaching the archipelago.

Greigia berteroi is an endemic species of Bromeliaceae on Robinson Crusoe Island, along with *Ochagavia elegans*. In contrast to the latter, however, *G. berteroi* is very rare, so much so that we have never found it during ten expeditions to the island. Ricci (2006), however, has reported its existence on Cerro Agudo. The genus contains only four species, and all are endemic to Chile. It is therefore extremely likely that the Juan Fernández endemic was derived from this mainland lineage. This includes the species *G. landbeckii*, *G. pearcei*, and *G. sphacelata* (Will and Zizka 1999). Skottsberg (1936) judged that *G. berteroi* and the two species first listed earlier were sufficiently different morphologically and anatomically to merit separation as a distinct genus, *Hesperogreigia*, which left *G. sphacelata* as the only species in *Greigia*. Recent authors (e.g., Smith and Downs 1979; Will and Zizka 1999), however, have not accepted this disposition. There has never been any quantitative morphological or molecular study of the entire genus, so the precise affinities of *G. berteroi* are not well understood. Will and Zizka (1999) mentioned that *G. berteroi* can be easily distinguished from the other species except for *G. pearcei*, which is more similar morphologically. It may be that birds provided the dispersal from this mainland lineage to Robinson Crusoe Island (Will and Zizka 1999), where anagenetic speciation gave rise to *G. berteroi*.

Gunnera (Gunneraceae) consists of about 40 species distributed in the southern hemisphere (Wanntorp and Wanntorp 2003). In the Juan Fernández Archipelago it consists of three endemic species, two on Robinson Crusoe Island, *G. peltata* (Fig. C76) and *G. bracteata*, and one on Alejandro Selkirk Island, *G. masafuerae* (Fig. C75). All reside in subgenus *Panke* (Wanntorp et al. 2003), which is characterized by having scales between leaves on the enlarged rhizomes. Morphological cladistic studies and flavonoid data presented by Pacheco et al. (1993) have shown clearly that the progenitor of the island species is *G. tinctoria* (or its immediate ancestor) in the Chilean continent. One obvious hypothesis is that the original immigrant arrived first on Robinson Crusoe Island and diverged cladogenetically into *G. peltata* and *G. bracteata*, followed by dispersal from out of *G. peltata* to the younger island and anagenetic speciation that yielded *G. masafuerae* (Pacheco et al. 1993). *Gunnera bracteata* appears closely related to *G. peltata*, as evidenced by extensive hybridization along the main trail in Valle de Villagra on Robinson Crusoe Island (Pacheco et al. 1991). There seems little doubt that *G. peltata* is very closely related to *G. tinctoria* (= *G. chilensis*) in the Chilean continent and most likely derived from it or its immediate ancestor. The controversial aspect of phylogeny among these species is the origin of *G. masafuerae*. Wanntorp et al. (2002) using ITS, *rbcL*, and *rps16*, examined nine species in subg. *Panke*, including *G. tinctoria* (= *G. chilensis*) and the three island species, which together form a monophyletic group, although only supported by 51% bootstrap with ITS data and 54% with combined ITS and plastid sequences. Unsurprising is that *G. peltata* and *G. bracteata* appear as sister species with 65% to 69% support, but more surprising is that *G. masafuerae* ties to *G. tinctoria* with 98% to 99% support. Ruiz et al. (2004) used different ITS and isozyme data to address these same relationships. These data reveal that the ITS divergence between *G. peltata* and *G. bracteata* is 0.0, that is, no difference at all. This provides still more corroborating evidence that these species are very closely related and suggests a cladogenetic (or anacladogenetic) event on Robinson Crusoe Island. Genetic identities from isozymes are high between each pair of species, with the very slightly higher 0.986 occurring between *G. masafuerae* and *G. peltata* in contrast to 0.957 between the former and *G. tinctoria*. Another set of comparative data, terpenoids, were investigated in *G. tinctoria*, *G. peltata*, and *G. masafuerae* (Bittner et al. 1994), which gives another view of relationships. Unfortunately, these data are also equivocal. Most of the terpene constituents are found in all three species, but two of them (lupeol acetate and lonona) are shared by *G. tinctoria* and *G. masafuerae*, and one of them (loliolide) is shared by *G. tinctoria* and *G. peltata*. Taking all facts into consideration, therefore, we hypothesize (Fig. 13.2L) that the common progenitor of all the endemic species of *Gunnera* in the archipelago was *G. tinctoria* or its immediate ancestor. However, two introductions established in the islands, the first one to Robinson Crusoe Island, whereby the immigrant split into two different lineages giving rise to *G. peltata* and *G. bracteata*. Another introduction occurred later to Alejandro Selkirk Island and resulted in the anagenetically derived *G. masafuerae*. The only other hypothesis that would be worth considering is that there was only one introduction to the islands, to Alejandro Selkirk Island first, followed by dispersal to the older island and subsequent cladogenetic speciation. This alternative seems less likely for several reasons. The fruits of this genus are fleshy and most likely have been dispersed by birds, such as thrushes, for example, *Turdus falcklandii* subsp. *magellanicus* (Chapter 12; see also Ridley 1930, p. 478), which means that flights must have been occurring from the continent during the first 2 million years of existence of the older island before the younger one was formed. This gave time, therefore, for the cladogenetic divergence to take place. *G. masafuerae* on the younger island is the only species of the genus found there, which suggests that it arrived more recently. It is doubtful that there would have been time for this lineage, after anagenetic speciation, to disperse back to the older island and then speciate cladogenetically.

Haloragis (Haloragaceae) is a genus of 26 species (Orchard 1975) distributed mainly in Australia but also occurring in New Zealand and a few of the Pacific islands, including the Juan Fernández Archipelago. The genus is not known from continental South America. Two species are endemic in the Juan Fernández Islands: *H. masatierrana* (Fig. C78) is confined to Robinson Crusoe Island, and *H. masafuerana* (with two varieties, var. *masafuerana* and var. *asperrima*) is restricted to Alejandro Selkirk Island. Orchard (1975), who monographed the genus most recently, confirmed the distinctions of Skottsberg (1921) for separating the two species, one on each island. Biogeographically, the genus must have immigrated from the western Pacific because it is unknown in continental South America. Close ties have been seen morphologically between the island endemics and *H. erecta* from New Zealand (Forde 1964; Orchard 1975). The only molecular phylogenetic investigation of the genus (Moody and Les 2007) examined only *H. masatierrana*, which with ITS and cpDNA *trnK* and *matK* tie very closely to *H. erecta* from New Zealand. It seems most certain, therefore, that *Haloragis* arrived to the Juan Fernández Archipelago from New Zealand, and after arriving in the islands, dispersal from one to the other island led to anagenetic divergence and speciation (Fig. 13.2M). The only question is to which island the initial colonizer came or whether immigrants arrived to both islands more or less at the same time. Because the molecular divergence between *H. masatierrana* and *H. erecta* is very low, and because the older island was available for colonization much earlier than the younger island, we suggest that the immigrant first arrived on Robinson Crusoe Island, followed by dispersal to the younger island. This would be another case of the progression rule, but with initial arrival coming from the much more distant western Pacific. One might wish to argue that because varietal distinctions have now occurred among populations of *H. masafuerana* on the younger island, this could indicate more time for such differentiation and hence place of first colonization. The morphological differences between these two varieties (Orchard 1975), however, deal with the presence or absence of pubescence and wings on fruit and also the size of fruit, which may be nothing more than normal populational variation within the species on the island. In this same vein, Danton (2014) has recently recognized two additional new varieties (var. *applanata* and var. *scabrida*) within *H. masatierrana*, which we do not accept at this time, pending molecular comparisons.

Luzula masafuerana (Juncaceae) is the single endemic species in the archipelago within a cosmopolitan genus of 115 species (Kirschner 2002). It falls within sect. *Alpinae*, which consists of 20 species in mountains of South America, the northern hemisphere, and in New Zealand (Kirscher 2002). Among these species, five (including

L. masafuerana) occur in Chile (Marticorena and Quezada 1985): *L. alopecurus*, *L. chilensis*, *L. leiboldii*, and *L. racemosa*. Skottsberg (1921) originally described the island taxon as a new subspecies of *L. racemosa*: subsp. *insularis*. In 1953, however, he elevated this to specific rank and chose a new epithet as *L. masafuerana*. The association of the endemic taxon with *L. racemosa* from the Chilean mainland suggests that the latter may be the progenitor of the former, but no specific investigations on this point have been made. Several molecular phylogenetic studies have been done now on the genus (Drábková et al. 2003, 2004, 2006), and the most complete sampling using 102 accessions has been done by Záveská Drábková and Vlček (2010) with nuclear and chloroplast markers. A new analysis of these data was subsequently completed by Bozek et al. (2012). As is often the case, no sample of *L. masafuerana* was included in any of these investigations. It seems likely that the endemic Juan Fernández species originated from a progenitor in Chile, probably *L. racemosa*, dispersed to the younger island and there speciated anagenetically.

Machaerina scirpoidea (Cyperaceae) is endemic to Robinson Crusoe Island, where it occurs commonly along streams or waterfalls. The genus is placed in tribe Schoeneae and consists of some 45 species in tropical and warm regions, especially Australia (Mabberley 2008). Several molecular phylogenetic analyses have been done on genera of Schoeneae, but the sampling of *Machaerina* has been minimal. Muasya et al. (2009) examined DNA sequences in 93 genera of Cyperaceae, but *Machaerina* was limited to *M. mariscoides* from Tahiti and an unidentified species from New Guinea. Verboom (2006) investigated relationships among genera of Schoeneae using three chloroplast markers, but only one sample of *Machaerina* sp. from New Guinea was included (the same used by Muasya et al. 2009). One message from this study was demonstration of patterns of intercontinental dispersal. Viljoen et al. (2013), using two nuclear and three chloroplast markers, sampled many of the genera of tribe Schoneae, including four species of *Machaerina*, but again, *M. scirpoidea* was not included. They also infer very frequent long-distance dispersal events throughout the southern hemisphere during evolution of the tribe. Based on morphology, Strong (1997) examined the five species of *Machaerina* occurring in South America, which included the Juan Fernández endemic. This is an odd pattern of disjunctions on the continent, with one species confined to the Guayana Highlands, three in mountains along the southeastern coast of Brazil, and one in the Juan Fernández Islands. If these taxa are related, then the genus again clearly shows good dispersal capability, and getting to the archipelago would have been unproblematic. Because most of the species of the genus occur in the Pacific area, one might infer that *M. scirpoidea* derived from that region. The key to the direction of dispersal to the islands has been indirectly suggested by Koyama (1972). He treated *Machaerina ficticium* from Brazil as a subspecies of *M. scirpoidea*, which then left the Juan Fernández taxon as an endemic subspecies, *M. scirpoidea* subsp. *scirpoidea*. Strong (1997), however, treats these two taxa as specifically distinct. Whether as an endemic subspecies or species, this is a strong inference that the Juan Fernández endemic derived from the South American continent and not the western Pacific. The distance involved with dispersal is obviously not a difficulty because many such long-distance events have taken place during evolution of the genus (Viljoen et al. 2013).

After arrival on Robinson Crusoe Island, anagenetic speciation resulted in the species seen today.

Margyricarpus digynus (Rosaceae) is endemic to Robinson Crusoe Island. The affinities of this species seem particularly clear because Skottsberg (1921) originally described the taxon as a subspecies of *M. setosus* (this is a later synonym of *M. pinnatus*, which has nomenclatural priority) (Kuntze 1898). Subsequently, Skottsberg (1951) elevated the taxon to specific status. *Margyricarpus pinnatus* is common in southern South America. The genus in a narrow sense consists of only two species (*M. digynus* and *M. pinnatus*), but it augments to eight species if *Tetraglochin* is included (Potter et al. 2007), with distribution in the Andes, southern Brazil, and Uruguay (Mabberley 2008). There seems little doubt in this case that the progenitor of *M. digynus* must have been *M. pinnatus*, followed by anagenetic divergence on the older island. Potter et al. (2007) summarized available molecular sequence data for numerous genera of Rosaceae, which included *M. pinnatus* and *M. cristatus* (= *Tetraglochin cristata*), but *M. digynus* has not yet been analyzed.

Myrceugenia schulzei (Myrtaceae) (Fig. C79) belongs to a genus of forty species confined to southern South America (Landrum 1981a; Murillo-Aldana et al. 2016). The island endemic is located on Alejandro Selkirk Island. Landrum (1981b), based on a morphological cladistic analysis, suggested a tie of this species with *M. exsucca* and *M. lanceolata*, both occurring in continental Chile. Data from flavonoids, isozymes, and ITS sequences (Ruiz et al. 1994, 2004) partially support Landrum's hypothesis. Murillo-A. et al. (2012), however, using two nuclear (ITS and ETS) and four plastid (partial *trnK-matK*, *rpl32-trnL*, *trnQ-50rps16*, and *rpl16*) sequences found a closer relationship to *M. colchaguensis*. Recent detailed population genetics studies using AFLPs and SSRs (López-Sepúlveda et al. 2013a) have reaffirmed this latter tie. From this continental progenitor, it seems clear that dispersal took place directly from the continent to the younger island followed by anagenetic speciation.

Nicotiana cordifolia (Solanaceae) has always been treated as being endemic to Alejandro Selkirk Island. Recently, Danton (2006) validly described a new subspecies, subsp. *sanctaclarae*, confined to Santa Clara Island, which lies just off the southwest coast of Robinson Crusoe Island. Two years earlier, Danton (2004) had published a lengthy description of this same taxon, with an excellent color plate but with indication of the name as a "nom. prov." (provisional name) and without designation of a type, both of which rendered the subspecies invalid in that publication. Subspecies *sanctaclarae* is mostly distinguishable by its pale yellow corollas, in contrast to the cream corollas that are often overlain with purple in subsp. *cordifolia* (but not always; see Goodspeed 1954, p. 347). If this new subspecies is, in fact, distinct (genetic and populational variation data would be most helpful), then a hypothesis needs to explain the distribution of the two subspecies within the archipelago. There can be little doubt that *N. cordifolia* is derived from a progenitor in South America, where much diversity in the genus occurs. Goodspeed (1954), in his impressive monograph of *Nicotiana*, suggested relationships with *N. raimondii* from the Andes of southern Perú or perhaps with *N. solanifolia* from coastal northern Chile, due in part to morphological similarity (cf. Goodspeed 1954, figs. 61, 63, and 64) but also the degree of chromosomal associations during meiosis and the viability and fertility of F_1 hybrids between these species in subgenus *Rustica* section *Paniculatae*. As might be expected in a genus that contains the economically significant tobacco, a number of taxonomic shifts and molecular investigations have recently been completed. Taxonomically, Knapp et al. (2004) maintained many of the sections of Goodspeed, but they did not use his subgenera. *Nicotiana cordifolia* remains in sect. *Paniculatae* with *N. solanifolia*, *N. raimondii*, and five other species. This is basically the same composition of species that Goodspeed had for this section, except that he also included *N. glauca* (which they place in sect. *Noctiflorae*). They also include *N. cutleri*, which was recently described by D'Arcy (1976). As for molecular investigations on *N. cordifolia* and relatives, Chase et al. (2003) and Clarkson et al. (2004) have used ITS and five plastid loci, respectively, for comparison; Komarnytskyy (2005) has examined AFLP relationships; and Khan and Narayan (2007) have presented RAPD comparative data. All these studies have yielded the same conclusion that *N. solanifolia* from coastal northern Chile is sister to the island endemic. Biogeographically, there appears to have been long-distance dispersal to the archipelago, apparently to the younger island, followed by anagenetic divergence during the past 1 to 2 million years. At some later time, dispersal also occurred to Santa Clara Island, after it was formed off the coast of Robinson Crusoe Island approximately 1 million years ago (Sanders et al. 1987), followed by some morphological changes. Other less parsimonious explanations are possible, of course, such as parallel introductions and evolution and/or original immigration and extinction on the older island.

Ochagavia elegans (Bromeliaceae) is an attractive species (Fig. C31) that is endemic to Robinson Crusoe Island. The genus is small, with only three other known species (*O. andina*, *O. carnea*, and *O. litoralis*), all endemic to continental Chile (Zizka et al. 2002). Monteiro et al. (2015) completed a morphological phylogenetic analysis of genera of the family and included *O. elegans* and *O. litoralis*, which formed a small, exclusive clade. Molecular phylogenetic studies have also examined *O. elegans*, but not involving all other species of the genus. Schulte et al. (2009), using one nuclear and five plastid loci, analyzed only the island endemic in comparison with other genera. A more comprehensive study for the entire family (Givnish et al. 2011) using eight plastid loci examined *O. elegans* and *O. carnea*, but not the other two species. The analyzed species formed a monophyletic group along with *Fascicularia bicolor*, which is known to be the most closely related genus (Zizka et al. 2002). Evans et al. (2015) also analyzed genera of subfamily Bromelioideae and included the same two species, *O. elegans* and *O. carnea*. This study used three chloroplast loci and found that *O. carnea* ties closely with *Fascicularia bicolor* and that *O. elegans* joins preferentially with *Deinacanthon urbanianum*. Schulte et al. (2005) examined *O. elegans* and *O. litoralis* with three chloroplast markers, and the latter joined closest with *Fascicularia bicolor*, with *O. elegans* joining next. The most comprehensive approach was with AFLPs that considered *O. elegans*, *O. litoralis*, and *O. carnea* (Horres et al. 2007). The results showed *O. litoralis* and *O. carnea* to be most closely related with *O. elegans*, their sister group. Although *O. andina* was not analyzed in this study, the data do suggest that *O. elegans* may have originated from the common ancestor of the continental species. In any event, there can be no doubt that immigration came from mainland Chile by long-distance dispersal (Horres et al. 2007), and anagenetic speciation resulted in *O. elegans*.

Peperomia (Piperaceae) is a massive genus of more than 1,600 species (Wanke et al. 2006) distributed in tropical parts of the world. Of this immense diversity, four species occur in the Juan Fernández Archipelago (Skottsberg 1921; Marticorena and Baeza 2001). Two of them are endemic to the islands: *P. margaritifera* restricted to Robinson Crusoe Island and *P. skottsbergii* found only on Alejandro Selkirk Island. Yuncker (1953) added a new variety var. *umbraticola* Yuncker ex Skottsberg within *P. margaritifera* on Masatierra, but it seems a minor variant, and we have not recognized it in this book. *Peperomia berteroana* subsp. *berteroana* (Fig. C27) occurs on both islands and is an endemic subspecies, with the other subspecies disjunct on Inaccessible Island in Tristan da Cunha in the Atlantic Ocean. Skottsberg (1946) and Christophersen (1968) treated these as closely related but distinct species, but Valdebenito et al. (1990b) combined them into one species at the level of subspecies. One native species, *P. fernandeziana* (Fig. C28), is known from both islands as well as the Chilean continent. Phenetic and cladistic morphological studies by Valdebenito et al. (1992a) have indicated that the endemic species of the genus in the islands form a monophyletic group but that the native *P. fernandeziana* has resulted from a separate origin and introduction. Yuncker (1953), in his treatment of the Argentinian, Bolivian, and Chilean species, also placed this latter species far from the other island species. Morphological features indicate that *P. margaritifera* on the older island shows the least derived condition in comparison with continental congeners (Valdebenito et al. 1992a). Particularly indicative is that the continental relatives are facultative epiphytes, and *P. margaritifera* in the archipelago is also of this nature. It is likely, therefore, that *P. margaritifera* evolved anagenetically first from an ancestor from the Chilean continent and then dispersed to the younger island resulting in *P. skottsbergii* (Fig. 13.2N). Explaining the origin of *P. berteroana* is a bit more challenging. This species is found on both islands, and it also ties morphologically most closely to subsp. *tristanensis* of the Tristan de Cuhna Archipelago (Valdebenito et al. 1990a) in a very odd disjunct distributional pattern between the Pacific and Atlantic oceans, doubtless due to long-distance dispersal by birds (Valdebenito et al. 1990b). *Peperomia berteroana*, despite its different vegetative appearance, is not out of place in the island group and fits well within a monophyletic concept (Valdebenito et al. 1992a). Also important is that this species possesses sulfated flavones (see Chapter 10), which also occur in *P. skottsbergii* on the same island and are not known elsewhere in the family (Valdebenito et al. 1992b). It is likely, therefore, that *P. berteroana* also originated on Alejandro Selkirk Island and subsequently dispersed to the older island as well as to Tristan da Cunha. Molecular data have been examined within *Peperomia* (Wanke et al. 2006; Naumann et al. 2011; Samain et al. 2011; Symmank et al. 2011), but none of the Juan Fernández endemics has been included in the analyses. Likewise, there is an interesting sampling of 45 species for leaf crystal patterns within the genus (Horner et al. 2009), but again, no island endemic was included.

Pernettya (Ericaceae) consists of 14 species (Sleumer 1985) distributed in Tasmania (two species), New Zealand (three species), subtropical Central and South America (four species), and temperate South America (five species). A number of workers have combined *Pernettya* into *Gaultheria* (Stevens 1971; Middleton and Wilcock 1990; Middleton 1990, 1991; Powell and Kron 2001; Teillier and Escobar 2013), but in this book we follow Luteyn (1995) in maintaining the two as separate genera (see Chapter 5 for more discussion). In the Juan Fernández Archipelago, *P. rigida* (Fig. C71) is endemic to both islands of the archipelago. Sleumer (1985) does not comment on the origin of *P. rigida*, but he places it with *P. myrtilloides* s.l. from Chile in his key. In the recent treatment of the genus for Chile (as *Gaultheria*), Teillier and Escobar (2013) place *P. rigida* (as *Gaultheria racemulosa*) in their key next to *G. poeppigii*, also from Chile and Argentina. Three molecular phylogenetic studies (Powell and Kron 2001; Bush et al. 2009; Fritsch et al. 2011) have been done on *Pernettya* (*Gaultheria*) and generic relatives, but none has included the island endemic. Available facts suggest that *P. rigida* derived by long-distance dispersal from an ancestor on the Chilean mainland to Robinson Crusoe Island and subsequently diverged anagenetically. Dispersal to the younger island must have taken place relatively recently because there is no obvious morphological divergence between populations on the two islands.

Plantago fernandezia (Plantaginaceae) is the only endemic member in the Juan Fernández Archipelago of a genus of approximately 240 species found typically in temperate and higher elevations in tropical regions of the world (van der Aart and Vulto 1992). Unsurprisingly, in the islands there are also the ubiquitous weeds, *P. major* and *P. lanceolata*, especially in and around the village of San Juan Bautista. The endemic species is a rosette small tree confined to the upper slopes of ridges on Robinson Crusoe Island and rather scarce. Johow (1896) suggested that this species has similarity with *P. princeps* from the Hawaiian Islands, which was echoed by Pilger in Skottsberg (1921, p. 171). Rahn (1996) conducted a morphological cladistic analysis of the entire family, which did include *P. fernandezia*. The results placed the endemic in a small clade with *P. principes* and *P. rupicola* from Rapa, but this was supported only by the single character state of elongated stems. Carlquist (1970) examined the wood anatomy of *P. fernandezia* in comparison with other insular species of the genus, especially *P. principes*, and they do have some anatomical similarities (e.g., vessels that are longer and wider in diameter than other species sampled). He cautions, however (Carlquist 1970, p. 359), that "Too often, the similarities in growth form are taken as evidence of phylogenetic relationship, and insular rosette-trees and rosette shrubs tend to be misunderstood on this account." Several molecular phylogenetic studies have been done on the genus, for example, Rønsted et al. (2002), Hoggard et al. (2003), Dunbar-Co et al. (2008), and Tay et al. (2010), but none has included material of *P. fernandezia*. Dunbar-Co et al. (2008) did focus on *P. principes* and other species from Hawaii, but *P. fernandezia* was not available for comparison. It is known that 16 species are native to continental Chile (Marticorena and Quezada 1985). Most of these species are placed in subg. *Albicans* sect. *Gnaphaloides* or subg. *Plantago* sect. *Oliganthos*, whereas *P. fernandezia* falls in this latter subgenus but in sect. *Plantago* (Rahn 1996). Rahn (1984) also revised seven species of sect. *Oliganthos* in southern South America, but no comment was made of any relationship to the Juan Fernández endemic. It should be mentioned that other authors have questioned his infrageneric classification based on morphological features (e.g., Tay et al. 2010). There is also another endemic species,

P. lundborgii, from San Ambrosio Island in the Desventuradas, north of the Juan Fernández Archipelago, but this species is placed in subg. *Albicans* sect. *Gnaphaloides* by Rahn (1996) and in subg. *Psyllium* by Ronsted et al. (2002). Species of the genus disperse relatively well, as evidenced by affinities between taxa in Australia and New Zealand (Tay et al. 2010) and presumably due to the sticky seed exudates (Tay et al. 2010) that might facilitate attachment to birds. There would be no great difficulty for propagules arriving in the Juan Fernández Islands, either from the South American continent or from the western Pacific. More sampling and analysis are needed, but from whichever direction the progenitor arrived in Robinson Crusoe Island, anagenetic speciation occurred thereafter.

Ranunculus caprarum (Ranunculaceae) (Fig. C82) is a rare species that occurs only on Alejandro Selkirk Island. In our two expeditions to the island, we found only one plant. Ricci (2006) cites three localities. Hörandl et al. (2005) have analyzed approximately 200 species of the genus in a worldwide survey, and this island endemic ties with ITS data to the clade XVI (Chilean-Argentinian lowland) closest to *R. maclovianus*, also known from Chile (Ruiz 2001). New analyses (Hörandl and Emadzade 2012; Hörandl 2014) with ITS and *matK*, *trnK*, and *psbj-petA*, as well as morphological and karyological data, reveal a close relationship of *R. caprarum* to *R. chilensis* and *R. peduncularis*, also found in Chile. It is likely, therefore, that the species dispersed from the Chilean mainland directly to Alejandro Selkirk Island and there speciated anagenetically.

Rhaphithamnus venustus (Verbenaceae) (Fig. C87) is an endemic species that ties to the only other known species of the genus, *R. spinosus*, in southern South America (Sun et al. 1996). Isozyme studies have compared levels of genetic variation between populations on the continent and those on the islands, and a lower level of variation in the island endemic has been recorded (Crawford et al. 1993c). More recent molecular genetics studies on both species (López-Sepúlveda et al. 2016) confirm the origin of the former from the latter via anagenetic progenitor-derivative speciation. For more details of this generic system, see Chapter 14.

Santalum fernandezianum (Santalaceae) (Fig. 9.5) is extinct, last seen by Carl Skottsberg in 1908 (Skottsberg 1910). Among islanders, it is almost a social game, especially with tourists, to talk about aromatic wood of the lost sandalwood hidden in some secluded ravine. Herbarium specimens exist because the species was first described by Federico Philippi in 1892, but the species is now presumed extinct. Collections only exist from Robinson Crusoe Island, but occasionally some islander has reported finding aromatic wood from Alejandro Selkirk Island. During our extensive expeditions of 1986 and 2011 on this island, we did not encounter any such evidence. Skottsberg (1921, p. 117) did list this species as being from Alejandro Selkirk Island, but it was known only from "semi-fossil wood" collected by P. Gutiérrez and G. Arredondo from Quebrada del Sándalo, Quebrada Sandalito, and Quebrada del Varadero. It is unknown if this was, in fact, a species of *Santalum* or, if so, whether it was the same species as on Robinson Crusoe Island or a different one. Because of the scanty evidence and no specimen of any kind from that island, we assume that *S. fernandezianum* was only known with certainty from Robinson Crusoe Island. Authentic wood samples do exist from Robinson Crusoe Island (e.g., Fig. 9.5B), however, such as that obtained by F. Philippi and sent for analysis to Kraus (1882) or that collected by Johow on January 30, 1892 and now housed in the herbarium of the University of Concepción (CONC), which was analyzed chemically and anatomically by Hoeneisen et al. (1998) and Baeza et al. (1999), respectively. Molecular phylogenetic studies on the genus have been done by Harbaugh and Baldwin (2007), and some inferences on relationships of the Juan Fernández endemic have been forthcoming. Because no living material exists to allow extraction of high-quality DNA, the only marker obtainable from a herbarium collection (*Johow s.n.*, over a century old) was the 3′ *trnK* intron. This marker showed a clear tie to Australian species, which suggests an introduction from the western Pacific to Robinson Crusoe Island, followed by anagenetic change. The entire genus, in fact, is distributed in the Pacific region. Other genera of the family occur in South America (Nee 1996) but not the genus *Santalum*.

Solanum fernandezianum (Solanaceae), endemic to Robinson Crusoe Island, falls within *Solanum* subg. *Potatoe* sect. *Etuberosum* (Contreras-M. and Spooner 1999). Two other species occur in this same section, *S. etuberosum* from central and south-central Chile and *S. palustre* (= *S. brevidens*) from southern Chile (plus two possible introductions near Viña del Mar) and adjacent Argentina. David Spooner and colleagues have examined these three species for relationships based on isozymes (Spooner et al. 1992), cpDNA restriction sites (Spooner et al. 1993), and RFLPs (Spooner et al. 1996). All data suggest that *S. fernandezianum* is a distinct species more or less equidistant from the other two, and therefore, it is not possible to determine which of the mainland relatives might be the progenitor species. It is also possible that the Juan Fernández endemic originated from the common ancestor of the two continental taxa. It appears that a long-distance dispersal event occurred to Robinson Crusoe Island from out of this ancestral complex followed by anagenetic evolution. Based on isozyme divergence, Spooner et al. (1992) estimated a divergence time of 1.1 million years.

Sophora (Fabaceae) contains approximately 45 species, and two endemic species of sect. *Edwardsia* exist in the Juan Fernández Archipelago: *S. masafuerana* on Alejandro Selkirk Island and *S. fernandeziana* (with vars. *fernandeziana* [Fig. C74] and *reedeana*) on Robinson Crusoe Island. These two species are very close morphologically (Skottsberg 1921), and ITS sequence and isozyme genetic identity between the two are 0% and 0.973 (Ruiz et al. 2004), respectively, confirming this close evolutionary affinity. Among the 19 species of sect. *Edwardsia*, *S. macrocarpa* from continental Chile has been suggested as a possible progenitor (Ruiz et al. 2004), and the ITS divergence between this species and *S. fernandeziana* is 1.3% and 0.814 genetic identity. Another continental species, *S. cassioides*, has also been compared to *S. fernandeziana*, and by the same measures of affinity, results give 0.0 ITS divergence but a much lower 0.756 genetic identity (Ruiz et al. 2004). Peña and Cassels (1996), in a morphological cladistic analysis of Chilean species of the genus, showed a close tie of *S. prostrata* to the island endemics. Peña et al. (1993), based on pollen data, suggested that *S. masafuerana* is also closely related to *S. cassioides* (they gave *S. microphylla* subsp. *macnabiana*, but this is now regarded as a synonym of *S. cassioides*) (Heenan 2001). From whatever continental progenitor in South America (Peña et al. 2000), the data conform most strongly to the hypothesis that the genus first arrived on Robinson Crusoe Island and speciated anagenetically there, followed by dispersal to Alejandro Selkirk Island with still another anagenetic speciation (Fig. 13.2O). Molecular phylogenetic studies in sect. *Edwardsia* have been done by Mitchell and Heenan (2002) and Heenan et al. (2004), but neither of the Juan Fernández endemics was included.

Spergularia (Caryophyllaceae) is a genus with approximately 60 species in North America, Europe, and southern South America (Mabberley 2008). In the archipelago, *S. masafuerana* is found only on Alejandro Selkirk Island, and *S. confertiflora* is endemic to both islands. Two varieties have been recognized within the latter species (Skottsberg 1921, 1953), var. *confertiflora*, known from both major islands and Santa Clara, and var. *polyphylla*, also occurring on the two major islands. Whether these varietal distinctions will hold up with detailed investigation remains to be seen. As far as we can determine, no molecular investigations exist involving these two endemic species of *Spergularia*. Molecular studies have examined intergeneric relationships in Caryophyllaceae (Fior et al. 2006; Greenberg and Donoghue 2011), but no broad sampling has yet been done within *Spergularia*. In the most recent monograph on the genus in North and South America (Rossbach 1940), *S. confertiflora* is placed in the sequence of species and together in the key with *S. villosa*, which occurs in southern Chile and is apparently native there (Rossbach 1940, 1943). We can only speculate at this point that the genus in the archipelago derived from southern South America, perhaps continental Chile, where 14 native species occur (Marticorena and Quezada 1985). From the original immigrant, most likely to the older island, an anagenetic speciation would have produced the original *S. confertiflora* lineage (Fig. 13.2P). Further dispersal after the younger island emerged from the sea would have resulted in another anagenetic speciation yielding *S. masafuerana*. More recently, two varieties in *S. confertiflora* diversified cladogenetically, and further dispersal has resulted in both varieties now occurring on the older as well as the younger island.

Ugni (Myrtaceae) is a genus of four species of tropical and warm America (Mabberley 2008), of which *U. selkirkii* (Fig. C81) is endemic to Robinson Crusoe Island. Three species occur in Chile (Landrum 1988), with *U. candollei* and *U. molinae* occurring on the Chilean mainland, the latter of which also occurs on Robinson Crusoe Island as an introduced species (Figs. 8.5 and 8.6). Unfortunately, *U. molinae* has been spectacularly successful in the archipelago, and it may be the most abundant invasive species. It can successfully invade slopes and ridges from 350 to 600 m, and it blends well into the landscape, not appearing out of place (Fig. 8.6). It has competed successfully with the endemic *U. selkirkii*, however, and therein lies a danger for the future. Based on morphology, it is unclear which of the two continental species might be closer to *U. selkirkii*. Landrum (1988) keys the endemic closest to *U. molinae*, but no comments are proffered regarding origins. Biffin et al. (2010) have analyzed fleshy-fruited genera of Myrtaceae with DNA sequences, but only *U. molinae* was sampled. Murillo-A. et al. (2012, 2013) have examined molecular relationships among species of *Myrceugenia* and related genera, but within *Ugni*, only *U. selkirkii* and *U. molinae* were included. These latter results are interesting in that with ITS and ETS, the two species separate into independent lineages, but with only chloroplast markers or when combined with the nuclear markers, they form a monophyletic group distinct from other genera. If, in fact, *U. selkirkii* is derived from *U. molinae*, it would be a curious case whereby the progenitor species has been introduced to the archipelago, presumably in modern times, and it now behaves invasively, posing a threat to the anagenetically derived endemic species.

Uncinia (Cyperaceae) is a genus of 60 to 70 species, with approximately 20 in the mountains and cooler regions of South America (Wheeler and Goetghebeur 1995). The Global *Carex* Group (2015) has recently submerged *Uncinia* into *Carex*, but we do not accept that disposition in this book. For more discussion on this point, see Chapter 5. Six species of *Uncinia* occur in the Juan Fernández Archipelago, of which four are endemic (Wheeler 2007): *U. aspericaulis*, *U. costata*, *U. macloviformis*, and *U. douglasii*. The first three are found only on Alejandro Selkirk Island, and *U. douglasii* is known from both islands. Wheeler (2007) has provided the most detailed study on these species, and he also treated *Carex* in the archipelago. He indicated that of the taxa of *Uncinia, U. douglasii* is very similar to *U. costata* and, in fact, that the former may have given rise to the latter. Wheeler (2007) also suggests that *U. aspericaulis* resembles *U. scabriuscula* from the mainland of South America. *Uncinia macloviformis* may be related to *U. macloviana*, but this might be only "outward similarity" (Wheeler 2007, p. 137). These suggestions regarding affinities are helpful, but they do not allow specific hypotheses to be produced with confidence. It does seem likely that these four species in the archipelago originated from South America rather than elsewhere. That all four species occur on Alejandro Selkirk Island could suggest that they evolved there, but the difficulty is that the relationships among these species are not at all clear. Wheeler's (2007) suggestion of a very close tie between *U. douglasii* and *U. costata* suggests that both diverged from a colonist to the younger island, with a back-migration to the older island in more recent times for *U. douglasii* (as happened with *Erigeron fernandezia*). Another alternative might be that there was a progression of arrival first at Robinson Crusoe Island with anagenetic speciation yielding *U. douglasii*, followed by dispersal again to Alejandro Selkirk Island and anagenetic speciation giving rise to *U. costata*. Because *U. douglasii* also now exists on the younger island, it would mean that this species would have dispersed westward again more recently. The available data do not allow rejection of either hypothesis. Apparently, *U. macloviformis* and *U. aspericaulis* are not particularly closely related to each other, and neither seems close to the other two endemic taxa, which might suggest two additional independent introductions. One of the difficulties of determining the phylogeny and biogeography of *Uncinia* is that the fruits, with their exserted hooked rachillae at maturity, are well adapted for bird dispersal (Guppy 1917, pp. 498–500). As evidence of this, two additional native species occur in the archipelago on the younger island, *U. tenuis* and *U. phleoides*, also having dispersed successfully from the South American mainland more recently without speciation.

Only two molecular studies exist that include material on *Uncinia* from the islands. Roalson et al. (2001) examined *Solbrig 3647* from Alejandro Selkirk Island via ITS and *trnT-L-F* intergenic spacer. This was cited as from *U. brevicaulis*, but more recently Wheeler (2007) listed this collection from the US-Chilean 1965 expedition as belonging to the endemic *U. douglasii*. The only other *Uncinia* investigated by Roalson et al.

(2001) was *U. uncinata* from Hawaii, which in combined analysis appears as the sister group with 99% jackknife support. Waterway and Starr (2007), using nuclear ITS, ETS-1f, and cp *trnL* intron, *trnL-trnF* intergenic spacer, and *trnE-trnD* intergenic spacer, analyzed selected species of *Carex*, *Kobresia*, *Cymophyllus*, *Schoenoxiphium*, and three species of *Uncinia*, as well as outgroups. Under *U. brevicaulis* in their table 1 are listed three accessions of this species that were examined for DNA data, the first two from Tristan de Cunha (erroneously cited under St. Helena) and the third from the Juan Fernández Islands (listed as "San Fernandez Island"). This latter has the same GenBank number as the *trnT-L-F* sequence originally determined by Roalson et al. (2001), which is actually *U. douglasii*. It is not possible to assess sequence divergence among the three listed accessions of *U. brevicaulis* because each voucher yielded a different sequence, and these were combined for the phylogenetic analysis. In the combined tree based on ITS, ETS, and *trnL-trnF*, *U. brevicaulis* exists as only a single short branch that ties first to *U. phleoides* (from Chile) and then to *U. uncinata* (from New Zealand and Hawaii), which were the only other two species in the analysis. Among the three taxa, all are supported with 100% bootstraps. These scanty data obviously do not provide clues to the specific origin of the island taxa, and more sampling is very much encouraged.

Urtica (Urticaceae) is a genus of 45 to 70 species mainly in temperate regions of the world (Geltman 1998; Taylor 2003). In the Juan Fernández Archipelago, there are two endemic species: *U. masafuerae*, restricted to Alejandro Selkirk Island, and *U. glomeruliflora*, found on both islands, although most commonly on the younger island. Few molecular studies on the genus exist. The most pertinent is that of Kim et al. (2015) as part of an intergeneric study within Urticaceae, who included nine species, but none of them is from the Juan Fernández islands. Charlotte Taylor (2003) on a morphological level has suggested that *U. masafuerae* is similar to *U. berteroana* of mainland Chile, where seven other presumably native species occur (Marticorena and Quezada 1985). Navas (1961) also cited this species from the Juan Fernández Archipelago, but Taylor (2003) placed this in doubt based on an inability to find and confirm the cited specimen (*Bertero 1623*). Because of the existence of numerous species of *Urtica* in southern South America (including Argentina) (Soraru 1972; Weigend and Luebert 2009), it appears likely that this represents another example of successful dispersal to and establishment on the younger island, followed by cladogenesis there and then a dispersal of *U. glomeruliflora* back to the older island, perhaps relatively recently (Fig. 13.2Q). Taylor (2003) points out, however, that this species has not been collected for some time on Robinson Crusoe Island and may, in fact, no longer be there.

Wahlenbergia (Campanulaceae, Campanuloideae) is another genus that contains five endemic species in the archipelago. It is also a large genus worldwide, with 267 species (Lammers 1996) centered in Africa (81% of species) and Australasia (13%). In the Juan Fernández Archipelago, *W. fernandeziana* (Fig. C69), *W. grahamiae*, and *W. berteroi* (Fig. C68) (also on Santa Clara) are restricted to Robinson Crusoe Island, whereas *W. masafuerae* and *W. tuberosa* occur only on Alejandro Selkirk Island. The species *W. larrainii*, recognized by Ricci and Eaton (1994), was treated as a minor variant of *W. fernandeziana* by Lammers (1996), and we concur with this view. Morphological phenetic and cladistic analyses of the endemic species of *Wahlenbergia* reveal a close affinity between *W. fernadeziana* and *W. grahamiae* on Robinson Crusoe Island and likewise also a close relationship between the two species *W. masafuerae* and *W. tuberosa* on Alejandro Selkirk Island. Most interesting is that *W. berteroi*, confined to Robinson Crusoe Island (and Santa Clara Island), appears morphologically most closely related to *W. masafuerae*. Lammers (1996) proposed that the original progenitor to the archipelago arrived first on the older island and diversified into the two species *W. fernandeziana* and *W. grahamiae*. After the younger island appeared, dispersal from *W. fernandeziana* led to the divergent cormose lineage of *W. tuberosa* on the higher elevations and *W. masafuerae* in the lower coastal environments. Back-dispersal to the older island from out of the lineage of *W. masafuerae* and subsequent anagenetic speciation produced *W. berteroi* (Fig. 13.2R). Isozyme data (Crawford et al. 1990), based on analysis of three species (*W. tuberosa* [cited as *W. masafuerae*], *W. fernandeziana*, and *W. berteroi*), are not in conflict with this hypothesis, but these authors suggested that *W. berteroi* might have diverged cladogenetically from *W. fernandeziana* on Robinson Crusoe Island, followed by dispersal to Alejandro Selkirk Island and further cladogenesis. Crawford et al. (1990) suggested that a possible progenitor for the Juan Fernández complex might have been similar to *W. peruviana* of the Andean region, but Lammers (1996) was of the opinion that these island taxa may have been derived from a lineage similar to *W. scopulicola* from northeastern Australia. The phylogenetic pattern in *Wahlenbergia* in the archipelago (Fig. 13.2R) is distinct in two ways. First, it represents the only known example of serial cladogenesis, whereby out of a cladogenetic lineage on the older island comes a dispersal to the younger island followed by another cladogenetic divergence. Second, it may be the only known example of a back-dispersal from the younger to the older island that led to anagenetic speciation. Another hypothesis to explain the presence of *W. berteroi* only on the older island could be that it originated cladogenetically on the younger island, dispersed to the older island perhaps even during historical time, and then was extirpated on the younger island. This is even less parsimonious, but it is not impossible. A molecular phylogenetic study is much needed.

Zanthoxylum (Rutaceae) is a genus of approximately 200 species in tropical regions of the world (Mabberley 2008). The two endemic species of the Juan Fernández Archipelago, *Z. mayu* (Fig. C84) and *Z. externa*, were originally published within *Fagara* (Skottsberg 1921), but this genus has now been submerged into *Zanthoxylum* by most workers (Rodríguez 2011, however, retains the genus). Morphological data (Brizicky 1962; Hartley 1966; Waterman 1975; Beurton 1994) reveal that *Fagara* is polyphyletic and fits comfortably within the generic limits of *Zanthoxylum*. Two molecular studies have examined intergeneric relationships within Rutaceae (Chase et al. 1999; Groppo et al. 2008), but neither of these included the island species. A recent molecular phylogenetic study of Hawaiian (and other Pacific) Rutaceae at the molecular level (Appelhans et al. 2014), using four nuclear and two plastid markers, analyzed 37 species of *Zanthoxylum*, and one of these was *Z. mayu* from Robinson Crusoe Island. In their analysis, the authors indicated that this island endemic ties with 99% Bayesian posterior probability but only 60% bootstrap support to *Z. riedelianum* and

Z. rhoifolium, both from tropical South America. It seems likely, therefore, that the origin of this island endemic (Fig. 13.2S) was from South America and that first colonization occurred on the older island, leading to the anagenetic evolution of *Z. mayu*. Dispersal to the younger island after it was formed yielded the second anagenetic endemic island species (*Z. externa*).

Ferns: Endemic Genus

A number of treatments have been done on the ferns of the Juan Fernández Archipelago. In addition to the historical works of Hemsley (1884), Johow (1892–93, 1896), and Christensen and Skottsberg (1920), more recent surveys have been published by Gunkel (1984), Rodríguez (1995), and Barrera M. (1997). These newer studies have focused primarily on documentating the occurrence of fern diversity within the archipelago and much less on relationships of island taxa with continental relatives. Most of the insights on relationships have come from taxonomic treatments plus a limited number of molecular phylogenetic investigations.

The only endemic genus of ferns in the Juan Fernández Archipelago is *Thyrsopteris*, with the single species, *T. elegans* (Kunze 1834) (Fig. 9.6), which occurs on both major islands. This unusual species of tree fern is so distinctive that it is now placed in its own family (Smith et al. 2006), Thyrsopteridaceae (formerly in Dicksoniaceae; Kramer 1990). Despite its unique position of being the only endemic fern genus in the islands, it is not rare on either island, being found frequently in moist forests above 500 m. One of the major interests in this genus is its presumed age. A number of earlier reports of fern fossils from the Jurassic and Cretaceous have been assigned to *Thyrsopteris* (reported in Seward 1910), but many of these have now been transferred to other genera (Berry 1911; Nishida and Nishida 1979). One report that appears more accurate, however, is of *T. antiqua* from the upper Cretaceous (ca. 70 million years ago) of Cerro Guido, Magallanes, Chile, which was described by Menéndez (1966). Nishida and Nishida (1979) described a new fossil genus, *Thyrsopterorachis*, from the Upper Cretaceous of Japan, that closely resembles *Thyrsopteris* in vascular patterns and histological features. It appears, therefore, that *Thyrsopteris* existed as a distinct genus in the northern and southern hemispheres, but through time, it dispersed to the Juan Fernández Archipelago during the past 4 million years, probably from a source in South America, and then became extinct in all continental regions. This pattern is the same as with *Lactoris*, discussed earlier, which is known to have had a broad distribution in the southern hemisphere but now exists refugially only in the Juan Fernández Islands. In these two cases, it makes no sense to talk about speciation in the archipelago because the species evolved much earlier and are surviving only refugially in the islands. Relationships of *Thyrsopteris* to other ferns are remote, as revealed with molecular phylogenetic studies on the tree ferns by Korall et al. (2006). Based on four cpDNA loci (*atpA*, *atpB*, *rbcL*, and *rps4*), the genus is isolated as sister group to species of *Plagiogyria* (Plagiogyriaceae), *Culcita* (Dicksoniaceae), and *Loxsomopsis* and *Loxsoma* (Loxsomataceae), a remoteness that recommends separate familial status for *Thyrsopteris*. The spore morphology is also quite different from that of other tree fern genera of the former Dicksoniaceae (Gastony 1981).

Ferns: Endemic Species

Argyrochosma chilensis (Pteridaceae) is another fern endemic that is found on both major islands. This species has had several generic placements. Christensen and Skottsberg (1920) treated it as belonging to *Pellaea* (*P. chilensis*), but Weatherby and Tryon (Tryon 1956) included it in the large *Notholaena* (*N. chilensis*), which consisted of 58 species. Since that time, many changes have taken place in generic circumscriptions, primarily by division of the genus into more supportable evolutionary groups (Windham 1986; Rothfels et al. 2008), hence resulting in fewer species in *Notholaena* and requiring some new combinations (Yatskievych and Arbeláez A. 2008). *Notholaena chilensis* has now been segregated with 15 others into the genus *Argyrochosma* (Windham 1987) based on morphology, cytology, and nucleotide data (Rothfels et al. 2008). As for origin, the Juan Fernández endemic was suggested by Christensen and Skottsberg (1920, p. 30) to be close ("Very near, or too near . . .") to *Pellaea* [*Argyrochosma*] *nivea* from Andean South America, a species also known from Chile (Marticorena and Quezada 1985) in the provinces of Parinacota and El Loa (Rodríguez 1995). Rothfels et al. (2008) examined three species of *Argyrochosma* with *atpA*, *rbcL*, and *trnG-trnR* sequences, but *A. chilensis* was not sampled, and hence this study provides no insight into the origin of the island endemic. Because the genus is entirely of the New World, however, it is nearly certain that the progenitor of *A. chilensis* would come from South America, and *A. nivea* is a good candidate. Dispersal would have taken place probably first to the older island, followed by anagenesis, and then dispersal to the younger island but without further speciation.

Arthropteris altescandens (Arthropteridaceae, following Liu et al. 2013) (Fig. C7) occurs in a genus of approximately 15 species that is distributed mainly in the Old World tropics as well as being absent in South America (Holttum 1966; Rodríguez 1995). Christensen and Skottsberg (1920) suggested close affinity of this endemic species to those of the Polynesian region, in particular, near *A. obliterata* from Samoa. A very recent molecular phylogenetic analysis of the genus did include material from *A. altescandens* taken from the collections *Crawford et al. 11126* and *Stuessy et al. 15379* (Liu et al. 2013). The closest relationship discovered using *rps4* and *rps4-trnS* IGS sequences and maximum likelihood is with *A. paucivenia* from Madagascar. There seems little doubt, therefore, that this is yet another example of a species arriving in the islands from the western Pacific, possibly even from eastern Africa. It probably first colonized Robinson Crusoe Island, followed by anagenetic speciation and subsequent dispersal to Alejandro Selkirk Island, but without further speciation there.

Two endemic species of *Asplenium* (Aspleniaceae) occur in the Juan Fernández Archipelago, and they both occur on both islands. *Asplenium macrosorum* was suggested by Christensen and Skottsberg (1920) to be related to *A. obtusifolium* and *A. semicordatum* from tropical America. These authors also commented that the specimens on both islands closely resemble each other but that those from Alejandro Selkirk Island are smaller in size and with narrower pinnae and with the terminal pinna being symmetrical. *A. stellatum* is the second endemic species of this genus, and Christensen and Skottsberg (1920, p. 24) commented that this species belongs to the "*A. lunulatum* assemblage," which would include *A. erectum* from South Africa and *A. pteropus* from tropical America, among others. It seems clear, therefore, that each of the endemic species of *Asplenium* ties to a different progenitor, probably from South America (Fig. 13.2T). Both probably arrived on the older island first, anagenetically speciated, and subsequently dispersed without further speciation to the younger island when it became available for colonization. Looser (1944) provided a synopsis of the Chilean species of the genus, and he concluded that the two endemic species are (p. 234) "... muy peculiares y sin relaciones con las especies chilenas continentales." Clearly, more work is needed on a morphological level to assess affinities among species in Latin America. Some molecular phylogenetic studies have been done in this genus (Murakami and Schaal 1994; Van den Heede et al. 2003; Schneider et al. 2004; Perrie and Brownsey 2005), but these have only scratched the surface of this large genus of approximately 700 species, and as is often the case, none of the island endemics has been sampled.

Blechnum (Blechnaceae) is a large genus of some 150 species worldwide and perhaps 50 in the Americas (Tryon and Tryon 1982). Although *Blechnum* (Blechnaceae) is the fern genus with the largest number of endemic species in the archipelago (*B. cycadifolium* [Fig. C10], *B. longicauda* [Fig. C11], *B. mochaenum* var. *fernandezianum*, and *B. schottii* [Fig. C12]), which would immediately suggest some pattern of cladogenesis among them, the evidence is strong that none of these is closely related to any of the others. The three endemic species are each placed in different subsections of sect. *Caudiculadas* of subgen. *Lomaria* (Gunkel 1984). Each represents a separate case of anagenetic speciation, three of them first on Robinson Crusoe Island and one of them (*B. longicauda*) on Alejandro Selkirk Island. All taxa are found on both islands, except for *B. longicauda*, which is restricted to the younger island. This example illustrates the importance of having phylogenetic understanding in order to interpret patterns of speciation in this (or any other) archipelago. As for affinities, all endemics appear to have been derived from South America (Fig. 13.2U). *Blechnum cycadifolium*, a very attractive treelike fern 1 to 2 m tall (Fig. C10), appears to be closely related to *B. magellanicum* from continental Chile. Hemsley (1884) suggested that it was the same as *Lomaria* (*Blechnum*) *magellanicum*, but Johow (1896) pointed out that he had seen both *B. cycadifolium* and *B. magellanicum* growing together in a botanical garden in Santiago, and both maintained their specific features in this common environment. Christensen and Skottsberg (1920), Looser (1947), and Gunkel (1984) also mention the close affinity between *B. cycadifolium* and *B. magellanicum*. Christensen (1910) earlier, in fact, had treated *B. cycadifolium* as a variety of the other species. Dispersal from a continental population of *B. magellanicum* and anagenetic speciation on Robinson Crusoe Island therefore would seem to be the most probable hypothesis for the origin of *B. cycadifolium*. Regarding *B. longicauda* (Fig. C11), Christensen and Skottsberg (1920) have suggested a relationship with *B. sprucei* from Ecuador, with dispersal apparently directly to the young island when it became available for colonization, followed by anagenetic speciation. These same authors also suggested that the endemic *B. schottii* (Fig. C12) might be closely allied with *B. attenuatum* (from Africa) or *B. meridense* (extending from the Yucatan Peninsula to Perú). It seems likely, therefore, that *B. schottii* evolved from a progenitor from northwestern South America and speciated anagenetically on Robinson Crusoe Island, later dispersing to the younger island when it formed without another round of speciation, perhaps more recently. The final endemic in *Blechnum* is *B. mochaenum* var. *fernandezianum*. This variety obviously relates to *B. mochaenum* var. *mochaenum* that occurs in continental Chile and Argentina (Rodriguez 1995; de la Sota 1972). Two other varieties have also been recognized within this species from Argentina, var. *achalense* and var. *squamipes* (de la Sota 1972). There seems little doubt, therefore, that the progenitor of the endemic island variety has been derived anagenetically on Robinson Crusoe Island from within *B. mochaenum* from southern South America, and it also dispersed to the younger island after it was formed, but without further divergence. A few molecular phylogenetic investigations in Blechnaceae have been completed (Shepherd et al. 2007; Gabriel y Galan et al. 2013; Perrie et al. 2014), but none of the Juan Fernández endemics has been included.

Dicksonia (Dicksoniaceae) is a genus of tree ferns with 20 to 25 species (Korall et al. 2006). There are two endemic species in the Juan Fernández Archipelago, one on each of the islands: *D. berteroana* (Figs. C13 and C14) is restricted to Robinson Crusoe Island, and *D. externa* (Figs. C15 and C16) is confined to Alejandro Selkirk Island. They appear to be closely related to each other based on the evidence that *Dicksonia externa* was originally published by Christensen and Skottsberg (1920) as var. *virgata* of *D. beteroana*, but Skottsberg later (1953) elevated it to specific status as *D. externa*. *Dicksonia berteroana* was suggested by Christensen and Skottsberg (1920) as being closely related to *D. lanata* from New Zealand, which, if accurate, would mean that the island lineage would have come from the western Pacific rather than South America. Seven species of the genus have been analyzed with nucleotide sequences (Korall et al. 2006), but none of these came from the islands. Sarah Noben is pursuing a doctoral thesis on relationships in the genus (e.g., Noben and Lehnert 2013), including molecular affinities, and island material has been provided to her for analysis. DNA was obtained only from *D. berteroana*, and its relationships appear to be closest to *D. arborescens* from St. Helena. These two species then join with others from the western Pacific (S. Noben, personal communication). Based on available evidence, therefore, we hypothesize that spores of *Dicksonia* dispersed from the western Pacific and arrived first in Robinson Crusoe Island, where it speciated anagenetically, followed by dispersal to the younger island shortly after it was formed, followed by another anagenetic speciation event (Fig. 13.2V). It would also have been possible for two more recent simultaneous introductions to both islands from the same western progenitor (or quick island-hopping after arrival to either one of the islands) and then divergence, which would be a cladogenetic type of speciation between the two islands, but this would seem less probable based on the morphological closeness between the two species and the shorter time available for such divergence to take place.

Hymenophyllaceae consist of approximately 600 species with a broad tropical distribution. Because of many recent taxonomic changes in this family, especially at the generic level, some comments are required before addressing the phylogeny of species in the Juan Fernández Archipelago (see Chapter 5 for more discussion). The endemic taxa of Hymenophyllaceae in the archipelago were treated by Christensen and

Skottsberg (1920) as follows: *Hymenophyllum rugosum* on both islands, *H. cuneatum* var. *rariforme* only on Robinson Crusoe Island, *Serpyllopsis caespitosa* var. *fernandeziana* on both islands, and *Trichomanes ingae* and *T. philippianum* both restricted to Robinson Crusoe Island. Diem and Lichtenstein (1959), in a study of the family from the south of Argentina and Chile, recognized these same taxa, as did Rodríguez (1995) in his treatment of ferns for the *Flora de Chile*, as did Marticorena et al. (1998) and Danton et al. (2006) in their catalogues of the flora of the archipelago. Ebihara and Iwatsuki (2007) and Ebihara et al. (2010) treated all species of *Hymenophyllum* from the Pacific region, but Juan Fernández was not included in this survey. The most recent treatment of *Hymenophyllum* for southern Argentina and Chile (Larsen et al. 2013), following the recent general revision of the family by Ebihara et al. (2006; see Chapter 5), includes *H. rugosum* and *H. cuneatum*, but *H. cuneatum* var. *rariforme* from the islands is not recognized as distinct from the continental variety, and *Serpyllopsis caespitosa* is moved into *Hymenophyllum* as *H. caespitosum* but not regarded as endemic. Ebihara et al. (2006) transferred *Trichomanes ingae* and *T. philippianum* into the genus *Polyphlebium* as *P. ingae* and *P. philippianum*. In summary, the present endemic members of the family as recognized in this book include *Hymenophyllum rugosum*, *Polyphlebium ingae* and *P. philippianum*.

As for affinities and modes of origin of these endemic species, the only molecular study of *Hymenophyllum* that included the Juan Fernández endemic *H. rugosum* was that of Hennequin et al. (2010) based on *rbcL*, *rbcL-accD* IGS, *rps4*, and *rps4-trnS*. These results showed a tie with *H. tunbridgense* from different localities (France, Scotland, Tanzania, and Madeira) but also Chile. Christensen and Skottsberg (1920) also mentioned this close relationship based on morphology. Regarding the *Trichomanes* complex, one of the endemic species, *T. ingae* (at that time in sect. *Lacosteopsis* but now as *Polyphlebium ingae*) was sampled for *rbcL* and analyzed by Ebihara et al. (2007), but they did not sample *T.* (*Polyphlebium*) *philippianum*. In their analysis, *T. ingae* was closest to *T. diaphanum* (from French Guiana), *T. hymenophylloides* (from Guadeloupe), and *T. borbonicum* (from Réunion), which would suggest immigration from northeastern South America or, less probably, eastern Africa. Christensen and Skottsberg (1920) commented that *T. ingae* belongs to a group of species centered around *T.* (*Polyphlebium*) *pyxidiferum*, which is spread in tropical America, the Caribbean, Africa, and Polynesia, and this species was also placed in *Polyphlebium* by Ebihara et al. (2006). With regard to *T.* (*Polyphlebium*) *philippianum*, Christensen and Skottsberg (1920) mentioned (p. 2) that this species "... is one of the most distinct species of the genus," with no comment about relationship with *T.* (*Polyphlebium*) *ingae*. Hence, based on available evidence, all these endemic species in the Juan Fernández Archipelago appear to have arrived independently and speciated anagenetically. This is perhaps not surprising because of the great long-distance capacity of ferns over the globe (Kramer 1993). Critical study of affinities among these two species and relatives in *Polyphlebium* would be very helpful.

Following the recent revision by Sundue et al. (2010), three species of *Megalastrum* (Dryopteridaceae) are endemic in the Juan Fernández Archipelago: *M. inaequalifolium* (Fig. C18), *M. glabrius*, and *M. masafuerae*. *Megalastrum inaequalifolium* was treated by Christensen and Skottsberg (1920) as belonging in *Dryopteris*, but Smith and Moran (1987) moved it into *Megalastrum*, a genus newly created by Holttum (1986). The species was viewed as consisting of two forms by Christensen and Skottsberg (1920), the typical form and forma *glabrior*. Rodríguez (1990) elevated both of these to varietal status based on differences in fronds, with var. *glabrior* being larger, darker green, almost glabrous, and sori that are also mostly glabrous but sometimes with one to two short hairs (from Rodríguez 1995). These varietal distinctions were accepted by Marticorena et al. (1998) and Danton et al. (2006). Some differences of opinion prevailed, however, on the distribution of the two varieties in the archipelago. Christensen and Skottsberg (1920) listed their forma *glabrior* as occurring on Alejandro Selkirk Island and the typical form as being "probably confined" to Robinson Crusoe Island. Rodríguez (1995) gave var. *glabrior* as occurring on both islands and the typical variety only from Robinson Crusoe Island. Marticorena et al. (1998) listed var. *inaequalifolium* as being only on Robinson Crusoe Island and var. *glabrior* as restricted to Alejandro Selkirk Island. Danton et al. (2006) gave var. *glabrior* as being from Alejandro Selkirk Island but var. *inaequalifolium* as occurring on both major islands. Greater clarity of the taxonomic and geographical situation in the islands is now provided by Sundue et al. (2010), with *M. inaequalifolium* being restricted to Robinson Crusoe Island. The form or variety *glabrior* has been elevated to specific status as *Megalastrum glabrius* with distribution on Alejandro Selkirk Island only from Quebrada Casas. The new species, *M. masafuerae*, based on *Solbrig et al. 3676*, is also confined to Alejandro Selkirk Island, but in Quebrada de las Vacas and the north branch of Quebrada Varadero. As for biogeographical origin, *Megalastrum* (*Dryopteris*) *inaequalifolium* was treated by Christensen and Skottsberg as belonging to ". . . a group of closely allied species distributed through tropical America" (1920, p. 19). Sundue et al. (2010) do not comment on origins of the island endemics, but certainly the species they treated in their revision from the circumaustral region have dispersed widely to different island systems. *Megalastrum spectabile* occurs in continental Chile in the Valdivian temperate forests, and this might be a relative. In any event, it seems most probable that the species endemic to the archipelago arrived first in Robinson Crusoe Island, the older island, followed by anagenesis to produce *M. inaequalifolium*. Dispersal from this new taxon to the younger island when it arose from the sea resulted in an established population that dispersed into different parts of the quebrada system, *M. glabrius* into Quebrada Casas and *M. masafuerae* into Q. de la Vacas and Q. Varadero, followed by speciation (Fig. 13.2W). It seems, therefore, that this may be the only known case of cladogenesis in ferns in the archipelago. It could also reflect parallel anagenesis from the progenitor on the older island to the younger island, but more study would be needed to determine whether such an alternative hypothesis might be feasible. Molecular phylogenetic investigations involving species of *Megalastrum* have not included these island taxa (Liu et al. 2007; Zhang et al. 2012; Labiak et al. 2014) and hence provide no insights into continental origins.

Ophioglossum fernandezianum (Ophioglossaceae) is found only on Robinson Crusoe Island. At the time of the original description, Christensen and Skottsberg (1920, p. 44) indicated that *O. fernandezianum* "mostly resembles" *O. ypanense* of Brazil and

Colombia but that it also "falls within the aggregation of forms called *O. nudicaule* by Baker." Lichtenstein (1944) continued this direction and formally placed *O. fernandezianum* into synonymy under *O. nudicaule* var. *nudicaule* ("var. *typicum*") with the comment that she was unable to find sufficient characters to justify separation of the two species. Kunkel (1965), Godoy (1989), and Rodríguez (1995), however, returned to the original perspective and listed *O. fernandezianum* as a good species, an opinion also followed by Ricci (1996), Marticorena et al. (1998), and Danton et al. (2006). Meza Torres et al. (2015a), however, point out that sculpturing of spores in *Ophioglossum* provides the best features for delimiting species within the genus, and in that context, *O. fernandezianum* is a "doubtful species." Although there is clearly some uncertainty on recognizing this species as endemic for the flora, we accept it here following majority opinion. More study is obviously needed. At the molecular level, 12 of the approximately 25 species of *Ophioglossum* (s.s.) have been analyzed for molecular relationships using *rbcL* and *trnL-F* sequences (Hauk et al. 2003), but *O. fernandezianum* was not included. Five species of *Ophioglossum* (s.s.) were examined with *matK* (Shinohara et al. 2013), but again the island endemic was not sampled. An abstract has been published on use of five cpDNA regions with 17 species in the genus (Lee and Hauk 2013), but the full results have apparently not yet appeared. Regarding origin, it is very clear that *O. fernandezianum* is very closely related to *O. nudicaule* from continental South America, where it occurs also in central-south Chile (Parra et al. 2012), and this is undoubtedly the ancestor of the island taxon. From wherever precisely this species originated, it was likely transported by wind via long-distance dispersal, a quality that has been documented convincingly for another species of the genus (*O. reticulatum*; Meza Torres et al. 2015b).

Pleopeltis ×*cerro-altoensis* (Polypodiaceae), endemic to Robinson Crusoe Island, has been described as new by Danton et al. (2015). This was suggested as an F_1 hybrid between two native species, *P. macrocarpa* and *P. masafuerae*. We agree that documentation is strong for the hybrid nature of *P.* ×*cerro-altoensis*, but we have concerns regarding taxonomy and distribution. *Pleopeltis masafuerae* was treated earlier by Rodríguez (1995) as *Polypodium masafuerae* Phil. Subsequently, de la Sota et al. (2007) moved it to *Pleopeltis*, and this has been followed by Smith and Tejero-Díez (2014), continuing a trend of breaking up and moving pieces of *Polypodium* into other genera (Otto et al. 2009). Despite this trend, we favor treating this species in *Polypodium* based on its lacking peltate scales (R. Rodríguez, personal communication). *Polypodium masafuerae*, as its name suggests, would be assumed to occur only on Alejandro Selkirk Island in the archipelago, but Rodríguez (1995) initially gave its distribution as extending to Antofagasta and southern Perú. He now regards these extra-archipelago reports as erroneous (R. Rodríguez, personal communication), which is the reason we have listed *Polypodium masafuerae* as endemic in Table 5.1. More to the point, we suspect that the reported hybrid may be a cross between *Pleopeltis macrocarpa* (Fig. C23) and *Polypodium intermedium* subsp. *intermedium* (Fig. C24), the latter of which is endemic to Robinson Crusoe Island. If this might be so, then it becomes another case of intergeneric hybridization in the archipelago (the other being ×*Margyracaena*, Rosaceae), although limits among these genera are somewhat tenuous. In addition, the cross would then have occurred between one native and one endemic species and both on Robinson Crusoe Island. Testing of this alternative hypothesis would be welcome.

Another case of anagenesis is found in the endemic *Polypodium intermedium* (Polypodiaceae). Taxonomically, the species has been divided (Christensen and Skottsberg 1920) into two subspecies, subsp. *intermedium* (their subsp. "typicum") on Robinson Crusoe Island (Fig. C24) and subsp. *masafueranum* on Alejandro Selkirk Island. Interesting is that within subsp. *masafueranum*, four varieties were also recognized by Christensen and Skottsberg (1920): var. *masafueranum* (not named by them but mentioned on p. 38 as "the type"), var. *obtuseserratum*, var. *basicompositum*, and var. *cambricoides*. These varieties were recognized principally by differences in leaf shape, especially details of the margins. Christensen and Skottsberg (1920) were struck with these many variations and made this comment (p. 38): "Masafuera is inhabited by a number of forms connected by transitions. As they grow side by side, sometimes two widely distant ones on the same branch, it is impossible to regard them as merely results of local influences. Either the species is composed of a number of units, elementary species, giving the impression of a more or less continuous series, or we have a few subspecies and their hybridogenous offspring." Rodriguez (1995), in his treatment of ferns for Chile (including the Juan Fernández Archipelago), recognized the two subspecies, but he regarded the differences among the varieties as weak characters that are mixed in populations on the island. Marticorena et al. (1998) and Danton et al. (2006) have likewise not recognized these varieties, and we agree with this perspective. As for molecular phylogenetic data, only two studies have been done on the genus, and those both were restricted to the *P. vulgare* complex in the northern hemisphere (Haufler et al. 1995; Sigel et al. 2014). From available information, therefore, *Polypodium intermedium* probably arrived from somewhere in South America and dispersed first to the older island, resulting anagenetically in what is now called subsp. *intermedium*, followed by dispersal to the younger island when it became available for colonization and with additional divergence at the subspecific level to yield subsp. *masafueranum* (Fig. 13.2X). That considerable morphological variation exists within populations may attest to an early stage of genetic differentiation within the subspecies.

Polystichum tetragonum (Dryopteridaceae) (Fig. C19) is endemic to both of the Juan Fernández major islands. It was treated as belonging to *P. vestitum* from New Zealand (Perrie et al. 2003) by Christensen and Skottsberg (1920), although they remarked (p. 21) that "It may be doubted whether the *aculeatum*-form [referring to *Aspidium aculeatum* Hemsl. placed in synonymy with *P. vestitum*] of Juan Fernandez really is identical with *P. vestitum*." Rodríguez (1995) treated the species as distinct, as did Ricci (1996), Marticorena et al. (1998), and Danton et al. (2006). The genus is a large group of 330 species with a wide distribution (Barrington 2014). Lu et al. (2007) analyzed 19 species from Asia, McHenry and Barrington (2014) examined 34 species from the Andes of South America, and Driscoll and Barrington (2007) investigated 50 species from the Hawaiian Islands and other regions, but no one has sampled the Juan Fernández endemic. Perrie et al. (2003) did examine *P. vestitum* and relatives from Australasia, but not with *P. tetragonum*. Because of the implied relationship with *P. vestitum* (Christensen and Skottsberg 1920), the best one can say at this point is that it appears that the island endemic originated from the western Pacific, first established on the available older island, anagenetically speciated, and then dispersed to the younger island when it was formed. Six continental Chile species of the genus exist, however (Rodríguez 1995), and these relationships have not yet been critically examined.

Three species of *Pteris* (Pteridaceae) occur in the Juan Fernández Archipelago, *P. berteroana* (Fig. C25), *P. chilensis*, and *P. semiadnata*, and all are found on both islands. *P. berteroana* is the only endemic, the others being native. The genus *Pteris* is large, with more than 250 species (Tryon et al. 1990). Several molecular phylogenetic studies have been completed. Prado et al. (2007) examined ten species for *rbcL* variation, Chao et al. (2014) examined 135 species with *rbcL* and *matK*, and Zhang et al. (2015) analyzed 119 species with six plastid regions, but none of these investigations included *P. berteroana*. Christensen and Skottsberg (1920) suggested relationships of *P. berteroana* with *P. endlicheriana* from Norfolk Island off the eastern coast of Australia in the western Pacific. Tryon and Tryon (1982), however, placed *P. berteroana* in the "*Pteris chilensis* Group" along with *P. chilensis*, *P. leptophylla*, and *P. tremula*. It would not be surprising, therefore, if *P. berteroana* was derived from the continental *P. chilensis*, although this is far from certain at this point. *P. tremula* is also a similar species, and it is known from the western Pacific (Looser 1945). Dispersal would have probably taken place first on Robinson Crusoe Island and, after anagenesis, further dispersal to the younger island when it arose from the sea. An interesting study on the development of the gametophyte of *P. berteroana* has been done by Mendoza et al. (1996), but no comment on specific relationships was offerred.

Rumohra (Dryopteridaceae) is a small genus of eight species of the southernmost regions (Labiak et al. 2014). Two species occur in Chile, *R. adiantiformis* from the continent and *R. berteroana* endemic to both major islands of the Juan Fernández Archipelago. The endemic has undergone a number of generic shifts, originally being described in *Aspidium* by Colla (1836), treated by Christensen and Skottsberg (1920) in *Polystichum*, and finally transferred into *Rumohra* by Rodríguez (in Duek and Rodríguez 1972). On the molecular level, Liu et al. (2007) began the analyses with comparison of *R. adiantiformis* with other genera using *rbcL* and *atpB* genes, but this did not include the endemic species. Labiak et al. (2014) analyzed four chloroplast markers in three accessions of the phenotypically variable (Boeger et al. 2007) *R. adiantiformis* from different geographical localities, as well as one of *R. berteroana* (*Danton 1964*). Unsurprisingly, *R. berteroana* nests within the accessions of *R. adiantiformis*. Although not all species of the genus have been compared, the available data do suggest that *R. berteroana* has been derived from dispersal from the continental progenitor *R. adiantiformis* to Robinson Crusoe Island, followed by anagenetic change and subsequent dispersal to Alejandro Selkirk Island when it became available for colonization.

One of the very few new species that have been described during our expeditions in the Juan Fernández Archipelago is *Sticherus lepidotus* (Gleicheniaceae; originally described as *Gleichenia lepidota* by Rodríguez-Ríos 1990), endemic to Alejandro Selkirk Island and found between 820 and 1,100 m. This species has recently been transferred into *Sticherus* (Rodríguez and Ponce 2007), a pantropical genus of approximately 95 species (Gonzales R. and Kessler 2011). Christensen and Skottsberg (1920. p. 43) recognized the distinctness of this taxon, but instead of describing it as new, they labeled it "*G.* cf. *litoralis* [*sic*]," indicating a relationship with this species from the Chilean continent. This new taxon seems very distinct from *G. littoralis*, also now transferred to *Sticherus* (Nakai 1950), and the latter most probably gave rise to the former by dispersal to Alejandro Selkirk Island directly from the mainland followed by anagenesis. Some molecular phylogenetic work has been done on *Gleichenia* (Perrie et al. 2007), but none has been published yet with *S. lepidotus*. Material of *S. lepidotus* was sent to Carlos Lehnebach back in 2006, and his preliminary analyses from *trnL-F* revealed (*in litt*) a clade with *S. lepidotus*, *S. quadripartitus*, *S. littoralis*, and *S. squamulosus*, as well as two species of *Sticherus* from New Zealand.

Phylogeny and Speciation

The detailed phylogenetic discussions in the preceding section now allow syntheses on sources of origin and speciation for all the endemic taxa of the Juan Fernández Archipelago. Although geographical origins of the flora will be discussed in more detail in Chapter 16, a brief overview resulting fom the previous analyses can be found in Table 13.6. Most important, and not at all surprising, is the strong contribution from the South American continent of 64% of the established endemic taxa. The western Pacific contributed 12% of the endemic flora, and interisland anagenetic speciation accounts for 18% of the total. For a finer-scale interpretation of geographical origins, see Chapter 16 (adapted from Bernardello et al. 2006). Very interesting is that the ferns and angiosperms do not differ in their geographical origins. In addition, far more species in the archipelago have originated from Robinson Crusoe Island than from Alejandro Selkirk Island, which seems reasonable due to the different geological ages of the two islands. The older island would also have been the closest source area for dispersal to the young island when the latter appeared from below the ocean.

Speciation will be discussed in context of cladogenesis, anagenesis, and hybridization, which are the three major speciation processes at work in the Juan Fernández Archipelago. Summary data from the previous disussions on the endemic taxa are provided in Table 13.7.

Cladogenesis

Cladogenesis is a splitting process (Fig. 13.1) within an island, which involves dispersal from the established immigrant population into new habitats, subdivision into isolated subpopulations, and subsequent morphological and genetic divergence. Generally this type of speciation is most common in large islands with strong environmental zonation, often correlated with island elevation. The Hawaiian Archipelago and the Canary Islands offer good examples where cladogenesis is common (Stuessy et al. 2006). In the Juan Fernández Archipelago, cladogenesis has taken

Table 13.6 Sources of Origin of Endemic Taxa of Vascular Plants in the Juan Fernández Archipelago

	Western Pacific	South America	Alejandro Selkirk Island	Robinson Crusoe Island	Unclear
Ferns	3 (12%)	17 (65%)	0 (0%)	5 (19%)	1 (4%)
Angiosperms	13 (12%)	70 (65%)	2 (2%)	17 (16%)	7 (6%)
Totals	16 (12%)	87 (64%)	2 (2%)	22 (16%)	8 (6%)

Table 13.7 Numbers and Percentages of Hypothesized Modes of Speciation in the Endemic Vascular Flora (134 Taxa) of the Juan Fernández Archipelago

	Anagenesis	Cladogenesis	Hybridization
Ferns	22 (88%)	2 (8%)	1 (4%)
Angiosperms	52 (50%)	51 (49%)	2 (2%)
Totals	74 (57%)	53 (41%)	3 (2%)
Species in endemic genera	8	22	1
Other endemic species	66	31	2

place in 41% of endemic taxa. A number of endemic genera have undergone cladogenesis within the islands, ranging in size from 2 to 11 endemic species. Five genera fall within this category: *Centaurodendron* (Asteraceae), *Cuminia* (Lamiaceae), *Dendroseris* (Asteraceae), *Megalachne* (Poaceae), and *Robinsonia* (Asteraceae). This occurrence may correlate with the age of these groups. With endemic genera, they are so designated because they are more divergent in morphological characters from continental relatives such that they are placed at a higher level in the taxonomic (informational) hierarchy. Time, as well as original environmental diversity, may have played a role in providing more opportunity for divergence as well as geographical speciation within the islands.

It is extremely interesting that there are very few examples of cladogenesis (only 8%) among endemic ferns in the Juan Fernández Archipelago (Table 13.7). Of the few examples of cladogenesis in ferns given in Stuessy et al. (1990), especially in *Asplenium* and *Blechnum*, now with more data and analyses of phylogenetic relationships, it becomes clear that most of these are instead cases of anagenesis. This low level of cladogenesis seems to be a peculiarity of the Juan Fernández Archipelago because recent investigations by Hennequin et al. (2014) on the ferns of the Mascarene Islands have shown that of the 46 endemic species, at least 16 have originated from anagenesis and up to 30 from cladogenesis. The Juan Fernández Islands are reasonably small, with limited ecological zonation, and Robinson Crusoe Island has undergone considerable modification over its ontogeny. These factors taken together may have been significant in leading to the low level of cladogenesis in the ferns seen today in the archipelago. Furthermore, differences in characteristics of ferns versus flowering plants may be responsible in part for differences in modes of speciation. Ferns tend to settle into relatively stable habitats, preserving phylogenetic niche conservatism (Patino et al. 2014). They also have a remarkable ability to disperse over long distances (Kramer 1993), and colonists can arrive from different regions and at different times. Continuing dispersal after establishment might interfere with isolation of populations that could lead to allopatric speciation within or between islands. Species from very different lineages, which are each able to establish in distinct ecological zones, are likely to be successful in such situations. They are genetically divergent and hence not suitable partners for interlineage hybridization in the islands. The only known case of cladogenesis in ferns occurs in the genus *Megalastrum* (Fig. 13.2W).

Adaptive Radiation

Adaptive radiation is the evolution of separate lineages within an island or archipelago into ecologically distinct habitats, usually accompanied by dramatic morphological divergence. This can be a confusing concept, overlapping with geographical speciation, nonadaptive radiation, explosive diversification, and so on. Givnish (2015) provides a good overview of these distinctions and stresses that adaptive radiation is speciation involving invasion of adaptive zones. This phenomenon often is associated with cladogenesis in oceanic islands. Several studies have been done on the genera *Dendroseris*, *Robinsonia*, and *Erigeron* to understand adaptive radiation on the Juan Fernández Archipelago. The species of *Dendroseris* have few and small populations, which makes it challenging to infer their ecological tendencies. However, *D. litoralis*, as the name suggests, seems adapted to open, sunny places on coastal rocks. *Dendroseris pruinata* is found in similar habitats but on cliff edges at higher elevations (300 to 500 m). In contrast, *D. micrantha* and *D. pinnata* are true forest denizens.

Robinsonia also shows similar tendencies, but again, interpretations are hindered by the habitat alterations over millions of years, including recent human disturbances. *Robinsonia gayana* occurs on exposed basaltic ridges. *Robinsonia evenia* and *R. thurifera* are found in the deep, moist forest, and *R. gracilis* is restricted to open forest on ridge tops above 700 m. Sanders et al. (1987) conducted analyses of associated species and soils at the localities of each of the species of *Robinsonia*, and little variation in these environmental factors was seen. The likely explanation for these results may relate to the changing landscape of Robinson Crusoe Island over the past 4 million years (see Chapters 3 and 7). Because the flora of this island exists compacted in a refugium, the original spatial and ecological differences among the species of *Robinsonia* may have become obscured.

The best place to assess adaptive radiation in the Juan Fernández Archipelago would be on the younger island, Alejandro Selkirk Island, which has undergone less geological and anthropogenic change. The only genus on this island that has adaptively radiated conspicuously is *Erigeron* (Asteraceae) (Valdebenito et al. 1992a). Population genetic studies (López-Sepúlveda et al. 2015) have shown the genus to be in an early stage of radiation: *E. rupicola* frequents the rocky coastal bluffs; *E. stuessyi* is confined to the deep, cool quebradas; and *E. fernandezia* is found on the more open upper portions of ravines. These three species have separated ecologically with some clarity. Species of the *E. ingae* complex, however, consisting of this species plus *E. turricola* and *E. luteoviridis*, are not always distinct genetically or morphologically (López-Sepúlveda et al. 2015). They appear to be diverging with morphological characters but do not yet have clear ecological distinctions.

Why some groups tend to radiate adaptively and others do not is a central question of island biology. The answer appears to lie largely with the nature of the progenitors. Groups that have been extraordinarily successful (i.e., speciose) in continental regions have tendencies for good dispersal, rapid generation times, and an ability to adapt to diverse environmental conditions (Takayama et al., submitted). When immigrants from these groups arrive in an island at an early stage of its ontogeny and ecological development, the colonists have a good chance of also radiating in the new environment. This is certainly the case with *Dendroseris* (originating from out of *Sonchus*), *Robinsonia* from *Senecio*, and *Erigeron* from out of the large complex of this genus in South America (Solbrig 1962). In other words, highly successful taxa in continental regions continue to produce more species in the new environments of oceanic islands.

Anagenesis

Anagenesis is transformational speciation by divergence from an established immigrant population to the islands (Fig. 13.1). For the flora as a whole (Table 13.7), 57% of the taxa have speciated anagenetically. Interesting is that the ferns show a much higher level of anagenetic speciation (88%) than do angiosperms (50%), again probably due to the greater dispersal ability of ferns to arrive in an island, diverge, and then disperse once more to another island. It is important to emphasize that there may have been more species in these genera originally, perhaps followed by extinction of one or more species as the island, especially Robinson Crusoe, altered over time. In the absence of any evidence to the contrary, however, the working hypothesis is that a single species within the archipelago indicates a single anagenetic origin.

Anagenetic speciation in the Juan Fernández Archipelago can be seen (Table 13.3) to have occurred in genera with patterns 1, 2 and 3, in which only single endemic species now exist. In addition, anagenetic speciation has occurred in the origin of some species within more speciose genera, such as *Dendroseris*, *Robinsonia*, *Peperomia*, *Gunnera*, and *Sophora*. In some cases, anagenetic speciation took place in Robinson Crusoe Island and only dispersed to Alejandro Selkirk Island without further speciation (e.g., *Rhaphithamnus*, *Drimys*). In these latter cases, it may suggest that the dispersal to Alejandro Selkirk Island occurred relatively recently such that not enough time existed to allow further differentiation. More commonly, and as would be expected over the course of 1 to 2 million years, is dispersal to the younger island and repeated speciation. This pattern is seen in, for example, *Dendroseris*, *Gunnera*, *Haloragis*, *Robinsonia*, and *Sophora*. *Dendroseris* is a particularly good example. Dispersal and anagenetic speciation have occurred three times in parallel, each line coming from a different subgenus (Sanders et al. 1987). In the other genera, single derived species have originated anagenetically.

Table 13.8 Sources of Origin for Anagenetically Derived Endemic Taxa of Vascular Plants in the Juan Fernández Archipelago

	Western Pacific	South America	Alejandro Selkirk Island	Robinson Crusoe Island
Ferns	3	16	0	2
Angiosperms	3	34	2	10
Totals	6	50	2	12

As for geographical origin, anagenetically derived species (Table 13.8) follow the general trend of major immigration from South America and, less so, from the western Pacific. Inter-island anagenesis has also been common, especially from the older to the younger island. It also is not surprising that many more species have anagenetically speciated on the older island rather than on the younger one, doubtless due to the existence of this island at least 2 million years prior to the appearance of the younger island and also because of its proximity to the major source region.

Hybridization

Although hybridization is a common feature in the floras of some island archipelagos, such as the Hawaiian Islands (Carr 2003), it is not common among the endemic species of the Juan Fernández Archipelago. Only 2% of the endemic taxa show some signs of hybridization (Table 13.7). It is impossible to know whether hybridization might have been more frequent when the islands were younger, but at the present time only a few cases are known, and all occur on Robinson Crusoe Island. We can confidently conclude, therefore, that reticulate modes of speciation in the archipelago are rare. Four cases, however, are known.

The most obvious case of interspecific hybridization is occurring between *Gunnera bracteata* and *G. peltata* in Quebrada Villagra. Morphological studies by Pacheco et al. (1991) using a morphological hybrid index have revealed hybridization and introgression between these two species along the main path down into the valley. The two species are ordinarily elevationally distinct, but human traffic between these two zones has probably encouraged interbreeding. The hybrids can survive with vegetative reproduction with no difficulty.

Three cases exist where hybridization has occurred and the hybrids have been given binomial names, and as such, they have been included in the inventory of the flora (Table 5.1). First, an intergeneric hybrid, ×*Margyracaena skottsbergii*, is presumably an F_1 hybrid between the introduced *Acaena argentea* and the endemic *Margyricarpus digynus*, clearly documented with RAPD data (Crawford et al. 1993a). This suggests that the genetic differences between the hybridizing genera are not substantial, but the evolutionary potential of this hybridization appears to be low. Few individuals have ever been collected and always from the same locality on the Centinela Ridge, Robinson

Crusoe Island. A second case of interspecific hybridization has occurred between two endemic species of *Eryngium*, *E. bupleuroides* and *E. inaccessum*, on Robinson Crusoe Island. The presumptive F_1 hybrid (named *E. ×fernandezianum*) is not common, and it is uncertain whether it represents a stabilized hybrid taxon or simply a few occasional spontaneous individuals. The third case of a named hybrid is the fern *Pleopeltis ×cerro-altoensis*, recently documented, described, and named by Danton et al. (2015). This was interpreted as interspecific hybridization between two native species, *P. macrocarpa* on Robinson Crusoe Island and *P. masafuerae* only known from Alejandro Selkirk Island in the archipelago. If this were the case, it would suggest that spores of *P. masafuerae* must have dispersed from the younger to the older island, germinated, and produced a gametophyte that allowed hybridization with another gametophyte of *P. macrocarpa* in the vicinity of Cerro Alto. Roberto Rodríguez, however, has suggested (personal communication) that one of the parents instead might have been *Polypodium intermedium* subsp. *intermedium*, which is endemic to Robinson Crusoe Island. More study is recommended.

Biogeographical Implications

The models presented in Fig. 13.2 show a number of points regarding biogeographical implications in the Juan Fernández Archipelago. Most important are how the data conform with (1) the progression rule and (2) degrees of divergence.

Progression Rule

The progression rule (Funk and Wagner 1995) is the concept that initial colonists to a newly formed island archipelago would arrive on the oldest island first and then progress (disperse) to newer islands as they are formed. The Juan Fernández Archipelago is well suited to this hypothesis because the older island is also the one closest to the continent. Robinson Crusoe Island was available for 2 to 3 million years before Alejandro Selkirk Island appeared. Evidence is strong in most cases for the ancestral species arriving first on Robinson Crusoe Island with subsequent dispersal to the younger island. This applies to many of the genera with detailed phylogenetic investigations shown in the models in Fig. 13.2 as well as to the entire endemic flora.

As with all rules, however, there are exceptions. *Erigeron* (Fig. 13.2J) appears to be in a young stage of adaptive radiation (López-Sepúlveda et al. 2015), and this has taken place on Alejandro Selkirk Island, suggesting that the colonizer first arrived on this younger island. It is always possible that *Erigeron* did arrive and become established on Robinson Crusoe Island but later became extinct as the island reduced in size and ecological diversity, but no evidence is available on this point. The recent description of a new species of *Erigeron* (*E. corrales-molinae*; Danton et al. 2014) on Robinson Crusoe Island could alter our biogeographical hypothesis. From inspection of the data presented in the protologue, we judge that this new species is simply a diminutive vegetative form of the relatively common and morphologically variable *E. fernandezia*.

Based on the pattern of molecular data (AFLPs and SSRs), *E. fernandezia* apparently arrived on Robinson Crusoe Island from Alejandro Selkirk Island, perhaps during historical time (López-Sepúlveda et al. 2015a).

Another exception to the progression rule is *Myrceugenia schulzei*, which is endemic to Alejandro Selkirk Island and ties closely to *M. colchaguensis* of the Chilean mainland. It was previously regarded (Stuessy et al. 1990) that *M. schulzei* was closely related to *M. fernandeziana* on Robinson Crusoe Island. Recent molecular phylogenetic studies on the genus (Murillo-A. et al. 2012), however, have shown clearly that *M. fernandeziana* on Robinson Crusoe Island is not at all related to *M. schulzei*. The former is best placed in its own genus, *Nothomyrcia*, related to *Blepharocalyx* (Murillo-Aldana and Ruiz 2011; Retamales and Scharaschkin 2015). *N. fernandeziana* therefore represents an endemic genus in the archipelago, whereas its affinities with *Blepharocalyx* are being more fully explored. *Myrceugenia schulzei* apparently colonized the younger island directly from the continent rather than having been derived from a progenitor from the older island.

Wahlenbergia (Fig. 13.2R) provides what appears to be a third exception. Five species are endemic to the Juan Fernández islands (Lammers 1996): *W. fernandeziana* and *W. grahamiae* are confined to Robinson Crusoe Island, and *W. masafuerae* and *W. tuberosa* are restricted to Alejandro Selkirk Island. The fifth endemic species, *W. berteroi*, however, is found on Robinson Crusoe Island (and Santa Clara). Phenetic and cladistic studies by Lammers (1996) and allozyme studies by Crawford et al. (1990) have shown that the two species now found on Alejandro Selkirk Island were apparently derived from a progenitor from Robinson Crusoe Island. Although the data are somewhat equivocal, the remaining species on Robinson Crusoe Island, *W. berteroi*, may have resulted from a back-dispersal from *W. masafuerae* on Alejandro Selkirk Island and subsequent anagenetic speciation. If this actually happened, it would be the only known case in the archipelago of such a phenomenon. See Chapter 15 for more discussion.

Other examples of introduction, establishment, and anagenetic speciation directly on Alejandro Selkirk Island are *Acaena masafuerana*, *Agrostis masafuerana*, *Blechnum longicauda* (Fig. 13.2U), *Cardamine krusselii*, *Carex stuessyi* (Fig. 13.2G), *Eryngium sarcophyllum* (Fig. 13.2K; now presumed to be extinct), *Euphrasia formosissima*, *Galium masafueranum*, *Gavilea insularis*, *Gunnera masafuerae* (Fig. 13.2L), *Luzula masafuerana*, *Nicotiana cordifolia*, *Ranunculus caprarum*, and *Urtica masafuerae* and *U. glomeruliflora* (Fig. 13.2Q). Of the 129 endemic species in the archipelago, the exceptions to the progression rule add up to 12%. The vast majority of the species do conform to the progression rule.

Divergence

Another important biogeographical dimension that can be assessed from analyses of phylogeny is divergence. One would predict that the degree of divergence between continental progenitors and island-endemic derivatives would be greater than between congeneric endemic island species. One may also predict that the degree of divergence between species on the younger island and the continent or between the islands should be

Table 13.9 Degrees of Divergence in Different Geographical Settings among Progenitor-Derivative Species Pairs in Five Genera of the Juan Fernández Archipelago

	Geographical relationships of progenitor-derivative species pairs			
Genus	AS-RC	Within RC	AS-Continent	RC-Continent
Berberis	*B. masafuerana–B. corymbosa*			*B. corymbosa–B. microphylla*
Isozymes	0.006 (0.936)			0.001 (0.99)
ITS	[a]			0.0
Dendroseris	*D. regia–D. pinnata* (or *D. berteroana*)	Average among 8 spp.		
ITS	0.0212 [0.0169]	0.0262		
Gunnera	*G. masafuerae–G. peltata*	*G. peltata–G. bracteata*	*G. masafuerae–G. tinctoria*	*G. peltata–G. tinctoria*
Isozymes	0.014 (0.986)	0.105 (0.900)	0.044 (0.957)	0.037 (0.963)
ITS	1.4	0.0	0.2	1.2
Robinsonia	*R. masafuerae–R. evenia*	Average among 3 spp.		
ITS	0.006	0.029		
Isozymes		Average among 4 spp. (0.625)		
Sophora	*S. masafuerana–S. fernandeziana*			*S. fernandeziana–S. macrocarpa*
Isozymes	0.027 (0.973)			0.206 (0.814)
ITS	0.0			1.3

Note: Isozyme divergence is given as Nei's genetic distance (identity in parentheses); proportion sequence divergence shown for ITS.
[a] Unsuccessful DNA extraction in *B. masafuerana*.
Sources: Data from Sang et al. (1994); Crawford et al. (1998); Ruiz et al. (2004); and Stuessy et al. (2005).

less than that between those on the older island and the continent. It should also be the case that the degree of genetic divergence is greater among species on the older island in comparison with those on the younger island. Not all genera have allowed such calculations, but enough have been done to suggest some general points. Crawford et al. (1998, 2001b), Ruiz et al. (2004), and Stuessy et al. (2005b) have calculated the genetic identities (from isozymes) and genetic distances (from nrITS) between and among some of the endemic taxa and their continental relatives, namely, in *Berberis*, *Dendroseris*, *Gunnera*, *Robinsonia*, and *Sophora* (Table 13.9).

It should be mentioned that data assembled from ITS and isozymes for the endemic *Myrceugenia fernandeziana* and *M. schulzei* cannot be shown accurately in Table 13.9 because the continental progenitors have changed due to new phylogenetic studies. For example, *Myrceugenia fernandeziana* was seen with isozymes and ITS to be dissimilar (0.68 distance) to *M. schulzei*, a low level that is now not surprising because they are not close relatives, and each ties to different continental progenitors (Murillo-A. et al. 2012). *M. fernandeziana* is now segregated into its own genus, *Nothomyrcia*, near *Blepharocalyx* (Murillo-Aldana and Ruiz 2011; Retamales and Scharaschkin 2015). *M. schulzei* still resides in the same genus, but more precise estimates of affinity place this closer to *M. colchaguensis* (Murillo-A. et al. 2012) rather than to *M. exsucca*, as believed previously (Ruiz et al. 2004).

The relative values of divergence from isozymes and ITS (Table 13.9) in these five genera allow some conclusions, even though the data are not abundant. Regarding divergence between progenitors in the continent and endemic species on the younger versus the older island, the average difference, respectively, is 0.044 for isozymes and 0.2 for ITS versus 0.081 and 0.833. Both sets of data support the idea of greater genetic divergence between the continent and the older island. As for divergence between the two islands versus divergence among species on the older island, the averaged data reveal 0.0148 isozymes and 0.3568 ITS versus 0.105 and 0.0184. The isozymes therefore also conform to the predictions of having greater genetic divergence between species on the older island, but the ITS data are not in conformity. The problem lies with *Gunnera*. Wanntorp and Wanntorp (2003) using ITS sequences showed that *Gunnera tinctoria* (as *G. chilensis*) was closest to *G. masafuerae* from the younger island than to the two species on the older island. This could suggest that there were two introductions to the archipelago (Fig. 13.2L). The relationship in ITS divergence between *G. masafuerae* and *G. peltata* is 1.4, which is much higher than between the two endemic species on Robinson Crusoe Island (0.0). Isozyme divergence between these two species, however, is lower (0.014). This may suggest that speciation between the two endemic species on the older island occurred more recently than the origin of *G. masafuerae* on the younger island. We do know that the two endemics on Robinson Crusoe Island, *G. peltata* and *G. bracteata*, hybridize extensively (Pacheco et al. 1991), and hence they are genetically very compatible. If this ITS value is ignored, the other values average to 0.0091, which is much lower than the 0.0184 among species on the older island. Having more comparative data would obviously be desirable in providing more robust tests of these biogeographical hypotheses.

14 Genetic Diversity and Divergence

Daniel J. Crawford, Patricio López-Sepúlveda, Koji Takayama, and Eduardo A. Ruiz

Importance and Nature of Genetic Diversity

Depending on the methods employed to detect and assess variation, the term "genetic diversity" can mean different things in different studies. A general distinction in diversity has often been made between quantitative characters and single-locus molecular markers commonly used to measure genetic variation in populations. Traits such as leaf size and shape, plant height, and flowering time are often controlled by multiple loci of small effect (Conner and Hartl 2004; Edwards 2015). While these types of characters may be important for adaptive evolutionary change (Karrenberg and Wilder 2008; Oakley 2015), it is not a trivial matter to determine their genetic basis in part because of environmental effects on the phenotype and interaction among loci (Allendorf et al. 2013, chap. 11, pp. 205–229; Harrisson et al. 2014; Edwards 2015). Single-locus molecular markers comprise a second general class of markers commonly used to assess genetic diversity. Many single-locus molecular markers are now available and will be considered herein because they are emphasized almost exclusively. Single-locus molecular markers are generally assumed to be neutral or near neutral (McKay and Latta 2002; Leinonen et al. 2008), but even if the loci themselves are largely neutral, they may be linked to loci under selection (Kirk and Freeland 2011; Garcia-Verdugo et al. 2015). Rapid technological developments in genome sequencing now provide an almost unlimited supply of molecular markers, something that was difficult to imagine 10 to 15 years ago (Frankham et al. 2010; Allendorf et al. 2013, chap. 4, pp. 54–76; Harrisson et al. 2014). The issue of whether population diversity and differentiation in quantitative traits and single-locus molecular markers exhibit similar patterns has not been resolved, and different analyses of published data have produced differing results (Pfrender et al. 2000; Reed and Frankham 2001; Merilä and Crnokrak 2001).

There is now considerable interest in identifying selectively important loci that facilitate adaptive evolution. Genomic methods, sometimes combined with ecological field or experimental studies, provide the potential to link adaptive traits with their genetic bases. This topic is beyond the scope of this chapter, but concise reviews and relevant studies may be consulted for general overviews (Harrisson et al. 2014; Eizaguirre and Baltazar-Soares 2014; Tiffin and Ross-Ibarra 2014). Genetic diversity is necessary for plant populations to adapt to and evolve in changing environments (Eizaguirre and Baltazar-Soares 2014; Edwards 2015); if a population lacks the requisite genetic diversity to meet environmental change, then it will decline and be vulnerable to extinction. In oceanic archipelagos, as in continental areas, a number of factors, both natural and the direct or indirect consequences of human activities, can lead to a reduction in population number and size (Stuessy et al. 2014; Garcia-Verdugo et al. 2015). These reductions in size and number of populations result in the loss of genetic diversity and eventually could lead to the extinction of populations and species (Kirk and Freeland 2011).

Although single-locus molecular markers are likely not the "stuff" of adaptive evolution, they are still of value for detecting cryptic genetic variation (Crawford and Stuessy 2016), quantifying genetic diversity within and among populations and species, estimating gene flow via both pollen and seeds, determining mating systems in natural populations, estimating effective population sizes, detecting hybridization, and as a tool for the recognition of units (populations) worthy of conservation concern (Kirk and Freeland 2011; Ekblom and Wolf 2014). The purposes of this chapter are to list the molecular markers available for the study of genetic diversity and to discuss their application to the study of genetic diversity in the context of the biology and evolution of plant species endemic to the Juan Fernández Islands.

Molecular Markers

Allozymes

Allozymes are different forms of an enzyme encoded by alternative alleles at a locus. The terms "isoenzymes" and "isozymes," strictly speaking, refer to forms of an enzyme encoded by different loci, but the terms are sometimes used in a more general sense to refer to different forms of an enzyme regardless of the genetic basis. Allozymes were the first molecular markers widely used in plant studies, following the publications of classic papers documenting their variability in *Drosophila* (Hubby and Lewontin 1966) and humans (Harris 1966). Gottlieb (1977) established the foundation for the use of enzyme electrophoresis in plants and elucidated the kinds of systematic studies that would take full advantage of electrophoretic data. Later he provided a critique of allozymes for studying genetic variation in plant populations (Gottlieb 1981a). In addition to Gottlieb's pioneering contributions (1971, 1977, 1981a), Crawford (1983, 1985, 1989) also discussed the advantages of allozymes for systematic and population studies. Briefly, allozymes provided the most exact equation between phenotype (colored bands in a gel) and genotype (loci and alleles) available at the time (beginning in the mid-1970s) because they were inherited as codominant characters. This makes it possible to determine allele frequencies in populations or species, and from the frequencies, various diversity and divergence statistics may be calculated (see below).

Among the major limitations of allozymes are the relatively few loci that can be resolved and the low level of allelic variation at loci (Coates and Byrne 2005). This can be a particular problem with rare island endemic species, where very low genetic variation may be detected at loci (Crawford et al. 1988, 1994, 2001b). Another potential limitation of allozymes is the need for fresh or frozen plant material, which can be a severe problem with plants from isolated archipelagos such as the Juan Fernández Islands. One solution to this problem was the establishment of an allozyme laboratory on Robinson Crusoe Island and Alejandro Selkirk Island so that freshly collected material could be processed and results obtained on the islands (Crawford et al. 2001b; Ruiz et al. 2004; Stuessy et al. 2005b). However, setting up a laboratory on the islands required considerable time, effort, and expense.

Various types of DNA markers have largely supplanted allozymes, but there is an extensive literature on correlations between allozyme variation/divergence and various ecological attributes (Hamrick 1989; Hamrick and Godt 1989, 1996, 1997; Hamrick et al. 1979, 1991; Pérez de Paz and Caujapé-Castells 2013). Allozymes have been used rather extensively in assessing genetic variation in island plants (e.g., Helenurm and Ganders 1985; Lowrey and Crawford 1985; Crawford et al. 1987, 1988, 1990, 1993b, 1994, 2001b; Francisco-Ortega et al. 1996, 2000; Pérez de Paz and Caujapé-Castells 2013). Thus, despite their decline in popularity, it is appropriate and necessary to discuss allozymes in this chapter.

PCR-Based DNA Markers

A wide variety of DNA markers is now available for measuring genetic diversity in plant populations and species; recent overviews are provided by Agarwal et al. (2008) and Henry (2013). The most commonly used markers are based on the polymerase chain reaction (PCR); that is, the PCR is used to amplify regions of the genome, and the amplified products are used as data to calculate various genetic statistics. The so-called arbitrarily amplified markers such as random amplified polymorphic DNA (RAPD), inter–simple sequence repeats (ISSRs), and amplified fragment-length polymorphisms (AFLPs) have frequently been used in population and systematic studies. The major advantage of these approaches is that they require no knowledge of the parts of the genome being included for study, and universal primers may be used to amplify the DNA. A limitation of ISSRs, and particularly RAPD markers, may be their less than perfect reproducibility (Agarwal et al. 2008). By contrast, because AFLPs use restriction fragment-length polymorphisms (RFLPs) combined with PCR, their reproducibility is considerably higher than RAPDs and ISSRs (Meudt and Clark 2007; Agarwal et al. 2008).

These three types of markers (AFLPs, ISSRs, and RAPDs) have both advantages and limitations for the study of genetic variation, all of which apply to plants from oceanic islands. One of their major advantages over allozymes is that many loci can be surveyed, and there is higher variation at the loci (Levsen et al. 2008). Another desirable attribute of these markers is that living material is not needed. Rather, small amounts of leaf material can be collected and dried on silica gel for later extraction of DNA. These are especially important advantages for the study of rare island plants, where only a few individuals and minimal amounts of material are available for some species. When working on isolated islands, it is easier to dry plant material in small quantities of silica gel for further use than to keep material fresh or to gather viable seeds, either of which is necessary for allozymes.

One of the limitations of RAPD, ISSR, and AFLP markers is that they, in contrast to allozymes, are inherited as dominant rather than codominant characters. This means that the presence of a band (amplified fragment) could be due to two dominant alleles (homozygous) or one dominant and one recessive allele (heterozygous) at a locus. This, in turn, makes it difficult to impossible to calculate allellic frequencies in a population or species, which precludes calculating some common genetic statistics with the ease with which it can be done for codominant data (see below).

Microsatellites or simple sequence repeats (SSRs) are gaining popularity as molecular markers. SSRs share one important attribute with allozymes: they are inherited as codominants (Coates and Byrne 2005). However, they have a big advantage over allozymes because more loci can be scored, and there are typically many more alleles per locus. SSR markers are similar to dominant PCR-based markers in that small amounts of dried plant material can be used as a source of DNA. So SSR markers combine some of the important advantages of allozymes and dominant PCR-based markers. The downside of SSRs has been the time and expense of designing locus-specific primers. However, it is becoming ever more feasible to develop microsatellite primers (Glenn and Schable 2005) with the use of next-generation sequencing methods. In fact, extensive studies of genetic diversity in Juan Fernández plants include the use of SSR data, with primers having been developed for a number of species from the archipelago using next-generation sequencing technologies (López-Sepúlveda et al. 2013b, 2015a; Takayama et al. 2011, 2012a, 2015a).

Analysis of Molecular Marker Data

Numerous statistics can be calculated from molecular marker data. Measures that have been employed for Juan Fernández plants are emphasized in the following discussion for the purpose of providing the background for the general discussion of genetic variation in plants of the archipelago. With the increased use of SSRs relative to allozymes and the numerous DNA makers being used, programs are now available to calculate a variety of measures of diversity and divergence. Caujapé-Castells et al. (2013) describe and provide a valuable resource for transforming data matrices from common formats such as Excel spreadsheets into formats that can be analyzed by many population genetic programs, and they provide references to many different programs.

Codominant Markers

Genetic Diversity

The analysis of codominant markers such as allozymes and microsatellites is more straightforward than that of dominant markers for the reason mentioned earlier: it is possible to determine the frequencies of alleles at loci in populations or species. One of the most popular statistics used to analyze data is Nei's gene diversity statistics

(Nei 1973, 1978). Allele frequencies within a population or species are used to calculate expected heterozygosity at each locus, assuming panmixis (random mating), and the values are averaged over loci. By way of illustration, assume that a population has two alleles at a locus with equal frequencies of 0.5. With random mating, the frequency of homozygotes would be $2 \times (0.5)^2$, or 0.5, because there will be homozygotes for each of the alleles. The frequency of heterozygotes would be one minus the frequency of homozygotes, or 0.5. Another calculation of expected heterozygosity would be $2 \times (0.5 \times 0.5) = 0.5$. The diversity within a species can be calculated by using allele frequencies for the species as a whole, or diversity could be determined for each of the populations and a mean value calculated over all populations. The difference between the total diversity for a species and the mean value for the populations represents the proportion of total species diversity existing between populations and is a measure of genetic differentiation between populations.

In the terminology of Nei (1973), and assuming that we are concerned with a species and its component populations,

H_T is total diversity for a species.
H_S is the mean diversity within the populations of the species.
D_{ST} is the diversity among populations, that is, $H_T - H_S$.
G_{ST} is the proportion of total diversity existing between populations, that is, D_{ST}/H_T.

The value for G_{ST} can vary from zero (all populations with the same alleles in the same frequencies) to one (populations sharing no alleles).

The expected proportion of heterozygotes H_E, as described earlier, is often calculated for populations. The observed heterozygosity H_O is simply the heterozygotes detected in a population.

Wright's F-statistic (Wright 1951) includes the F_{ST}-statistic, which is similar to G_{ST} and is a measure of the proportion of total diversity in a population (or a species) that occurs among subpopulations (or among populations of a species). The total diversity may involve populations, groups of populations, species, and so on, and the subunits can be subpopulations, populations in particular areas, and so on, or populations of a species.

Wright's F-statistic can also be used to calculate the inbreeding coefficient F_{IS}, which is given by

$$F_{IS} = 1 - H_O/H_E.$$

The inbreeding coefficient is a measure of the departure from Hardy-Weinberg proportions, with a positive value indicating a deficit of heterozygotes and a negative value an excess of heterozygotes. This is intuitive because an excess of observed heterozygotes relative to expected heterozygosity would make the value of H_O/H_E greater than 1 and F_{IS} would be negative, whereas if H_E is greater than H_O, F_{IS} will be positive.

Hardy-Weinberg equilibrium (HWE) or random mating (see earlier) for loci in a population can be tested from the allelic frequencies at each locus. Several other genetic diversity estimates are often calculated for populations and species. One of them is average number of alleles per locus (NA) for all the loci examined; allelic richness

(AR) is a measure similar to NA, but it uses various statistical techniques to compensate for different sample sizes from different populations.

Genetic Divergence/Structure

Nei's (1973) genetic identity or genetic distance is used to calculate divergence between populations or species. The genetic identity is calculated as the probability of the identity of two randomly chosen genes from different populations or species compared with the probability of identity within each of the populations or species. The value at any locus can range from 0 (no alleles in common) to 1 (same alleles in same frequencies).

Assume that one population has only allele *a* at a locus and a second population has only allele *b*. This means that the probability of identity of two randomly chosen alleles within each of the populations is 1; that is, they must be identical because there is only one allele in each population. However, the probability of identity of two randomly chosen alleles from each of the populations is 0 because they share no alleles in common. By contrast, if both populations had the same single allele, then the probability of identity within and between the populations is the same; that is, the genetic identity is 1.

Thus, if for any locus x_i and y_i are the frequencies of the *i*th allele in populations *X* and *Y*, respectively, then Nei's genetic identity *I* for that locus is given by

$$I = \sum x_i y_i / (\sum x_i^2 \sum y_i^2)^{0.5}.$$

The identity over all loci is calculated as the arithmetic means of $\sum x_i^2$, $\sum y_i^2$, and $\sum x_i y_i$.

Nei's standard genetic distance *D* is equal to –ln *I*, and *D* can range from 0 to infinity. *D* is interpreted as the mean number of codon substitutions per locus, with a correction for multiple hits.

Other genetic divergence parameters sometimes calculated include number of private alleles (NPA), which, as the designation implies, are alleles present in only one population or species. The number of locally common alleles (NLCA) are alleles present in a frequency of 5% or greater in 25% or fewer of the populations sampled.

Dominant Markers

As indicated earlier, with dominant data it is not possible to determine directly the frequency of alleles within populations, but a number of methods are now available to calculate genetic diversity and divergence using the data.

Genetic Diversity

One measure of diversity is the total number of different phenotypes, or individuals with a unique combination of bands. Another diversity parameter is the average gene diversity over loci in each population and species (AGDOL), which is the probability that two homologous band sites, randomly chosen, are different. Other genetic diversity estimates by populations and species include percentage of polymorphic bands (PPB), which simply refers to the percentage of bands that are variable. This measure will be sensitive to sampling because increased sample size increases the likelihood of detecting variation. Another parameter that can be considered for dominant data is the total number of bands (TNB) detected in a population or species. The *Shannon diversity index* (SDI) also can be estimated:

$$HSh = -\sum p_i \log_2 p_i$$

over all loci where p_i is the frequency of the ith band in a population.

Genetic Divergence/Structure

Several measures of genetic divergence and structure are often calculated. One is the number of private bands (NPB). As the name suggests, it refers to bands detected exclusively in a population, in two or more populations that are grouped according to certain criteria such as a particular geographical area, or in a species. Another statistic is the "rarity index" (RI), sometimes referred to as "frequency-down-weighted marker values" (DWs). This is calculated for each population as the number of occurrences of a particular band divided by the number of occurrences of that particular band in the total data set (Schönswetter and Tribsch 2005). As mentioned earlier, a measure of genetic differentiation similar to G_{ST} is F_{ST} of Wright's F-statistic, and it may be used for dominant as well as nondominant markers. An analysis of molecular variance (AMOVA) is a statistical method for partitioning total genetic variation, typically in a species, into components within and among populations. It can be used to partition variation at various hierarchical subdivisions of populations or groups of populations.

A number of methods are available for using genetic diversity/divergence statistics from dominant data to group individuals or populations in some graphical form; these methods fall into two general categories. One is a band-based approach in which different similarity coefficients are used to calculate similarities (or distances), usually between individuals within or among populations. The mean difference among individuals in a population is a measure of diversity within the population. Similarity measures such as the Jaccard or Dice coefficients emphasize the presence of shared bands over shared absence of bands, the basic rationale being that there could be different reasons for the absence of a band. That is, absence could be homoplasious.

The Jaccard coefficient is given by

$$a/(a+b+c)$$

where a is the number of bands shared by two individuals, and b and c are bands present in each of the individuals.

The Dice coefficient is similar and is given by

$$2a/(2a + b + c)$$

and gives more weight to bands shared by two individuals.

The second general method of analysis is based on allele frequencies, and as with codominant data, it is a population-level approach. As indicated earlier, Nei's diversity statistics require allele frequency data, and with dominant data, it is not possible to distinguish (in most cases) whether presence of a band is due to homozygosity or heterozygosity at a locus. Thus calculation of allele frequencies is not a straightforward issue. One approach is to assume Hardy-Weinberg equilibrium for a population, and then the frequency of band absence (presumably a homozygous recessive) can be used to calculate the frequency of the two alleles. For outcrossing species, this may be a valid assumption, but for highly selfing species, this would not be the case. More detailed discussions of the statistical analyses of dominant data are given by Bonin et al. (2007).

Molecular Marker Studies of Island Plants: Population and Evolutionary Questions

As indicated earlier, molecular markers commonly used in population genetic studies have been generally assumed to be neutral or near neutral (McKay and Latta 2002; Leinonen et al. 2008). This means that marker differences between populations or species probably do not represent adaptive traits, although they could be linked to such traits. Neutral molecular markers have proven valuable as indicators of genetic variation in populations and divergence between populations, even if they are not the basis of adaptive divergence (Hamrick 1989; Hamrick and Godt 1989, 1996, 1997; Hamrick et al. 1979, 1991). Neutral markers may also be used as a rough estimate of historical gene flow between populations. For example, a crude approximation of gene flow has been calculated from G_{ST} by the equation

$$G_{ST} \approx 1/(4Nm + 1)$$

where N is the effective population size, and m is the migration rate between populations.

The primary purpose for using molecular markers is to determine the level and apportionment of genetic diversity within and among populations. The proportion of diversity among populations is a measure of genetic divergence or distance among populations; the higher the interpopulation diversity, the greater is the genetic divergence. These data can be critical for island species consisting of very few populations because markers can provide a first, though far from perfect, indication of differences between populations and an initial guideline for the conservation of populations. Markers can be used to assess divergence among populations on different islands of an archipelago or in different habitats or different areas on an island and have the potential for identifying cryptic species.

There is an immense literature on correlations between life-history traits or other attributes of species with genetic diversity in plants (Hamrick 1989; Hamrick and Godt 1989, 1996, 1997; Hamrick et al. 1979, 1991). It would be of interest to determine whether the general correlations for plants as a whole detected by Hamrick and collaborators hold for populations of insular endemic species. If this were the case, then relative levels of allozyme divergence could be used to estimate gene flow between populations and to infer some key attributes of species such as mating system (outcrossing or highly selfing). All these observations would be useful in formulating strategies for the conservation of genetic diversity in a species.

Molecular markers may also be of conservation value in another way. It is common practice to grow rare species in cultivation from seed collected from natural populations and then to reintroduce the plants into nature. Allozyme divergence between populations from different areas could indicate the possibility of divergence in ecologically important, yet cryptic, adaptive traits. Such data would dictate that progeny from cultivation be returned to the same source area as their parents.

Molecular markers may be used to infer the mating systems of plants, that is, to determine the levels of selfing and outcrossing in a population (Ritland 2002; Koelling et al. 2012). This is very valuable information for island plants (Sun and Ganders 1988; Nielsen et al. 2000; Crawford et al. 2010), especially in formulating conservation strategies for rare species. These types of studies are rare for island plants and have not yet been attempted for Juan Fernández species.

Genetic Diversity/Structure in Plants of the Juan Fernández Islands

Studies of genetic diversity and structure in Juan Fernández plants over the past three decades have involved two phases. The initial series of studies using primarily allozyme markers (and to a much lesser extent two PCR-based DNA markers) were done during the 1980s and 1990s and barely into the twenty-first century. The second series of studies has resulted in papers published from 2013 to the present and has employed SSRs instead of allozymes as codominant markers and AFLPs instead of RAPD or ISSRs as dominant markers (see literature cited). The two series of studies differed in more than the markers employed, with the earlier efforts focusing more on general surveys of the flora. A major purpose of these studies was to see whether there were correlations between genetic diversity/structure and ecological/life-history traits and species rarity (numbers and sizes of populations) (Crawford et al. 2001b). The more recent investigations have concentrated on extensive sampling of populations of selected species, both in the islands and their relatives on the continent, with the primary purpose of comparing genetic diversity/structure with modes of speciation in the Juan Fernández Archipelago. The results of earlier studies will be integrated into and compared with results of later investigations when appropriate. In other instances, it will be appropriate to consider the new data separately from that of earlier studies.

Diversity in Endemic Species

Crawford et al. (2001b) reported and summarized allozyme diversity for twenty-nine endemic species from the Juan Fernández Islands, including summaries of prior studies (Crawford et al. 1987a, 1988, 1990, 1992, 1993c, 1994, 1998), supplementation of data from the prior studies, and first reports for some species. The major purpose of Crawford et al. (2001b) was to determine whether the level and apportionment of allozyme diversity in populations of native and endemic species of the Juan Fernández Islands are associated with plant life history or other attributes detected in the extensive reviews of Hamrick and collaborators for flowering plants (Hamrick 1989; Hamrick and Godt 1989, 1996, 1997; Hamrick et al. 1979, 1991). Ten different groups, some based on the categories from Hamrick's studies, were identified, and tests of significance between the categories were done for various diversity statistics. Significant differences were found in one or more of the diversity measures (proportion of polymorphic loci per species and Nei's genetic diversity at the population or species level) for species with (higher diversity first in each comparison) large populations (two or more populations with more than 100 individuals) versus one or two populations; small scattered populations (10 or more with fewer than five plants) versus one or two populations; and large populations versus all other species.

Breeding system is a life-history trait that has been most important in explaining allozyme diversity in plants (Hamrick 1989; Hamrick and Godt 1989, 1996, 1997; Hamrick et al. 1979, 1991). However, using the conservative sequential Bonferroni analysis, Crawford et al. (2001b) detected no significant differences between wind-pollinated (largely outcrossing) and highly selfing species in any of the diversity categories. It should be stated, however, that the lower diversity in selfing species compared with both wind- and wind/bird-pollinated as well as insect/bird- and wind-pollinated species combined may be biologically significant given the low *p* values for these comparisons and the conservative nature of the Bonferroni analysis (Crawford et al. 2001b, table 3). For dioecious species, which must be outcrossing, the conservative statistical test shows no difference with other categories of breeding system; however, dioecious taxa have about twice the average diversities found in nondioecious species, and the low *p* values again suggest that the differences may be of biological significance (Crawford et al. 2001b, table 3).

The results for allozyme diversity in the Juan Fernández Islands may be compared with the comprehensive review and analysis by Pérez de Paz and Caujapé-Castells (2013) for allozyme diversity in the Canary Islands. As in the Juan Fernández Islands, there was higher diversity in taxa with larger population sizes, species that are self-incompatible (highly outcrossing) or have mixed mating systems, compared with selfing species. There was a significant positive association between polyploidy and higher allozyme diversity, likely a reflection of duplicate allozyme loci and higher heterozygosity at loci. The association between ploidy level and allozyme diversity was not tested in the Crawford et al. (2001b) study, but the two largest genera examined, *Dendroseris* and *Robinsonia*, are polyploids based on both chromosome number (Sanders et al. 1983; Spooner et al. 1987) and "extra" allozyme loci (Crawford et al. 1987, 1992) compared with what would be expected with diploids (Gottlieb 1982). Species of *Dendroseris* have generally much lower allozyme diversity than those of *Robinsonia*, and there is variation among species in each of the genera (Crawford et al. 2001b). It is likely that the very small population sizes and the self-compatibility in at least one species (Anderson et al. 2001b; Bernardello et al. 2001) are at least partially responsible for the lower diversity in *Dendroseris*. In contrast, most species of *Robinsonia* have larger populations than *Dendroseris*, and all species of *Robinsonia* are dioecious and thus obligately outcrossing, both attributes of which are associated with higher diversity.

Genetic Differentiation among Populations of Endemic Species

No significant differences were found between any of the groups in the proportion of genetic diversity residing among populations as estimated from G_{ST} (Crawford et al. 2001b). One of the consistent results from the literature has been the higher G_{ST} values obtained among populations for highly selfing species compared with mixed mating or outcrossing species (Hamrick and Godt 1989, 1997), but G_{ST} could not be calculated for five of the six selfing species because they were monomorphic at all allozyme loci (Crawford et al. 2001b). It is possible that if the five selfing species could have been included in the analyses, significantly higher G_{ST} values would have been obtained for the selfing category versus other breeding systems. Pérez de Paz and Caujapé-Castells (2013) found that self-compatibility was associated with higher G_{ST} values in Canary Island plants.

Genetic Diversity in Island Endemics and Continental Congeners

Stuessy et al. (2005a) and Ruiz et al. (2004) compared allozyme diversity for Juan Fernández endemics and continental congeners. Two of the major criteria for including genera in those studies were occurrence on both islands and some evidence for the closest continental relatives of the island taxa. Total genetic diversity was higher in the continental species than in insular endemics in all the genera. Species diversity for continental and island species is given for each genus (two continental species of *Sophora* were studied): *Gunnera* (0.102, 0.032), *Myrceugenia* (0.214, 0.091), *Rhaphithamnus* (0.103, 0.028), and *Sophora* (0.093 and 0.067, 0.038). A subsequent study of *Myrceugenia* (López-Sepúlveda et al. 2013b) showed comparable levels of diversity in one island and one continental species. However, the results of the two studies using different markers differed in several ways, and valid comparisons between the two are not possible. The value reported by Stuessy et al. (2005b) is for allozyme diversity between *M. fernandeziana* from Robinson Crusoe and its then presumed closest continental species *M. exsucca*. By contrast, López-Sepúlveda et al. (2013b) compared *M. schulzei* from Alejandro Selkirk with its presumed closest continental relative *M. colchaguensis*. There can be little doubt that the results of López-Sepúlveda et al. (2013b) provide a more valid comparison of island and continental species because it is now known from the molecular phylogenetic study of Murillo-A. et al. (2012) that *M. fernandeziana* from Robinson Crusoe is very distantly related to *M. exsucca*.

In fact, *M. fernandeziana* is not closely related to any other species of *Myrceugenia* (and should be recognized as distinct at the generic level), making the comparisons of Stuessy et al. (2005b) tenuous at best. Murillo-A. et al. (2012) showed strong support for *M. schulzei* and *M. colchaguensis* as sister species, providing confidence in their results by showing comparable levels of SSR and AFLP diversity in the insular and continental species.

The genus *Rhaphithamnus* (Verbenaceae) has one species in continental South America and one species in the Juan Fernández Islands; the extensive molecular phylogenetic study of Verbenaceae by Marx et al. (2010) provides definitive evidence that the two species are sister taxa that are not closely related to other members of the family. The recent study of López-Sepúlveda et al. (2016) employing SSR and AFLP markers is concordant with the earlier allozyme study of Crawford et al. (1993c) in showing reduced diversity in the island compared with the continental species. Lastly, the results of Stuessy et al. (2005b) cited earlier for *Gunnera* and *Sophora* should perhaps be taken with some caution because of the lack of strong evidence that the Juan Fernández endemics and the continental species with which they were compared do indeed represent sister taxa.

The lower allozyme diversity in Juan Fernández endemics compared with their continental relatives is a pattern seen in other endemics (de Joode and Wendel 1992; Frankham 1997), and indeed endemics in general harbor lower levels of allozyme diversity than more widespread taxa (Hamrick and Godt 1989, 1996). The hypotheses most commonly advanced for the lower diversity in island plants include dispersal of a small subset of the parental species to the island and drift and inbreeding in small island populations (de Joode and Wendel 1992). However, the generality that plants endemic to oceanic islands have lower diversity than their closest continental relatives has been called into question by later studies of Juan Fernández plants, and those results are discussed below.

Comparative Diversity in Allozymes and PCR-Based Dominant Markers

Although allozymes were used almost exclusively in the earlier studies to assess genetic diversity in Juan Fernández endemics (Crawford et al. 2001b), several species were studied with dominant PCR-based DNA markers, and results from different markers may be compared. *Lactoris fernandeziana* is arguably the most fascinating endemic species in the Juan Fernández Archipelago, and it is discussed in other chapters of this book. The species is restricted to a number of mostly small populations on Robinson Crusoe. The species was once considered possibly extinct (Skottsberg, in Carlquist 1964). Later estimates placed the total number of plants as low as ten (Lammers et al. 1986), and the most accurate, though still rough, estimates range from several hundred to perhaps over a thousand individuals (Crawford et al. 1994; Bernardello et al. 1999).

Brauner et al. (1992) used 16 RAPD primers to examine 27 plants from 15 populations. The primers amplified 106 bands that could be reliably scored. Fifteen of the 27 plants had one or more variant bands, and in only one instance was the same variant detected in plants from two different populations (Brauner et al. 1992). The overall polymorphism (41 variants of a total of 2,532 bands scored for all primers and individuals) was a very low 1.6%. The low level of polymorphism could be due in part to the small number of plants sampled relative to the sizes of populations: for seven of the 15 populations, less than 20% of the estimated total number of plants were sampled.

Crawford et al. (1994) surveyed 83 plants from 12 populations for allozyme variation and detected no variation among any of the plants at 22 presumptive loci. Allozymes are apparently less sensitive than RAPD markers for assessing population diversity and structure in *L. fernandeziana*. Finally, in a third study, Crawford et al. (2001a) used three ISSR primers to examine genetic variation in 89 plants from 16 populations of *Lactoris fernandeziana*. Sixty-two loci were scored, with each unique band interpreted as diallelic for the presence or absence of the band. Pairwise similarities were calculated and used to generate distances (1 − similarity). Average similarities were calculated for plants in the same and different populations. Mean distances between individuals within populations varied from 0.00 to 0.10, and for plants from different populations the range was 0.02 to 0.18. The neighbor joining tree based on the distance values had plants from 10 of the 16 populations forming exclusive groups, all with bootstrap support above 60%. This indicates that over 60% of the populations were distinct from all other populations examined, something that was not detected with either allozyme or RAPD markers. There was no correlation between geographical and ISSR distance for populations.

The results from this study demonstrate that ISSR markers, even when using only three primers, detect much more variation than did allozymes or RAPD markers. The increased variation detected with ISSRs allowed several inferences to be made about genetic diversity in *L. fernandeziana*. Most of the diversity resides among rather than within populations, so there is population differentiation. Second, ISSR population distance is not correlated with geographical distance. These two findings have important conservation implications because they suggest that all populations must be conserved to maintain genetic diversity in the species. Also, the lack of geographical correlation indicates that populations from the same area are just as important for conservation as populations from different areas of Robinson Crusoe Island.

Crawford et al. (1987a) examined allozyme variation in six species of *Dendroseris*, with subsequent data and analyses given in Crawford et al. (1998, 2001b). Two to six populations of these very rare species (Stuessy et al. 1998b, 1998d) were studied, and in all cases except one, fewer than 10 plants comprise each population (Crawford et al. 1987a). No allozyme diversity was detected in the two species *D. berteroana* and *D. pruinata*, whereas the other four species had total diversities ranging from 0.02 to 0.071 (Crawford et al. 2001b). Esselman et al. (2000) surveyed RAPD marker diversity in seven species and 26 populations of *Dendroseris*, with the majority of the populations the same as those included in the allozyme studies. Some RAPD diversity was found in all species of *Dendroseris*. Esselman et al. (2000) calculated RAPD locus diversity using the Shannon-Weaver statistic (see earlier), and these values ranged from 0.003 to 0.022.

Despite the rather low diversity, every individual of five species was separable by the array of RAPD bands; that is, each plant had a different multilocus genotype. Of the remaining two species, five plants of one species had three genotypes, and in the other species, four different arrays of bands occurred among the 13 individuals.

Esselman et al. (2000) compared relative levels of diversity for allozymes and RAPD markers among species of *Dendroseris*. Using Spearman rank correlations, it was determined that relative diversity rankings for the two markers were not correlated. However, the two species with the highest allozyme diversity, *D. litoralis* and *D. micrantha*, also exhibited the highest RAPD band diversity, and *D. berteroana* exhibited no allozyme diversity and has the second lowest RAPD band diversity of any of the species.

With both *Lactoris* and *Dendroseris*, the PCR-based DNA markers ISSR and RAPD were more efficient than allozymes in detecting variation. In the case of *Lactoris*, ISSR markers generated much more variation than RAPD markers while using fewer primers. The results for allozymes, while limited, are not surprising in view of the earlier discussion of the advantages of PCR-based DNA markers over allozymes, including the ability to incorporate more loci and with higher variation at the loci. The higher variation seen with ISSR than RAPD markers illustrates that the results will likely depend on the PCR-based markers employed.

Second-Phase Studies: Different Molecular Markers and Testing New Hypotheses for the Level and Apportionment of Genetic Diversity in Endemic Species

The primary purpose of the first phase of genetic diversity/divergence studies using molecular markers, primarily allozymes, was to test the correspondence between diversity and a suite of traits that had been found for a wide variety of flowering plant taxa. The second phase of studies, as indicated earlier, used different molecular markers. The primary purpose of these studies was to test hypotheses of genetic diversity with different modes of speciation in the Juan Fernández Archipelago and speciation in the context of the ontogeny of the Juan Fernández Archipelago (Stuessy 2007; Stuessy et al. 2006).

There are several components to the Stuessy model of anagenetic and cladogenetic speciation that predict levels of genetic diversity in island endemics compared with their closest continental relatives (progenitors or sister taxa) and with each other. As indicated earlier, the long-held paradigm is that island populations typically have lower genetic diversity because the colonization of islands by one or a few propagules will contain diminished diversity compared with their source populations; this is often referred to as the *founder effect* (Nei et al. 1975; Templeton 2008). The severity of the genetic bottleneck associated with colonization and the time to recovery will depend on how long the population remains small following establishment (Nei et al. 1975). Stuessy and collaborators (Stuessy 2007; Stuessy et al. 2006) argue that diversity comparable with that found in the source populations could develop through processes such as mutation and recombination in an island lineage as populations increase in size and number and diverge from the original colonizer (Fig. 14.1). Under this scenario, any initial genetic

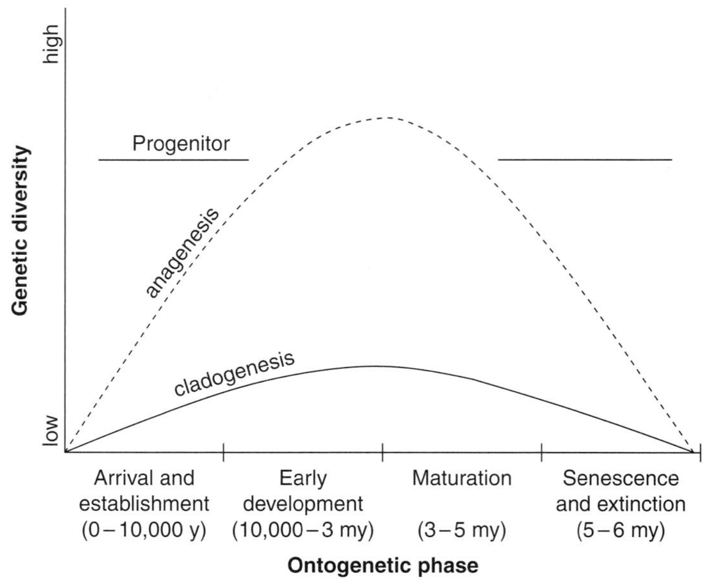

Figure 14.1 Model of contrasting levels of genetic diversity within species with anagenesis (no divergence into distinct lineages) versus cladogenesis (lineage divergence and speciation) during ontogeny of an island. (Simplified and modified from Stuessy 2007.)

bottlenecks associated with colonization would likely go undetected (Stuessy et al. 2012). Eventually, the island plants may be recognized as a distinct species; this is termed "anagenetic speciation," and the process is "anagenesis" (Stuessy 2007; Stuessy et al. 2006, 2014c; Takayama et al. 2015b). If, instead of a single species evolving following colonization, there is lineage splitting to produce several species, it is termed "cladogenetic speciation," and the process is "cladogenesis" (Stuessy et al. 2006). With cladogenesis, genetic diversity would presumably be partitioned among the species, and thus diversity within each of the species would be lower than in a single species originating from anagenesis (Fig. 14.1). There could also be secondary loss of species diversity during the ontogeny of an island. As the island becomes reduced in size and available habitats decrease in size and number due to erosion and subsidence, reduction in population size and number would erode genetic diversity (Fig. 14.1). Stuessy (2007) mentioned the difficulty in determining whether reduced genetic diversity in a species is the result of dispersal and establishment (primary) or decline in accrued diversity with island age and development (secondary) (Fig. 14.1). Several lineages in the Juan Fernández Archipelago can serve as model systems to test the hypothesis of Stuessy (2007) and Stuessy et al. (2012) on the impact of anagenetic and cladogenetic speciation on genetic diversity in island plants and their continental ancestors (sister taxa).

Anagenetic Speciation and Genetic Diversity

Three Juan Fernández genera suitable for examination of genetic diversity and anagenesis are *Myrceugenia* (Myrtaceae), *Drimys* (Winteraceae), and *Rhaphithamnus*

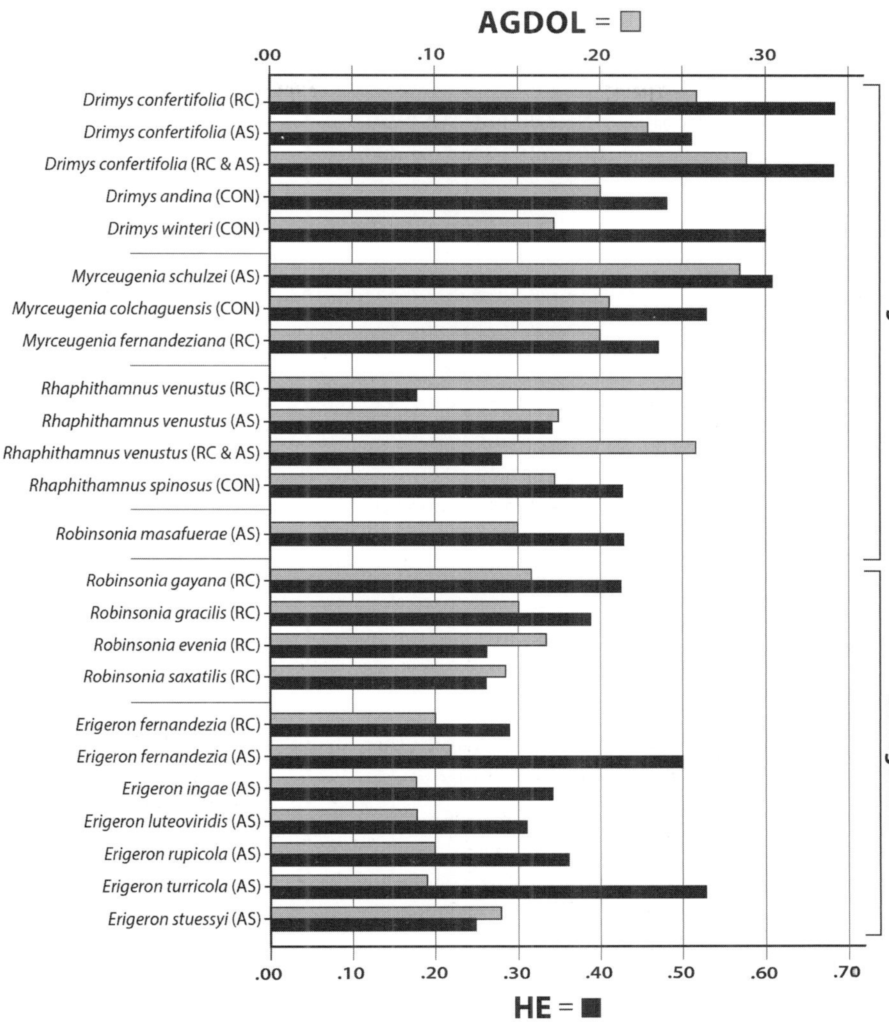

Figure 14.2 Summary of average genetic diversity over loci (AGDOL) in AFLP markers (grey bars) and expected heterozygosity H_E at loci (black bars). All are average values. (Data from Takayama et al. 2015.)

(Verbenaceae). Genetic diversity in the insular *M. schulzei* endemic to Alejandro Selkirk Island is comparable with, and indeed slightly higher than, its closest continental species *M. colchaguensis* (López-Sepúlveda et al. 2013b) (Fig. 14.2). Likewise, SSR and AFLP diversity in the Juan Fernández endemic *Drimys confertifolia*, which occurs on both Robinson Crusoe and Alejandro Selkirk Islands, is similar to that estimated in the continental *D. andina* and two varieties of *D. winteri* (López-Sepúlveda et al. 2015b) (Fig. 14.2). The closest relative of the endemic *D. confertifolia* has not been identified with confidence, and the continental sampling was somewhat limited (one population of *D. andina* and two populations of one variety of *D. winteri*), making comparison of diversity between insular and continental taxa somewhat less rigorous than for *Myrceugenia*. The insular species *Rhaphithamnus venustus* has genetic diversity comparable with its continental progenitor (sister) species with AFLP markers and lower SSR diversity than the continental species (López-Sepúlveda et al. 2016) (Fig. 14.2); a similar pattern was seen with allozymes (Crawford et al. 1993c).

Looking at results from another island, two anagenetically derived species of *Acer* (Sapindaceae) from Ullung Island, Korea, have lower diversity in SSR and/or AFLP markers than their most closely related species (Pfosser et al. 2002; Takayama et al. 2012b, 2013). It was suggested that the slightly lower diversity in the two Ullung Island endemic *Acer* species was the result of genetic bottlenecks associated with colonization of the island, the young age (1.8 my) of the island and the persistently small population sizes on the islands. In contrast to *Acer*, the Ullung endemic *Dystaenia takesimana* (Apiaceae) exhibited significantly higher AFLP diversity than its sister species from Japan (Pfosser et al. 2005). Pfosser et al. (2005) considered several factors that could account for the higher insular diversity than its progenitor species. Pfosser et al. (2005) suggested that there was an initial founder effect with reduced AFLP diversity and that through time mutation and recombination replenished diversity as population numbers and sizes increased.

Results for anagenetically derived species from both Juan Fernández and Ullung illustrate that there may be either reduced or increased diversity in insular species compared with species from their source areas. This indicates that within the concept of anagenetic speciation, various factors (discussed below) may influence the genetic diversity in populations and species. As mentioned earlier, lower diversity in island species could be the result of either genetic bottlenecks associated with colonization and establishment on an island or loss of genetic diversity as populations decrease in number and/or size because of one or a combination of factors, both natural and the result of human activities. As examples, the size of an island will decrease in area and relief through time due to subsidence and erosion, and human impact includes the destruction of native habitats (Fig. 14.1). In the Stuessy et al. (2012) model, higher or comparable diversity in island compared with continental species is the result of the generation of diversity during radiation in the archipelago as the derivative species increases in population number and size and possibly with some gene exchange between populations. Another potential cause for comparable diversity in island taxa is the avoidance of a severe genetic bottleneck during colonization and establishment; that is, the populations must avoid remaining small for many generations. The initial diversity carried by the colonizer(s) to an island could have a large influence on diversity in the insular species, and several variables may be involved. If the colonization involves only one diaspore, then it matters whether the source population is highly selfing or mixed mating/highly outcrossing because the former would be highly homozygous and the latter would likely have high genome wide heterozygosity. The ploidy level of the diaspore could also be a factor, with neopolyploids, especially allopolyploids, having higher heterozygosity than diploids due to the duplication of loci (Crawford et al. 2009). A single dispersal episode could include more than one diaspore when the dispersal unit is a multiseeded fruit or an inflorescence with several fruits, such as in Asteraceae

(Crawford et al. 2009). Indeed, a single dispersal event with multiple diaspores of an outcrossing colonizer could potentially carry considerable genetic diversity. There is no evidence to suggest that the Juan Fernández species are other than the result of single colonization episodes, and this appears to be the situation for the vast majority of flowering plant lineages on remote oceanic archipelagos (e.g., Sang et al. 1994, 1995; Puppo et al. 2014).

It is apparent from the preceding discussion that a robust test of the hypothesis proposed by Stuessy and collaborators (Stuessy 2007; Stuessy et al. 2006, 2012) of how anagenetic speciation could influence genetic diversity would require knowing, among other things, the ontogenetic stage of an island when colonization occurred, the genetic diversity included in the colonizing event, and fluctuating population numbers and sizes over the ontogenetic history of an island. Obviously, most of these things cannot be known with any degree of certainty for any island species because they are all historical in nature. However, comparative data, especially molecular markers, may provide some insights into these questions. Identification of the closest extant relative of an island species is typically most robustly documented by resolving the two as sister taxa in a molecular phylogeny. High sequence similarity between an island species and its closest relative would be indicative of recent divergence, whereas higher sequence divergence would suggest an older lineage split and the possibility that the colonizing ancestor of island species arrived early in the ontogeny of an island. Another indication of recent island colonization is the distribution of novel or private alleles (or bands) with molecular markers such as SSRs and AFLPs. If the island species is of recent origin, then it would be expected to have either the same or, more likely, a subset of the markers found in its closest relative. As a result, alleles in the closest relative that were not included during colonization of an island will be seen as "private." In contrast, there would be few, if any, private alleles in the island species because of insufficient time for origin of new variation via mutation (Meloni et al. 2015). If, however, increased diversity generated during growth in population size and number during early ontogeny of an island was then lost secondarily later in the ontogeny of an island (Fig. 14.1), then one might expect private alleles in both the island species and its relative because of mutations of novel alleles in each.

Data for the genus *Rhaphithamnus* (Verbenaceae), with one species (*R. spinosus*) in South America and another (*R. venustus*) endemic to the Juan Fernández Islands, indicate that it is an example of anagenetic speciation in which lower diversity in the insular species, especially with SSR markers (Fig. 14.2), is the reflection of a genetic bottleneck associated with colonization. As mentioned briefly earlier, a molecular systematic study with extensive taxon sampling of Verbenaceae by Marx et al. (2010) provided compelling evidence that the two species are sister taxa. On the basis of identical ITS sequences for the two species and very low or non-existent divergence at several plastid loci, Marx et al. (2010) suggested that colonization must have been recent. López-Sepúlveda et al. (2016) documented lower SSR and AFLP diversity in *R. venustus* and many private alleles (bands) in the continental *R. spinosus* compared with their low or nonexistence in the island species, a pattern suggesting recent divergence (Meloni et al. 2015).

Two other anagenetically derived Juan Fernández species mentioned earlier, *Drimys confertifolia* (Winteraceae) and *Myrceugenia schulzei* (Myrtaceae), have comparable or higher diversity than their closest relatives (Fig. 14.2). Marquínez et al. (2009) provided strong evidence that *D. confertifolia* is sister to a clade consisting of *D. andina* and *D. winteri* from Chile and Argentina, the two species used by López-Sepúlveda et al. (2015b) for genetic diversity comparison with the insular species. However, divergence time between *D. confertifolia* and its sister species was estimated at 9.0 to 10.8 Mya from combined nuclear ITS and plastid sequences (Marquínez et al. 2009). This estimate predates the origin of the Juan Fernández Archipelago, which is approximately 4 Mya (Stuessy et al. 1984). Murillo-A. et al. (2012) demonstrated that *M. schulzei* from Alejandro Selkirk Island is sister to *M. colchaguensis*, with the divergence during the Pliocene, some 2.6 to 5.3 Mya. This estimated divergence time largely predates or at best coincides with the earliest origin of Alejandro Selkirk Island some 1.0 to 2.0 Mya (Stuessy et al. 1984). López-Sepúlveda et al. (2013) found comparable levels of diversity in the two species. In contrast to *Rhaphithamnus*, the molecular data suggest that *Drimys* and *Myrceugenia* were early colonizers of the Juan Fernández Archipelago. If this is the case, then there are two scenarios to explain their diversity, one being that the colonizers carried considerable diversity to the island, and there was no strong or prolonged bottleneck. Another possibility is that the initial bottleneck is now obscured by the generation of diversity during evolution on the island, as envisioned by the Stuessy et al. (2012) model. The number of private AFLP bands and private SSR alleles are comparable in the insular *Drimys confertifolia* and its two closest relatives in continental South America, suggesting that a portion of the diversity in the island species is the result of novel mutations (Meloni et al. 2015).

A word of caution is in order when comparing private bands/alleles in two species because the results could be affected by sample size. For example, in the study of *Drimys* by López-Sepúlveda et al. (2015b), sampling of continental taxa included only one population of *D. andina*, two populations of *D. winteri* var. *winteri*, and eight populations of *D. winteri* var. *chilensis*, whereas 31 populations from the island species were sampled. Additional sampling from the continent could have revealed additional bands/alleles not detected on the island; that is, the number of private alleles could increase with sampling. Low-frequency bands/alleles are particularly sensitive to sampling error. The situation for private bands/alleles in *Myrceugenia* is similar to that in *Drimys*, with the insular species having a comparable, but somewhat lower, average percentage of private alleles per locus with SSR markers. A total number of eight private AFLP bands were detected in the 13 populations (129 individuals) of insular species, and seven bands were found in the two populations (19 plants) of the continental *M. colchaguensis*. These observations suggest that novel bands/alleles have been generated in the insular species. With the aforementioned caveats in mind, the totality of data suggest that *Drimys* and *Myrceugenia* are examples of the model of anagenetic speciation in which genetic diversity has been generated following an initial genetic bottleneck (Stuessy et al. 2012).

No examples have been detected in Juan Fernández plants where higher genetic diversity in insular taxa likely reflects diversity in the colonizer(s) and not the result of recovery from a bottleneck. As discussed earlier, *Dystaenia takesimana*, endemic to

Ullung Island, has higher AFLP diversity than its sister species from Japan (Pfosser et al. 2005). The authors suggest that the species lost diversity during colonization and establishment and that it accumulated higher diversity through mutation, recombination, and drift. However, these workers observed that all of the 130 fragments scored occur in both species; that is, there are no private bands in either species. This means that at least with regard to the bands used in the analyses, none was lost from the progenitor species during colonization. However, if new bands were created via mutation in the insular species, as suggested by Pfosser et al. (2005), those bands would have been detected only in the insular species as private bands. These observations suggest that the high diversity in the island species results from minimal loss of diversity during colonization rather than an initial loss of diversity followed by an increase in AFLP diversity in the island. Furthermore, Pfosser et al. (2005) make the important point that sampling in Ullung Island included large populations, with most of the diversity occurring within populations. The authors suggest that *Dystaenia takesimana* on Ullung Island is a large population with gene flow among the subpopulations. This is exactly what the model of anagenetic speciation envisions as important in generating diversity by recombination. In contrast to population structure and sampling in the island, sampling of populations of the progenitor species *D. ibukiensis* in Japan included small, geographically isolated populations, a situation that would be expected to have lower diversity than large metapopulations.

In addition to anagenetic speciation involving source areas outside an archipelago (usually continental sources), anagenesis could occur following dispersal between islands within an archipelago. *Rhaphithamnus* and *Drimys* are ostensibly two examples of dispersal from the older Robinson Crusoe Island to the younger Alejandro Selkirk Island without speciation. *Drimys* has lower AFLP and SSR diversity in Alejandro Selkirk than in Robinson Crusoe (Fig. 14.2) and has all the signatures of founder effect on the younger island (López-Sepúlveda et al. 2015b). In *Rhaphithamnus*, AFLP diversity is much lower on Alejandro Selkirk Island than on Robinson Crusoe Island, but there is higher diversity in several measures of SSRs in the younger island (López-Sepúlveda et al. 2016). It is important to note, however, that there are no private alleles on Alejandro Selkirk, indicating that all alleles on Alejandro Selkirk are a subset of those found on Robinson Crusoe, indicative of founder effect. There can be little doubt that *Robinsonia masafuerae* (Asteraceae-Senecioneae), which is endemic to Alejandro Selkirk Island, has evolved from a colonizer from Robinson Crusoe Island, where several endemic species of *Robinsonia* occur (see below). Further, molecular and morphological data demonstrate that *R. evenia* is the closest extant relative of *R. masafuerae* (Sanders et al. 1987; Sang et al. 1995). The AFLP diversity is just slightly lower in *R. masafuerae* than in *R. evenia*, but SSR diversity is considerably higher, and the differences between the two markers are striking (Fig. 14.2). For example, *R. masafuerae* has only one private AFLP band, and *R. evenia* has 39 unique bands, but the situation is completely reversed for SSR alleles, where *R. evenia* has no private alleles and *R. masafuerae* has several novel alleles (Takayama et al. 2015a). The reasons for the contrasting patterns with the two markers are obscure, though it should be noted that only one individual was sampled in four of the five populations, and five plants were included in the other population, for a total of nine plants. It may be that the different results for the two markers are the result of the very small sample size for *R. masafuerae*.

Cladogenetic Speciation and Genetic Diversity in the Juan Fernández Archipelago

The splitting of lineages via cladogenesis will partition genetic diversity among species, and the individual species will have lower diversity than anagenetically derived species (Fig. 14.1). With the one exception in *Robinsonia*, all comparisons of genetic diversity between anagenetically and cladogenetically derived species are among and not within genera. Species resulting from cladogenesis in the Juan Fernández Archipelago for which there are diversity data come from the two genera *Erigeron* (López-Sepúlveda et al. 2015a) and *Robinsonia* (Takayama et al. 2015a).

Takayama et al. (2015b) reviewed comparative diversity in Juan Fernández species resulting from anagenesis and cladogenesis and provided statistical analyses of data from AFLP and SSR markers. It is beyond the scope of this chapter to discuss the results in detail; Takayama et al. (2015b) should be consulted for the details. The most meaningful results in which speciation mode is considered were presented by Takayama et al. (2015b) as a two-way ANOVA. At $p < 0.05$, the only diversity measure in which anagenetically dervived species are higher is H_O with SSR. However, the species derived by anagenesis are higher and approach significance (*p*-values in parentheses) for AFLP average genetic diversity over loci (0.063). With SSRs, expected heterozygosity (0.085) and allelic richness (0.051) are very nearly significantly higher. None of the other statistics with either marker approached a significant difference for speciation mode. Statistics aside, the average species diversities are higher in species originating by anagenesis than for those evolving by cladogenesis for all statistics with both markers (Fig. 14.2) (Takayama et al. 2015b). Another important result of the diversity analyses mentioned by Takayama et al. (2015b) is that diversity in Juan Fernández species resulting from anagenesis, with the notable exception of *Rhaphithamnus venustus*, is comparable to the total diversity over all species derived by cladogenesis in *Robinsonia* and in *Erigeron*. This is what would be expected under the models of anagenesis and cladogenesis, in which total diversity in a lineage that has undergone cladogenesis has been partitioned into the component species, whereas the total diversity accumulates within a single species during anagenesis (Fig. 14.1). Pérez de Paz and Caujapé-Castells (2013) noted that lineages on the Canary Islands with two or fewer species tended to have higher species diversity than those in which radiation had been more extensive. This observation is concordant with what would be expected under the models of anagenetic and cladogenetic speciation.

The partitioning of genetic diversity within and among species during cladogenesis and speciation will depend on how lineage splitting occurs. Speciation is discussed in Chapter 15, and the reader may wish to consult that chapter for more background information. However, a brief consideration of the biogeography of speciation will suffice for the present discussion. With peripatric speciation, there is a budding off of a small population at the periphery of a larger population or the dispersal of propagules from the larger population to establish a smaller population outside the normal dispersal range of the larger population. Under this scenario, which is similar to the founding of populations in an archipelago, there would initially be lower diversity in the small populations, and the diversity in the ancestral population would be little, if any, diminished. By contrast, with the classical model of allopatric (or vicariant) speciation, there could be a more equitable partitioning of genetic diversity between the two populations (or population systems).

Robinsonia represents the best group in the Juan Fernández Archipelago to examine genetic diversity in species within the context of the probable mode of speciation. *Robinsonia gayana* and *R. thurifera* are sister species in a molecular phylogeny, with identical ITS sequences (Sang et al. 1995), and the two species are similar morphologically (Sanders et al. 1987). All data indicate recent divergence between *R. gayana* and *R. thurifera*, with the former species much more abundant than the latter. Allozyme diversity is almost 5.5 times higher in the more common *R. gayana* than in the rarer *R. thurifera* (Crawford et al. 1992), which is what would be expected for a progenitor-derivative species pair (Gottlieb 1971, 1973, 1977, 2003; Crawford 2010) where there has been a budding off of a peripheral population (*R. thurifera*) carrying a subset of the diversity in the progenitor (*R. gayana*) population. The study of Takayama et al. (2015a) included only one individual of *R. thurifera* for AFLP and SSR analyses, precluding valid comparison of the results with the allozyme data. The recently described and very rare *R. saxatilis* (Danton 2006) is similar morphologically and with molecular data to *R. gayana* (Figs. 14.2 and 14.3C, D) (Takayama et al. 2015a). This species has not been included in a molecular phylogenetic study. In concordance with the hypothesis that *R. saxatilis* is a derivative species of *R. thurifera*, *R. gayana*, or a common ancestor of both, the AFLP and SSR diversity in the five plants from one population of *R. saxatilis* is much lower than in *R. gayana* (Fig. 14.2) (Takayama et al. 2015a). Another example of peripatric speciation in *Robinsonia* is *R. masafuerae*, which, as indicated earlier, all evidence suggests evolved from *R. evenia* (Sanders et al. 1987; Sang et al. 1995; Takayama et al. 2015a, 2015b) (Fig. 14.2). The two species differ by only three mutations in ITS sequences, which makes them much less divergent from each other than they are from any other species (Sang et al. 1995); *R. masafuerae* was not included in the allozyme study of Crawford et al. (1992b). Depending on the statistic, *Robinsonia masafuerae* has AFLP diversity comparable with or lower than *R. evenia*, which is what might be expected given the presumed origin of the species. However, as indicated earlier, *R. masafuerae* has higher SSR diversity than *R. evenia* in all measures (Fig. 14.2) (Takayama et al. 2015a). This is the opposite of what would be expected and runs counter to the AFLP data. Takayama et al. (2015a) suggested that the diversity in *R. masafuerae* has accumulated over time during anagenetic speciation. However, there are at least two factors that seem to be at odds with this hypothesis. First, the comparable or lower AFLP diversity in *R. masafuerae* contrasts sharply with SSR diversity, which suggests that diversity increased during anagenetic speciation with one marker (AFLP) but not the other (SSR). Second, *R. masafuerae* is a rare species that occurs in very small populations, which would not be conducive to the generation

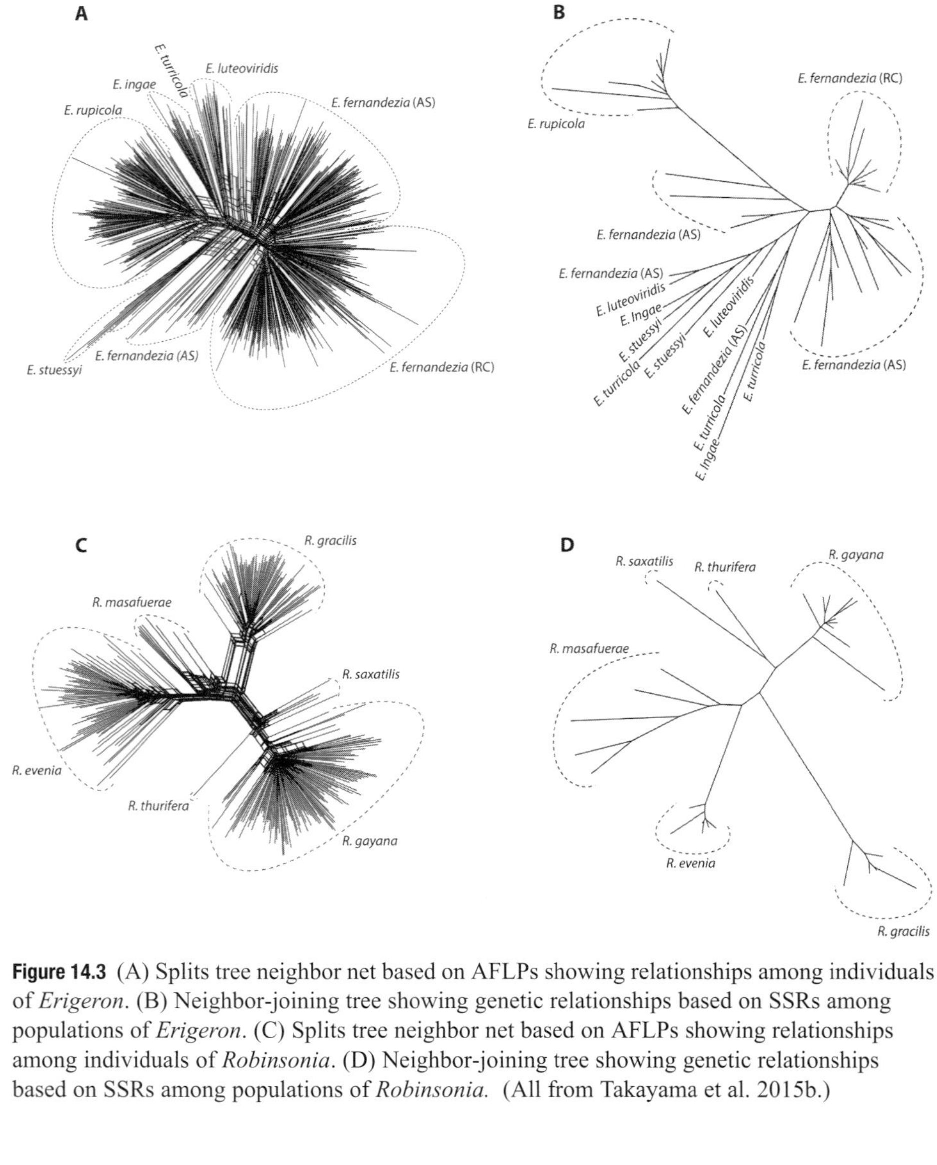

Figure 14.3 (A) Splits tree neighbor net based on AFLPs showing relationships among individuals of *Erigeron*. (B) Neighbor-joining tree showing genetic relationships based on SSRs among populations of *Erigeron*. (C) Splits tree neighbor net based on AFLPs showing relationships among individuals of *Robinsonia*. (D) Neighbor-joining tree showing genetic relationships based on SSRs among populations of *Robinsonia*. (All from Takayama et al. 2015b.)

of diversity and the recovery from bottlenecks associated with colonization and establishment on Alejandro Selkirk Island. There are no readily apparent reason(s) for the higher SSR diversity in *R. masafuerae* than in *R. evenia*. Equally enigmatic are the very low levels of SSR diversity in *R. evenia*. The low SSR diversity is particularly puzzling because this species, along with *R. gayana* and *R. gracilis*, are the three most common species of *Robinsonia*. Also, the AFLP diversity in *R. evenia* is comparable to or higher than that in the other two common species (Fig. 14.2), making it difficult to imagine how population/genetic/biological factors could account for such differences with essentially the same sample of individuals for the two markers (Takayama et al. 2015a).

The genus *Erigeron* (Asteraceae-Senecioneae) offers another suitable study for genetic diversity and cladogenesis in the Juan Fernández Archipelago. *Erigeron* differs from *Robinsonia* in that cladogenesis has occurred recently or is occurring on the younger Alejandro Selkirk Island rather than on the older Robinson Crusoe Island. This is likely reflected by the fact that some species are not as well defined morphologically as in *Robinsonia* (López-Sepúlveda et al. 2015c; Takayama et al. 2015a, 2015b). Within *Erigeron*, there are three morphologically well-defined groups. The three rare species *E. ingae*, *E. luteoviridis*, and *E. turricola* from the so-called alpine zone are difficult to distinguish morphologically. *Erigeron rupicola* and *E. stuessyi* are similar morphologically, with the former species more common than the latter and occurring on coastal rocks at sea level up into the ravines. The rarer *E. stuessyi* also occurs on rocky ledges but is restricted to cool, deep ravines. Lastly, *E. fernandezia* is broadly distributed from 100 to 1200 m in altitude and commonly occurs in rocky areas at middle elevations (Valdebenito et al. 1992a; López-Sepúlveda et al. 2015a). This species also occurs on Robinson Crusoe Island, where it is particularly common on disturbed sites, and it appears to be a rare example of back-migration from the younger Alejandro Selkirk Island (Valdebenito et al. 1992a; López-Sepúlveda et al. 2015a). Takayama et al. (2015a) suggested that the three named alpine species may consist of a complex of populations that are in the early stages of speciation. If this is the case, then there still may be some gene flow among them, and it is possible that despite the rarity of each of the species, diversity would be relatively high in populations of the complex because of gene exchange among them. That is, the diversity in the lineage has not been completely partitioned among the species. While results vary depending on the marker (AFLP or SSR) and the statistic, diversity in the three rare alpine species is comparable with that found in the other species of *Erigeron* (Fig. 14.2) (López-Sepúlveda et al. 2015a), which is concordant with, though not strong evidence in favor of, the hypothesis that the species are in early stages of divergence (Takayama et al. 2015b).

Diversity in the two closely related, morphologically similar species *E. rupicola* and the very rare *E. stuessyi* could be interpreted from two different scenarios. If, as with the three alpine species, there is gene flow among populations, then a comparable level of diversity could be expected in the two species. In contrast, if *E. stuessyi* is a peripheral isolate of *E. rupicola* and gene flow between them no longer occurs, then one might expect lower diversity in the rare species and perhaps some reduced diversity in *E. rupicola*, depending on how the total diversity from the original populations was partitioned into the two species. The results for AFLP and SSR markers give very different measures of diversity in the two species, with *E. stuessyi* having higher AFLP diversity by all measures, whereas SSR diversity is higher in *E. rupicola* (Fig. 14.2) (López-Sepúlveda et al. 2015a). If only AFLP markers were used, then the interpretation would likely be that the two species are similar to the three alpine species. That is, there is gene flow between the species, with the high diversity in the rare species a reflection of gene flow with the more common species. None of the populations of *E. rupicola* approaches the diversity in the one population of *E. stuessyi* examined for AFLP markers (López-Sepúlveda et al. 2015a). However, if SSR markers alone were employed, the lower total diversity in *E. stuessyi* would suggest that it is a recent derivative of the more common *E. rupicola*. However, there is a wide range of SSR diversity among the populations of *E. rupicola* (e.g., expected heterozygosity ranging from 0.10 to 0.34), which raises the possibility that *E. stuessyi* may have originated from a low-diversity population of *E. rupicola*. As with other cases cited earlier, the reason(s) for the differing pattern of AFLP and SSR diversity is (are) obscure.

The common and widespread *E. fernandezia*, which occurs on both islands, has higher total diversity than the other Juan Fernández *Erigeron* species in both AFLP and SSR, with diversity lower in *E. fernandezia* on Robinson Crusoe Island than on Alejandro Selkirk Island (Fig. 14.2). The higher total diversity in the species likely is a reflection of the larger and greater number of populations compared with the other endemic species of *Erigeron*. The lower diversity on the older Robinson Crusoe Island is concordant with the species having originated on Alejandro Selkirk Island with subsequent dispersal to Robinson Crusoe, with the lower diversity the result of colonization and establishment on Robinson Crusoe (Takayama et al. 2015b; López-Sepúlveda et al. 2015a).

Apportionment of Genetic Diversity and Divergence among Populations and Species in the Juan Fernández Archipelago

The relatively high proportion of total genetic diversity (mean G_{ST} of 0.338) among populations of species of Juan Fernández plants using allozyme markers suggests that the preservation of maximum genetic diversity within species requires the conservation of multiple populations. As a way of illustration, with a high G_{ST} value of 0.60, six populations would be needed to capture 95% of the diversity in the species; in contrast, with a G_{ST} of 0.20, two populations will contain 95% of the diversity (Hamrick et al. 1991). While the mean value provides a general overview of G_{ST} values for Juan Fernández species, it is important to keep in mind that values among species range from the lowest to highest possible, that is, from 0.00 to 1.00. Furthermore, it is not clear whether 20% of the 30 species examined with values above 0.41 reflect the general pattern of allozyme variation for the enzymes assayed because even one locus could have a large effect on the G_{ST} value. Additionly, high G_{ST} values could be a true reflection of the scattered, isolated nature of the populations or be an artifact of inadequate sampling of inaccessible populations. Francisco-Ortega et al. (2000) reported a mean and range of G_{ST} values for populations of plants in the Canary Islands similar to those calculated for Juan Fernández plants. In an important paper, Caujapé-Castells (2010) provided a caveat for interpreting G_{ST} values in island plants by showing that inadequate sampling within populations could inflate G_{ST} values and overestimate the roles of restricted gene flow among populations and the level of differentiation among populations. Caujapé-Castells (2010) cautioned that while direct comparison of the mean G_{ST} value of 0.280 reported by Francisco-Ortega et al. (2000) for Canary Island endemics with the mean of 0.179 estimated by Hamrick and Godt (1997) for endemic outcrossing species in general is correct in a numerical sense, it may not reflect the biological situation for the Canarian flora. The analyses of Caujapé-Castells (2010) and the reviews of Francisco-Ortega et al. (2000) and Crawford et al. (2001b) were all based on allozyme data.

No significant differences, again using the rather conservative sequential Bonferroni technique, were found between any of the Juan Fernández groups in the proportion of allozyme diversity residing among populations (G_{ST}) (Crawford et al. 2001b). Although no significant differences were detected, it should be noted that species occurring in small, scattered populations had high G_{ST} values. One of the consistent results from the literature has been the higher G_{ST} values obtained among populations of highly selfing species compared with species with mixed mating or outcrossing species (Hamrick and Godt 1989, 1997), but G_{ST} could not be calculated for five of the six selfing species because no diversity was detected in them (Crawford et al. 2001b). It is possible that if the five selfing species could have been included in the analyses, significantly higher G_{ST} values would have been obtained for the selfing category versus other breeding systems. In this regard, Pérez de Paz and Caujapé-Castells (2013) found that one of the variables that most influences high G_{ST} values in Canary Island plants is self-compatibility.

The later Juan Fernández studies using AFLP and SSR markers reported the percent of total genetic variance residing within and among populations of species as a measure of genetic differentiation among populations. For three abundant tree species, the proportion of total variance in the species occurring among populations is relatively low with both SSR and AFLP markers: *Myrceugenia fernandeziana* (4.5%, 10.4%), *M. schulzei* (8.5%, 16.4%), and *Drimys confertifolia* (5.09%, 5.79%) (López-Sepúlveda et al. 2013b, 2015b). These results indicate that little of the genetic variance within each of these species occurs among the populations. Crawford et al. (2001b) calculated a G_{ST} value of 0.232 from allozymes for 18 populations of *Myrceugenia fernandeziana*, which suggests a higher proportion of total genetic variance among populations compared with the other two markers.

There are three species of *Robinsonia* for which percent of total genetic variance among populations using SSR or AFLP markers and proportion total allozyme species diversity (G_{ST} value) may be compared (Crawford et al. 2001b; Takayama et al. 2015a). It is important to note that the number of populations sampled and the number of individuals sampled per population were comparable for the two studies. The values for SSR, AFLP, and allozyme markers, respectively, are given for each of the three species: *R. evenia* (10.9%, 2.3%, 0.319), *R. gayana* (5.2%, 5.8%, 0.395), and *R. gracilis* (12.9%, 4.0%, 0.207). As in species of *Myrceugenia* and *Drimys*, both SSR and AFLP markers show that genetic variance resides largely within populations of *Robinsonia* species, with little of the total variance among populations. Also, as in *Myrceugenia fernandeziana*, higher among-population diversity is seen with allozymes than with the other two markers. In fact, the differences between allozymes and the other two markers are more pronounced in *Robinsonia* than in *M. fernandeziana*. The discrepancy between markers is best exemplified in *R. gayana*, where less than 6% of the variance is among populations with SSRs and AFLPs, whereas allozymes estimate that nearly 40% of total species diversity resides among populations. Clearly, very different pictures of genetic variation in *R. gayana* emerge in the Crawford et al. (2001b) allozyme study and the report of Takayama et al. (2015a) using SSRs and AFLPs. The same inconsistencies emerge when comparing different species of *Robinsonia*, where *R. gracilis* would be seen as having a much higher level of among-population genetic variance than *R. gayana* if SSR markers were analyzed, whereas quite the opposite would be seen with allozymes where the G_{ST} value for *R. gracilis* is little more than half the value for *R. gayana*. The reasons for the higher among-population diversity seen with allozymes compared with AFLPs and SSRs are not readily apparent. One possibility is that the higher mutation rates for AFLPs and SSRs than for allozymes influence the values of differentiation seen for the same populations. While difficult to estimate, mutation rates for SSRs are more than three orders of magnitude (1,000 times) greater than for allozymes (Allendorf et al. 2013, pp. 233, 234). Wang (2015) points out that if mutation rates are greater than migration or, more precisely, mutation rates times population size are greater than migration, then there would be an underestimation of population differentiation.

In *Erigeron*, the percent of total genetic variance among populations within species is higher than in the preceding three genera. With AFLP markers, four of the five species have 17% to 26% of total variance among populations, with only one species lower at 6% (López-Sepúlveda et al. 2015a). The percent of genetic variance among populations is considerably higher with SSR than with AFLP markers, ranging from 25% to 61% for five species (López-Sepúlveda et al. 2015a). Using allozyme markers, a G_{ST} of nearly 0.300 was calculated for 15 populations of *E. fernandezia* from Robinson Crusoe Island (Crawford et al. 2001b), which is just below the mean for all Juan Fernández species examined.

The causes for the higher among-population variance in *Erigeron* than in the other genera using AFLP and SSR markers are not known, but one potential cause is the small population sizes in *Erigeron* compared with the other species. Both drift and biparental inbreeding in the small populations could result in both lower diversity in populations and increased differences in the presence and frequencies of alleles among populations. This seems a feasible explanation for *Myrceugenia* and *Drimys*, both of which have large populations compared with *Erigeron*. However, some populations of *Robinsonia* species are small and scattered as in *Erigeron*. Also, it was suggested earlier in this chapter that gene flow may still be occurring among populations of the same and even different species of *Erigeron*. If this were the case, then it might be expected that variance among populations would be low, and the percent of total variance within populations would be increased. It is also noteworthy that the estimates of genetic variance among populations are generally higher with SSR than AFLP markers for all groups. This may be due to the fact that 9 to 12 SSR loci were used whereas several hundred AFLP markers were employed, and the higher values for SSR could be the result of differences in allele frequencies or presence/absence of alleles at very few loci in different populations.

Divergence among species has been estimated using AFLP and SRR markers in the two Juan Fernández genera *Erigeron* and *Robinsonia* (López-Sepúlveda et al. 2015a; Takayama et al. 2015a). Because *Robinsonia* radiated on the older Robinson Crusoe Island and the species are morphologically distinct (Sanders et al. 1987), higher genetic divergence among its species (as measured by F_{ST}) would be expected than the divergence among the species of *Erigeron*, where divergence has and is occurring on the younger Alejandro Selkirk Island (Takayama et al. 2015a, 2015b; López-Sepúlveda et al. 2015a). With AFLP markers, the mean F_{ST} value for pairwise comparison of all species is 0.269 for *Erigeron* and 0.522 for *Robinsonia*, which is what would be expected for the latter genus from the older island. However, with SSR markers, the mean divergence for *Erigeron* species is slightly higher than it is for *Robinsonia* (0.428 versus 0.403). The reasons for the difference between the two markers is not known, but one possibility seems likely. Higher F_{ST} values can result from differences in allelic frequencies and presence/absence of alleles, both of which may be generated by drift in small populations. This could be the situation in *Erigeron* because of the few small populations that comprise certain species. In addition, the difficulty of sampling from natural populations because of accessibility could further exacerbate the issue; total sampling for four of the species included 25 or fewer individuals, and six of the ten populations included 19 or fewer plants (López-Sepúlveda et al. 2015a).

The levels of genetic divergence among species in the two genera may be viewed within the context of relationships within each of the genera inferred from other data. No explicit molecular phylogenetic study has been done on Juan Fernández *Erigeron*, but several distinct groupings have been suggested, primarily from morphological and ecological considerations (Valdebenito et al. 1992a; López-Sepúlveda et al. 2015a). As indicated earlier, the three groups include the three rare species *E. ingae*, *E. luteoviridis*, and *E. turricola*; *E. rupicola* and *E. stuessyi*; and *E. fernandezia*. Mean pairwise AFLP and SSR F_{ST} values are lower for the three species *E. ingae*, *E. luteoviridis*, and *E. turricola* than they are for comparisons between groups, which is what might be expected. However, *E. rupicola* and *E. stuessyi* have higher F_{ST} values with both markers than do most other comparisons between groups, which is surprising given that the two species are separated with some difficulty on the basis of morphology (Valdebenito et al. 1992a; López-Sepúlveda et al. 2015a). Genetic relationships based on AFLP genetic distances among individuals and on SSR distances among populations of *Erigeron* are shown in Fig. 14.3A, B. The three ostensibly closely related *E. ingae*, *E. luteoviridis*, and *E. turricola* do not group together with either marker (Fig. 14.3A, B). In addition, the morphologically similar *E. rupicola* and *E. stuessyi* do not group together (Fig. 14.3A, B). The morphologically distinct *E. fernandezia* is more or less divergent from the other two groups depending on which groups are compared and which markers are used (Fig. 14.3A, B). Thus there is no consistent concordance between marker divergence and accepted relationships in *Erigeron*.

Prior data from allozymes and ITS sequences for *Robinsonia* (Crawford et al. 1992b; Sang et al. 1995) offer a more rigorous systematic/phylogenetic framework than is available for *Erigeron* for interpreting AFLP and SSR divergence. However, only four species were included in the allozyme survey, and the ITS study did not include the more recently described *R. saxatilis*. The low SSR divergence between *R. gayana* and *R. thurifera* is also reflected in low allozyme and ITS sequence divergence, with the sequences identical (Crawford et al. 1992b; Sang et al. 1995). A similar pattern is seen with AFLP markers, where *R. gayana* and *R. thurifera* have the lowest divergence of any species pairs. Both SSR and AFLP markers indicate that the recently described *R. saxatilis* (Danton 2006) is closest to *R. gayana* and *R. thurifera* rather than to

R. evenia, as suggested by Danton (2006) (Fig. 14.3C, D). The mean divergence for species in sect. *Robinsonia* is low (0.217), whereas it is higher for species in sect. *Eleutherolepis* (0.427); the mean divergence between species in two different sections is 0.456, whereas the mean for all species in the genus is 0.522. Overall, species divergence is lowest in AFLP and SSR markers between species viewed as closely related from other data, and these relationships are seen in trees generated from the two markers (Fig. 14.3C, D).

Genetic Diversity in Juan Fernández Plants: Summary and Conservation Implications

The two pervasive patterns seen in molecular marker studies of Juan Fernández endemic plants are the low level of diversity in populations and species and the generally high proportion of species diversity residing among populations. Although there are conflicting studies and views about the association between neutral genetic diversity (e.g., molecular markers such as allozymes, AFLPs, and SSRs), fitness, and extinction (Grueber et al. 2008; Bouzat 2010), very low population diversity in molecular markers provides at least a crude guide to populations that are worthy of special conservation concern because they may be more vulnerable to habitat alterations caused by various factors, including environmental changes (Edwards 2015; Harter et al. 2015). Pérez de Paz and Caujapé-Castells (2013) caution that while they found correlations between population size and allozyme diversity in Canary Island plants, there are also many studies of plants from several archipelagos where there are no such correlations. These authors present a thoughtful, wide-ranging discussion that emphasizes the necessity of using genetic markers as part of a multifaceted study of any given species; especially important are components of the reproductive biology of the species (see Chapters 12 and 15). The important point of their discussion is that despite correlations between marker diversity and abiotic and biotic factors that may be detected for island floras, there are many exceptions for individual species.

The high apportionment of genetic diversity among populations indicates that the preservation of maximum genetic diversity within species requires the conservation of multiple populations. As indicated earlier, the higher the G_{ST} value, the greater are the number of populations needed to capture 95% of diversity in a species. The G_{ST} values obtained for the same species and similar population sampling using different markers may vary, the reasons for which are usually not apparent, but they may be related to mutation rates of the markers relative to migration rates (Wang 2015). Despite the uncertainty about the causes of different values, from a conservation perspective, it is imperative to gather biological data (e.g., seed set, age structure, pollen fertility within populations and morphological and habitat differences among populations) for any species for which high G_{ST} values are obtained with any given marker. High among-population genetic diversity could be a reflection of low level or absence of historical gene flow, meaning that the populations are isolated genetically and subject to such processes as inbreeding depression (reduced fitness of progeny due to self-fertilization or mating among genetically similar individuals) and loss of genetic diversity by chance due to genetic drift (Bouzart 2010 and references therein). Populations with unique alleles or multilocus genotypes should be of special interest for the study of possible morphological, geographical, or habitat differences. This is so because even if the different markers themselves are not of adaptive value, they may be linked to adaptive features, and the markers may reflect the reproductive isolation of the populations.

If molecular marker diversity is very low in populations and other data suggest problems such as inbreeding depression, then the sometimes contentious issue arises of whether to replenish the genetic diversity of a population ("genetic rescue"). This can be done by introducing plants from other natural populations of the same species, from living plants in botanical gardens, or from seed banks (Edmands 2007; Weeks et al. 2011; Frankham 2015). A basic issue in decision-making is the balance between the harmful effects of inbreeding depression within a population and outbreeding depression resulting from the introduction of plants that are so distantly related that the resulting hybrids have low fitness (Edmands 2007; Weeks et al. 2011; Frankham 2015). Unfortunately, inbreeding and outbreeding depression are not known prior to making a decision on genetic rescue. Molecular markers such as those discussed in this chapter cannot by themselves provide strong guides as to whether or what plants should be introduced into natural populations. However, Caujapé-Castells et al. (2008a, 2008b) provide an elegant example of how markers, when combined with other observations of natural populations such as low seedling survival and declining population sizes, can inform decisions about introduction of plants from different populations for the preservation of genetic diversity in rare species in the Canary Islands.

Some of the correlations found between life-history attributes and patterns of allozyme diversity detected in reviews of plants (Hamrick 1989; Hamrick and Godt 1989, 1996, 1997; Hamrick et al. 1979) were not found in Juan Fernández plants. A possible reason for the lack of some correlations is the very low allozyme variation detected in species and/or inadequate intra- or interpopulational sampling in some species.

Results from AFLP and SSR markers show general support for the predictions of genetic diversity under the models of anagenetic and cladogenetic speciation (Stuessy 2007; Stuessy et al. 2006, 2012). Interpretations of the results are especially informative when combined with morphological observations and molecular phylogenetic studies. However, results for some Juan Fernández plants indicate that depending on the molecular marker employed, different inferences would be made about patterns of genetic diversity and divergence. As discussed in this chapter, many biological, abiotic, and historical factors may have an impact on genetic diversity in plant species, and this is particularly true in a small, isolated system such as the Juan Fernández and other oceanic archipelagos. This means that understanding the major impact(s) on the genetic diversity and structure seen in insular species is a challenging task; Caujapé-Castells (2011) and Pérez de Paz and Caujapé-Castells (2013) provide comprehensive analyses and discussions of the complexity of inferring genetic diversity and differentiation of genetic diversity in the Canary Islands. Many of the points they make apply to the Juan Fernández Archipelago and to oceanic archipelagos in general.

The studies of genetic diversity in Juan Fernández plants over several decades have provided important insights into patterns of genetic variation within and among populations of species and among congeneric species. In addition to the inherent scientific value of these studies, the results also provide some general guidelines for the conservation of some of the most interesting and imperiled plants on Earth.

Future Studies of Juan Fernández Plants

The kinds of studies of genetic diversity in Juan Fernández plants using molecular markers that could be undertaken in the future are nearly unlimited. Indeed, future investigations will be limited only by the logistics of working on inaccessible populations on the remote islands, especially, Alejandro Selkirk Island. At the beginning of this chapter, some general comments were presented on the potential for determining the genetic basis of ecologically important traits in island lineages by using a combination of genomic methods and ecological studies (Kover and Mott, 2012). While such studies are highly desirable because they would provide insights into genetic diversity in island plants relevant to evolution, diversification, and conservation of insular endemics, they would involve large multidisciplinary endeavors. As emphasized by Frankham (2010), Pérez de Paz and Caujapé-Castells (2013), Edwards (2015) and many others, identifying the ecologically important traits, including the genetic basis of the traits, is critical to implementing effective conservation planning.

In reality, the most feasible studies over the next several years would seem to be to expand the use of hypervariable markers such as SSRs and AFLPs as well as the many markers that may be obtained from currently available genomic techniques to other species with minimal to nonexistent allozyme variation. For example, Crawford et al. (2001a) found no allozyme variation in eight of the 30 species (26.6%) examined. Of special interest in this regard are highly selfing species such as *Chenopodium sanctae-clarae* and *C. crusoeanum* where lack of variation precluded calculating G_{ST} values. Given the results for self-pollinating flowering plants in general (Hamrick 1989; Hamrick and Godt 1989, 1996, 1997; Hamrick et al. 1979) and for the Canary Islands (Pérez de Paz and Caujapé-Castells 2013), one would expect high differentiation among populations. Markers such as SSRs and AFLPs, unlike allozymes, do not require fresh material or viable seeds from which progeny can be grown. Rather, material can be processed easily in the field and DNA extracted later in the laboratory. In some of the rarest species consisting of fewer than 25 total individuals known from very few populations (Stuessy et al. 1998b), a concerted effort should be made to sample every individual. This will assure that allelic richness (number of alleles at each locus), genotypes or genets (different multilocus genotypes), and heterozygosity (genetic diversity) will be detected.

Simple sequence repeats, being highly variable and inherited as codominants, are ideal for studies of mating system. Surveys of the reproductive biology of Juan Fernández plants (Anderson et al. 2000, 2001; Bernardello et al. 1999, 2001; see also Chapter 12) have provided insights into breeding systems (self-compatible or self-incompatible), but there are no data on the levels of outcrossing in natural populations, that is, on mating systems. Information on mating systems would be valuable for understanding the population genetics structure of species and for inferring levels of gene flow between populations. One potential limitation on such studies is obtaining progeny from maternal plants, but with variable markers and newer programs for analysis (Koelling et al. 2012), lower sample sizes are required. The program of Koelling et al. (2012) also allows for an estimation of inbreeding depression, information that could be valuable in making conservation decisions. If there is an indication of high inbreeding depression in a given population, consideration should be given to introducing individuals from other populations into that population (Edmands 2007; Weeks et al. 2011). The inbreeding coefficient for populations of the genera examined with SSR markers shows a general and often significant deficiency of heterozygotes, which could arise from various levels of self-pollination or biparental inbreeding, that is, mating between more closely related individuals rather than random mating. Studies combining molecular markers and field work provide the potential for understanding the reproductive biology of Juan Fernández plants, and this knowledge, in turn, would be critical in formulating strategies for the conservation of the marvelous endemic plants of the archipelago.

15 Speciation

Daniel J. Crawford, Tod F. Stuessy, Koji Takayama, Patricio López-Sepúlveda, Gregory J. Anderson, and Gabriel Bernardello

Speciation is arguably the most fundamental process in plant evolution because it leads to evolutionary lines that represent the basic units of plant classification, the species, and is the means of diversification within lineages (Rieseberg and Brouillet 1994; Coyne and Orr 2004). In addition, the plant diversity for a given area is usually presented in terms of species richness, and thus species enter prominently into discussions of conservation (Steele and Pires 2011). Despite the central role of the origin of species in plant evolutionary biology, there is no one universally accepted species concept for plants (The Marie Curie SPECIATION Network 2012) or a comprehensive, refined understanding of the process(es) of speciation. Indeed, the lack of a single species concept in plants and the extensive biological/philosophical debates and discussions of the species concepts have historically retarded the study of speciation (Rieseberg and Willis 2007). Species concepts have been reviewed extensively elsewhere and thus will not be considered in detail here (de Queiroz 1998, 2007; Coyne and Orr 2004, pp. 447–72 and therein; Hausdorf 2011; Shapiro et al. 2016; see also Chapter 5). Most important, as argued by de Queiroz (1998, 2007), despite the different concepts of species, there is general agreement that species are independently evolving metapopulation lineages. Thus speciation can be discussed from the perspective of the factors promoting divergence into independent lineages. Even this perspective is complex given that the elucidation of lineages may be difficult, especially when different criteria suggest incongruent species boundaries or when certain criteria indicate divergence between populations but others do not. Along the same line of thought, Lowry and Gould (2016) discuss speciation as process, and the species concept adopted in any particular case will depend on where along the continuum the process has progressed. As indicated by Bacon et al. (2012) and others, these issues may be especially problematical in island plants, where divergence and speciation are often recent. Recent divergence may not have provided sufficient time for completion of the speciation process in the sense that there has not been sorting out of the attributes (e.g., morphological, physiological, molecular) commonly associated with "older" continental species. Perhaps most important, because the final stages of the "completion" of speciation often involve the evolution of strong postzygotic isolating factors, the biological species concept will not apply to island plants in many cases because fertile interspecific hybrids can often be synthesized (see below). However, lack of completion of the process in island plants also offers significant advantages for studying speciation. The geologically young and recently colonized habitats presented by islands present the unusual, if not unique, opportunity to study populations/species at different stages along the speciation continuum. In addition to speciation via divergence, so-called primary speciation, species may originate by hybridization between two differentiated populations (usually recognized as species) at the same ploidy level to give stabilized, novel phenotypes that are isolated from each of the parental species. This is most commonly called "homoploid hybrid speciation" (Rieseberg 1997; Mallet 2007; Abbott et al. 2010, 2013; Yakimowski and Rieseberg 2014). An increase in ploidy level is also a mechanism of speciation and may involve chromosome doubling within species (autopolyploidy) or doubling following interspecific hybridization (allopolyploidy) (see Soltis et al. 2014). In this chapter, the emphasis is on primary speciation because it is the most frequent, if not exclusive, mode of speciation in Juan Fernández plants.

Primary Speciation

Regardless of the species concept, there can be little doubt that the essence of speciation is the reduction in gene flow between populations or subpopulations such that divergence can occur by selection and/or drift (The Marie Curie SPECIATION Network 2012). Thus the study of speciation is essentially elucidating or inferring the barriers to gene exchange. At the most rigorous level, this is not an easy task for several reasons. First, identification of all the barriers and their relative strengths in promoting reproductive isolation is not a trivial undertaking (Ramsey et al. 2003; Kay 2006; Martin and Willis 2007; Gavrilets 2014; Sobel and Chen 2014). Even if all barriers were identified, it is challenging, if not impossible, to determine the order in which they appeared over time (evolved) and thus which barriers were key to speciation and which evolved after gene flow had already been reduced and isolation was already effectively complete (Templeton 1982; Coyne and Orr 2004, p. 57).

Barriers to Gene Flow

Barriers to gene flow have traditionally been classified by the point at which they act during the life history of the plant (as in Stebbins 1950, chap. 6; Levin, 1978, 2000, p. 11; and Grant 1981, chap. 9). The more recent classification of Levin (2000) is, with slight modification, shown in Table 15.1. Isolating factors are often broken into two broad categories depending on whether they act prior to or after pollination (pre- versus post-pollination) or either before or after zygote formation (pre- versus postzygotic). Historically, postzygotic barriers to gene flow have been more commonly studied in plants because less effort is required than would be needed to elucidate the factors accounting for the ecological/spatial separation of species in nature. Studies of postzygotic barriers have typically involved determining the vigor and fertility (often pollen fertility) of synthetic interspecific hybrids (Stebbins 1958; Grant 1981, chap. 11; Levin 2000, pp. 44–52). More recently, there has been increased interest in isolating factors that act before pollination; especially prominent is the concept of ecological

Table 15.1 Factors Reducing Gene Flow

Ecological
A. Habitat divergence
B. Temporal divergence: flowering at different times
C. Floral divergence: floral attributes attract different pollinators
D. Reproductive mode: self-fertilization and asexual reproduction
Genomic: Pollen is transferred, but gene exchange is limited.
A. Cross-incompatibility: pollen not able to send tubes to ovules of another plant
B. Reduced fitness of first-generation hybrids:
1. Hybrid nonviability and weakness
2. Hybrid floral isolation: pollen not efficiently transferred by parental pollinators
3. Hybrid sterility
C. Hybrid breakdown (reduced fitness) in advanced or backcross generations.

Source: After Levin (2000).

speciation (Rundle and Nosil 2005; Hendry et al. 2007; Lowry et al. 2008; Nosil et al. 2009; Givnish 2010; Sobel et al. 2010). These factors include ecogeographical isolation, differences in flowering time (phenology), and divergence in floral traits that serve to reduce movement between two species. Suffice it to say that factors acting both before and after pollination are the topics of current research in plant speciation (e.g., Fishman et al. 2013; Sobel 2014; Stathos and Fishman 2014; Sobel and Streisfeld 2015). There are very few comprehensive studies of the strengths of the different barriers between two sister species. The study by Ramsey et al. (2003) is a notable exception: they reported multiple barriers to gene flow between sister species of *Mimulus*, with prezygotic, and especially habitat divergence, being particularly strong. Also, because these prezygotic factors act early in the life history of the plants, they are particularly effective and reduce the relative contributions of later-acting factors to the total reduction in gene flow.

Geographical Modes of Speciation

Historically, modes of primary speciation have been classified by biogeographical criteria, although not without considerable discussion and debate. For example, Coyne and Orr (2004, p. 3) recognized that "Among all the scientifically tractable questions about speciation, the most hotly contested concerns its biogeography." Allopatric speciation is the most widely accepted geographical mode of speciation in large part because it is the most intuitive. Once two populations are separated by a distance that exceeds the dispersal capabilities of the plants, the populations can accrue differences through time without the homogenizing effects of gene flow (Fitzpatrick et al. 2009). Different types of allopatry have been recognized, with the "classical" or vicariant mode occurring when populations are separated by some type of barrier (e.g., rise of mountains, continental drift, extinction of intermediate populations) (Fig. 15.1A). Allopatry can also be achieved by habitat divergence without geographical separation, in which selection drives spatial separation.

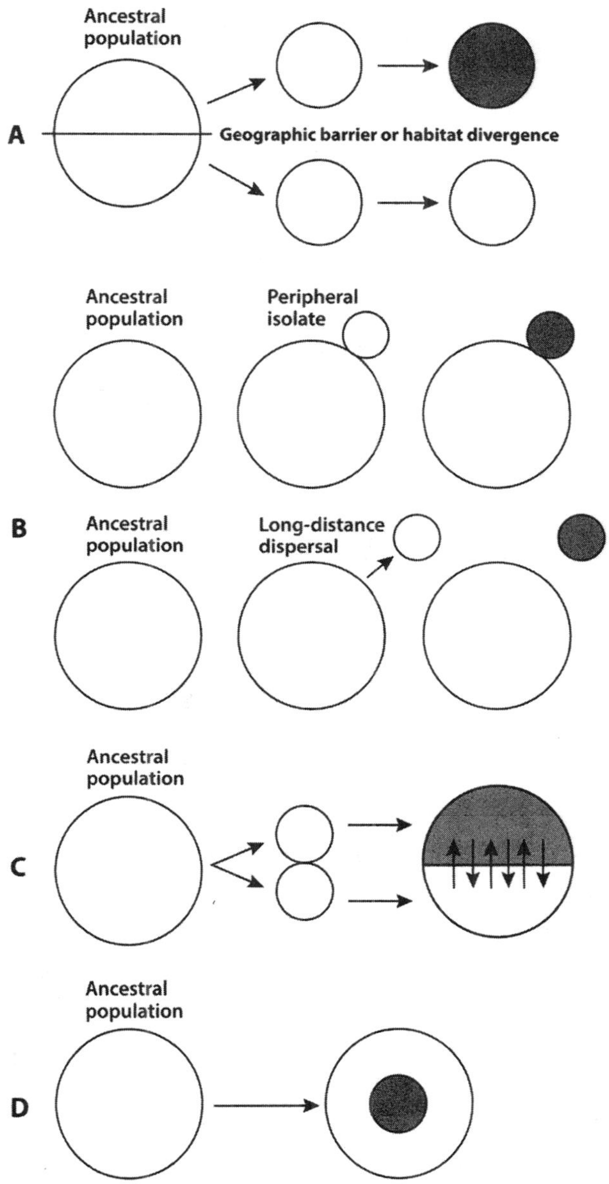

Figure 15.1 Geographical modes of speciation. (A) Classical allopatric mode. (B) Peripatric mode by either "budding off" to produce peripheral isolates (top) or with long-distance dispersal (bottom), such as to oceanic islands. (C) Parapatric mode in which two populations diverge in the face of gene flow. (D) Sympatric speciation in which populations diverge despite overlapping geographical ranges.

Peripatric speciation is a particular variant of the classical allopatric mode (i.e., the vicariant mode described earlier) wherein the diverging population is much smaller than the larger parental population. This may occur by one population budding off the periphery of the other in some continental setting or via long-distance dispersal of colonists beyond the range of the parental source population (Fig. 15.1B). The latter is intrinsic to island organisms, has been described in many instances (e.g., Carlquist 1974, chaps. 1 and 2; Bernardello et al. 2006), and is the focus of this chapter and this book.

Parapatric speciation is another mode of speciation in which reproductive isolation evolves between populations despite some gene flow among them (Coyne and Orr 2004, pp. 111–123) (Fig. 15.1C). "Clinal" and "stepping-stone" are the two most cited examples of parapatric speciation. In the former, populations occur more or less continuously over different environments, with populations variously adapted to local habitats despite some gene flow among them. In the latter model, there are more or less distinct populations, but there is some level of gene flow among them.

Sympatric speciation is by far the most controversial and contentious geographical mode of speciation (Butlin et al. 2008; Fitzpatrick et al. 2008; Gavrilets 2014). In fact, one of the problems is the lack of a widely accepted definition (Templeton 1981; Bolnick and Fitzpatrick 2007; Fitzpatrick et al. 2008). One concept emphasizes geographical separation, and the other views sympatric speciation within a population genetic context (Fig. 15.1D). The primary distinction between the two concepts is that geographic/spatial distribution of populations is the basis of the geographical concept of sympatry, whereas the foundation of the population genetics concept is gene flow. By way of explanation, plants in two subpopulations diverging at both the same place and time are sympatric under both concepts (Fitzpatrick et al. 2008). In contrast, populations could diverge and occupy spatially distinct niches occurring in a mosaic over the same local area. In this situation, the populations are sympatric biogeographically, but if plants in the different niches are not mating randomly, they are not sympatric from a population genetics perspective. Another scenario posits that two populations are geographically disjunct, but individuals from the two populations are just as likely to mate with each other as are plants within each area. Geographically, these populations are not sympatric, but they are genetically sympatric because they meet the random-mating criterion. In a recent review of speciation models, Gavrilets (2014, p. 747), citing Fitzpatrick et al. (2008, 2009), makes the point that "Some authors now argue that sympatric speciation is virtually impossible to demonstrate in an uncontroversial way and that, in any case, testing whether a particular case fits a particular definition of sympatric speciation is less informative than evaluating the biological processes affecting divergence . . .."

Studying Speciation

At the broadest level, the study of speciation may be divided into experimental and comparative studies of closely related (ideally, sister or progenitor-derivative) species. The experimental approach includes pollination studies, essential for understanding gene flow, and furthermore, it can be important to elucidate prezygotic barriers by studying pollen germination and pollen tube growth following cross-pollination (Anderson 1979; Bernardello et al. 1999) and postzygotic isolating factors by making crosses between species and examining their viability and fertility (examples from island studies include Gillett and Lim 1970; Carr 1985; Carr and Kyhos 1986; Lowrey 1986; Crawford et al. 2009). One approach to assaying pre-pollination barriers to gene flow that is less common for insular floras is transplantation. Transplanting one species to the habitat of another species and measuring fitness components in the foreign habitat obviously provide insights into the potential role of habitat divergence in reducing gene flow between species (e.g., Givnish and Montgomery 2014). While experimental studies of speciation are to be desired for island plants, they can be challenging and time-consuming, especially ecological investigations of barriers to gene flow (see below).

Comparative studies of speciation involve determining differences among species and then inferring contrasting features that may be associated with reduction in gene flow and thus speciation (Crawford and Stuessy 1997). This approach, while perhaps less rigorous, is more widely employed than experimental approaches simply because it is much more feasible. Certain caveats must be kept in mind when employing the comparative method. It is critical that the correct species be compared, that is, progenitor-derivative species pairs or sister species that evolved from a common ancestor (Crawford and Stuessy 1997; Crawford 2010). The former situation, a progenitor-derivative pair, is presumably preferable to a sister relationship for comparison because the extant progenitor shares many features with the derivative, whereas sister species presumably diverged from a common hypothetical ancestor. Another important concern, alluded to earlier, is whether character differences between species are associated with speciation or evolved after speciation was complete (Templeton 1982; Coyne and Orr 2004, p. 57). This is difficult to determine, but if species have recently diverged or are in the process of diverging, then there should be few observed differences, and those that do occur are more likely to have been associated with divergence. This we consider to be one of the most attractive elements of island plants for speciation studies.

Island Plants as Model Systems for Studying Speciation

Island plants have many advantages for the study of plant speciation, whether using experimental or comparative methods (Crawford et al. 1987; Crawford and Stuessy 1997; Warren et al. 2014; Fernández-Palacios et al. 2015). As mentioned earlier, island endemics are relatively "young," as judged by the fact that they occupy, and presumably originated on, islands of defined ages, usually fewer than 5 million years (Stuessy 2007). Another line of evidence supporting recent speciation in islands is the low level of molecular divergence often detected among congeners (e.g., Helenurm and Ganders 1985; Francisco-Ortega et al. 1996; Mort et al. 2010). Recent divergence is an advantage for both experimental and comparative studies of speciation. Species within an endemic lineage often differ by conspicuous morphological and ecological (habitat) features. However, because they are of recent origin and have not had sufficient time to accumulate incompatibilities, they are often able to hybridize and produce viable, fertile progeny in the F_1 and later generations (Gillett and Lim 1970; Lowrey 1986; Crawford et al. 2009). Divergence in morphological/pollination/ecological characters combined with interfertility may allow for inferences regarding the genetic architecture of character differences between species. In addition, these hybridization studies provide insights into the potential contributions of postpollination and postzygotic factors as barriers to gene flow. From a comparative perspective, if speciation is recent, then the aforementioned problem of distinguishing features associated with speciation from those that have accumulated subsequent to divergence becomes more tractable than with older lineages. Many oceanic islands do not occur alone but, like the Juan Fernández Archipelago, are part of an archipelago consisting of two or more islands. The distribution of species on the same and different islands within an archipelago provides insights into possible processes and patterns of speciation, especially whether divergence has occurred within an island (*intraisland radiation*) or subsequent to dispersal of colonizers from one island to another (*interisland dispersal*). This issue is particularly relevant in discussions of speciation in the Juan Fernández Archipelago because it consists of two major islands.

Potential Limitations of Island Species for Speciation Studies

Despite the many advantages of island endemics for studies of speciation, they sometimes pose challenges as model systems. One limitation using molecular tools is the difficulty in some lineages of obtaining a completely resolved, highly supported phylogeny with DNA sequence data. Lack of resolution is caused by low variation in the regions frequently employed for resolving relationships within genera (Kim et al. 1996a; Ganders et al. 2000; Archibald et al. 2006; Mort et al. 2010). The low variation could preclude the identification of sister species for both experimental and comparative studies. In addition, a completely resolved phylogeny is fundamental to inference of biogeographical relationships, including the roles of interisland dispersal and intraisland radiations in speciation (Kim et al. 1996a; Francisco-Ortega et al. 1997; Baldwin 2003). However, the use of next-generation sequencing methods is providing the variation necessary to provide high resolution of phylogenetic relationships in groups that had been poorly resolved from Sanger sequencing of several regions from the nucleus and plastids (Wessinger et al. 2016), including the genus *Tolpis* (Asteraceae) in the Macaronesian Archipelago (Mort et al. 2015). The many markers provided by next-generation approaches likely will result in much better resolution of island lineages, thus making them even more attractive plant systems for studying speciation. While island endemics are in theory ideal for experimental studies of speciation, the high frequency of perennials with long generation times (versus annuals or short-lived perennials) and the associated longer time to sexual maturity are significant limitations. Two notable examples from the Juan Fernández Islands are the genera *Dendroseris* and *Robinsonia* (Sanders et al. 1987) and several species of *Solanum* from other archipelagos (Anderson et al. 2014, 2015). And, as noted earlier, transplant studies are difficult because many of the species are quite rare, and permission would not be routinely granted to manipulate such rare species. However, granting permission to researchers of recognized institutions with sound projects could be important because the data produced could eventually be of immense value in formulating strategies for protecting the species.

Speciation in Juan Fernández Plants

As we described earlier, there are two basic ways to study speciation, the experimental and comparative. The second method of comparing attributes of the species for inferring the process of speciation will be employed; essentially no experimental data exist for Juan Fernández plants. The many expeditions to the islands by T. F. Stuessy and collaborators over the past three decades (see Chapter 2) provide the data from which inferences of speciation may be made. Stuessy et al. (1990) analyzed the patterns of phylogeny for Juan Fernández endemics, and those results, along with information on phylogeny in Chapter 13, will be extensively incorporated into the discussion. Information from the review by Stuessy et al. (1998e) of the isolating mechanisms in Juan Fernández endemics will likewise be incorporated into the discussion. Lastly, the extensive studies of G. J. Anderson and G. Bernardello on the reproductive biology of Juan Fernández plants provide important insights into speciation (Anderson et al. 2000a, 2000b, 2001; Bernardello et al. 1999, 2001, 2004; see also Chapter 12).

Single-Island Endemics: Geographical Speciation in the Juan Fernández Archipelago

When considered within the broadest historical and biogeographical contexts, speciation in islands is typically initiated by dispersal of one or a few colonists and thus is peripatric. If only one species has evolved in the island setting, then speciation has been strictly peripatric. However, if there is divergence and speciation within an archipelago, other geographical modes of speciation are relevant. The most common distribution pattern for flowering plants in the Juan Fernández Archipelago is single-island endemics. That is, following colonization, there has been divergence between the island populations and their continental progenitors (Stuessy et al. 1990). Stuessy et al. (1990) calculated that 31% of the genera (lineages) have one species on the older Robinson Crusoe Island, which is nearer to the primary source area of continental South America than is the more distant Alejandro Selkirk Island. As will be discussed below, present knowledge indicates that this figure is higher (40% for angiosperms; see Table 13.3) than previously reported. A lower percentage (<16%) of the flora is endemic to the younger, more distant Alejandro Selkirk Island.

The frequent occurrence of only one endemic species in a genus in the archipelago means that peripatric speciation is a common geographical mode in the Juan Fernández Archipelago. Even though peripatric speciation is often viewed as primarily a budding off from the periphery of the ancestral population, it has also been defined to include the long-distance dispersal of one or a few propagules from the parental population (Coyne and Orr 2004, chap. 3). Coyne and Orr (2004, chap. 3) make the distinction between more classical allopatric (or vicariant) speciation on the basis of the relative sizes of the two populations. Mayr (1954) apparently was the first to formulate the concept of peripatric speciation, and it has been widely accepted as a mode of geographical speciation in plants. For example, Grant (1963, pp. 456–459; 1981, pp. 155–160) and Lewis (1966) discussed the concept under the name *quantum speciation*. Genetic drift could become a factor in divergence and speciation in the small peripheral isolates, especially in the founding of new populations following long-distance dispersal such as colonization of the Juan Fernández Archipelago or other isolated oceanic archipelagos (Crawford et al. 1987; Crawford and Stuessy 1997; Whittaker and Fernández-Palacios 2007, pp. 167–172).

With the peripatric mode of speciation for Juan Fernández plants, reproductive isolation from the continental source populations for insular species is presumably complete. Given the distance of the islands from the source areas and the small size and number of islands, it seems unlikely that there has been dispersal from the continental source area after the initial colonizers of a lineage first arrived. In addition, the lack of back-dispersal seems a reasonable assumption, and the fact that there is no evidence for back-dispersal in the long-studied flora of Chile provides further support for the lack of dispersal from the Juan Fernández Archipelago back to the continent. It is not known whether there are any other barriers to gene flow that would isolate the island endemics from their continental progenitors should they be in sympatry. In some instances, such as *Lactoris fernandeziana*, the ancestor is no longer extant, and the question of interbreeding is moot. However, for other genera, such as *Rhaphithamnus*, with an extant continental relative, the question is more relevant. The morphological and molecular divergence between the insular endemic and the continental congener in *Rhaphithamnus* is minimal (Crawford et al. 1993b; Sun et al. 1996), and it is possible that the two species could hybridize if brought together. Some would argue, for example, Coyne and Orr (2004, pp. 93–94), that speciation is not complete if the species could not pass the so-called test of sympatry. This, however, seems too restrictive because of the unlikelihood that they will ever become sympatric in nature. De facto, the Coyne and Orr definition of speciation implies acquisition of internal barriers to crossing. However, there is no question that the Juan Fernandez species and insular species from around the world function as independent evolutionary lineages separate from their proposed continental (or other island) relatives, with virtually no possibility of intercrossing.

The present distribution of endemic species in the Juan Fernández Archipelago demonstrates convincingly that peripatric speciation has been a common geographical mode of speciation in the archipelago, whether from continental source areas to the islands or interisland dispersal within the archipelago. In addition, there are several examples of peripatric speciation in the Juan Fernández Archipelago via interisland dispersal where there apparently has been diversification into two or more species on a single island with subsequent dispersal to and speciation on the other island. The most unambiguous examples come from two of the larger genera in the Juan Fernández Archipelago, both members of Asteraceae, as discussed in Chapter 13. In *Dendroseris*, there have been three independent dispersals from different lineages on Robinson Crusoe Island to Alejandro Selkirk Island with divergence to a distinct species in each case (Sanders et al. 1987; Sang et al. 1994). In *Robinsonia*, there was dispersal and divergence to produce one endemic species (Sanders et al. 1987; Sang et al. 1995).

In other instances it may not be a trivial matter to determine which of the biogeographical scenarios is correct: given sometimes short spans of geological time, it may not be easy to distinguish whether there were independent introductions from continental source areas to each of the islands or dispersal to an island was followed by interisland dispersal. As discussed below, phylogenetic relationships among island and continental taxa provide the best evidence for choosing between the two biogeographical scenarios. Stuessy et al. (1990) listed five flowering plant genera in which there is one species endemic to each of the islands, these include *Berberis*, *Haloragis*, *Myrceugenia*, *Sophora*, and *Zanthoxylum*. It is now known that the two species of *Myrceugenia* are not closely related, and the species on Robinson Crusoe Island is recognized as the distinct genus *Nothomyrcia* (Murillo-Aldana and Ruiz 2011; Murillo-A. et al. 2012; see also Chapter 13). Thus these two species provide an example of peripatric speciation involving independent dispersals from continental source areas to both Robinson Crusoe and Alejandro Selkirk. While evidence is not as strong as for the *Myrceugenia/Nothomyrcia* example, a similar pattern of independent dispersal is likely for the two species of *Zanthoxylum*, which were formerly treated as members of *Fagara* (see Chapter 13). There is less compelling evidence that the two species of *Sophora* are the result of separate introductions, but the bulk of evidence suggests that this is the case (see Chapter 13). The situation for *Berberis* is also not clear; Landrum (1999) indicates that comparative morphological observations do not provide convincing data for judging whether or not the two endemics originated from a common or independent dispersal event. The situation is also unclear for *Haloragis*. Danton (2014) recently published two new varieties of *Haloragis masatierrana*, one from Robinson Crusoe Island and the other from the tiny island of Santa Clara, which is very near to Robinson Crusoe Island. A judgment as to whether these varieties represent entities in the process of diverging from their parental populations must await the results of further studies. *Chenopodium* is a particularly interesting genus, with one species each on Robinson Crusoe, Alejandro Selkirk, and Santa Clara. The relationships of these species have not been rigorously tested, but regardless, these are all examples of peripatric speciation, whether involving continental-island or interisland dispersal.

Two or More Endemics on Single Islands: Larger Lineages and Selected Examples from Smaller Lineages

As indicated earlier, peripatric speciation has been an important factor in speciation in the Juan Fernández Islands, and whether there are other barriers to gene flow is presumably of little consequence because the geographical distance between species prevents or greatly reduces gene exchange. Other isolating factors could have originated subsequent to geographical divergence. The footprint of the geographical scale of divergence and speciation on islands may be obscured by several processes, and the Juan Fernández Archipelago is no exception. One factor could be extinction of species within lineages, leaving a spatial gap, which could result in wider geographical/spatial separation of extant species than was originally present during initial adaptive radiation and divergence. This would suggest that speciation had a stronger spatial/geographical component than was in fact the case. In contrast, as the areas of islands decrease during maturation by processes such as erosion and subsidence, the spatial separation among species within a lineage could diminish so that present distributions would suggest that speciation occurred at a finer spatial scale than was actually the case. The areas of Robinson Crusoe and Alejandro Selkirk have decreased since they initially emerged above the surface of the Pacific Ocean. This is especially pronounced for Robinson Crusoe Island, which has lost perhaps 95% of its area since formation (Stuessy et al. 1984; Sanders et al. 1987). The loss of area is estimated to be 28% for the younger Alejandro Selkirk Island. A further ramification of loss of surface area is the likely increased potential for species extinction. Extinction has certainly been a part of the evolutionary history of the Juan Fernández flora, with evidence from historical times (see Chapter 9). Another set of factors about which we have no information but which could have played important roles in the speciation process consists of those associated with the amount of island land mass exposed or not during the vagaries of the Late Quaternary sea-level changes (e.g., Weigelt et al. 2016).

The following discussion of multiple speciation events within lineages on single islands will consider possible isolating factors (Table 15.1) in addition to commenting on the possible role of geography. First consideration will be given to the three largest genera in the Juan Fernández Archipelago, all members of Asteraceae, though not closely related to each other: *Dendroseris* (Cichorieae: Sonchinae), *Robinsonia* (Senecioneae: Senecioninae), and *Erigeron* (Astereae: Conyzinae).

Dendroseris is the largest genus in the archipelago with 11 species; eight species from the three subgenera (two or three species per subgenus) occur on Robinson Crusoe Island, and one species from each subgenus is endemic to Alejandro Selkirk Island (Sanders et al. 1987). In addition to documenting the monophyly of the genus, Sang et al. (1994) produced sufficient resolution of phylogenetic relationships (with ITS sequence data) (Fig. 15.2) to allow an assessment of factors that could have promoted divergence and speciation of sister species. For instance, by all measures, *Dendroseris litoralis* and *D. marginata* are very closely related species in subg. *Dendroseris* (based on morphology and DNA sequences). They are ecologically and altitudinally differentiated: *D. litoralis* occurs at low elevations along the coast, and *D. marginata* is restricted to exposed cliffs at higher elevations (Stuessy et al. 1998e). In subg. *Phoenicoseris, D. berteroana* and *D. pinnata* are supported as sister taxa (Sang et al. 1994) (Fig. 15.2) and are the only two members of the subgenus on Robinson Crusoe Island. The two species are easily distinguishable morphologically and ecologically, with *D. berteroana* restricted to tree/fern forests at higher altitudes and *D. pinnata* occurring only on open wind-swept ridges at higher elevations; the two species are clearly separated spatially by the different habitat preferences (Sanders et al. 1987; Stuessy et al. 1998e). In the third subgenus *Rea*, *Dendroseris micrantha* and *D. pruinata* are strongly resolved as sister taxa (Sang et al. 1994) (Fig. 15.2), and they are morphologically distinct. Like the two previous species pairs, they are also distinct ecologically and altitudinally: the former species occurs widely in forests at middle elevations, whereas the latter species occurs in open areas, either at low elevations near the coast or on open,

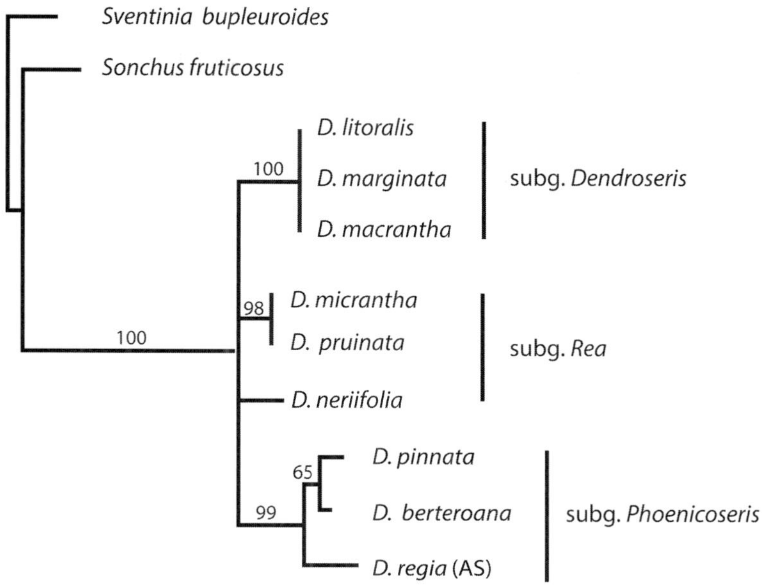

Figure 15.2 Phylogeny of *Dendroseris*. Strict consensus tree of the four most parsimonious trees revolved by ITS sequences. *D. regia* is endemic to Alejandro Selkirk Island (AS); other species occur only on Robinson Crusoe Island. Bootstrap support percentages (1,000 replicates) are given above branches. (Modified from Sang et al. 1994.)

windy cliffs at higher elevations (Sanders et al. 1987; Stuessy et al. 1998e). Keeping in mind the caveats discussed earlier about factors that could erase the biogeographical scale of speciation on an oceanic island, the present distribution of extant species of *Dendroseris* on Robinson Crusoe Island suggests that spatial/ecological/altitudinal factors were likely involved in divergence and speciation.

Anderson et al. (2001) and Bernardello et al. (2001) showed that *D. litoralis* is self-compatible, visited by hummingbirds, and it may have a mixed mating system. It is doubtful whether the transition to self-compatibility could serve as an isolating factor, as might occur if the species were highly selfing (Table 15.1) (Levin 2000; Brys et al. 2013; Hu 2015). Chromosome numbers are known for most species of *Dendroseris*; all have a gametic number of $n = 18$ (reviewed by Stuessy and Crawford 1998; see also Chapter 11), indicating the lack of either polyploidy or aneuploidy. This does not, however, rule out the possibility of karyotypic polymorphisms that could reduce the vigor or fertility of interspecific hybrids. Unfortunately, no such information is available. However, some anecdotal evidence suggests that there are likely no strong postpollination isolating mechanisms among species of *Dendroseris*. A plant growing along the main street of the village of San Juan Bautista in 2010 was distinct from any known endemic species and appeared intermediate between *D. litoralis* and *D. pruinata* in flower size and color as well as leaf shape. Assuming that this plant is a hybrid, we suggest that one parental species has orange corollas (*D. litoralis*) and the other white corollas (*D. pruinata*), whereas the putative hybrid has yellow flowers. Both species are commonly cultivated in the village and grow in close proximity. The pollen fertility of the proposed hybrid was high (91%) (D. J. Crawford, unpublished data), and the plant was vigorous and robust with many floral buds. Given that the two putative parental species occur in different subgenera and are not closely related phylogenetically (Sang et al. 1994) (Fig. 15.2), it seems highly likely that hybrids between sister species would also be vigorous plants with high pollen fertility.

The size and color of floral parts distinguish members of subg. *Dendroseris* from the other subgenera in being larger and orange in color rather than smaller and whitish (cf. Figs. C41 and C48). As indicated earlier, hummingbirds visit *D. litoralis*, whereas only occasional flies or moths were seen on *D. neriifolia* (Anderson et al. 2001b; Bernardello et al. 2001). Anderson et al. (2001b) and Bernardello et al. (2001) suggested that hummingbirds are the likely pollinator of *D. litoralis*, and this is probably the case also for the other very rare members of the subgenus. The flies and moths are not considered likely pollinators of any significance. If true, this suggests that the species within the subgenus are not isolated by pollinators, and there could be gene exchange among the species of subg. *Dendroseris* via sharing of the same bird pollinators.

The genus *Robinsonia* has the second highest number of species in the Juan Fernández Islands, but comparison of traits of the extant species offers few clues to mode(s) of speciation. All species are dioecious and thus are obligately outcrossing. They all have the same chromosome number of $n = 20$, so polyploid or aneuploid speciation is unlikely (Stuessy and Crawford 1998; see also Chapter 11). The species are putatively wind pollinated because no floral visitors were observed on three species from Robinson Crusoe Island (Anderson et al. 2001; Bernardello et al. 2001). There is no evidence on whether post-pollination isolating barriers exist among species.

The totality of data indicate that *R. gayana* and *R. thurifera* are more closely related to each other than either is to any other species on Robinson Crusoe Island (Crawford et al. 1987a; Sang et al. 1995; Takayama et al. 2015) (Figs. 14.3C, D and 15.3). Based on limited sampling of plants for molecular marker analyses, the more recently described *R. saxatilis* (Danton 2006) shows a close relationship to *R. gayana* and *R. thurifera* (Takayama et al. 2015) (Figs. 14.3C, D and 15.3). Although *R. gayana* and *R. thurifera* are known to occur together, they differ in the former being more, but not totally, restricted to drier, exposed ridges below 500 m, whereas the latter species grows on more mesophytic sites under a tree-fern canopy (Sanders et al. 1987; Stuessy et al. 1998e; Takayama et al. 2015). As stated by Sanders et al. (1987, p. 211), "*Robinsonia gayana* comes closest to having a restricted, xerophytic habitat on exposed rocky crevices. But occasionally, taller and laxer individuals establish in the ecotone with the tree ferns and forest within one or a few meters of *R. thurifera* and *R. evenia*." These authors further observed that *R. gayana* may also grow very close to *R. gracilis*, which typically is in open scrub and on ridges. These observations show quite clearly that several different combinations of *Robinsonia* species may come into close contact, including the most closely related taxa. Sanders et al. (1987) also noted, as did Carlquist (1974), that *R. thurifera* flowers later than *R. gayana*, a factor that could be effective in reducing gene exchange between the two species. However, Sanders et al. (1987) suggested that later flowering was likely the response to becoming sympatric

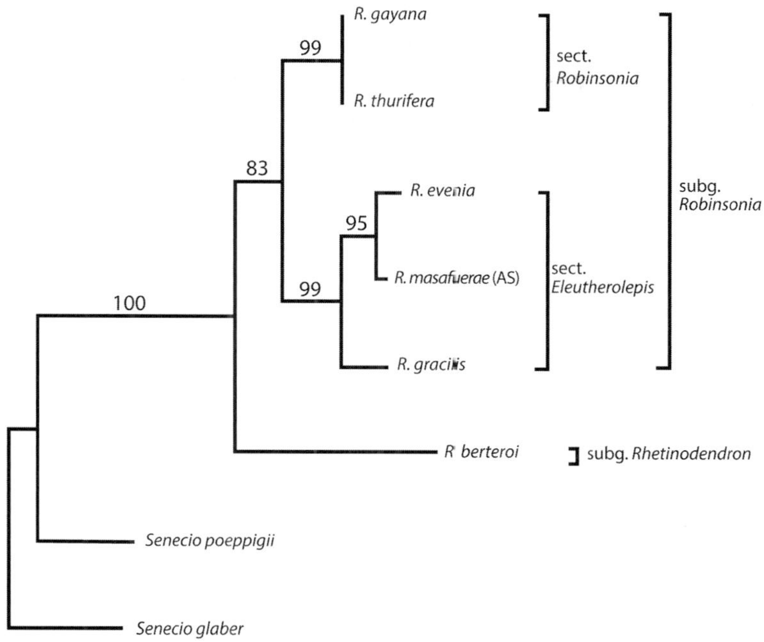

Figure 15.3 The most parsimonious tree for *Robinsonia* from analyses of ITS sequences. *R. masafuerae* is endemic to Alejandro Selkirk Island (AS); all other species occur only on Robinson Crusoe Island. Numbers above branches are bootstrap support percentages (1,000 replicates). (Modified from Sang et al. 1995.)

with its sister species, *R. gayana*. This could indeed be a case of reinforcement, as suggested by Sanders et al. (1987), but it could also have evolved from later-flowering peripheral isolates of *R. gayana*, with reduced gene flow that facilitated divergence into a new species.

A number of molecular markers (Crawford et al. 1992a; Sang et al. 1995; Takayama et al. 2015) have shown that, in general, all species of *Robinsonia* are distinct genetically, and the species maintain their morphological integrity despite different combinations of species in flower occurring together in nature on Robinson Crusoe Island. This point is illustrated most dramatically by the results of Takayama et al. (2015a), in which extensive sampling of individuals of *R. evenia* and *R. gracilis* co-occurring at four different sites in different areas of the island are distinct in two sets of molecular markers (Fig. 14.3C, D). The factor(s) that are ostensibly preventing the origin and/or establishment of interspecific hybrids is(are) unknown. There is no reason to think that wind pollination (Anderson et al. 2001) would function in any way to prevent gene flow between two species in flower in the same locality, and as noted earlier, no biological pollinators were observed on any of the flowers (note from Chapter 12: the biological pollinators in the Juan Fernández Archipelago are essentially limited to hummingbirds). The present distribution of the species may not give an accurate picture of the geographical scale at which the original speciation events occurred because the changing landscape on Robinson Crusoe Island during which the dramatic reduction (95%) of

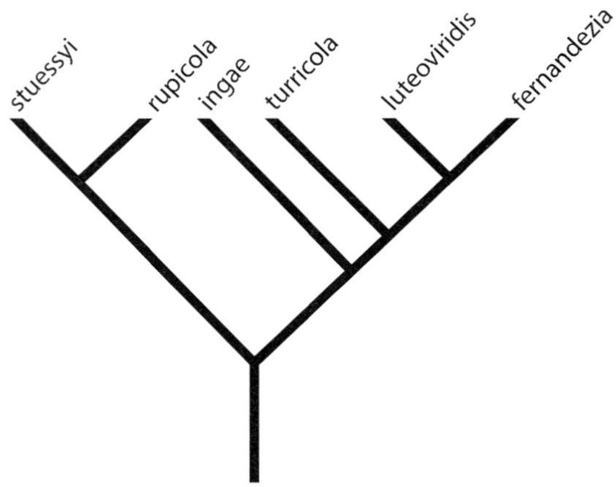

Figure 15.4 Phylogenetic hypothesis of relationships among *Erigeron* species endemic to the Juan Fernández Archipelago based on morphological characters. Six species from continental South America were used as outgroups. *E. fernandezia* occurs on both islands, while others are restricted to Alejandro Selkirk Island. (Modified from Valdebenito et al. 1992a.)

island area over geological time could have resulted in the species being "crowded" close together with the diminishing area of the island and/or the possible temporary expansion of ranges associated with the Late Quaternary climate-induced sea-level changes (Weigelt et al. 2016). That is, the original divergence may not have occurred in sympatry, although there is no evidence to refute this. Even if the original divergence was facilitated by spatial/habitat divergence, the factors that now allow the species to "pass the test of sympatry" (Coyne and Orr 2004) remain obscure. Two species of *Robinsonia* have gone extinct, *R. berteroi*, the only known member of subg. *Rhetinodendron*, and *R. macrocephala*, the only known representative of sect. *Symphyochaeta* of subg. *Robinsonia* (Danton and Perrier 2005; Ricci 2006; Danton et al. 2006). Whether these single species representing distinct lineages in *Robinsonia* were the last remnants of much larger lineages with species extinctions before historical times can never be known; there is no fossil record (other than pollen) (Haberle 2003) on this volcanic archipelago and hence no former evidence of *Robinsonia*. There could also have been prehistorical species extinctions in the remaining two sections of subg. *Robinsonia*.

The genus *Erigeron* is found on both islands, but in contrast to *Dendroseris* and *Robinsonia*, it has diversified on the younger Alejandro Selkirk Island, with one species also occurring on both islands (Valdebenito et al. 1992a). Also, unlike the other two genera, no rigorous molecular phylogenetic study has been done for *Erigeron*, although Valdebenito et al. (1992a) produced a hypothesis of relationships based on morphological characters (Fig. 15.4). Noyes (2000), as part of a large-scale phylogeny of *Erigeron*, included two Juan Fernández endemics, *E. fernandezia* and *E. rupicola*, the former being the one species occurring on both Robinson Crusoe and Alejandro Selkirk. The two species do occur together in the ITS phylogeny and provide some support for all endemic taxa being the result of a single dispersal event. The phylogeny of Valdebenito et al. (1992a) showed two basic groups, one consisting of *E. rupicola* and *E. stuessyi* and the other consisting of *E. fernandezia*, *E. ingae*, *E. luteoviridis*, and *E. turricola* (Fig. 15.4). Flavonoid compounds were variable among populations of the same species and were of very limited utility in distinguishing the species, with instances where populations of different species were more similar than populations of the same taxa (Valdebenito et al. 1992a; see also Chapter 10). A recent study by López-Sepúlveda et al. (2015a) used microsatellites (simple sequence repeats) and AFLPs (see Chapter 14) to assess genetic diversity and divergence within and among populations of *Erigeron*. However, with the exception of the two species *E. rupicola* and *E. fernandezia*, only two or three populations of each species were sampled, a reflection of their rarity. AFLP markers are concordant with other prior assessments in showing a close relationship between *E. rupicola* and *E. stuessyi* (Fig. 14.3A, B) (López-Sepúlveda et al. 2015a), and these markers generally sort individuals into their respective species based on morphology (Fig. 14.3A). There was considerably lower resolution with microsatellites. While there was reasonably good grouping of *E. rupicola* and *E. fernandezia* on Robinson Crusoe Island and most of *E. fernandezia* on Alejandro Selkirk Island, there was intergradation among the other species (Fig. 14.3B). In contrast to prior views, *E. rupicola* and *E. stuessyi* are not resolved as closely related. Overall, microsatellites portray a picture of species that do not sort out genetically, and populations assigned to different species occur intermixed (Fig. 14.3B). Recall that the data from flavonoids give a similar picture for *Erigeron* (Valdebenito et al. 1992a; see Chapter 10).

All Juan Fernández species of *Erigeron* have a chromosome number of $n = 27$. This number presumably indicates hexaploidy on a base of $x = 9$ (Valdebenito et al. 1992a; see Chapter 11), and the fact that all species have the same number indicates that change in ploidy level has not been a factor in speciation in the Juan Fernández Archipelago. Nothing is known about the breeding/mating system or pollinators of *Erigeron*, although the very low seed set in the field and on herbarium specimens suggests that the species are self-incompatible and thus highly outcrossing. As mentioned earlier, given the lack of biotic pollinators on the Juan Fernández Archipelago besides the important role that hummingbirds play (Anderson et al. 2000, 2001b; Bernardello et al. 2001), anemophily is the likely method of pollen distribution. In addition, four of the species are protandrous (Bernardello et al. 2001), which would promote outcrossing. The two species *E. rupicola* and *E. stuessyi*, which may be sister species, occur at lower elevations on Alejandro Selkirk Island and are separated spatially from other species except for the rare occurrence of *E. rupicola* with *E. fernandezia*, a species that occurs over a wide range of altitudes. *Erigeron rupicola* and *E. stuessyi*, although occurring at similarly low elevations, are differentiated ecologically, with *E. rupicola* growing on dry coastal rocks that are subjected to salt spray at low elevations; *E. stuessyi* is found on the cool, moist walls of the ravines (Valdebenito et al. 1992a; López-Sepúlveda et al. 2015a). The two species are morphologically similar in being small rosette herbs with short flowering stalks and small flowering capitula, but *E. stuessyi* differs from *E. rupicola* by having thinner leaves and longer flowering stalks. The different habitat preferences of the two species may serve to reduce or eliminate gene flow between them, and this may have been a factor in the initial divergence of populations in the two areas.

Erigeron fernandezia occurs on the two major islands of the Juan Fernández Archipelago but likely originated on Alejandro Selkirk Island and eventually dispersed to Robinson Crusoe Island, where it is relatively common but often occupies disturbed areas such as along trails (Chapter 14). The species is morphologically distinct and much more common than its congeners on Alejandro Selkirk Island, occurring primarily in areas dominated by grasses, from less than 100 m to over 1,200 m. It is likely isolated from other species primarily by its habitat.

The three remaining species are problematical and include *Erigeron ingae*, *E. luteoviridis*, and *E. turricola* (the *E. ingae* complex). The species are difficult to distinguish on morphological characters, and when Skottsberg (1921) described them as new, he failed to provide convincing characters for distinguishing them and even expressed some reservations about the distinctiveness of the taxa. Solbrig (1962) noted the difficulty in identifying the species, and he synonymized *E. turricola* and *E. luteoviridis*. In field collections made on expeditions to Alejandro Selkirk Island (led by T. F. Stuessy), it was difficult to assign identifications to some of the individuals obtained. Neither AFLP nor microsatellite data cluster the species together into distinct groups, this being especially true for microsatellites (López-Sepúlveda et al. 2015) (Fig. 14.3A, B). All of this suggests that the group is in a dynamic evolutionary condition, with consistent genetic lines not yet having stabilized. That is, speciation may be in progress and has not been "completed" (Nosil et al. 2009). All three taxa of the *E. ingae* complex occur in the high-elevation "alpine" habitats of Alejandro Selkirk Island, and there are no apparent differences in the habitats where they are found.

Erigeron on Alejandro Selkirk Island likely represents a better system than either *Dendroseris* or *Robinsonia* on Robinson Crusoe Island for studying the processes of divergence and speciation because it appears to contain taxa in different stages of divergence, with *E. rupicola* and *E. stuessyi* representing a low-elevation, fairly distinct lineage that diverged from a common ancestor with subsequent divergence into different habitats. The widely distributed *E. fernandezia* seems to represent a distinct lineage, but López-Sepúlveda et al. (2015a) showed that populations on the two islands are divergent (Fig. 14.3A, B). The remaining three strictly high-elevation "species" may provide the best opportunity to study early stages of divergence, but such studies would be a challenge due to the rarity and inaccessibility of the taxa. Detailed studies of the components of the habitat, such as was done by Sanders et al. (1987) for *Dendroseris* and *Robinsonia* on Robinson Crusoe Island, could provide clues to factors isolating the populations. It may also be that the lack (other than hummingbirds) of biotic pollinators and the imprecision of wind in transferring pollen over even short distances could serve to isolate populations.

Depending on the taxonomic treatment, *Wahlenbergia* (Campanulaceae) includes two or three species endemic to Robinson Crusoe Island and one or two restricted to Alejandro Selkirk Island (Lammers 1996). We have seen no effective pollinators on *Wahlenbergia*. However, the species on Robinson Crusoe Island are self-compatible and likely autogamous when the stigmata reflex and touch the style, on which pollen has been placed by the pollen brush (Anderson et al. 2000, 2001b; see Chapter 12). A very interesting exception to that syndrome occurs with *Wahlenbergia berteroi* where pollination is accomplished with the aid of the wind, though not like the typical wind distribution of pollen from stamens to stigmata. The scenario is analogous to that of a bell and clapper. In this species, the stigmata do not reflex to touch the style. However, there is pollen covering the inner surface of the corolla (deposited by the pollen brush), and the nearly constant wind in the cliff-side habitats of this species moves the flowers in such a way that the stigmata (the analogue of the clapper in a bell) contact the inner corolla (the analogue of the body of a bell) and pollen resides there (Anderson et al. 2000).

Available evidence suggests that *Wahlenbergia* in the Juan Fernández Archipelago is the result of a single introduction because they share a chromosome number of $n = 11$ (see Chapter 11) and a distinctive combination of morphological characters (Lammers 1996). Lammers (1996) carried out phenetic and cladistic analyses of morphological characters to recognize three species (*W. fernandeziana*, *W. grahamiae*, and *W. berteroi*) on Robinson Crusoe Island and two (*W. masafuerae* and *W. tuberosa*) on Alejandro Selkirk Island and suggested an initial colonization of Robinson Crusoe, dispersal to Alejandro Selkirk, with divergence and speciation, and then back-dispersal to Robinson Crusoe to account for the presence of a species (*W. berteroi*) that occurs in a clade with the two species endemic to Alejandro Selkirk (see Chapter 13) (Fig. 13.2R). Whether indeed this represents the best estimate of relationships for Juan Fernández *Wahlenbergia* remains an open question until intensive molecular phylogenetic studies are undertaken. If the Lammers' scenario is correct, then *W. berteroi* likely originated following dispersal of *W. masafuerae* or a common ancestor of *W. berteroi* and *W. masafuerae*. Alternatively, *W. berteroi* could have originated on Alejandro Selkirk Island and become extinct there following dispersal to Robinson Crusoe Island. Finally, it is possible that the Lammers' (1996) hypothesis is not the best estimate of phylogeny and that *W. berteroi* could in fact have originated on Robinson Crusoe Island.

With the recognition that the issue of relationships is not settled, it will be assumed that the latter scenario is correct and that *W. berteroi* originated on Robinson Crusoe Island, and the focus will be on potential factors other than interisland dispersal for isolation and speciation. If *W. berteroi* originated on Robinson Crusoe Island, then spatial/ecological divergence likely was a factor because this species typically grows on rocky, open areas at low elevations near the sea, though it may be found as high as 300 m (Lammers 1996). In contrast, the other one or two species on the island are found at higher elevations. Anderson et al. (2000, 2001b) and Bernardello et al. (2001) also demonstrated that *Wahlenbergia* on Robinson Crusoe Island are highly self-compatible and capable of self-pollination. This breeding system can also obviously maintain separation and promote divergence (Hu 2015).

Two of the species recognized by Lammers (1996), *W. grahamiae* on Robinson Crusoe and *W. tuberosa* from Alejandro Selkirk, have not been universally accepted as distinct from *W. fernandeziana* and *W. masafuerae*, respectively, and could be considered as either morphologically "cryptic" species or simply reflect intrapopulational variation. Lammers (1996) indicated that the two species on Alejandro Selkirk are separable spatially and ecologically, but in three instances, mixed collections of the two species are cited among the specimens examined. Likewise, three mixed collections were cited for *W. fernandeziana* and *W. grahamiae*, and Lammers (1996) suggested that the mixed collections as well as two other collections could contain hybrids between the two species. Detailed studies at the population level are to be desired to determine whether the "species" pairs on each of the islands represent secondary intergradation (i.e., hybridization) between distinct entities or natural variation within the same basic gene pools.

The genus *Centaurodendron* is endemic to Robinson Crusoe Island and is generally recognized as including two species. Although *Yunquea* was described as a monospecific genus endemic to Robinson Crusoe Island, it is highly likely that it and the two species of *Centaurodendron* all originated from the same colonizing event (A. Susanna, personal communication). All three species are extremely rare, being known from a few, very small isolated populations, and they rarely flower (Skottsberg 1921, 1956, 1958). The two species of *Centaurodendron*, when they do flower, apparently do so at different times (Skottsberg 1956). It is apparent that the three species are now isolated spatially, and the low frequency of flowering, as well as different flowering times for two of the species, further reduces the potential for gene flow, even if the species were not spatially isolated. Bernardello et al. (2001) cite a report by Brooke (1987) that hummingbirds visit one species of *Centaurodendron*, but whether it is an effective pollinator is unknown. A likely scenario for divergence and speciation in this lineage is that following initial colonization and establishment, chance dispersal of fruits initiated disjunct populations with subsequent divergence.

Ruiz et al. (2000) studied *Cuminia* on Robinson Crusoe Island using morphological and ITS sequence data and concluded that two species, *C. eriantha* and *C. fernandezia*, should be recognized. Both species occur in similar habitats and in close proximity but never in mixed populations, and they remain distinct in morphology and ITS sequences (Ruiz et al. 2000). The two species share the same chromosome number of $n = 22$ (Sanders et al. 1983; see Chapter 11). The species are visited by hummingbirds, which are the likely pollinators (Anderson et al. 2001; Bernardello et al. 2000; see Chapter 12). There are no apparent mechanisms that reduce or prevent gene exchange between the two species that could allow them to exist in what must be described as sympatric populations. Ruiz et al. (2000) suggested that the two species may have diverged when Robinson Crusoe Island was more ecologically diverse than it is presently (or perhaps when Robinson Crusoe Island was larger based on the Late Quaternary glaciations) (e.g., Weigelt et al. 2016) and that their present sympatry is the result of secondary contact brought on by the smaller size and diminished habitat diversity on the island. This may well be true, but it does not explain the apparent ability of the two species to exist in sympatry without recognized hybrids. The answer may lie in one or more postpollination factors (Table 15.1).

Hybridization and Speciation

Natural hybridization, both ancient and contemporary, between endemic and native species has been shown in different groups of flowering plants in various archipelagos (Nielsen et al. 2003; García-Verdugo et al. 2013; Jones et al. 2014). The reasons frequently cited for the presence of natural interspecific hybrids in insular lineages are the absence of strong post-pollination barriers and natural and human disturbance bringing together species typically isolated by spatial/ecological factors (Gillett and Kim 1970; Borgen 1976; Carr and Kyhos 1981; Ganders and Nagata 1984; Nielsen et al. 2003; van Hengstum et al. 2012). The stabilization of hybrids may result in the origin of new species at the same ploidy level as the parents, a process known as "homoploid hybrid speciation" (Rieseberg 1997; Abbott et al. 2010, 2013; Yakimowski and Rieseberg 2014). Whether the few reports of homoploid hybrid speciation are the result of its uncommon occurrence in nature or the difficulty in detecting it, or possibly a combination of both, remains an open question (Mallet 2007; Schumer et al. 2014; Yakimowski and Rieseberg 2014).

Several examples of possible homoploid hybrid species have been reported for oceanic island plants. In the Hawaiian silversword alliance (Asteraceae), Baldwin (2003) interpreted incongruence between placement of several species in nuclear and plastid phylogenies as suggestive of the hybrid origin of several species. Also in Hawaii, Howarth and Baum (2005) provided data from several lines of investigation, including DNA sequences, to infer the hybrid origin of two species of *Scaevola* (Goodeniaceae). There is also strong molecular evidence for two homoploid hybrid species of *Argyranthemum* (Asteraceae) in the Canary Islands (Brochmann et al. 2000; Fjellheim et al. 2009).

There are very few possible examples of natural hybridization and no compelling evidence of homoploid hybrid speciation in the Juan Fernández Archipelago. The most conspicuous example of interspecific hybridization in the Juan Fernández Archipelago involves two species in the genus *Gunnera* endemic to Robinson Crusoe Island (Pacheco et al. 1991a). The two parental species, *G. bracteata* and *G. peltata*, typically occur at slightly different elevations, with the former from 350 to 500 m and the latter from 400 to 600 m. Skottsberg (1921) noted the occurrence of numerous individuals displaying a complex array of characters intermediate between the parental species in a disturbed area following a trail along a transect in Quebrada Villagra. In contrast, few intermediates were found in undisturbed areas, even where the parental species were in close proximity (Pacheco et al. 1991a). The putative parental plants show no obvious reduced vigor and appear to be doing well along the trail. This suggests, but clearly does not prove, that the open areas along the trail provide suitable habitats for the hybrids where they do not compete with their parents. Based on morphology, the hybrid zone consists of a complex of different types of hybrids, with plants having hybrid indices ranging from values similar to each of the parental species to values intermediate between the parents. Thus there is no indication of stabilized hybrid variants that could represent incipient hybrid speciation.

The genus *Eryngium* (Apiaceae) has four described endemic species in the Juan Fernández Archipelago, with one, *E. sarcophyllum*, endemic to Alejandro Selkirk Island. The species on Robinson Crusoe Island include *E. bupleuroides*, *E. ×fernandezianum*, and *E. inaccessum*. Calviño et al. (2008a) showed that the three species on Robinson Crusoe Island form a strongly supported monophyletic group. The species on Alejandro Selkirk Island is assumed to be extinct (Danton et al. 2006). All species on Robinson Crusoe are quite rare, especially *E. ×fernandezianum* and *E. inaccessum* (Stuessy et al. 1998b, 1998d). Skottsberg (1921) suggested that *E. ×fernandezianum* is a hybrid between *E. bupleuroides* and *E. inaccessum* and provided drawings and a table showing that it is intermediate between the other two species in a number of morphological features. He also commented that the putative hybrid species grows in close proximity to the presumed parental species. He concluded, however, that "Still no proofs [*sic*] that it is a bastard can be given at present" (Skottsberg 1921, p. 161). Whether *E. ×fernandezianum* is of hybrid origin has still not been documented, for instance, with molecular markers, beyond the original morphological data presented by Skottsberg (1921).

Perhaps the most unusual case of possible hybridization in the Juan Fernández Archipelago involves the two genera *Acaena* and *Margyricarpus* (Rosaceae) on Robinson Crusoe Island. *Acaena argentea* is an invasive weed, while *Margyricarpus digynus* is now a rare species endemic to Robinson Crusoe Island. The putative intergeneric hybrid, *×Margyracaena skottsbergii*, is also quite rare. Skottsberg (1921) recognized that *×M. skottsbergii* was likely of hybrid origin. Because the chromosome number of only one of the species (*Acaena argentea*) has been determined (see Chapter 11), it is not known whether all taxa are at the same ploidy level, and thus this is a potential case of homoploid hybrid speciation. Crawford et al. (1993a) used RAPD markers to show that the two plants of *×Margyracaena* from two different localities along the same path separated by 100 m in altitude have additive profiles, with bands detected only in each of the parents now combined in the presumed hybrid species. Skottsberg (1921) observed that *Margyricarpus digynus* is common in one and "not uncommon" in another area of Robinson Crusoe Island, as well as present in several other areas. However, the species is now much rarer; the plant used by Crawford et al. (1993a) was cultivated in the CONAF garden, its origin unknown. Field work since Crawford et al. (1993a) has resulted in the discovery of naturally occurring *M. digynus* on two rocks off the coast of Robinson Crusoe Island, but apparently the other parental species, *A. argentea*, despite being a weed, does not occur in either place (Danton et al. 2006). Skottsberg (1921) reported the hybrid from four localities and indicated that it was quite rare and growing on disturbed or open ground. It is not known whether the hybrids are fertile; Crawford et al. (1993a) indicated that only sterile plants or plants with immature fruits are known. In summary, the hybrid *×Margyracaena skottsbergii* likely does not have a bright evolutionary future in terms of becoming a stable, sexually reproducing lineage distinct from both of its parental species. The rarity of one of the parental species in nature does not bode well for the generation of new hybrids. Likewise, the loss of hybrid individuals over historical time and the question of their fertility and fecundity cast doubt on the future success of hybrid types. Lastly, the endemic parent and the hybrid may well suffer from competition with the other parental species, the weedy, invasive *Acaena argentea*.

The most recent possible example of hybridization and speciation was reported by Danton et al. (2015). They described the fern species *Pleopeltis ×cerro-altoensis* (Polypodiaceae), which is endemic to Robinson Crusoe Island and known from only a small population. Danton et al. (2015) suggested that the new species was a first-generation hybrid between two species native to the Juan Fernández Archipelago, *Pleopeltis macrocarpa* and *P. masafuerae*. The former species is native to both islands of the Juan Fernández Archipelago, while the latter is endemic to Alejandro Selkirk Island. Danton et al. (2015) presented morphological data concordant with the hybrid origin of the new species, but there is some question as to whether their hypothesis of the parental species is correct. Roberto Rodríguez (personal communication) suggests that the reported hybrid may be a cross between *Pleopeltis macrocarpa* and *Polypodium intermedium* subsp. *intermedium*. If this were the case, then the two parental species would both occur on Robinson Crusoe Island, whereas if *Pleopeltis macrocarpa* and *P. masafuerae* were the parents, then the species would most likely be an interisland hybrid because the latter parent occurs only on Alejandro Selkirk Island and the hybrid is known only from Robinson Crusoe Island. An alternative, though less parsimonious, explanation would be that the hybrid originated on Alejandro Selkirk Island with subsequent dispersal to Robinson Crusoe Island, and the hybrid has either not been discovered on Alejandro Selkirk Island or has gone extinct there. With the Rodríguez hypothesis, the new species would be an intergeneric hybrid, although generic limits for *Polypodium* and *Pleopeltis* are not clear and are continually being evaluated (e.g., de la Sota et al. 2007; Otto et al. 2009; Tejero-Díez 2014). Regardless of the parental species, the morphology and rarity of *Pleopeltis* ×*cerro-altoensis* provide support for its hybrid origin. It would be highly desirable to use molecular markers to provide insights into the origin of the new species.

In summary, none of the very rare examples of interspecific hybridization in the Juan Fernández Archipelago show any indication that they have resulted or will result in the formation of stabilized hybrid derivatives potentially worthy of designation as distinct species. Rather, they appear to represent instances where hybrids have become established on open or disturbed sites, and in two instances, the hybrids are extremely rare and appear to be the result of gene exchange between equally rare parental species. There are several factors that may limit the frequency of interspecific hybridization in the Juan Fernández Archipelago compared with other archipelagos such as the Canary and Hawaiian Islands. The first is the very large percentage of species with only one endemic species on an island, thus limiting any opportunity for gene exchange in many groups. Furthermore, when two or more congeneric endemic species are found on an island, they are often very rare, occurring in a few isolated, very small populations. Periodic natural disturbances such as volcanic activity and landslides that may bring species into contact and provide suitable open habitats for hybrids in other archipelagos do not occur in the Juan Fernández Archipelago. As with most archipelagoes, there is no good evidence to measure the impact of the likely significantly increased area and then its subsequent loss as a result of glacial maxima in the Late Quaternary (Weigelt et al. 2016) on the origin and establishment of hybrids in the Juan Fernández Archipelago. Lastly, disturbances from human activities are not as severe in the Juan Fernández Archipelago as in the Canary and Hawaiian Islands. There are no known examples of polyploid speciation in situ in the Juan Fernández Archipelago.

Speciation in the Juan Fernández Archipelago: An Overview and Perspective

Stuessy et al. (1998e) calculated the number of single species of a genus endemic to the archipelago (i.e., species that diverged from a continental progenitor without further speciation in the islands), the number of genera with one endemic species on each island, and the number of speciation events associated with dispersal to another island following the evolution of two or more species on the source island. These tabulations indicate that geographical isolation via continental-island or island-island dispersal accounts for about 70% of the endemic species of flowering plants on the Juan Fernández Archipelago. In the nearly two decades since that review, there have been no additional data suggesting that this estimate requires significant modification. Stuessy et al. (1998a) pointed out that there could be additional factors isolating these now geographically distinct species, but, to our knowledge, there are still no data to address this question.

Stuessy et al. (1998e) suggested that habitat divergence on islands could be a factor for slightly less than 20% of speciation events in the archipelago. As pointed out earlier, only sister species provide valid comparisons of habitat divergence as a possible isolating barrier facilitating divergence and speciation. This study does not alter the estimate of Stuessy et al. (1998e) for the role of habitat divergence, with notable examples, mentioned earlier, from *Dendroseris*, *Robinsonia*, and *Gunnera*.

There are few examples of congeneric species flowering at different times (temporal divergence; Table 15.1); Stuessy et al. (1998a) reported five possible examples (ca. 7%) where this mechanism could have been a factor in divergence. However, in two of the examples cited, one from the genus *Uncinia* (Cyperaceae) and the other from *Chenopodium* (Chenopodiaceae), there are other factors that could have been involved in divergence. In the case of *Uncinia*, one of the species is native and the other endemic. Whereas flowering time may now be a factor in reducing gene flow between the two species, the native species likely originated in the continental source area (assuming that they did not originate on the island with back-dispersal to the continent), and therefore it cannot be inferred that flowering time was involved in speciation. As pointed out by Stuessy et al. (1998e), two of the *Chenopodium* species are separable geographically, with one on Robinson Crusoe Island and the other known only from Santa Clara Island. In addition, both species are highly self-pollinating, which could be a factor in isolation (see below).

Stuessy et al. (1998e) commented on the calculations of Skottsberg (1928) for the pollinators of the endemic flora, suggesting that his estimates of bird pollination were too low and of insect pollination too high. Subsequent studies by Anderson et al. (2000a, 2000b, 2001) and Bernardello et al. (2001, 2004) have verified the significance of bird pollination and have shown that there are no nonbird pollinators. In fact, the potential example of mechanical floral isolation involving differences in pollinators cited by Stuessy et al. (1998e) in *Wahlenbergia* is questionable because both species are wind pollinated (Anderson et al. 2000a, 2001; Bernardello et al. 1999, 2000, 2001). Whether the differences in corolla color between the two *Cuminia* species affect hummingbird preference for *C. eriantha* is an open question (Anderson et al. 2001; Bernardello et al. 2001). There are no reports for *C. fernandezia*, which has more lightly colored corollas than *C. eriantha*. In comparison with the most extensively studied archipelagoes (e.g., the Galápagos, Hawaiian, and Canary Islands), a striking conclusion derived from our extensive field studies (and critical reviews of the literature) is that isolation by pollination does not play the same extensive role in speciation for the Juan Fernández flora. The reason for this has been pointed out many times in other chapters; other than the spectacular hummingbird scenarios, there are no dedicated biotic pollinators (see the caveats in Chapter 12). Pollen transfer for many species, with a morphology implying biotic pollination (that led to speculations by Skottsberg of such), is via autogamy/geitonogamy and/or wind.

While the majority of endemic species examined in the Juan Fernández Archipelago are self-compatible, relatively few of them appear to be highly autogamous because they have mechanisms promoting outcrossing (Anderson et al. 2001; Bernardello et al. 2001; see Chapter 12). One possible example is from the genus *Megalachne* (Poaceae), where one species (*M. masafuerana*) from Alejandro Selkirk was reported as cleistogamous by Skottsberg (1921), cited in Stuessy et al. (1998e). Two other species (*M. berteroana* and *M. robinsoniana*) occur on both islands, and Danton et al. (2006) noted a possibly undescribed taxon on Alejandro Selkirk Island, but the situation is in need of study. Additional studies are to be desired to determine the role of cleistogamy in speciation. As noted earlier, there is one species of *Chenopodium* endemic to Robinson Crusoe Island and another restricted to the small island of Santa Clara off the coast of Robinson Crusoe Island. Assuming that all species are the result of one colonization event, it appears that dispersal and subsequent divergence, rather than autogamy alone, most likely facilitated the origin of the three species.

Unfortunately, there are no data for intrinsic genic (postpollination) factors that could reduce gene flow and facilitate divergence and speciation in Juan Fernández plants. While results from other archipelagos (Gillett and Lim 1970; Lowrey 1986; Crawford et al. 2009) might suggest that these factors would not be highly developed in the Juan Fernández Archipelago, this obviously cannot be assumed without data.

Part VII

Biogeography

Biogeography is the distribution of organisms in space and time. Investigations on processes of evolution that have taken place in oceanic archipelagos cannot be accomplished effectively without having an appreciation of the spatial context of these processes. Because evolution usually proceeds slowly, the geological history of islands provides a backdrop when attempting to reveal evolutionary mechanisms. Interpretation of biogeography of the vascular flora of oceanic islands therefore involves many aspects of understanding the abiotic and biotic environments both now and millions of years previously. In the Juan Fernández Archipelago we have emphasized understanding of two fundamental aspects of biogeography. The first is to learn the origin of the native and endemic vascular flora, particularly the flowering plants. Where did the flora come from, and how did it get there? To sit under an endemic tree and have a quiet lunch, such as under a large individual of the endemic *Drimys confertifolia*, immediately brings to mind the question of how this species first came to the islands and how it evolved after arrival. It is nearly impossible for a biologist to avoid these important questions. In Chapter 16, Bernardello and Anderson examine the angiosperm flora of the Juan Fernández Archipelago and provide an excellent summary of where the immigrants have come from. The results indicate southern South America as the primary source, which correlates with its proximity to the islands. The second dimension of biogeography that we have stressed involves the use of the structure of the islands and its flora for insights on the modeling of species diversity. This approach, often referred to as "island biogeography" after the influential book by MacArthur and Wilson (1967), attempts to examine the factors that play a role in species levels in oceanic islands in the hope of providing general answers. Much emphasis on the sizes of islands and the distance from the source area has been given over the decades, and rates of immigration and extinction have also been examined. Because of the simple structure of the Juan Fernández Archipelago, and due to knowledge of the sources of much of the flora, it provides an excellent opportunity to attempt to model the factors that have been responsible for this specific diversity. Chapter 17 deals with these points, and even in an extremely simple island system, the perspectives and calculations become somewhat complex. This is due to the dynamic and constantly changing nature of the island system, as well as evolutionary processes taking place within the islands, all of which have an impact on overall levels of species diversity.

16 Plant Origins and Dispersal

Gabriel Bernardello and Gregory J. Anderson

Oceanic islands have played a key role in the understanding of the evolution and dispersal of organisms on the Earth (Crawford and Stuessy 1997; Whittaker 1998; Emerson 2002; Cowie and Holland 2006). The extraordinary faunas and floras of oceanic islands containing unique assemblages of species, and high levels of endemism have long been of interest to naturalists, particularly following the publication of Darwin's (1859) and Wallace's (1895) books that highlighted these fascinating small areas of the world as remarkable laboratories of evolution. Sherwin Carlquist's books from 1965 and 1974 increased interest in island plant diversity, evolution, and ecology. The addition of a firm timeline (the emergence of the islands) for evaluation of evolutionary processes has conferred a huge advantage over study of continental systems. When the arrival of a biota can be dated, the divergence from nearest relatives can be assessed with a degree of accuracy not usually available with continental organisms (e.g., Carlquist 1974; Barrett 1995). Understanding evolution on oceanic islands must be founded on substantial field-based data integrated with sound principles of dispersal and biogeography. There is still much to be learned on these subjects both about and from these waif biotas (Darlington 1957; Thorne 1963; Carlquist 1974; Adsersen 1995; Crawford and Stuessy 1997; Cronk 1997; Emerson 2002; de Queiroz 2005; Midway and Hodge 2012; Garnock-Jones 2014).

The biota of oceanic islands frequently bears a different assemblage of plant families than equivalent mainland areas (Carlquist 1965, 1974; Adsersen 1995; Grant 1998; Midway and Hodge 2012). Most islands are small, and many are distant. These factors, combined with a wide array of dispersal capabilities among continental species, result in island floras that possess a nonrepresentative sample of the species from the potential source continents. This disharmony in composition of an insular flora with its continental source flora is considered prime evidence for the filter effect of long-distance dispersal (Carlquist 1974). The filter that long-distance dispersal imposes is related to the size of disseminules, desiccation resistance, cold resistance, and other features that vary among taxonomic groups. The composition of island floras also depends ultimately on the existence of appropriate dispersal vectors (Carlquist 1974).

Clearly, chance also plays a key role in determining which species arrive, in what order (i.e., some species might become established if they are initial pioneers, whereas others will only flourish when certain other biotic elements are present), when, and in what numbers (Carlquist 1981). In addition, difficulties of both establishment and reproduction (Baker 1967; Stebbins 1957; Whittaker 1998; Anderson et al. 2001; Bernardello et al. 2001; Midway and Hodge, 2012) will further influence the composition of island communities by favoring some types of colonists over others. Thus establishment is not simply defined by dispersal of propagules to islands but also by the availability of an open niche (in all the contexts of that term) in which seeds of new migrants can establish a viable reproducing population before challenges that might lead to their extinction are experienced (Kingston et al. 2003; Crawford et al. 2011; Anderson et al. 2015).

The fantastic flora of the Juan Fernández Archipelago contrasts with its terrestrial fauna, which is far less impressive. There are no native mammals, amphibians, or reptiles, and the insect fauna is small (with no known native insect pollinators; see Chapter 12) (Kuschel 1952; Wilson 1973; Ingram et al. 2006). The total land and sea bird diversity of the Juan Fernández Archipelago is very small. In the most recent comprehensive survey of birds, Lönnberg (1921) listed fewer than 30 species total. This includes eight native-breeding land birds and six breeding sea birds (the rest of Lönnberg's list included the following: three species of accidental visitors, five species of South American "visitors," five species of roving sea birds, and two species introduced). If this extant fauna is taken to represent that available as original dispersal agents, opportunities for dispersal have been relatively few. The endemic land bird species notably include the Juan Fernández firecrown (*Sephanoides fernandensis*), which is the only endemic hummingbird known on an oceanic island (Colwell 1989; Roy et al. 1998). Among the native birds, there is also a second sister hummingbird species (also found on continental South America). Both of these hummingbirds are, of course, important for pollination, not dispersal (Bernardello et al. 2001). The other endemic land birds are the rayadito (*Aphrastura masafuerae*) from Alejandro Selkirk Island and the Juan Fernández tit-tyrant (*Arlairetes fernandezianus*). The native bird species shared with continental Chile include the short-eared owl (*Asio flammeus*), the green-backed firecrown hummingbird (*Sephanoides sephanoides*), the austral thrush (*Turdus falcklandii*), and the austral blackbird (*Guracus curaeus*) (Hahn 1996).

Floral visitors are absent or rare in the island forests (Skottsberg 1928; Anderson et al. 2001; Bernardello et al. 2001), other than the two hummingbird species. Approximately 10% of the extant flora is hummingbird pollinated (see Chapter 12). The diet of the hummingbirds includes nectar from 14 autochthonous plant species (Bernardello et al. 2000, 2004; Anderson et al. 2001). It is estimated that around 46% of the flora is wind pollinated, whereas the pollination of the remaining percentage of the flora is unknown (Bernardello et al. 2001; see also Chapter 12). Among the usual classes of insects, there are virtually no native species that regularly visit flowers as pollinators. That is, there are no pollinating bees (with the exception now of a species we found that is described as a new species and is very likely a recent Chilean immigrant) (Engel 2000) or pollinating butterflies, moths, flies, or beetles (Anderson et al. 2001; Bernardello et al. 2001).

The origin of the angiosperm flora of the Juan Fernández Archipelago has been the subject of considerable discussion (i.e., Skottsberg 1925, 1934, 1956; van Balgooy 1960, 1971) and was recently reviewed by Bernardello et al. (2006). In this chapter we analyze the most probable places of origin and the methods of arrival of the first angiosperm colonists to the Juan Fernández Islands. The penultimate section, as indicated in its title, "Summary of Foundational Data and Observations," presents a summary of the data on which the arguments in the following two sections ("Plant Origins" and "Mechanisms of Arrival") are based.

Plant Origins

Species were assigned to one of six classes according to their place of origin: (1) Chile (the coastal country on the nearest continent and most probable source), (2) Neotropics (lowland tropical zones), (3) South America (remaining parts), (4) Pacific area, (5) Pacific area (Australia), and (6) Pacific area (New Zealand). The works by Skottsberg (1956) on the Juan Fernández Islands, van Balgooy (1960, 1971) on plant geography of the Pacific, Carlquist (1974) on islands in general, Porter (1983) on the Galápagos Islands, and Wagner et al. (1990) on the Hawaii Islands were used as guides to make determinations on the possible origins and methods of arrival of the Fernandezian species. In addition, phylogenies of the island species and their likely closest colonizing relatives were used (see Chapter 13 and citations therein).

Dispersal mechanisms for arrival of the first colonists were tabulated into the four most common categories: (1) bird internal (seeds or fruits eaten and carried in the digestive tract), (2) bird external (either mechanically attached to feathers by viscid/sticky or bristly/barbed seeds or fruits or embedded in mud on feet), (3) wind (air flotation), and (4) water (oceanic drift). Native and endemic angiosperm species on the archipelago were scored for their area of origin, method of arrival, fruit type and length, dispersal unit length, and dispersal type (Table 16.1).

Eighty-two percent of the species are postulated to have come from South America (e.g., species of *Colletia*, *Chenopodium* [Fig. C70], and *Nicotiana*), as judged from the distribution of closest relatives and the respective areas of the families. This includes 12% that are likely indigenous to Chile (e.g., species of *Drimys* [Fig. C29], *Gunnera* [Figs. C75 and C76], *Mimulus*, *Peperomia* [Figs. C27 and C28], *Pernettya* [Fig. C71], and *Spergularia*), where the genera are well represented with analogous species, and 10% to the Neotropics more generally (e.g., most species of Asteraceae, *Juania* [Fig. C30], and *Ugni* [Fig. C81]). Pacific colonizers (in general) account for the other 18% (e.g., species of *Dendroseris* [Figs. C36–C52], *Dichondra*, *Hedyotis*, *Nertera*, *Ranunculus* [Fig. C82], and *Santalum*). Making up about half of this figure (i.e., 4% each) are species associated with Australia (e.g., *Wahlenbergia* [Figs. C68 and C69]) and New Zealand (e.g., *Calystegia*, *Coprosma*, and *Haloragis* [Fig. C78]). At the specific level, almost all native (*not* endemic) species are shared with Chile.

Among the families that have numerous endemic genera, Asteraceae (four genera and 27 species) and Poaceae (two genera and six species) stand out; most constituent genera from these two families are derived from South American colonists. Genera with several endemic species are *Gunnera* (three species) (Figs. C75 and C76) and *Peperomia* (four species) (Figs. C27 and C28) with their ancestral colonists from Chile, *Chenopodium*

Table 16.1 Autochthonous Angiosperm Flora Checklist of the Juan Fernández Islands (with Family in Bold and Genus and Species in Italics) Indicating Probable Place of Origin, Method of Arrival of the First Colonizers, Fruit Type, Fruit Length, Dispersal Unit Length, and Current Dispersal

Family, genus	No. of species	Species: Native	Species: Endemic	Place of origin	Method of arrival	Fruit type	Fruit length (mm)	Dispersal unit length (mm)	Dispersal
Apiaceae									
Apium	2	*A. australe*	*A. fernandezianum*	Pacific area	Bird internal	Schizocarp	2	2	Anemochory
Centella	1	*C. asiatica*		S. America	Bird internal/ bird external	Schizocarp	2.5	2.5	Endozoochory
Eryngium	4		*E. bupleuroides*	S. America	Bird external	Schizocarp	2	2	Epizoochory
			E. ×fernandezianum						
			E. inaccessum						
			E. sarcophyllum						
Arecaceae									
Juania[a]	1		*J. australis*	Neotropics	Bird internal	Drupe	15	15	Autochory
Asteraceae									
Abrotanella	1	*A. linearifolia*		S. America	Bird external	Achene	1.5	1.5	Autochory
Centaurodendron[a]	2		*C. dracaenoides*	Neotropics	Bird external	Achene	5	5	Autochory
			C. palmiforme						
Dendroseris[a]	11		*D. berteroana*	Pacific area	Bird external	Achene	4	4	Autochory
			D. gigantea						
			D. litoralis						
			D. macrantha						
			D. macrophylla						
			D. marginata						
			D. micrantha						
			D. neriifolia						
			D. pinnata						
			D. pruinata						
			D. regia						

Erigeron	5		*E. fernandezia*	Neotropics	Bird external/ wind	Achene	2	2	Anemochory
			E. ingae						
			E. luteoviridis						
			E. stuessyi						
			E. rupicola						
Gamochaeta	1	*G. chamissonis*		S. America	Bird external	Achene	1	1	Anemochory
Lagenophora	1	*L. hariotii*		Neotropics	Bird external	Achene	2.5	2.5	Autochory
Robinsonia[a]	8		*R. berteroi*	S. America	Bird external	Achene	3.5	3.5	Autochory
			R. evenia						
			R. gayana						
			R. gracilis						
			R. macrocephala						
			R. masafuerae						
			R. saxatilis						
			R. thurifera						
Taraxacum	2	*T. fernandezianum*		S. America	Bird external/ wind	Achene	3	3	Anemochory
		T. subspathulatum							
Yunquea[a]	1		*Y. tenzii*	S. America	Bird external	Achene	6	6	Autochory
Berberidaceae									
Berberis	2		*B. corymbosa*	S. America	Bird internal	Berry	8	8	Endozoochory
			B. masafuerana						
Boraginaceae									
Selkirkia[a]	1		*S. berteroi*	S. America	Bird internal	Nutlet	25	25	Epizoochory
Brassicaceae									
Cardamine	3	*C. chenopodioides*	*C. kruesselii*	S. America	Bird internal	Silique	35	2	Autochory
		C. flaccida							
Bromeliaceae									
Greigia	1		*G. berteroi*	S. America	Bird external	Dry berry	17	17	Autochory
Ochagavia	1		*O. elegans*	S. America	Bird external	Dry berry	10	10	Autochory
Campanulaceae									
Lobelia	1	*L. anceps*		S. America	Bird external	Capsule	7	0.7	Autochory

Table 16.1 (cont.)

Family, genus	No. of species	Species: Native	Species: Endemic	Place of origin	Method of arrival	Fruit type	Fruit length (mm)	Dispersal unit length (mm)	Dispersal
Wahlenbergia	5		*W. berteroi*	Pacific area (Australia)	Bird external	Capsule	4	0.5	Anemochory
			W. fernandeziana *W. grahamiae* *W. masafuerae* *W. tuberosa*						
Caryophyllaceae									
Spergularia	2		*S. confertifolia* var. *confertifolia* *S. confertifolia* var. *polyphylla* *S. masafuerana*	Chile	Bird internal/ bird external	Capsule	4	1.5	Autochory
Chenopodiaceae									
Chenopodium	3		*C. crusoeanum* *C. nesodendron* *C. sanctae-clarae*	S. America	Bird external	Utricle	2	2	Anemochory
Convolvulaceae									
Calystegia	1	*C. tuguriorum*		Pacific area (New Zealand)	Bird internal	Capsule	8	2	Autochory
Dichondra	1	*D. sericea*		S. America	Bird internal	Capsule	3	2	Autochory
Cyperaceae									
Carex, *Cyperus*, *Eleocharis*, *Machaerina*,	18	*Carex phalaroides* *Cyperus*	*Carex berteroniana* *Carex fernandezensis*	S. America	1 Bird internal, 4 bird external, 2	Achene	2	2	Anemochory, endozoochory

Oreobolus, *Scirpus, Uncinia*		*eragrostis* *Cyperus reflexus* *E. fuscopurpurea* *O. obtusangulus* *S. cernuus* *S. nodosus* *U. phleoides* *U. tenuis*	*Carex stuessyi* *M. scirpoidea* *U. aspericaulis* *U. costata* *U. douglasii* *U. macloviformis*		bird internal/ bird external				
Ericaceae									
Empetrum	1	*E. rubrum*		S. America	Bird internal	Drupe	9	9	Endozoochory
Pernettya	1		*P. rigida*	Chile	Bird internal	Berry	8	8	Endozoochory
Escalloniaceae									
Escallonia	1		*E. callcottiae*	S. America	Bird external	Capsule	4	1	Anemochory
Euphorbiaceae									
Dysopsis	1		*D. hirsuta*	S. America	Bird internal	Schizocarp	1	1	Active ballistic
Fabaceae									
Sophora	2		*S. fernandeziana* var. *fernandeziana* *S. fernandeziana* var. *reedeana* *S. masafuerana*	S. America	Water	Legume	60	8	Autochory
Gunneraceae									
Gunnera	3		*G. bracteata* *G. masafuerae* *G. peltata*	Chile	Bird internal	Drupe	3	3	Autochory
Haloragaceae									
Haloragis	2		*H. masafuerana* var. *masafuerana* *H. masafuerana* var. *asperrima* *H. masatierrana*	Pacific area (New Zealand)	Bird internal	Nutlet	3	3	Autochory
Iridaceae									
Herbertia	1	*H. lahue*		Chile	Bird internal	Capsule	2	3.5	Autochory
Libertia	1	*L. chilensis*		Chile	Bird internal	Capsule	10	2	Autochory

Table 16.1 (cont.)

Family, genus	No. of species	Species: Native	Species: Endemic	Place of origin	Method of arrival	Fruit type	Fruit length (mm)	Dispersal unit length (mm)	Dispersal
Juncaceae									
Juncus	5	*J. capillaceus*		S. America	Bird external/ wind	Capsule	4	0.8	Anemochory
		J. imbricatus *J. pallescens* *J. planifolius* *J. procerus*							
Luzula	1		*L. masafuerana*	S. America	Bird external	Capsule	3	1.5	Anemochory
Lamiaceae									
Cuminia[a]	2		*C. eriantha* *C. fernandezia*	S. America	Bird internal	Nutlet	12	4	Autochory
Lactoridaceae									
Lactoris[a]	1		*L. fernandeziana*	S. America	Wind	Follicle	3	0.7	Anemochory
Loranthaceae									
Notanthera	1	*N. heterophylla*		S. America	Bird external	Drupaceous	6	6	Epizoochory, endozoochory
Myrtaceae									
Myrceugenia	1		*M. schulzei*	S. America	Bird internal	Berry	9	9	Endozoochory
Myrteola	1	*M. nummularia*		S. America	Bird internal	Berry	8	8	Endozoochory
Nothomyrcia[a]			*N. fernandeziana*	S. America	Bird internal	Berry	8	8	Endozoochory
Ugni	1		*U. selkirkii*	Neotropics	Bird external	Berry	8	8	Endozoochory
Orchidaceae									
Gavilea	1		*G. insularis*	S. America	Wind	Capsule		0.05	Anemochory
Orobanchaceae									
Euphrasia	1		*E. formosissima*	S. America	Bird external	Capsule	3	0.8	Anemochory
Phrymaceae									
Mimulus	1	*M. glabratus*		Chile	Bird external	Capsule	4	0.5	Anemochory
Piperaceae									
Peperomia	4	*P. fernandeziana*	*P. berteroana* subsp. *berteroana* *P. margaritifera* *P. skottsbergii*	Chile	Bird external	Drupe	1	1	Epizoochory

Plantaginaceae									
Plantago	3	*P. australis* *P. firma*	*P. fernandezia*	Neotropics	Bird external	Capsule	3	1.5	Anemochory
Poaceae									
Agrostis, *Chusquea*, *Danthonia*, *Megalachne*,[a] *Piptochaetium*, *Podophorus*[a]	9	*D. chilensis* var. *chilensis* *D. malacantha* *Piptochaetium bicolor*	*A. masafuerana* *C. fernandeziana* *M. berteroana* *M. masafuerana* *M. robinsoniana* *Podophorus bromoides*	S. America	2 Bird internal, 9 bird external	Caryopsis	4	4	Anemochory, endozoochory
Ranunculaceae									
Ranunculus	1		*R. caprarum*	Pacific area	Bird internal/ bird external	Achene	7	7	Autochory
Rhamnaceae									
Colletia	1		*C. spartioides*	S. America	Water	Drupe	10	10	Autochory
Rosaceae									
Acaena	1		*A. masafuerana*	S. America	Bird external	Achene	1	1	Epizoochory
×*Margyracaena*	1		×*M. skottsbergii*	S. America	Bird external	Drupe	6	6	Autochory
Margyricarpus	1		*M. digynus*	S. America	Bird external	Drupe	6	6	Autochory
Rubus	1	*R. geoides*		S. America	Bird internal	Aggregate	4	18	Endozoochory
Rubiaceae									
Coprosma	2		*C. oliveri* *C. pyrifolia*	Pacific area (New Zealand)	Bird internal/ bird external	Drupaceous	20	20	Autochory
Galium	1		*G. masafueranum*	S. America	Bird internal	Schizocarp	2	2	Endozoochory
Hedyotis	1	*H. salzmannii*		Pacific area	Bird internal/ bird external	Capsule	5	5	Endozoochory
Nertera	1	*N. granadensis*		S. America	Bird internal	Drupe	3	3	Endozoochory
Rutaceae									
Zanthoxylum	2		*Z. externum* *Z. mayu*	S. America	Bird internal	Follicle	3	3	Endozoochory

Table 16.1 (cont.)

Family, genus	No. of species	Species: Native	Species: Endemic	Place of origin	Method of arrival	Fruit type	Fruit length (mm)	Dispersal unit length (mm)	Dispersal
Salicaceae									
Azara	1		*A. serrata* var. *fernandeziana*	S. America	Bird internal	Berry	8	8	Endozoochory
Santalaceae									
Santalum	1		*S. fernandezianum*	Pacific area	Bird internal	Drupe	4	4	Endozoochory
Solanaceae									
Nicotiana	1		*N. cordifolia* subsp. *cordifolia* *N. cordifolia* subsp. *sanctaclarae*	S. America	Bird external/ wind	Capsule	10	0.3	Anemochory
Solanum	2	*S. pentlandii* subsp. *interandinum*	*S. fernandezianum*	S. America	Bird internal	Berry	6	6	Endozoochory
Urticaceae									
Boehmeria	1		*B. excelsa*	S. America	Bird internal/ bird external	Achene	1.5	1.5	Anemochory
Parietaria	1	*P. debilis*		S. America	Bird internal	Achene	1	1	Anemochory
Urtica	2		*U. glomeruliflora* *U. masafuerae*	S. America	Bird external	Achene	1	1	Anemochory
Verbenaceae									
Rhaphithamnus	1		*R. venustus*	S. America	Bird internal	Drupe	15	15	Autochory
Winteraceae									
Drimys	1		*D. confertifolia*	Chile	Bird internal	Follicle	7	2	Autochory

Note: Characteristics are generally considered at the generic level. Empty cell in *Gavilea* indicates missing information. Where there is more than one species per genus, the general character state or average (e.g., size), is given; if some species are very different, the individual character states are given.

[a] Endemic genera. Varieties are included only when the variety is the only representative of the species occurring in the archipelago. The place of origin, methods of arrival, and dispersal are all presumed.

(three species) (Fig. C70) and *Eryngium* (four species) (Fig. C34) from South America, and *Wahlenbergia* (five species) (Figs. C68 and C69) from Australia. A number of genera have two endemic species each: *Berberis*, *Cuminia* (Fig. C77), *Sophora* (Fig. C74), *Urtica*, and *Zanthoxylum* (Fig. C84) are all derived from South America, *Spergularia* from Chile, and *Haloragis* and *Coprosma* from New Zealand.

At the generic level, about 77% of the Juan Fernández genera are shared with Chile. Notable exceptions are *Boehmeria* (Fig. C86), *Coprosma*, *Zanthoxylum* (Fig. C84), *Haloragis* (Fig. C78), ×*Margyracaena*, and *Santalum*. Fifteen percent of the Fernandezian angiosperm genera are endemic (see Chapter 5).

Biogeographically, the Juan Fernández Islands have been considered a separate floristic region, generally placed within the Neotropical kingdom (e.g., Engler 1882; Turrill 1959; Thorne 1963; Mattick 1964; Takhtajan 1969; Good 1974). Later, Takhtajan (1986) took a different view, considering the flora most similar to that of the Chile-Patagonian region, thus including the Fernandezian region within his Holoantarctic kingdom. Our analysis of the literature positing origins of the Fernandezian species clearly suggests that the overall affinities of the Juan Fernández flora are with the South American continent as a whole (Skottsberg 1934, 1956; van Balgooy 1960, 1971; Takhtajan 1986). Asteraceae, a worldwide family with its cradle in South America (Bremer 1994; Stuessy et al. 1996), is a particularly notable family with several genera and species endemic to the archipelago.

According to Carlquist (1974), relicts in the strictest sense are scarce on oceanic islands. If primitive forms have migrated to islands and the mainland remnants have become extinguished recently, the insular representatives appear as relicts. The Juan Fernández Archipelago is outstanding because it has two relicts: *Lactoris fernandeziana* (Fig. C26) and *Thyrsopteris elegans* (Fig. 9.6), regrettably both considered endangered (Stuessy et al. 1998b, 1998c, 1998d; see also Chapters 5 and 9). *Lactoris fernandeziana* is a paleoherb or an Archaeangiosperm that constitutes a monotypic dicotyledoneous family; it is now exclusively confined to Robinson Crusoe Island, but around 70 to 90 million years ago it was more widely distributed in the southern hemisphere (Stuessy et al. 1998a; Macphail et al. 1999b). Today there are only a few populations isolated in the highest reaches of the island, but fortunately, each contains dozens of plants (Bernardello et al. 2001). The second relict, *T. elegans*, is a monotypic genus of ferns in Thyrsopteridaceae that was distributed in Mesozoic forests 80 to 170 million of years ago in both hemispheres (Moran 1993), but now lives solely on the two large Juan Fernández Islands (Robinson Crusoe and Alejandro Selkirk), where it is an important component of the upper montane forest communities (Greimler et al. 2002a; see Chapter 6).

Some endemic taxa are likely extinct (see Chapters 5 and 9): the Fernandezian sandalwood *Santalum fernandezianum* (the last tree was photographed in 1908; Fig. 9.5A) (Skottsberg 1910; Stuessy et al. 1998b, 1998c), the monotypic endemic grass genus *Podophorus bromoides* (Baeza et al. 2002), *Eryngium sarcophyllum, Robinsonia macrocephala*, and *R. berteroi* (but see note to Table 5.1). ×*Margyracaena skottsbergii* is extinct in the wild.

Mechanisms of Arrival

Birds seem by far to have been the most important dispersal agents of colonizers to the archipelago (ca. 90%). Close to half (46%) of the bird-dispersed species likely reached the islands externally via seeds or fruits mechanically attached to feathers by barbs and awns (e.g., Poaceae, *Acaena*), by viscid appendages (e.g., *Peperomia*), or simply embedded in mud on feet. The other significant portion of species (35%) likely was carried internally as bird-ingested seeds or fruits. The method of bird dispersal for the remaining 9% of the bird dispersed species is uncertain. The two other long-distance dispersal agents are (1) wind (2%) for species with small seeds (e.g., Orchidaceae, *Nicotiana*, or spores for the pteridophytes and bryophytes, not tabulated) or fruits with appendages that promote flying (e.g., some Asteraceae) or (2) oceanic drift (2%; only *Colletia* and *Sophora*). Finally, 6% of the species were carried either by wind or by birds.

Given that hermaphroditic-flowered species are the most frequent on the archipelago, few telling correlations were expected between the method of arrival and the sexual systems. Figure 16.1A shows that all the dispersal methods are represented among the hermaphroditic-flowered species. Similarly, but not surprisingly given the notable lack of pollinators, there were no associations between the presumed pollination type of the colonizers and the method of arrival. Additionally, comparisons of the current pollination system (known for 56% of the flora) (Bernardello et al. 2001) with the presumed method of arrival (Fig. 16.1B) show little association. This is again not surprising given that anemophily and ornithophily are the only pollination syndromes documented. Association with bird arrival is so pervasive that it would be surprising if most of the bird-pollinated species had not also arrived by bird dispersal. Similarly, species currently considered wind-pollinated mostly arrived carried by birds. The fundamental point is that given that there is no logical reason to associate methods of dispersal and pollination syndromes, there is no corresponding association between method of arrival and pollination system. A comparison supports this contention: most wind- and bird-pollinated genera arrived by birds, rarely by wind (Fig. 16.1B). Thus the method of dispersal is, not surprisingly, independent of the method of pollination.

Most species with bright-colored flowers (Fig. 16.1C), that is, yellow, white, red, and blue, seem to have arrived through avian transport. Similarly, the smaller, green-flowered species also mostly arrived by bird dispersal. And the same is true for growth form; most species that are annual herbs (Fig. 16.1D) likely also arrived via avian transport. But, here again, the lack of significant diversity in dispersal type, with 90% of the species presumed to be originally bird dispersed, means that there is not much variability to parse among flower color categories or growth forms.

As discussed earlier, migration to islands is chiefly governed by chance (Carlquist 1981; Martin 1981; Turner et al. 1998), although habitat availability is a significant factor determining the composition and size of a flora (Kingston et al. 2003). The disproportionate occurrence of certain attributes in the flora (e.g., dry fruits, small to medium-sized dispersal units) may be evidence for dispersal-mediated species selection. Given the fine sieve associated with establishment on islands, it would be surprising if

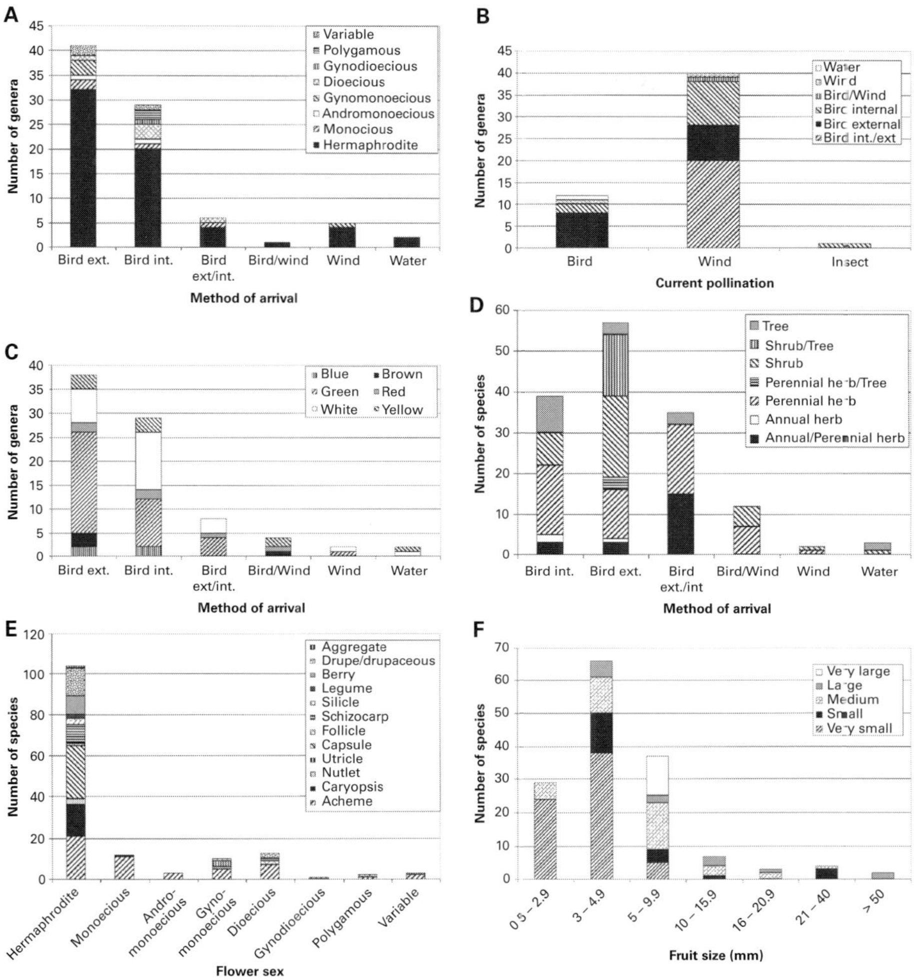

Figure 16.1 Histograms comparing the frequency of the following in the Juan Fernández angiosperm flora: (A) method of arrival with sexual system; (B) method of arrival with current pollination system; (C) method of arrival with flower color; (D) method of arrival with habit; (E) fruit type with flower sex, (F) fruit size with flower size. (From Bernardello et al. 2006, fig. 2; reprinted with permission of the *Canadian Journal of Botany*.)

the species with a propensity for more effective (long-distance) dispersal were not in the majority of island plants.

Among the several agents of long-distance initial dispersal, land or sea birds seem to have been the most important vector in bringing plant colonizers to this archipelago (around 90% of the species). Long-distance bird transport of fruits and seeds, either externally or internally, is generally regarded as the most common long-distance dispersal mechanism (cf. Carlquist 1974; Cox and Moore 1993; Whelan et al. 2008). In particular, for the high islands of the Pacific, migratory birds are generally suggested as the primary method of arrival for most groups of angiosperms (Ridley 1930; Carlquist

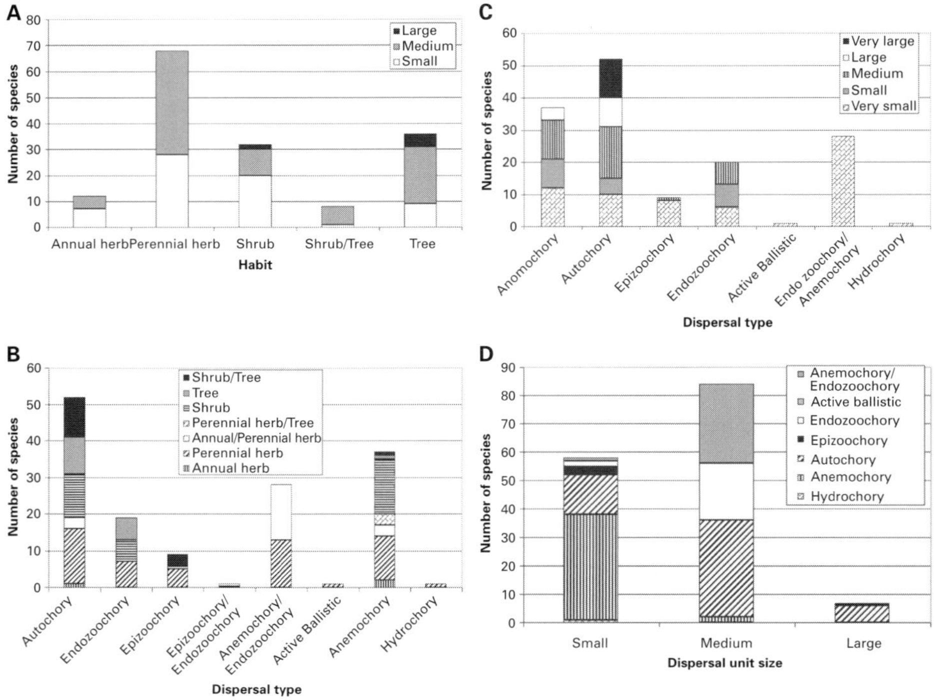

Figure 16.2 Histograms comparing the frequency of the following in the Juan Fernández angiosperm flora: (A) habit with dispersal unit size; (B) dispersal types with habit; (C) dispersal types with flower size; (D) dispersal types with dispersal unit sizes. (From Bernardello et al. 2006, fig. 3; reprinted with permission of the *Canadian Journal of Botany*.)

1974; Porter 1976; Ono 1991). Carlquist (1974) supported the primacy of bird dispersal for the Juan Fernández Archipelago, a conclusion supported by our data as well. The one interesting caveat is that there are only a few bird species currently associated with these small and isolated islands. We employed the word *currently* because, of course, we do not know what the bird fauna was like millions of years ago when the early colonization of the islands was taking place.

The other primary initial dispersal agent emerging from our analysis is air flotation. Carlquist's (1974) general estimates for islands of the eastern Pacific and Polynesia, as well as those for the Juan Fernández Archipelago, attribute less importance than we do to anemochorous dispersal. Finally, oceanic drift would have been negligible, although it has been proposed as significant for several Indian Ocean and other Pacific Ocean islands (Guppy 1890; Carlquist 1974; Murray 1986; Green 1999).

Carlquist's comprehensive work (1974) allows us to make comparisons with other archipelagos, in particular, the Galápagos Islands. The proportions for different arrival modes are analogous percentages, except for drift and air flotation, where Carlquist (1974) proposes higher percentages for the Galápagos than we think evident on Juan Fernández. The Juan Fernández Islands are closer to the South American continent than the Galápagos, on which basis we might expect passive mechanisms such as drift to account for more dispersal. However, the patterns of ocean currents (the Humboldt Current flows north from the Antarctic along the west coast of South America to southern Ecuador, then west; Akin 1991) and the prevailing winds (trade winds circulate counterclockwise around a high-pressure center located at about the Juan Fernández Islands, turning west near the equator; Akin 1991; see Chapter 4), provide obvious explanations for the differences found.

Fleshy fruits are adapted for vertebrate dispersal (van der Pijl 1982), are very common in most tropical and subtropical rain forests, and are moderately common in some north temperate forests and some south temperate moist forests (Willson et al. 1989). Given that internal dispersal by birds is the most common long-distance dispersal mechanism for oceanic islands (Ridley 1930; Carlquist 1974; Burrows 1994; Lord 1999), it might be expected that the Juan Fernández flora would contain a high proportion of species possessing fleshy fruits. However, our data show that fleshy-fruited forms account for only 20% of the flora. The explanation does not seem to lie in the nature of the primary source flora, in that data from the flora for mainland temperate forest species of Chile and Argentina show that approximately 42% bear fleshy fruits (Donoso Zegers 1993). It is not clear why there are not more species on the Juan Fernández Archipelago with fleshy fruits.

Establishment also requires successful sexually reproducing populations. This suggests that the arrival of single propagules will mostly not result in successful establishment. Given that a bit more than one-third of the bird-dispersed species are endozoochorus and that hydrochory (the other category usually considered beneficial for multiple-propagule dispersal) is considered minimal at best, establishment issues are of some interest. For hermaphroditic, self-compatible species, multiple independent bouts of dispersal of single individuals may be sufficient for establishment (Baker 1967), but for dioecious or self-incompatible species, more than one disseminule or dispersal event is usually needed. Or there must be some kind of "leaky" system. There is increasing evidence of leaky systems both in self-incompatible ("leaky self incompatibility" or "partial self compatibility") and in dioecious taxa ("leaky dioecy") (e.g., Crawford et al. 2015; Anderson et al. 2015, respectively; also see Chapter 12 on reproductive biology for further discussion of these topics). In this context, perennial plants have been alleged to possess an advantage over annuals because their longevity increases the likelihood of "finding" a mate, achieving successful outcrossing, and securing sufficient pollination to produce seeds to establish and maintain a species (Wallace 1895; Böhle et al. 1996). Data available on the Juan Fernández flora (Bernardello et al. 2001) support these hypotheses: the flora is typically composed of perennials, most species are hermaphroditic, and the vast majority of species we have studied (see Chapter 12) are self-compatible.

Current levels of endemism in oceanic islands are in large part the consequence of the ease with which certain highly dispersible groups can cross ocean gaps (McGlone et al. 2001). In addition, the presence of certain taxa in island systems from a given area may indicate an ancestral capability for long-distance dispersal. For instance, Asteraceae, mainly wind dispersed, have four endemic genera in the Juan Fernández Archipelago and the Galápagos Islands (Wiggins and Porter 1971) and seven in Hawaii (Wagner et al.

1990). Also, the wind- and bird-dispersed Cyperaceae and Poaceae have several endemic species in these three archipelagos.

The Brassicaceae, Boraginaceae, Caryophyllaceae, Convolvulaceae, Euphorbiaceae, Fabaceae, Iridaceae, Juncaceae, Lamiaceae, Myrtaceae, Rhamnaceae, Rubiaceae, and Verbenaceae are found in all three archipelagos, suggesting significant capacity for long-distance dispersal and establishment. However, members of the families Boraginaceae, Convolvulaceae, and Euphorbiaceae are underrepresented in the Juan Fernández Archipelago based on comparisons with the Galápagos and Hawaiian Archipelagos and with the source flora.

Interestingly, the highly dispersible family Orchidaceae (with exceedingly small seeds, apparently dispersed as easily as the spores of often-ubiquitous ferns, and a family that is present in Hawaii and the Galápagos Islands) is rare in the Juan Fernández Archipelago, with only one species endemic (and that one is restricted to the more distant and younger Alejandro Selkirk Island). The absence of insect pollinators or that the symbiotic fungi frequently needed for their germination may not be present may provide explanations for the virtual absence of orchids on the Juan Fernández Archipelago. Curiously, the families Amaranthaceae, Cucurbitaceae, and Apocynaceae that are widespread in the South American flora are not present on the Juan Fernández Archipelago, though there are representatives in the Galápagos and Hawaiian Islands.

The following nonendemic genera, shared by the Juan Fernández and Galápagos Islands, suggest that they too are highly dispersible: *Centella* (Apiaceae), *Dichondra* (Convolvulaceae), *Nicotiana* (Solanaceae), *Parietaria* (Urticaceae), *Pernettya* (Ericaceae), and *Chenopodium* (Chenopodiaceae). In addition, *Zanthoxylum* (Rutaceae), *Lobelia* (Campanulaceae), *Peperomia* (Piperaceae), *Plantago* (Plantaginaceae), *Ranunculus* (Ranunculaceae), and *Solanum* (Solanaceae) are shared by these two archipelagos and the remote Hawaiian Archipelago as well. However, these latter genera are very different from each other in many respects, including likely dispersal method. Thus the insights they might provide come from study of individual cases, not from the group.

Summary of Foundational Data and Observations

Information on fruit features was taken from Hemsley (1884), Johow (1896), and Skottsberg (1921, 1951) and supplemented with personal observations in several expeditions to the archipelago (cf. Bernardello et al. 2001). The works by Gay (1845–1854), Reiche (1896–1911), and Moore (1983) were also consulted. The variables measured or scored included fruit length, dispersal unit size length, and fruit type (as schizocarpic fruit, achene, berry, nutlet, silique, capsule, drupe/drupaceous, legume, follicle, aggregate, or caryopsis). Measurements were taken of the dispersal units (whole fruit or only the seeds). The fruit size categories were defined as small, 0 to 2.0 mm; medium, 2.5 to 10 mm; large, 11 to 30 mm; and very large, 40 mm and above. We selected the categories by using the data for all fruit sizes and by non-overlapping intervals.

Categories of dispersal of the current flora basically follow van der Pijl (1982) and are based on our own observations of the species of the archipelago. Species were assigned to one of the following six classes based on the characteristics of their fruits and seeds: (1) endozoochory (dispersed by fruit-eating birds), (2) epizoochory (dispersed passively by birds that carry fruits or seeds attached to feathers), (3) hydrochory (water dispersal), (4) anemochory (wind dispersal), (5) autochory (autonomous passive dispersal), and (6) active ballistic (by tension in dead hydroscopic tissues).

To compare these traits with other features, such as flower sex, flower color, pollination of the colonizers, and current pollination system (either observed or inferred from morphology), data were taken from Bernardello et al. (2001).

There are various features of fruits that provide some insight into the dispersal of the present flora. Most species (80%) have dry fruits; achenes are the most common (34%, chiefly Asteraceae and Cyperaceae), followed by capsules (18%, e.g., Campanulaceae, Convolvulaceae, and Juncaceae), caryopses (10%, Poaceae), and schizocarpic fruits (7%, e.g., Apiaceae, Euphorbiaceae, and Rubiaceae). The most common reproductive class, hermaphroditic-flowered species, have all fruit types (Fig. 16.1E), but in the group with more unusual sexual systems, monoecious, andro- and gynomonoecious, and dioecious species, achenes predominate.

Fleshy fruits are comparatively uncommon and are represented by two types: drupe or drupaceous (13%, e.g., Gunneraceae, Piperaceae, and Rhamnaceae) and berry (10%, e.g., Ericaceae, Myrtaceae, and *Solanum*). Fleshy fruits are present exclusively in perennial species, whereas achenes and caryopses are also found in annual herbs.

An analysis of the size of the dispersal units shows that the vast majority of the species have medium (57%) or small (38%) units, and only 5% have large units (e.g., *Juania*, *Rhaphithamnus*, and *Coprosma*). In general, the larger the flower, the larger the fruit (Fig. 16.1F); that is, very small and small flowers produce fruits that mainly range from 0.5 to 5 mm, whereas larger flowers tend to be associated with average/medium-sized fruits. In terms of dispersal unit size compared with the habit of the species (Fig. 16.2A), large and medium-sized units are mainly present in shrubs and trees.

Four classes of dispersal typify most of the current flora. A substantial proportion of species (35%) have autochorous dispersal, with anemochory (25%) and ornithochory (20%, including epi- and endozoochory) following in magnitude. Cyperaceae and Poaceae, considered wind or bird dispersed, account for 18% of the flora. Active ballistic are unusual; only *Dysopsis hirsuta* is dispersed in this way, whereas there are no hydrochorous species. Given that species with different dispersal modes are very diverse in their fruit types, no correlations can be drawn, except for the aggregate fruits and berries that are, as expected, among the endozoochorous species.

Both anemochorous and autochorous species are mainly perennial (Fig. 16.2B), and they have medium to large (Fig. 16.2C) unisexual flowers. Anemochorous species, as expected, generally have small dispersal units (Fig. 16.2D). However, species with large dispersal units are mainly autochorous, and medium-sized ones are either autochorous or ornithochorous. Most autochorous species have predominantly brightly colored flowers, whereas those that are anemochorous bear dull-colored flowers.

Interestingly, the four endemic genera of Asteraceae that possess achenes with reduced dispersal ability stand in contrast with most of the members of the family with strong dispersal ability. Effectively, *Centaurodendron*, *Dendroseris*, *Robinsonia*, and *Yunquea* species have bristles that are reduced or break easily. In addition, in *Centaurodendron* and some *Robinsonia* species, the involucral bracts tend to retain the achenes within the capitula, whereas the achenes of *Yunquea* and several *Dendroseris* species have irregular shapes and rugose surfaces, features that would reduce their dispersibility even further.

Correlating the geographical origin of the species with current means of dispersal, South American species are mainly dispersed by wind and birds or have autochorous dispersal. Anemochorous species are also mostly South American.

Current Dispersal in the Flora and Invasives

An assessment of the *current* dispersal mechanisms extant for the Fernandezian flora shows that the majority of the species have abiotic dispersal, chiefly autochory and anemochory. This fact may reflect either the ancestral dispersal capability of the colonizers or the absence of a conspicuous fauna to disperse seeds and fruits. The few land and sea bird species associated with these islands (Lönnberg 1921; Brooke 1987) disperse 20% of the plant species in two ways: actively by carrying seeds and fruits in their digestive tracts (endozoic) or passively by adhesion to their external body surfaces (ectozoic). Unfortunately, the birds also disperse the most aggressive noxious introduced invasives, such as *Rubus ulmifolius* (Figs. 8.3 and 8.4), *Aristotelia chilensis* (Figs. 8.1 and 8.2), and *Ugni molinae* (Figs. 8.5 and 8.6) (Smith-Ramírez et al. 2013). The invasives have severely affected the native vegetation, leading to a significant decrease of endemic plants in the montane forests and native shrublands (Dirnböck et al. 2003; see also Chapter 8). Unfortunately, the newly arrived and potentially dangerous alien species *Lantana camara* and *Lonicera japonica* (Swenson et al. 1997) are also bird dispersed. In addition to the damage that these aggressive invasives cause in terms of displacing the autochthonous species, there is a perhaps even more pernicious aspect of their presence. That is, the fruits of these invasives are often preferred by the native bird dispersers. This preference leads to decreased native plant dispersal and simultaneous enhancement of the dispersal of the introduced invasives, spreading the latter further and faster and at the expense of the natives.

A notable feature of many island colonizers, both animals and plants, that have become adapted to oceanic islands is the post-establishment loss of dispersal ability (Darwin 1855; Zimmerman 1948; Carlquist 1965, 1974; Roff 1990; Eliasson 1995). Obviously, dispersal mechanisms and dispersal ability are likely to be selected against because those that retain good dispersal mechanisms are more likely to have more propagules lost to the surrounding sea than those with poor dispersal. However, retention of good dispersal ability of endemic species might constitute a fixation of a character suite, dispersal ability, that served the plants not only in initial colonization but also in subsequent colonization in multi-island (e.g., the Canary or Hawaiian Islands) stepping-stone (in age and distance) archipelagos. This contention is supported by a recent study on the Canary Islands (Vazačová and Münzbergová 2014) in which the comparison of dispersal ability of endemics and closely related nonendemics failed to provide evidence to support the hypothesis regarding the loss of dispersal ability of endemics versus nonendemic species pairs (with the exception of endozoochory). The comparison of 27 pairs of endemic and nonendemic species suggested that a reduction of dispersal ability of island species may not be as general as formerly proposed. For the species on the Canaries that the authors studied, a number of endemic species had the same or better dispersal ability than their nonendemic congeners.

Although most Fernandezian species seem to have retained the dispersal mode of their ancestral immigrants, the four endemic genera of Asteraceae (see earlier) are a noteworthy exception. This family is known for having lost dispersibility in other archipelagos (Carlquist 1966, 1974; Cody and Overton 1996). The primary dispersal mode of its members is anemochory, but there is a trend toward diaspores with lower dispersibility in oceanic island species. In the particular case of the Juan Fernández Islands, Carlquist (1966, 1974) suggested that the first Asteraceae colonizers adapted to the wet forest and then subsequently lost dispersibility chiefly due to this ecological shift.

Conclusion

In many ways the diversity of the Juan Fernández flora is impressive given the relative youth of the islands, the remoteness of the archipelago, and the small bird fauna associated with them. The native flora is small enough that we could study it all either in the field or from specimens. Thus we are able to work with the whole flora of an archipelago and have produced the first comprehensive analysis of dispersal and origins of an island flora. Not surprisingly, and in concert with many previous studies, the most common source is the closest, and the most common dispersal modes are wind and bird. In contrast to some other studies, however, is the fact that animal dispersal is more limited, perhaps analogously to the limited biotic pollination, both situations induced by a relatively young, distant, and small archipelago, with consequently fewer animal dispersers. The biota of the Juan Fernández Archipelago is interesting from many vantage points, including as a paradigm for establishment and persistence in the face of strong distance, size, and age "filters."

17 Modeling Species Diversity

Walter A. Sontag, Jr., Tod F. Stuessy, and Ulf Swenson

Oceanic islands represent excellent places for investigating processes of evolution and development of organismic diversity (Warren et al. 2014). Islands offer special challenges because of their dynamic formation, erosion, and subsidence overlain by colonization, establishment, and survival of new biota over millions of years. Although contiguous land masses comprise by far most of the world's terrestrial areas, insular species diversity and endemism are much higher than would be expected by just the limited space available. This makes islands important for attempting to understand species diversity. Especially fascinating are small oceanic islands far from any mainland, such as is the case with Robinson Crusoe and Alejandro Selkirk Islands of the Juan Fernández Archipelago.

An attempt to understand mechanisms shaping species diversity in islands was undertaken by MacArthur and Wilson in their island equilibrium theory, the full extension of which was published in 1967. Unknown to most scientists, however, about twenty years before (1948), Eugene Gordon Munroe, an inspired entomologist (Vives Moreno 2008), had tackled the mystery of insular species richness and independently designed a similar concept in his dissertation, publishing only later in an obscure journal (Munroe 1963; see Lomolino et al. 2010). The MacArthur-Wilson model, which emphasizes sizes of islands and distance from source areas, has turned out to be highly stimulating for ecological modeling but insufficient for precisely explaining island species diversity. Well known, even to these authors, is that the theory lacks information on the critical dimensions of ecology, age of islands, speciation within the archipelago, and reproductive characters of species.

Recently, more dynamic models of estimating island species diversity have been suggested independently by Stuessy (2007) and Whittaker et al. (2008). Similar in their rationale, both take into account the geological and physical history of islands. These considerations render it more challenging, therefore, to analyze quantitatively the diversity of species on small islands far from mainland sources. Stuessy (2007) addressed biological diversity with particular reference to the Juan Fernández Archipelago. This island group is both geographically extremely remote, without contact with other land masses during its existence, and highly restricted in size and number of islands. Whittaker et al. (2008) focused in detail on present species diversity in different organismic groups in several archipelagos. Both approaches have turned the rather simple MacArthur-Wilson model into a much more differentiated rationale, especially by adding the factor of time. In this chapter we begin by reviewing the dimensions of the MacArthur-Wilson model, and then we examine the new approaches with reference to the development of species diversity of angiosperms in the Juan Fernández Archipelago.

The Equilibrium Model

A very important step for understanding the biogeographical development of oceanic islands was the equilibrium model of MacArthur and Wilson (1967). According to this theory, an equilibrium of species diversity is established by the interplay of immigration on the one hand and extinction on the other. The parameter driving the interaction of these phenomena is, in general, the open ecological niches (cf. Whittaker et al. 2010) that result from geographical isolation (e.g., Lack 1947; Levin 2004). MacArthur and Wilson (1967) specifically added two geographical parameters that influenced immigration rate and chances of survival: (1) distance from the immigration source (normally a continent) and (2) size of the available area to be occupied (i.e., the island).

Basics of the Model

The MacArthur-Wilson equilibrium model includes a number of assumptions. The immigration rate of new species to an island decreases continuously with increasing numbers of species that are already present on the island. Also, immigration is thought to stop when all species from the source area have settled on the island. The more species that are established on the island, the more they are subject to extinction. The model also provides predictions based on geographical relations. Islands located near the mainland should show higher immigration rates than remote islands, provided that species numbers of the source area are the same. Further, the extinction rates are suspected to be higher on smaller islands than on larger ones due to smaller population sizes and resource pools and more direct competition. Near islands therefore should contain more species than more distant islands of the same size, and large islands should carry more species than smaller islands with the same distance to the mainland (bearing comparable species numbers). The MacArthur-Wilson model does not predict that the species community is identical over time even at an equilibrium state; rather, it assumes a turnover of species composition. Many examples have been provided to confirm these predictions in a general way (see Lomolino et al. 2005). In a major island study on Bahamian web spiders, the traditional distance and area relations were confirmed as well as the relations of immigration to island area and extinction to island distance (Toft and Schoener 1983). Furthermore, a series of creative experiments was done by Simberloff and Wilson (1970), in which they sterilized small mangrove patches in the Florida Keys (United States), followed by examining the pattern of immigration and extinction, especially of the arthropods, on these small islands. The gathered data correlated well with their theoretical expectations, but the time scale, obviously, was very short.

Application of a test of the equilibrium model on the recent Juan Fernández Archipelago plant community is consistent with expected species numbers and island distances. Because ferns have such a high dispersal capacity, they are not included in these calculations. It can be noted, for example, that 49 taxa of ferns occur on Alejandro Selkirk Island and 47 are found on Robinson Crusoe Island (including those on Santa Clara Island). The former island is situated 181 km further westward from the latter and supports many fewer native and endemic species of angiosperms (65 versus 96, respectively; derived from Table 5.1). This general tendency is clearly what the MacArthur-Wilson model suggests based on distance from the mainland, with the actual proportional decrease in native and endemic species diversity on Alejandro Selkirk Island 32%, which is close to the expected level of 27%. But despite this reasonable fit to the actual species diversity, use of distance alone does not attempt to deal with the actual factors responsible for species diversity in the archipelago.

Problems with the Model

The few factors involved in the MacArthur-Wilson model are not sufficient to deal with the complexity and potential abilities of organisms nor the variable interactions involved with natural events (Heaney 2007; Schoener 2010). The model lacks (1) the idea of the enormous variance of ecological differentiation within islands, (2) aspects of anagenetic and cladogenetic speciation once populations are established on an island, and (3) the potential exchange and interaction of populations between different islands within an archipelago. Perhaps the main omission of the MacArthur-Wilson theory is that it does not consider the changing nature of islands over time. Because of this dynamic, there never can be a *permanent* equilibrium established in any oceanic island system – only a short, temporary one – before the island begins to erode and subside.

Ecological parameters. Ecological factors were not included in the MacArthur-Wilson model simply because they were too complex to model effectively. Taking the Juan Fernández Islands as an example, Greimler et al. (2002a) listed fifteen plant assemblages presently existing on Robinson Crusoe Island (see Chapter 6). No less than eleven of these are well-defined plant communities, with the remaining four representing mixed assemblages of vegetation types. Tremendous ecological differentiation due to elevation, wind, and precipitation has to be expected even on small islands. Over long time scales lasting millions of years, successional phenomena in habitats and communities will occur.

Examples of deviation from strong species-area relations have been analyzed by Losos and Parent (2010) in *Anolis* lizards in the West Indies. *Anolis* is limited to the Americas and includes 361 currently recognized living species. More than 40% of these are restricted to the West Indies, whereas the other members are found on the surrounding mainland. Losos and Parent (2010) compared the distribution of the genus among dozens of islands over the West Indies. They were able to distinguish different species-area relations between four subsets of islands defined by size class, land-bridge character, and historical background. Ecological aspects, however, played a major role in explaining additional amounts of variation: numerous ecomorph species and unique habitat specialists contributed to species richness. All these taxa, however, live in the larger islands, which links different habitats (including possible microhabtitats) positively with island size.

Speciation within the archipelago. The potential for forming new species from immigrants after arrival on an island was also not taken into account in the MacArthur-Wilson model. Those authors were well aware of the importance of this phenomenon, however, as discussed later in their book. They did not include it in their model because it is group specific and also related to habitat heterogeneity.

Adaptive radiation obviously takes place in many islands. Many organisms of totally different systematic groups have confirmed this mode of speciation. Striking evidence for this evolutionary drive has been presented from the Hawaiian and Galápagos Archipelagoes (e.g., Fleischer and McIntosh 2001; Grant and Grant 2010). In the Hawaiian Archipelago, various genera of Drosophilidae comprising nearly 1,000 species evolved from a single colonist 25 Mya (Magnacca et al. 2008), as did the silversword alliance (Asteraceae), although it contains fewer species (Barrier et al. 1999). In the Galápagos Archipelago, several bird and reptile genera or species complexes have originated from single ancestors (e.g., *Geochelone* species complex [Beheregaray et al. 2004], *Nesomimus*, and four genera of Geospizinae [Grant and Grant 2008]). Speciation is clearly encouraged by the ecological complexity of an island (Losos and Parent 2010) because distinct geographic habitat-structuring stimulates reproductive isolation and selection.

Interisland considerations. The MacArthur-Wilson model also left out the possible interplay between communities of neighboring islands. Considering an island as a sort of small territory or mainland suggests possible organismic exchange between these areas. For complex archipelagos, such as the Galápagos Islands, interisland considerations are obviously complicated and certainly significant.

New Dynamic Biogeographical Analyses

In recent years, new geological and organismic data have made it possible to develop a more sophisticated approach to estimating insular biodiversity. Two principal new approaches have been developed, and they deal with the ontogenetic history of an island and the evolutionary course of its ecosystem. Stuessy (2007), in the four-phases genetic model, focused on the genetic consequences of different speciation types (i.e., cladogenesis and anagenesis) (Stuessy et al. 1990) on islands of different elevations (and habitats). Whittaker et al. (2008), in the general dynamic model, applied a strictly mathematically based model on different organism-island complexes, combining the factors of immigration, speciation, and extinction.

The Four-Phases Genetic Model

The four-phases genetic model of Stuessy (2007) was based heavily on research experience in the Juan Fernández Archipelago, but it is applicable to any oceanic island (Fig. 17.1). The structure of the four-phases genetic model is based on the ontogenetic development of the physical and organismic aspects of islands, with emphasis on the

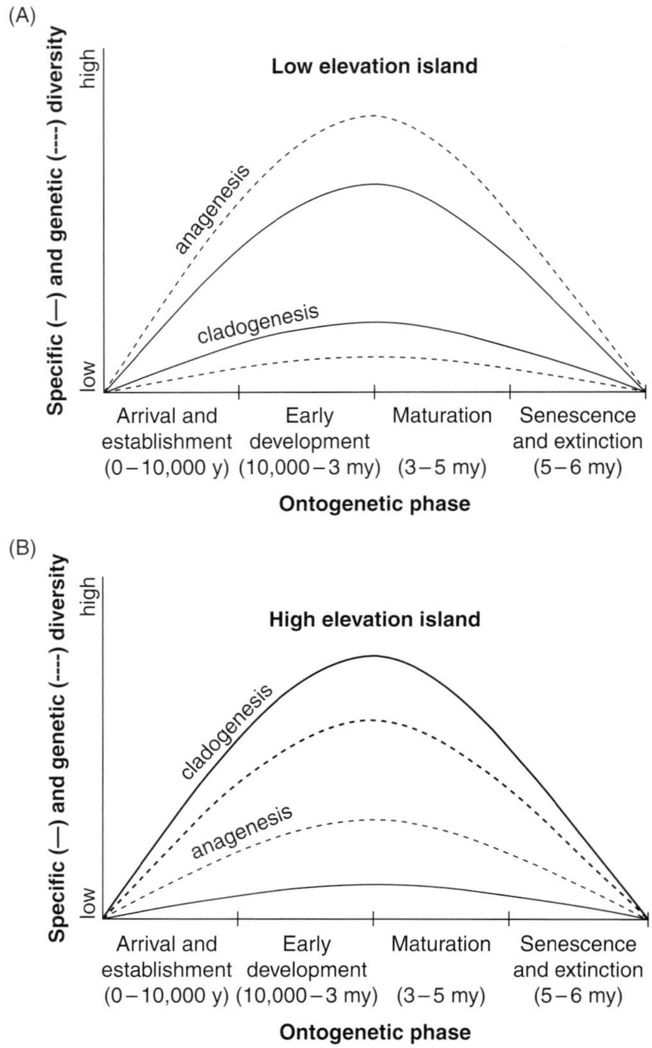

Figure 17.1 Representation of the four-phases genetic model of island biogeography: (A) low-elevation island; (B) high-elevation island. (From Stuessy 2007, p. 123.)

plant communities. The four phases of ontogeny of the oceanic island floras are (1) arrival of propagules and establishment during the first few thousand years, (2) early development of the floral communities (10,000–3 Mya), (3) maturation stage (3–5 Mya), and (4) senescence followed by extinction (5–6 Mya). The tremendous interference caused by human impact must also be considered. Due to the dramatic role of humans (e.g., Ono 1998), the present anthropogenetic period (the Anthropocene) (Crutzen and Stoermer 2000; Caro et al. 2011) could even be listed as a distinct stage for all islands because it interferes so strongly with previous evolutionary processes.

Phase 1: Arrival and establishment of plant species. At the beginning, an island is bare of terrestrial organisms. Thus dispersal from other localities is the first step of colonization. Despite immigration modes of floating, rafting, or lofting, a high percentage of arrivals results from transport by birds (Carlquist 1974; see also Chapter 16). Insular dimensions and distance from continents play a role in the probability of receiving immigrants in the initial phase. One single founding event versus a sequence of arrivals over a period of hundreds or thousands of years is also significant.

Also important in the initial phase is the quality of the first organisms. Certain seed and fruit types may better cope with poor soil qualities than others. Self-pollinating and self-compatible immigrants and those able to reproduce vegetatively may have better chances of survival. Perennial plants appear more predestined to colonize successfully. Individuals with broad genetic variation will also possess an advantage compared with those that are more genetically limited. Polyploidy may permit adaptation to a broader range of physiological challenges. Most strikingly, the early insular communities strongly deviate from those of source areas; only taxa that are successful dispersers reach the new islands, and they assemble into vegetation types often quite different from those of the mainland (said to be "disharmonic") (Carlquist 1974).

Phase 2: Early development. This stage is characterized by multiplication of individuals and populations and by speciation. Topographical structure such as stratigraphical properties and numbers and sizes of available niches (often connected to island size) strongly influence this part of the biogeographical development. Cladogenetic speciation is correlated with ecological diversity (Stuessy et al. 2006). Generally, the higher and more environmentally diverse an island is, the more rich with species it will become. There is also strong competition among evolving populations in this phase. Adaptations to the local environments mean a high risk because island areas are restricted, and environmental changes may quickly lead to extinction. The gene pool may be meager due to low numbers of individuals.

Speciation modes in oceanic islands are reduced in comparison with continental regions (see Chapter 15). Immigrants represent limited genetic variation in comparison with progenitor populations. This starting point goes both for cladogenetic and for anagenetic processes (Stuessy et al. 1990). Small changes in the genome and habitat diversity may lead to the development of morphologically distinct lineages (species), each with low genetic differentiation. Such mode of speciation is called "cladogenetic." If islands are more ecologically homogeneous, genetic variation will slowly accumulate over a long period, leaving a species with a broad genetic spectrum that is morphologically distinct from its ancestor. This kind of speciation is labeled "anagenetic" and is typical of monotonous, flat islands such as Ullung Island (Korea), where at least 88% of the speciation processes are anagenetic (Stuessy et al. 2006). See Fig. 17 1 and Chapters 13 and 15 for details of these modes of speciation and patterns of phylogeny.

Phase 3: Maturation. In this phase, continuous erosion and subsidence cause shrinking of the terrestrial surface (see Chapter 3). Erosion is largely due to meteorological and oceanic forces (wind and waves, temperature variations, rainfall, and so on) (Murphy et al. 2016), resulting in loss of niche spaces. Competition within and between populations increases, and species richness starts to thin out. During maturation, species diversity may be reduced by 25% or more (cf. Stuessy et al. 1998a, 2005a). Reduction of genetic variation takes place in both cladogenetically and anagenetically evolved species

(Fig. 17.1). Beyond the extinction of populations, a strong tendency to hybridization of the remnant populations can add to the loss of (pure) plant species.

Phase 4: Senescence and extinction. In the senescence stage, erosional processes and island subsidence continue, leading to eventual disappearance of the island (Ramalho et al. 2013). This period is rather short (1–2 my). Species diversity declines. The addition of new species resulting from new arrivals of individuals is expected to occur rarely, at best. Finally, all terrestrial species disappear completely.

Human impact. Human impact on the biodiversity of oceanic islands has been dramatic in past centuries. Even before seafaring in the fifteenth century, the strong influence of native tribes on island communities was documented, for example, in the South Sea and Hawaiian Archipelago (cf. Ziegler 2002). Since land discoveries and colonization by Europeans, an even more dramatic change has occurred, resulting in drastic reductions of island biota. On Easter Island (166 km^2 and perhaps the world's most continuously inhabited and most remote place) (McCall 2009), native plant species have widely disappeared. After drastic reduction by earlier Polynesian immigrants, most of the remaining native flora in the past century has disappeared due to modifications of the island into pasture for sheep (McCall 2009). The Hawaiian Archipelago shows similar losses of native vegetation and species richness (Ziegler 2002; Swenson et al. 2016).

The situation in the Juan Fernández Archipelago has been exacerbated by the introduction of a great number of plants due to human activities during historical times. Since the early botanical expeditions in 1823 by Mary Graham to Robinson Crusoe Island (see Chapter 2), the overall number of introduced plant species has averaged about one species per year (Matthei et al. 1993; cited by Swenson et al. 1997). Robinson Crusoe Island has been subject to much stronger human impact (Skottsberg 1954; Sanders et al. 1982; Wester 1991, Swenson et al. 1997; see also Chapter 7). This island has housed permanent colonists since the early 1700s, presently with a settlement (San Juan Bautista) of approximately 600 inhabitants (Greimler et al. 2002b).

The General Dynamic Model (GDM) and the ATT^2 Model

The general dynamic model (GDM) of oceanic islands, as discussed by Whittaker et al. (2010) (Fig. 17.2), incorporates the classical immigration/speciation-extinction dynamics plus the idea of ecological processes supported by empty niches and the historical time scale of islands. The model establishes a positive connection between speciation rates (*S*), single-island endemics (SIEs), and extinction rates (*E*) in island systems (Emerson and Kolm 2005a, 2005b).

The historical island perspective assumed by the general dynamic model reaches from youth and maturity to old age and eventual complete loss. This insular life cycle plays an important role in the ecological changes and evolutionary processes of oceanic archipelagos (e.g., Peck 1990; Peck and Kukalova-Peck 1990; Stuessy et al. 1998, 2005a, 2006; Gillespie 2004). Furthermore, the history of hotspot islands should clearly differ from that of complex island arc areas, where close physical and/or historical connections among islands may have existed. Also, catastrophic and erratic incidents may play

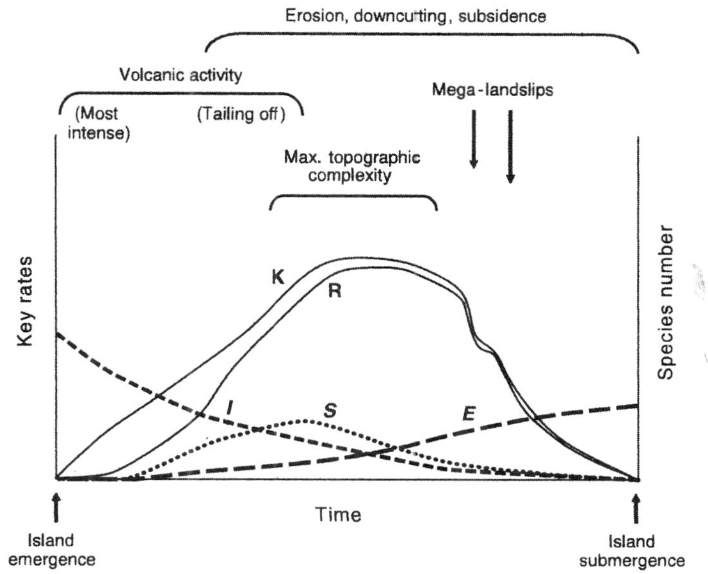

Figure 17.2 Representation of the general dynamic model of island biogeography. *I*, immigration rate; *S*, speciation rate; *E*, extinction rate; *K*, potential carrying capacity; *R*, realized species richness. (From Whittaker et al. 2008, p. 981.)

a major role. The simplified model of Whittaker et al. (2008), log(area) + time − time^2 (ATT^2), implies that the biological carrying capacity (i.e., biomass and number of individuals across all species) should be achieved with maximum area and elevational range, with maximum heterogeneity of environment, and thus maximum opportunity for within-island allopatry. In the model, immigration and speciation rates also differ in relation to distance between neighboring islands and dispersal capacities of organisms (Whittaker et al. 2008). In many cases, islands are incorporated within a dynamic archipelago, which makes them both species receivers and species donors.

Having these principles in mind, a general dynamic theory was developed with testable predictions and with a focus on single-island endemics. Species diversity as defined by this model is based on time (age) parameters ($T + T^2$) and logarithmic space parameters, and, in addition, a distinct diversification index that takes into account the number of single-island endemics and the number of genera within which single-island endemics occur. The biotas tested covered plants, various arthropod groups, and snails in the Canary, Hawaii, Galápagos, Marquesas, and Azores Archipelagos between 1,039 and 16,397 km^2 in size (Whittaker et al. 2008). The ATT^2 model describing the richness of species was significant statistically in all the 14 cases from these five island groups (Whittaker et al. 2008). Beyond this general statement, the model, characterized by a humped time ontogeny, that is, a sharp drop in the species curve in late midlife of island ontogeny, clearly outmatched several alternative models with less specific assumptions. With respect to the Azores, however, the application of a simpler model with a nonhumped time assumption resulted in a better fit. Difficulties in applying the

ATT[2] tests are exacerbated by insular fusion and separation as well as catastrophic events, such as the Roque Nublo ash flow on Gran Canaria some 3.5 million years ago, which occurred about 11 million years after the presumed emergence of the island. Other challenging methodological problems include fusion and separation of Lanzarote and Fuerteventura (Canary Islands), the Maui-Nui complex in the Hawaiian Archipelago (presently consisting of four separate islands), the fusion of three older massifs by the Teide formation on the island of Tenerife (Canary Islands) some 2 million years ago, and the formation of São Miguel (Azores) composed of five differentially aged sections in a mosaic-like pattern (Whittaker et al. 2008). Despite these difficulties, Whittaker et al. (2008) argue that compared with the MacArthur-Wilson approach, their general dynamic model represents a more comprehensive theoretical framework for understanding the evolutionary biogeography of oceanic islands. Additional discussions of the model can be found in Borregaard et al. (2016).

Analyses of Species Diversity in the Juan Fernández Archipelago

Because of its simplicity, the Juan Fernández Archipelago offers an excellent opportunity to attempt to understand the factors controlling species diversity in oceanic islands. They are of volcanic origin and small enough to represent strictly localized and relatively clear evolutionary processes, contrasting with what might happen in large islands. There are only two major islands, and they are oriented in an east-west line from the South American continent, which is the principal source area for immigrants (see Chapter 16). The islands are now about the same size (50 km^2). Their geological ages, however, are different (Stuessy et al. 1984), with the further island also being the younger (1–2 Mya; Alejandro Selkirk Island) and the closer island the older (4 Mya; Robinson Crusoe Island). The biogeographical probability of a propagule arriving in the archipelago, therefore, is vastly higher for the nearest island than for the further and younger island. It is this structure that allows strong inferences on the nature of species diversity in this archipelago, perhaps more favorably than in any other oceanic system of the world. Furthermore, the numerous phylogenetic and genetic studies (see Chapters 13 through 15) on many endemic genera and species add strength to the formulation of hypotheses. The basic approach should be to first work out the factors that can explain species diversity on Robinson Crusoe Island and then apply these same factors to predicting diversity on Alejandro Selkirk Island. Because we know the diversity on both islands, this provides a strong test of the method.

A total of 151 native and endemic angiosperm species have been documented for the Juan Fernández Islands (Table 5.1). One-hundred and four of these species are endemic, and another 42 are native. Moreover, there are 12 endemic genera, and one family is restricted to the archipelago (Lactoridaceae). Considering diversity on the two main islands, Robinson Crusoe Island (including Santa Clara Island due to its very recent separation from the main island; see Chapter 3) contains 96 species (this includes five shared with Santa Clara Island and one that is restricted to this small island), and

Alejandro Selkirk Island has 66 species. Twenty-two species are shared between the two major islands.

The challenge, then, is to determine the factors that have influenced development of this plant diversity. Relying on the equilibrium model of MacArthur and Wilson (1967), one could just take the present-day distance from the Chilean mainland for assessing the immigration frequency and the size of the islands for estimating the extinction rates, which in this case is the same (50 km^2). As we have seen, this approach, without regard to differential time scales (or other factors), is not a bad representation of the current species diversity. It is hard to believe that the different ages of the two main islands, one being two to four times older than the other, did not have consequences for the number of colonizing species and composition of the flora. The hypothesized geological dynamics of the Juan Fernández Archipelago (pointed out in detail in Chapter 3) have resulted in a change in terrestrial size in both islands from origin to modern times by an estimated 95% for Robinson Crusoe Island and 28% for Alejandro Selkirk Island (Stuessy et al. 1998a). Hence an enormous dynamic has taken place, leaving only a tiny portion of the original island in the case of Robinson Crusoe Island.

An important factor for immigration, settlement, and diversification of species is the original size of the island. By analogy to the Hawaiian Islands, as suggested by Macdonald et al. (1983; see also Chapter 3), Sanders et al. (1987) calculated the former dimensions for the Juan Fernández Archipelago based on the existing submarine erosional surfaces at 200 m below sea level. For the older Robinson Crusoe Island, estimates yielded 1,092.5 km^2, and for the much younger Alejandro Selkirk Island, the total area was estimated at 90.1 km^2. Thus, provided that these assessments are correct, Robinson Crusoe Island would have been reduced to less than a twentieth of its largest extensions, whereas Alejandro Selkirk Island would only have lost about half of its former area. Also, both islands probably lost substantial portions of their vertical dimensions, with Robinson Crusoe Island perhaps having been originally 3,000 m high (now 915 m).

In addition to these physical aspects, the geographical origin of the flora can be used to help assess biogeographical history. Judging by the systematic relationships, about 82% of the angiospermous flora came from South America (Bernardello et al. 2006; see also Chapter 16), and among these, 10% are from the Neotropics and 12% from Chile. The rest of the angiosperm species have been contributed from pantropical (Pacific) areas (18%), including Australia (4%) and New Zealand (4%). Another benchmark assessment is the suggestion by Stuessy et al. (1997) of angiosperm species diversity loss of about 25% to 30% on Robinson Crusoe Island. This is only an estimate based on consideration of the loss of surface area of Robinson Crusoe Island over the past 4 Mya.

Reliable estimates on overall speciation events within and between islands can be derived from several phyletic studies on the archipelago's flora (see Chapter 13). At least 35 cladogenetic origins took place on Robinson Crusoe Island (including Santa Clara Island) and another 10 on Alejandro Selkirk Island. The origin of species from neighboring island populations, that is, anagenetic speciation processes between the two islands, is assumed for 12 known species (see Table 13.8), 10 from Robinson Crusoe Island and two from Alejandro Selkirk Island.

Some principal assumptions of the rationale and method to be used are as follows: Arrivals and colonizations of an island should be influenced by the distance from the source area (mainland), decreasing with the distance. Extinction should be related to the respective size of an island, which declines, leading to a negative correlation between species number and island size (extinction rate). Between two islands of an archipelago (one older than the other), there is impact on the younger island by settlement through invasion of neighboring populations, the rate of which is influenced by between-island distance and number of existing species on the older island at the time of emergence of the younger island. The speciation type (i.e., anagenetic versus cladogenetic) between and within the islands will also affect species diversity.

Explaining Species Diversity on Robinson Crusoe Island

There are 151 native and endemic species of angiosperms on both of the Juan Fernández Islands. Robinson Crusoe Island (plus Santa Clara Island) contains 96 species, and Alejandro Selkirk has 65; they share 22. These figures are not by themselves satisfactory for estimating and predicting species diversity in the archipelago, however, because the islands have changed considerably since they were formed.

To provide an estimate of species diversity on Robinson Crusoe Island in the Juan Fernández Archipelago, we must (1) assume a number of species from the source area (South America), (2) estimate the colonization probability of a propagule arriving on the island, which is correlated with distance from the mainland, (3) add species to the totals for Robinson Crusoe Island to compensate for the presumed loss of diversity from extinction over its 4 million years of existence, (4) subtract speciation events occurring within Robinson Crusoe Island, and (5) subtract anagenetic speciation events resulting from dispersal from Alejandro Selkirk Island.

The specific calculations of original species diversity on Robinson Crusoe Island are as follows: At present, there are 96 native and endemic species on this island. First, we assume that 25% of the species have been lost to extinction on Robinson Crusoe Island. This is a very rough estimate in view of the presumed 95% loss of surface area (Sanders et al. 1987), especially during the first 3 million years prior to emergence of Alejandro Selkirk Island. We must add, therefore, 25% more species to the presently existing 96, giving a total of 128. Second, we need to subtract the number of speciation events (cladogenesis) having originated on the island because these species have nothing to do with dispersal and original establishment. Based on phylogenetic considerations (see Chapter 13), we assume 35 such events have taken place within Robinson Crusoe Island. This reduces the species diversity to 93. Third, we need to subtract the number of species (two) that have been anagenetically derived from Alejandro Selkirk Island, yielding 91 species.

Having this adjusted species total for Robinson Crusoe Island, we can attempt to structure a model of predicting species diversity. To determine probability of colonization, we need a value for species numbers in the major source area, namely, South America. Because no precise numbers exist for the entire flora of South America, for simplicity we assume that all propagules have come from the Chilean mainland, where statistics of the flora have been provided (Marticorena 1990). It is realized that only 12% of the flora of the islands has actually been derived from this region (see Chapter 15), but what is needed is a base value from which colonization probabilities can be estimated. The total number of angiosperm species in continental Chile is known as 4,975 (Marticorena 1990). The dispersal distance would be approximately 900 km, the distance of Robinson Crusoe Island from the Chilean mainland when the island was formed over the hotspot, which is now west of Alejandro Selkirk Island under the Friday Seamount (Astudillo M. 2014). Needed, therefore, is a colonization coefficient probability (*Cp*) based on this distance. We estimate that for every 100 km of ocean, *Cp* gets lowered by 0.640 (or 64%). This has been calculated by taking the total species diversity in the source area (4,975) and determining what level of *Cp* multiplied nine times for 900 km would give the known 91 species. The *Cp* value for 900 km, which was the original distance of Robinson Crusoe Island from the continent, is 0.0115; for the distance between the two islands (181 km), it is 0.4534; and for the distance of Alejandro Selkirk Island from the continent when it formed over the hotspot (900 km), it is also 0.0115.

Prediction of Species Diversity on Alejandro Selkirk Island

Having developed a model for estimating the diversity of species on Robinson Crusoe Island, we can now attempt to infer species diversity on Alejandro Selkirk Island. Sixty-five species are presently known on Alejandro Selkirk Island. From this we need to subtract the species that have speciated cladogenetically on the island (10), yielding 55 adjusted species. Within this number, we know that 10 species have evolved anagenetically from Robinson Crusoe Island (see Chapter 13). Because we know the diversity on Alejandro Selkirk Island, we can see how well the theory predicts the actual species diversity.

Initially, it is necessary to estimate the original size of the two islands as target areas because this was the situation when most of the colonization occurred. Sanders et al. (1987) suggested that Robinson Crusoe Island may have lost 95% of its terrestrial surface in the past 4 million years; Alejandro Selkirk Island would have lost 28%. These estimates can be taken to determine probability (*Tp*) of the islands as target regions for incoming diaspores. The target probability of the two islands now is 1.0 because they are presently the same size. Four million years ago, the *Tp* between the original size of Robinson Crusoe Island (1,092.5 km^2) and that of the younger Alejandro Selkirk Island (69.2 km^2), when it was formed 1 million years ago, is 0.063.

With respect to the extinction rate, there may be a big difference between the two islands. The older Robinson Crusoe Island has been drastically reduced in size, which has not been the case for Alejandro Selkirk Island. For the latter island, therefore, the relative extinction rate after the initial establishment of plant communities is thought to be much more limited or even zero due to relatively little change in its dimensions.

We might also examine dispersal to the archipelago and consider the impact that island size as a target area could possibly have had on different agents. For propagules arriving on an island via water or wind, island size is certainly the limiting biogeographical factor (e.g., Buckley and Knedlhans 1986). The larger the perimeter of an island (water related) and to some extent also its elevation (wind related), the higher is the probability that the diaspore will arrive safely. Bird-dispersed propagules (either externally or internally) (McAtee 1947) offer another perspective. If a bird were to fly above an ocean, and if it sought land, it would seek any island, no matter its size; to some extent, the larger the island, the higher is the probability that it would be observed. The size of islands for bird-dispersed propagules would be less critical than it would be for seeds or fruits dispersed by water or wind. Fifty-five bird species, including landbirds and seabirds, have been recorded from the Juan Fernández Archipelago, about one-third of them representing regular breeders and the others being visitors (Hahn et al. 2009). The archipelago has recently been estimated to maintain about 1 million breeding pairs of petrels (*Pterodroma* sp.) (Brooke 1987). Bernardello et al. (2006; see also Chapter 16) most recently estimated that 90% of the species of Juan Fernández have arrived by birds. Despite the different dispersal mechanisms represented by water and wind, their low percentage in the flora (10%) allows us to ignore this distinction in the calculations.

We can calculate the diversity of species on Alejandro Selkirk Island as follows: Species diversity = [(number of species in source area, or 4,975 species) × *Cp* for 900 km (0.0115) × *Tp* relative to Robinson Crusoe Island 4 mybp (0.063)] + [number of anagenetically originated species from Robinson Crusoe Island (10)] + [number of adjusted species on Robinson Crusoe Island (91) × *Cp* for 181 km (0.4534)]. These calculations yield a total of 54.9 species predicted for Alejandro Selkirk Island, and the actual adjusted number is 55. Recall that the equilibrium model calculations based only on distance of the islands from the mainland gave an expected value that was somewhat higher (see earlier in this chapter).

Because of the uncertainty of some of these calculations, especially level of loss of species diversity on Robinson Crusoe Island over 4 million years, arrival at the exact level of species diversity on Alejandro Selkirk Island can only be regarded as coincidental. Nonetheless, the approach does suggest an alternative model, instead of using only area of island and distance from source area, which can explain species diversity reasonably well. Perhaps more important than the specific calculations is that the example serves as an illustration of using factors that seem more relevant for affecting species diversity. Size of island and distance from the mainland alone are not by themselves able to explain species diversity in oceanic islands; they can serve only as a rough surrogate for the actual factors involved.

Conclusions

Tod F. Stuessy, Daniel J. Crawford, Carlos M. Baeza, Patricio López-Sepúlveda, Josef Greimler, Eduardo A. Ruiz, Koji Takayama, Gregory J. Anderson, Gabriel Bernardello, and Ulf Swenson

Island biology is not a new field. Many excellent books have been written in recent decades that summarize much of what we have learned from analysis of different plant and animal groups in diverse oceanic archipelagos (e.g., Carlquist 1974; Wagner and Funk 1995; Grant 1998; Stuessy and Ono 1998; Cody 2006; Thornton 2007; Whittaker and Fernández-Palacios 2007; Gillespie and Clague 2009; Caujapé-Castells et al. 2010; Losos and Ricklefs 2010; Bramwell and Caujapé-Castells 2011). It may seem presumptuous to suggest that our investigations in the Juan Fernández Archipelago have yielded any new general insights on island biology. There is no question that we have learned a great deal about this particular archipelago, but we also believe that a number of important general concepts have been revealed through these investigations.

A principal reason why the Juan Fernández Archipelago has served well for asking and attempting to answer evolutionary and biogeographical questions is its setting. The two main islands are at present nearly equal in size (ca. 50 km^2), but they are of different geological ages and positions with respect to primary sources of colonists. Robinson Crusoe Island is the closest to the Chilean continent, and it is the oldest at approximately 4 million years of age (Stuessy et al. 1984; see also Chapter 3). The more isolated island, Alejandro Selkirk, at 1 to 2 million years old, is 181 km further west, a substantial distance, but in a nearly direct east-west line with Robinson Crusoe Island and the South American continent, which is the major source area for propagules. This means that the probability of a colonist arriving first in Robinson Crusoe Island is vastly greater than in Alejandro Selkirk Island. There is no other oceanic archipelago in the world that is so favorably situated geographically and geologically for projecting order of colonization. Many other fascinating archipelagos exist (e.g., Hawaii, Galápagos, Canaries), but they are all more complex than the Juan Fernández Islands. This geological and geographical structure is what has facilitated the ease of generating hypotheses and their testing over more than three decades. Nor have the Juan Fernández Islands undergone catastrophic natural disturbances, such as massive landslides or renewed volcanic activity, since initial formation of the two islands.

Another favorable condition for gaining insights on the evolution and biogeography of the flora of the Juan Fernández Islands is the small endemic flora. This has allowed phylogeny, biogeography, and processes of evolution for the entire flora to be developed during our investigations. No great taxonomic surprises have been lurking in the background to confound our understanding of evolution. This can sometimes be a formidable problem, as with the Hawaiian flora. One of the major problems there had been the lack of a consistent species concept, which made accurate evolutionary inferences nearly impossible. The *Manual of the Flowering Plants of Hawai'i* (Wagner et al. 1990) straightened this problem out. In the Juan Fernández flora, thanks to the excellent efforts of Hemsley (1884), Johow (1896), Skottsberg (1921, 1953a), Marticorena et al. (1998), Danton and Perrier (2006), and now Danton and Perrier (2017), the floristic inventory is well understood (see Chapter 5). This has made it possible to carry out evolutionary investigations with some sense of order and reason.

Historical and Vegetational Analyses

One important concept that has been emphasized in numerous studies from our laboratories is the fundamental roles that geology and history have played in the islands (Stuessy et al. 2005a; see also Chapters 1 and 3). History has many dimensions, beginning with the ontogeny of the islands. Oceanic islands are relatively short-lived geologically, many surviving to 6 million years and then disappearing under the ocean due to subsidence and erosion. When islands are of different ages, different kinds of impacts have occurred, making it more challenging to compare processes of evolution from one island to another. In this context, processes of speciation might be more clearly revealed on species groups of the younger Alejandro Selkirk Island, which has undergone much less physical modification than the older Robinson Crusoe Island (Sanders et al. 1987; Stuessy et al. 2005a). However, the present distribution of the species on the latter island, in fact, may reveal little about the original processes that led to divergence.

Considerable effort has been spent on providing new vegetational surveys of both Robinson Crusoe and Alejandro Selkirk Islands (Greimler et al. 2002a, 2013; see also Chapter 6). Due to the absence of aboriginal people in the archipelago (Haberle 2003), a special opportunity exists for examining vegetational patterns and their interpretations. In many island systems, indigenous peoples already had an impact on the ecosystems prior to the arrival of Europeans. In the Juan Fernández Islands, humans only began to influence the archipelago after its discovery in 1574 (Woodward 1969; Medina 1974). Importantly, much historical documentation in the form of captains' and pilots' logs and diaries over 400 years has allowed a relatively good understanding of activities involving the vegetation of the islands (Stuessy et al. 1998b; see also Chapter 7).

Previous detailed descriptions of vegetation were developed primarily by Skottsberg (1953b) during his expedition in 1916–17. Our detailed analyses of the vegetation on both islands (Greimler et al. 2002a, 2013) have been based on 106 sampled plots (relevés) on Robinson Crusoe Island and 90 on Alejandro Selkirk Island, which have provided quantitative data for more precise classification of vegetation. Because Skottsberg also completed relevés in selected places in the archipelago, we have been able to resample some of these areas and make comparisons of vegetational change over the past century (Sanders et al. 1982; Greimler et al. 2002a). Notable have been the disappearance of native vegetation and increase in exotic and invasive species on Robinson Crusoe Island, particularly at the lower elevations.

The two islands are very different in their vegetation and physical appearance (Greimler et al. 2002a, 2013). The different ages, geomorphologies, and erosional patterns have had a strong impact on the assembly and spatial distribution of the vegetation. It is more homogeneous on Alejandro Selkirk Island, with its huge tableland dissected by deep canyons, in contrast to the highly eroded Robinson Crusoe Island, with many narrow ridges connecting the summits and the single high peak (El Yunque). These vegetational analyses again emphasize the importance of understanding the geological history as the islands have changed through time. Because of the greater subsidence and erosion of Robinson Crusoe Island, the vegetation there is more compacted, and the flora survives in what can be best regarded as a refugium. This important concept has major implications for interpreting modes of speciation and understanding patterns of genetic diversity within and among populations.

Another geological element for which we have little evidence but assume some impact is Late Quaternary glaciation (Weigelt et al. 2016). The floras of all islands were subjected to a drop in sea level at glacial maximum, estimated to be 122 m. The attendant increase in land mass, followed by the subsequent raising of sea level to current levels, would have affected the distribution, speciation, and extinction of the flora (Weigelt et al. 2016). Volcanic islands in deep water, however, such as Juan Fernández, would have experienced reduced impact because less surface area was exposed and/or lost.

Endangered Status of the Endemic Flora and Conservation Challenges

It is saddening to learn that so many of the endemic species of the Juan Fernández Archipelago are endangered at some level. The archipelago was designated a Chilean national park in 1935 and a UNESCO biosphere reserve in 1977. Despite these protections, statistics for the vascular flora reveal that 36% of the species are vulnerable, 41% endangered, 9% critically endangered, and 4% already extinct (see Chapter 9). The obvious conclusion is the need for conservation of these endemic taxa. This is not a new perspective; many authors have commented on this same point (Johow 1896; Muñoz Pizarro 1974; Ricci 1996, 2006; Danton et al. 1999; Danton 2004; Biodiversa 2009; Arana 2010; Vargas et al. 2011), and we have also added our own pleas (Sanders et al. 1982; Swenson et al. 1997; Stuessy et al. 1998d; see also Chapter 9). Underlining the importance of conservation in the archipelago is its designation of being one of the world's 12 most threatened national parks by the International Union for the Conservation of Nature and Natural Resources (IUCN). We hope that having a modern synthesis on the biogeographical and evolutionary significance of the plants of these islands, as presented in this book, stresses the point even more strongly.

One of the challenges to preserving the native and endemic flora of the archipelago is control of invasive species. Introduction of alien plants and animals has led to invasive species causing significant alteration to natural habitats (Matthei et al. 1993; Swenson et al. 1997; Greimler et al. 2002b; Danton and Perrier 2006; see also Chapter 8). Introduced taxa arrived in the archipelago more than 400 years ago with visits of the first sailing ships. By the end of the twentieth century, the number of introduced and naturalized taxa (227) exceeded the number of native and endemic species (Swenson et al. 1997). Dispersal of the seeds by animals (birds and rabbits) and wind, as well as rapid clonal growth of a few taxa, has helped them to spread effectively. Two shrubs (*Ugni molinae* and *Aristotelia chilensis*) from the South American continent and one European shrub (*Rubus ulmifolius*) are the most aggressive invasives on Robinson Crusoe Island (Dirnböck et al. 2003). *Aristotelia chilensis* has become invasive on Alejandro Selkirk Island, but much less so at this time. The South American ground herb *Acaena argentea* is extremely abundant and has replaced natural vegetation in many areas on Robinson Crusoe Island, but it has had only a minor impact on Alejandro Selkirk Island (López-Sepúlveda et al. 2013a), even though it has been observed there for nearly 100 years (Skottsberg 1921). Among the introduced animals, goats are a plague especially on Alejandro Selkirk Island, and rabbits have caused much damage to the vegetation of Robinson Crusoe Island (Biodiversa 2009). The growing impact of the invasive shrubs, herbs, and grasses, together with pressure from introduced animals and increased human activity, has promoted dominance of introduced plants and the homogenization of the flora (Sanders et al. 1982; Swenson et al. 1997; Greimler 2002a, 2013; López-Sepúlveda et al. 2013a).

Character Divergence in Flavonoids

In addition to interest in unraveling phylogeny and speciation for the endemic vascular flora, our studies have also attempted to understand the evolution of secondary plant metabolites (flavonoids). We have examined the taxonomic utility and phylogenetic implications of flavonoid compounds from plants of the Juan Fernández Islands and compared the flavonoid chemistry of the island lineages with that of their closest continental relatives (Chapter 10). As far as we are aware, no similar comprehensive analyses are available for any other oceanic island system. Flavonoids, in general, serve as useful markers for distinguishing species within genera of the archipelago, and this is particularly notewothy in the two largest endemic genera, *Dendroseris* and *Robinsonia* (Asteraceae) (Pacheco et al. 1985, 1991a). A notable exception is the genus *Erigeron* (Asteraceae) with several morphologically similar species on the younger island, where flavonoids fail to distinguish the species (Valdebenito et al. 1992a). The evolutionary transitions in flavonoid chemistry were inferred from phylogenies generated from published DNA data (Crawford et al. 1992b, 1993a, 1998; Sang et al. 1994, 1995; López-Sepúlveda et al. 2015a), most of them published only recently (see Chapter 13). There are few published examples of gains or losses of flavonoid classes between island plants and their continental relatives (for an earlier review, see Bohm 1998). In both *Dendroseris* and *Robinsonia*, mapping flavonoids onto molecular phylogenies reveals gains and losses of compounds and the origin of novel compounds during their evolution in the archipelago. The transitions in flavonoids occur at different stages of the biosynthetic pathway of the compounds in these two genera, from early steps in the pathway that produce different classes to later stages involving methylation and glycosylation. The most significant insight provided by flavonoids is the documentation of a close relationship between a species of *Peperomia* in the Juan Fernández Islands and one on Tristan de Cunha Island in the Atlantic Ocean (Valdebenito et al. 1990a, 1992b).

Chromosome Number Stasis during Island Speciation

Chromosome counts have been routinely made over the years from the endemic and native angiosperm flora of the Juan Fernández Islands, resulting in 48% of the dicots now having been examined (Sanders et al. 1983; Spooner et al. 1987; Sun et al. 1990; Stiefkens et al. 2001; Kiehn et al. 2005; see also Chapter 11). This has allowed us to assess the importance of chromosomal change during speciation within the archipelago. Although initial colonists of many plant groups arrived at already higher polyploid levels (66% of the flora), including the largest lineages, *Dendroseris*, *Erigeron*, and *Robinsonia*, there is no evidence of new cycles of polyploidy during evolution within the archipelago. Furthermore, there is no evidence of any change in chromosome number in any lineage within either island. This is in sharp contrast to the many changes in chromosome number that routinely take place in continental plant lineages. General comparisons with other archipelagos (Ono 1975; Ono and Masuda 1981; Carr 1998; Stuessy and Crawford 1998) have led to the concept that change in chromosome number is unusual within rapidly evolving lineages within oceanic islands. The reason for this may be that changes in chromosome number may be maladapted and selected against within the confines of narrow ecological niches. Even in species evolving anagenetically, the challenges of adaptation to the new island environment may be sufficient to minimize survival of any deviant chromosome numbers.

Reproductive Systems within the Endemic and Native Flora

Determination of reproductive systems in any flora is requisite for understanding the nature of initial colonists, for revealing isolating mechanisms and modes of speciation, and for interpreting patterns of genetic variation within and among populations. Reproductive data are also fundamental for developing proper conservation programs, especially in view of the fragile nature of many of the endemics of the Juan Fernández Islands. The aspects we have considered in detail have been pollination systems and breeding systems (Skottsberg 1928; Sun et al. 1996; Anderson et al. 2000a, 2000b, 2001; Bernardello et al. 1999, 2000, 2001, 2004; see also Chapter 12). These studies represent one of the most comprehensive analyses of reproductive features for an entire island flora. The only other island system in which reproductive aspects have been inventoried as comprehensively is the Bonin (Ogasawara) Islands (belonging to Japan) (Abe 2006).

Floral visitors are uncommon among the endemic and native species of the Juan Fernández Archipelago. Only on 11 plant species have insects (e.g., flies, moths, ants, and beetles) been documented. Other than a recently observed, and newly described, bee species, there are no other native bees on the Juan Fernández Islands (Anderson et al.

2001; Bernardello et al. 2001; see also Chapter 12). Given the infrequent, irregular, and imprecise nature of native insect association with flowers in the archipelago, there is no certainty that *any* species is truly insect pollinated. However, about 9% of the extant flora is currently hummingbird pollinated (involving one endemic and one native species of hummingbird). About 55% of the species offer nectar rewards, and only 2% offer pollen rewards. And despite a number of larger flowers, flowers with color, and those that offer nectar, we hypothesize that 47% of the flora is wind pollinated.

We infer that most of the colonizers to the archipelago were ancestrally either insect or wind pollinated. There is association between a number of current floral features and the hypothesized pollination of colonizers. We suggest that to a large extent flower size, shape, and color in the flora may reflect pollination syndromes of colonizers rather than of extant pollination. Most of the species in the Juan Fernández Archipelago have small flowers. Green is the most frequent flower color, followed by white and yellow. The presence of nectar in a number of species reflects ancestral, not contemporary, pollination relationships. The lack of alternative means of biotic pollination seems to have led in a number of instances to anemophily (wind pollination); in essence, wind pollination constitutes a default pollination system.

As for breeding systems, most species are hermaphroditic, 9% are dioecious, and 9% are monoecious. About 40% of the species are protandrous or protogynous. Experimental studies of compatibility of a number of species indicate that more than 8% of the cosexual species are self-compatible. Although most species studied are indeed self-compatible, their level of autogamy is low. Selfing mediated via geitonogamy (fertilization of a flower by pollen from another flower on the same plant) is the most frequent mechanism of pollen transfer, hence providing flora-wide support that island colonizers tend to be self-compatible ("Baker's rule") (Baker 1974; Crawford et al. 2009). Where outcrossing occurs, it is achieved mainly through dioecy and self-incompatibility, promoted by dichogamy (maturation of carpels and stamens at different times within a flower) in the hermaphroditic flowers and facilitated by wind pollination.

Phylogeny of Endemic Ferns and Flowering Plants

The simplicity of the Juan Fernández Archipelago allows modeling of the phylogenetic patterns and processes in all endemic species (Stuessy et al. 1990; Ruiz et al. 2004; Stuessy et al. 2005b; see also Chapter 13). A maximum of nineteen patterns of distribution exist among the endemic flora. There is a significant difference between the patterns of ferns versus flowering plants. Once ferns arrive in the archipelago, they tend to disperse and colonize both islands, likely due to the ease of tiny wind-dispersed reproductive spores; flowering plants often remain on only one island. The phylogenetic hypotheses for the entire flora facilitate analyses of genetic diversity and speciation in Chapters 14 and 15. This is one of the most comprehensive phylogenetic surveys for the endemic flora of any oceanic archipelago. Results show 57% of the endemic vascular flora having originated anagenetically and 41% cladogenetically, with 2% by hybridization. Degrees of divergence are greater between continental relatives and selected endemics on Robinson Crusoe Island than between the continent and Alejandro Selkirk Island, which would be expected based on the different ages of the islands.

The vast majority of lineages have diversified on Robinson Crusoe Island with subsequent dispersal to and evolution on Alejandro Selkirk Island. In other words, biogeographical parsimony prevails. Another way to frame this concept is by reference to the "progression rule" (Funk & Wagner 1995), which states that colonization typically occurs first on the oldest island in an archipelago. One conspicuous exception to this rule in the Juan Fernández Islands is *Erigeron*, which must have come from continental South America (Solbrig 1962; Noyes 2000) and established on the younger and more distant island and there adaptively radiated into six species (López-Sepúlveda et al. 2015a). A few other genera, such as *Coprosma* and *Haloragis*, doubtless colonized from the western Pacific area, but even so, these may have arrived first on Robinson Crusoe Island because for at least 2 to 3 million years it was the only island in the archipelago. Comprehensive phylogenetic studies on these two genera should be able to resolve these biogeographical alternatives.

Degrees of Genetic Variation within and among Populations of Endemic Species

In Chapter 14 we discussed different types of genetic diversity and methods to assess diversity, factors that are correlated with or shape the levels and patterns of diversity (see also Stuessy et al. 2014), as well as the use of diversity measures to guide conservation strategies. Several types of molecular markers have been employed in studies of Juan Fernández plants over several decades, initially allozymes, then RAPDs (Brauner et al. 1992; Esselman et al. 2000), ISSRs (Crawford et al. 2001a), and more recently, simple sequence repeats (SSRs) and amplified fragment-length polymorphisms (AFLPs) (López-Sepúlveda et al. 2015b; Takayama et al. 2015a).

Allozymes were used to assess genetic diversity within and among populations of species and to estimate divergence among congeneric species (Crawford et al. 1987a, 1988, 1990, 1992b, 1993c, 1994, 2001b). The results are, in general, similar to allozyme studies of plants from other archipelagos: low diversity within populations and a relatively high proportion of total species diversity distributed among populations and low divergence of congeneric species when compared with published summaries for species in general (Crawford et al. 2001b). Species with particular life-history or ecological traits correspond with differences in patterns of allozyme diversity reported for plants as a whole (Hamrick et al. 1991; Hamrick and Godt 1997). Francisco-Ortega et al. (2000) presented comparable analyses of genetic diversity/divergence among groups with different traits for the Canary Islands, and more recently, Pérez de Paz and Caujapé-Castells (2013) presented a critical review of allozyme diversity in plants of this same archipelago.

The SSR and AFLP markers were focused more sharply on comparing the relative levels of genetic diversity in the continental ancestors of island lineages with their insular relatives under the models of anagenetic and cladogenetic speciation (Stuessy et al. 2006; López-Sepúlveda et al. 2013b, 2014, 2015a, 2015b, 2016; Takayama 2015a, 2015b). The results are generally concordant with the predictions of the models, with single-island species having genetic diversity comparable with or higher than their progenitor species (anagenesis), whereas speciation within island lineages (cladogenesis) results in reduced diversity in each species compared with single-island species, ostensibly because the diversity generated in the insular lineage is partitioned among the species. This provides the first comprehensive overview of genetic diversity with anagenetic and cladogenetic speciation in an oceanic archipelago.

Modes of Speciation

We have synthesized data and observations made over several decades to produce an overview of the modes of speciation in the Juan Fernández Archipelago (Chapter 15). Allopatric speciation, that is, divergence following geographical isolation, accounts for about 70% of the species endemic to the islands (Stuessy et al. 1998e). The majority of allopatric speciation events involve divergence between an ancestral colonizer and its progenitor species in a continental source area, with no subsequent speciation in the islands. A much rarer form of allopatric speciation involves dispersal between islands in the archipelago. Habitat divergence on single islands likely accounts for about 20% of the speciation events. Factors that may have initiated or facilitated speciation on single islands include differences in flowering time, pollinator shifts, and very high or obligate self-fertilization, all of which are very rare in this archipelago (e.g., only bird pollinators, hence no biotic pollinators among which to shift). It is not known whether there are postzygotic isolating barriers among species because no experimental hybridizations have been conducted. Naturally occurring interspecific hybrids are very rare in these islands, and there is no evidence indicating that they are stabilized and isolated from their parental taxa, that is, worthy of taxonomic recognition. Three exceptions exist. An intergeneric hybrid (×*Margyracaena*) exists between a native and introduced species (Crawford et al. 1993a), but the hybrid is only known from very few plants and may be extinct in nature. *Eryngium* ×*fernandezianum* was originally described by Skottsberg (1914) as a good species, but he later interpreted it as an interspecific hybrid between *E. inaccessum* and *E. bupleuroides*. It is unknown whether this is a stabilized hybrid species or represents only occasional F_1 hybrids. The recently described fern *Pleopeltis* ×*cerro-altoensis* has been proposed as a hybrid between a native species on Alejandro Selkirk Island and a native species on Robinson Crusoe Island (Danton et al. 2015; see discussions in Chapters 13 and 15), but again, it is unclear how stabilized this hybrid is. Hybridization and apparent introgression also occur between *Gunnera peltata* and *G. bracteata* in Villagra Valley on Robinson Crusoe Island (Pacheco et al. 1991b), but these have not been accorded a binomial. There is no evidence of polyploid speciation in the archipelago, although several of the most successful radiations (largest lineages) appear to have originated from polyploid colonizers. With the notable exception of the Hawaiian Islands (e.g., Wagner and Funk 1995), there have apparently been no other syntheses of modes and mechanisms of speciation in an entire flora of an oceanic archipelago.

Plant Origins and Dispersal for the Entire Angiosperm Flora

Although previous estimates exist for geographical origins of the flora (e.g., Skottsberg 1956), modern phylogenetic insights have done much to refine our views of these continental relatives (Chapter 13). Most plant colonists to the Juan Fernández Archipelago came from South America (82%), including from Chile (12%) and the Neotropics (10%) (Chapter 16). The remaining species in descending order of importance have come from North America, New Zealand, and the Pacific area.

It is possible to infer modes of dispersal for the native and endemic flora based on observed morphological characteristics of continental and island taxa (Bernardello et al. 2006; see also Chapter 16). In the current flora, the majority of the species have dry fruits. In monoecious, andro- and gynomonoecious, and dioecious species, achenes predominate. Most species have medium to small dispersal units, and generally, the larger the flower, the larger is the fruit. Large and medium-sized dispersal units are common in shrubs and trees. Dispersal to the islands can be categorized into propagules arriving by internal bird dispersal (endozoochory), external bird dispersal (epizoochory), wind (anemochory), and water (hydrochory). Bird dispersal is estimated to be 90% for the entire vascular flora in the Juan Fernández Islands and hence is the principal mode of arrival to the islands. Within this category, 46% would have arrived externally, 35% internally, and the remaining 9% is uncertain. Sea and land birds were the most important initial long-distance dispersal agents. Monoecious, andro- and gynomonoecious, dioecious, and polygamous species were mainly carried by birds. Arrival via wind accounts for 2% of species (those with light seeds or obvious adaptations, such as the pappus in Asteraceae), with an equal percentage arriving via water surface. Approximately 10% of the species have arrived by mechanisms that cannot be determined precisely. Anemochorous (wind-dispersed) and autochorous (self-dispersed) species are mainly perennial and have medium to large unisexual flowers. Anemochorous species have small dispersal units and dull-colored flowers, whereas large dispersal units and brightly colored flowers are frequent in autochorous species. Medium-sized dispersal units are represented in autochorous or ornithochorous (bird-dispersed) species.

A Dynamic Approach to Determining Species Diversity

One of the central questions in island biology has been to determine which factors predict levels of species diversity. The equilibrium model of MacArthur and Wilson (1967) provided stimulating views of the importance of sizes of islands and distances from the major source area for immigrants. Species diversity was viewed as reaching equilibrium through opposing factors of immigration and extinction. This theory has now been largely replaced by the new general dynamic model proposed in detail by Whittaker et al. (2007, 2008, 2010; see also Borregaard et al. 2016) and as developed in parallel by our own investigations (Stuessy et al. 1998e; Stuessy 2007). This new approach involves taking the island ontogeny into account and modeling speciation and changing genetic diversity within this context. This book applies these new ideas to the flora of an entire archipelago. Calculations are provided as a means of assessing species diversity, taking into account concepts of island change over time (Chapter 17). As we have noted, the advantage of this spectacular little archipelago is its simplicity, which allows factors to be modeled for the older island and then to test these concepts by application to the younger island. Important to consider, in addition to size of island and distance from source areas, are ages of islands, dispersal capabilities of colonists, patterns of erosion and subsidence, and intra- and inter-island speciation. This point has been underlined by Steinbauer et al. (2016), who showed the importance of topography for speciation in islands.

Epilogue

Over the 35 years of our investigations on the evolution, biogeography, and conservation of the vascular flora of the Juan Fernández Archipelago, a number of insights on island biology have resulted. Some of these contributions have already been summarized in a number of general reviews (e.g., Crawford et al. 1987b; Crawford and Stuessy 1997; Baldwin et al. 1998; Stuessy and Ono 1998), but this book goes well beyond previous syntheses and presents one of the most comprehensive views of historical-biological aspects of the endemic flora of any oceanic archipelago. The concepts of evolution and biogeography presented here provide models for tests in other oceanic archipelagos.

This book also establishes a new research and conservation platform from which future investigations on the islands can more profitably be initiated. We hope that it will serve as a fundamental resource to support the Chilean government's search for international funds for stronger conservation initiatives. This book may also be helpful to support upgrading the archipelago from "tentative" (since 1994) to "official" status on the UNESCO World Heritage List. Over these three and one-half decades, we have seen a dramatic increase in sensitivity to and action by the Corporación Nacional Forestal (CONAF) toward conservation of the endemic flora of the islands. Much still needs to be done, but progress is steadily being made toward preservation of the species and vegetation of this remarkable archipelago.

Appendices

Appendix 1: Maps of Islands Showing Locations Mentioned in the Text

Key: C. = Cordón (ridge); Q. = Quebrada (ravine); Portezuelo = Mirador de Selkirk (Selkirk's Lookout). See Figs. C88 and C104 for color vegetation maps of these same islands.

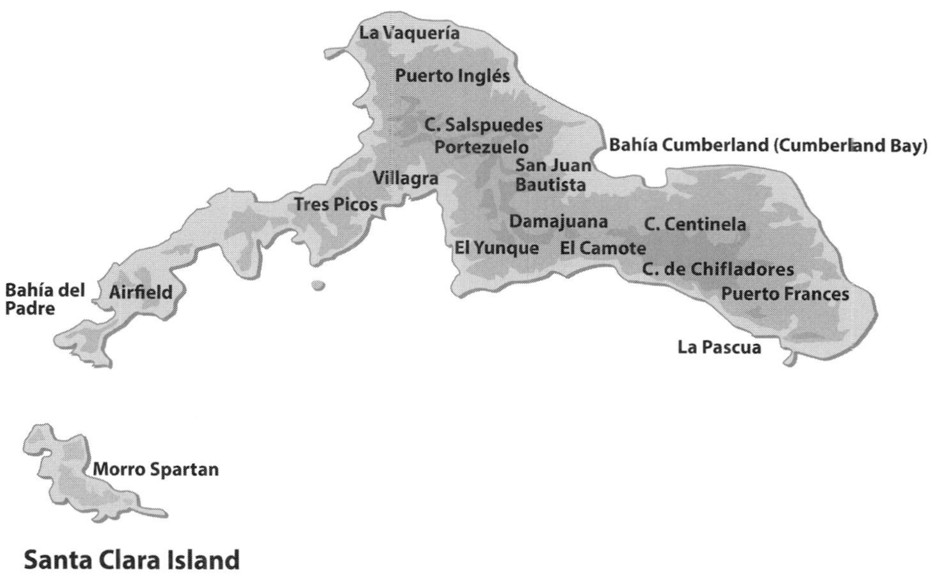

Map 1 Robinson Crusoe (= Masatierra or Más a Tierra) and Santa Clara Islands

Map 2 Alejandro Selkirk Island (= Masafuera or Más a Fuera)

Appendix 2: Nomenclatural Novelties and Notes

New Species

Erigeron stuessyi Valdebenito was described in a Ph.D. thesis by Hugo A. Valdebenito (1989, Ohio State University) but never validly published. We have used the name as a *nomen nudum* in several publications for more than twenty years (beginning with Valdebenito et al. 1992 and most recently with López-Sepúlveda et al. 2015). It is finally validly published here, having been extracted and somewhat modified from the thesis of Valdebenito (1989, pp. 61–66).

Erigeron stuessyi Valdebenito, sp. nov. – Type: CHILE: Juan Fernández Archipelago: Alejandro Selkirk Island, Cordón Atravesado down into Quebrada Vacas, east side of upper part of steep valley, 950 m, 25 Jan 1986, *A. Landero & E. Ruiz 9247* (holotype, OS; isotype, CONC).

Herbae suffruticosae perennes caespitosae, 9–20 cm altae. Folia rosulata, spathulata, apice acuta, pubescentia, margine integra vel serrato-dentata, 5–7 cm longa, 0.7–1.2 cm lata. Capitula laxe paniculata 3–6 floribus, pedunculis tenuibus villosis 4–6 cm longis. Involucrum turbinatum, 7–8 mm altum et 9–10 mm diametro. Bracteae involucrum 4 mm longae, 1 mm latae, lanceolatae, acutae, dorso dense hirsutae; bracteae interiorae subglabrae, margine dense ciliatae. Flosculi radiorum 35–42; ligulae albae tridentatae 4 mm longae, tubo 2 mm longo. Flosculi discorum 40–48, hermaphroditi; corolla tubulosa, 4–5 mm longa, 5-lobulata, lutea. Styli 1–1.5 mm longi, ad apicem attenuati pubescentes. Pappus albus, 2.1 mm longus. Achaenia compressa 1 mm longa, marginibus pubescentibus. Chromosomatum numerus $n = 27$ (Spooner et al. 1987, sub *E. rupicola*).

Additional specimens examined: Juan Fernández Archipelago, Alejandro Selkirk Island, Quebrada Casas, wet walls of the quebrada, 218 m, *P. Pacheco & E. Ruiz 6401* (OS, CONC); 250 m, 15 Jan 1986, *E. Ruiz & T. Lammers 8019* (OS, CONC).

Erigeron stuessyi is most similar to *E. rupicola*, from which it differs in vestiture, leaf color, leaf and capitulum size, and stigmatic branches of the disk florets. *Erigeron stuessyi* has pubescent leaves, bracts, and stems, which contrast strongly with the absence of pubescence in *E. rupicola*. Further, in herbarium collections, *E. stuessyi* has dark green leaves, whereas in *E. rupicola*, they appear light brown. In *E. stuessyi*, the leaves are 5–7 cm long, with margins entire, occasionally dentate, and with apex acute. In contrast, *E. rupicola* has smaller leaves (3.5–5 cm long), with margins entire and the apex obtuse. Capitula are arranged in a panicle in *E. stuessyi*; they are solitary in *E. rupicola*. The involucre is 7–9 mm in diameter and 4.5–5.5 mm tall in *E. rupicola*, whereas in *E. stuessyi* it is 9–10 mm in diameter and 7–8 mm tall. Stigmatic branches are triangular and with collecting hairs covering only the tip in *E. rupicola*, whereas in *E. stuessyi* they gradually taper toward the tip and are completely covered by the hairs (see Figs. C56 and C57 for a visual comparison).

As for ecology and distribution, the two species differ, with *E. rupicola* adapted to coastal rocks and mouths of quebradas from near sea level to 300 m, whereas *E. stuessyi* occurs in the darker and more humid canyon walls of Q. Casas and Q. Vacas (as far as known) from 200 to 1,200 m. *Erigeron rupicola* is relatively common along the eastern coast of Alejandro Selkirk Island, but *E. stuessyi* is more narrowly restricted to these two eastern quebradas. Both species flower at about the same time in January and February. See Chapters 13 through 15 for discussions of the evolution of the *Erigeron* complex on Alejandro Selkirk Island.

This species is named in honor of Tod F. Stuessy, professor emeritus of The Ohio State University, United States, and the University of Vienna, Austria.

New Combination

Zanthoxylum externum (Skottsb.) Stuessy, comb. nov. Basionym: *Fagara externa* Skottsb. Nat. Hist. Juan Fernandez (Botany) 2(2): 143. 1921. For discussion, see Chapters 5 and 13.

Notes on Names of Genera and Species

A nomenclatural issue has appeared that affects the proper spelling of one of the endemic species of *Erigeron* (Asteraceae), in fact, the most common one. Most authors, including in our own publications, have been using the name *Erigeron fernandezianus* (Colla) Solbrig. This is apparently an error, the correct name being *Erigeron fernandezia* (Colla) Harling, as we explain below. Two issues affect this name, the first being the gender of the generic name and the second being the spelling of the specific epithet.

As for gender, Linnaeus was the first to describe *Erigeron* (1753), and he treated it as neuter; epithets he used reflected this, for example, *E. philadephicum*, *E. alpinum*, as did other workers into the nineteenth century (e.g., Candolle 1836). The generic name *Erigeron* is, however, derived from the Ancient Greek *ἐρι-* (*eri-* = "truly or early") and *γέρων* (*geron* = "old man") (referring to the appearance of the white hairs of the fruit soon after flowering). Because γέρων is masculine, other specialists treated the generic name as masculine (e.g., Britton 1901). A Recommendation going back to the *Cambridge Rules* (Briquet 1935) was that "generic names formed from two or more Greek or Latin words should take the gender of the last." Because of this, *Erigeron* was given as an example (to Rec. 75A) of a name that should be masculine in the 1975 edition of the *International Code of Botanical Nomenclature* (*Leningrad Code*) (Stafleu et al. 1978). This became more clearly mandatory when such specifically voted examples were clearly marked as such in the *Tokyo Code* (Greuter et al. 1994, Art. 62 *Ex. 1), stipulating that *Erigeron* should be masculine "... for which botanical tradition has reestablished the classical gender despite another choice by Linnaeus." This example and wording are maintained in the most recent edition, now the *ICN* (McNeill et al. 2012) but without "botanical" before "tradition."

Confusion in the epithet relates to interpretation of the gender of the generic name and an attempt to correct the original spelling of the basionym or perhaps even lack of attention to detail when making combinations. This species was described by Candolle in the *Prodromus* (1836) as *Erigeron fruticosum* DC., and this name was adopted by Hemsley (1884) and Johow (1896) but treated as masculine as *E. fruticosus* DC. Both authors cited the earlier *Terranea fernandezia* Colla (1835) in synonymy, but they did not make any new combination. There appears no doubt that Colla published prior to Candolle (TL-II; Stafleu and Cowan 1976). Skottsberg (1921), another major contributor to floristics of the Juan Fernández Archipelago, also used *E. fruticosus* DC. for this species, but he did not cite *Terranea fernandezia* in synonymy.

More recently, awareness of the earlier applicable name for this species of *Erigeron* led to two proposals of new combinations. In a revision of the South American species of *Erigeron*, Solbrig (1962, 23 Nov) combined *Terranea fernandezia* Colla (but referring to it as *T. fernandeziana* Colla) into *Erigeron*, doing so incorrectly as *Erigeron fernandezianus* (Colla) Solbrig. It is not known whether this was simply an error or if he was attempting to correct what he considered a spelling mistake in the original epithet. IPNI (online) gives "Terranea (sphalm. fernandeziana) fernandezia," suggesting an error to be corrected. The *ICN* (McNeill et al. 2012) does recommend (60D.1.) use of "-(a)nus" for an epithet derived from a geographical name, but this is not a rule and cannot be used to contravene the original spelling in this case because the epithet of a species name "may be taken from any source whatever" (Art. 23.2), and "*fernandezia*" is a perfectly formed Latin noun derived from the name Fernández. Prior to Solbrig's publication, Harling (1962, 18 Jun) also made the combination from *Terranea* into *Erigeron*, but in this case he published it as "*Erigeron fernandezius*" (Colla) Harling. However, because "*fernandezia*" is a noun, originally published in apposition to *Terranea*, it retains its own gender and termination irrespective of the gender of the generic name (Art. 23.5). Consequently, Harling's spelling must be corrected, and the correct name is *Erigeron fernandezia* (Colla) Harling (in Acta Hort. Berg. 20: 108. 1962) ("*fernandezius*").

Literature Cited

Aagesen, L. 1999. Phylogeny of the tribe Colletieae, Rhamnaceae. *Bot. J. Linn. Soc.* 131: 1–43.

Aagesen, L., Medan, D., Kellermann, J., and Hilger, H. H. 2005. Phylogeny of the tribe Colletieae (Rhamnaceae): a sensitivity analysis of the plastid region *trnL-trnF* combined with morphology. *Plant Syst. Evol.* 250: 197–214.

Abbott, R., Albach, D., Ansell, S., Arntzen, J., Baird, S. J. E., et al. 2013. Hybridization and speciation. *Evol. Biol.* 26: 229–246.

Abbott, R. J., Hegarty, M. J., Hiscock, S. J., and Brennan, A. C. 2010. Homoploid hybrid speciation in action. *Taxon* 59: 1375–1386.

Abe, T. 2006. Threatened pollination systems in native flora of the Ogasawara (Bonin) Islands. *Ann. Bot.* 98: 317–334.

Adhikari, B., Milne, R., Pennington, R. T., Sarkinen, T., and Pendry, C. A. 2015. Systematics and biogeography of *Berberis* s.l. inferred from nuclear ITS and chloroplast *ndhF* gene sequences. *Taxon* 64: 39–48.

Adsersen, H. 1995. Research on islands: classic, recent, and prospective approaches. In P. M. Vitousek, L. L. Loope, and H. Adsersen (eds.), *Islands: Biological and Ecosystem Function* (pp. 7–22). New York: Springer.

AFP. 2010. Tsunami kills 5, 11 missing. *The Straits Times*, Feb. 28 (online).

Agarwal, M., Shrivastava, N., and Padh, H. 2008. Advances in molecular marker techniques in plant science. *Plant Cell Rep.* 27: 617–631.

Aguirre de Maino, I. M. (ed.). 1981. *Chile a Color: Biografías*, vol. 1. Santiago: Editorial Antártica.

Agurto, B. R. 1943. Sobre el valor económico de las Islas de Juan Fernández. Licenciado Thesis, Facultad de Ciencias Jurídicas y Sociales, Universidad de Chile, Santiago.

Ahrendt, L. 1961. *Berberis* and *Mahonia*, a taxonomic revision. *Bot. J. Linn. Soc.* 57: 1–410.

Akin, W. E. 1991. *Global patterns: Climate, Vegetation, and Soils*. Norman: University of Oklahoma Press.

Allan, R., Lindesay, J., and Parker, D. 1996. *El Niño Southern Oscillation and Climatic Variability*. Collingwood, Australia: CSIRO Publishing.

Allendorf, F. W., Luikart, G., and Aitken, S. N. 2013. *Conservation and the Genetics of Populations* (2nd edn.). West Sussex: Wiley-Blackwell.

Al-Shehbaz, I. A. 1988. The genera of Arabideae (Cruciferae, Brassicaceae) in the southeastern United States. *J. Arnold Arbor.* 69: 85–166.

Alston, R. E., and Turner, B. L. 1963. *Biochemical Systematics*. Englewood Cliffs, NJ: Prentice-Hall.

Amundsen, K., and Warnke, S. 2012. *Agrostis* species relationships based on *trnL-trnF* and *atpI-atpH* intergenic spacer regions. *Hort. Sci.* 47: 18–24.

Anderson, A., Haberle, S., Rojas, G., Seelenfreund, A., Smith, I., et al. 2002. An archaeological exploration of Robinson Crusoe Island, Juan Fernandez Archipelago, Chile. In S. Bedford, C. Sand, and D. Burley (eds.), *Fifty Years in the Field: Essays in Honour and Celebration of Richard Shutler Jr.'s Archaeological Career* (pp. 239–249). Auckland, Private Bag: New Zealand Archaeological Association (Monograph 25).

Anderson, G. and Bernardello, G. 2012. Robinson Crusoe principle: the pollination biology of the flora of the Juan Fernández Archipelago and how the species "made do" with what they had. Columbus, OH: Botany 2012 (Abstr.).

Anderson, G., Bernardello, G., and Engel, M. S. 2001a. Conservation implications of a newly discovered bee species on Isla Robinson Crusoe, Chile. *Conserv. Biol.* 15: 803–805.

Anderson, G., Bernardello, G., Opel, M., Santos Guerra, A., and Anderson, M. 2006. Reproductive biology of the dioecious Canary Islands endemic *Withania aristata* (Solanaceae). *Am. J. Bot.* 93: 1295–1305.

Anderson, G., Bernardello, G., Stuessy, T., and Crawford, D.J. 2001b. Breeding system and pollination of selected plants endemic to Juan Fernández Islands. *Am. J. Bot.* 88: 220–233.

Anderson, G., Crawford D. J., and Bernardello, G. 2013. The reproductive biology of island plants: the Juan Fernández and Canary Islands (facilitated by the work of Arnoldo Santos-Guerra). *Vieraea* 41: 109–121.

Anderson, G. J. 1979. Dioecious *Solanum* of hermaphroditic origin is an example of a broad convergence. *Nature* 282: 836–838.

Anderson, G. J., Anderson, M. K. J., and Patel, N. 2015. The ecology, evolution and biogeography of dioecy in the genus *Solanum*: with paradigms from the strong dioecy in *Solanum polygamum* to the unsuspected and cryptic dioecy in *Solanum conocarpum*. *Am. J. Bot.* 102: 471–486.

Anderson, G. J., Bernardello, G., López, P., Crawford, D. J., and Stuessy, T. 2000a. Reproductive biology of *Wahlenbergia* (Campanulaceae) endemic to Robinson Crusoe Island (Chile). *Plant Syst. Evol.* 223: 109–123.

Anderson, G. J., Bernardello, G., López, P., Stuessy, T., and Crawford, D. J. 2000b. Dioecy and wind pollination in *Pernettya rigida* (Ericaceae) of the Juan Fernández Islands. *Bot. J. Linn. Soc.* 132: 121–141.

Anderson, G. J., Bernardello, G., and Santos-Guerra, A. 2014. Reproductive biology of *Solanum vespertilio* (Solanaceae), a zygomorphic, heterantherous, enantiostylous, and andromonoecious rare Canary Islands endemic. *Plant Syst. Evol.* 301: 1191–1206.

Andrus, N., Tye, A., Nesom, G., Bogler, D., Lewis, C., et al. 2009. Phylogenetics of *Darwiniothamnus* (Asteraceae: Astereae): molecular evidence for multiple origins in the endemic flora of the Galápagos Islands. *J. Biogeogr.* 36: 1055–1069.

Anonymous. 1978. Isla Robinson Crusoe, from Surveys by the Chilean Navy, map. No. 22492. Defense Mapping Agency Hydrographic Center, Washington, DC.

APG (Angiosperm Phylogeny Group). 1998. An ordinal classification for the families of flowering plants. *Ann. Missouri Bot. Gard.* 85: 531–553.

APG (Angiosperm Phylogeny Group). 2003. An update of the Angiosperm Phylogeny Group classification for the orders and families of flowering plants: APG II. *Bot. J. Linn. Soc.* 141: 399–436.

APG (Angiosperm Phylogeny Group). 2009. An update of the Angiosperm Phylogeny Group classification for the orders and families of flowering plants: APG III. *Bot. J. Linn. Soc.* 161: 105–121.

APG (Angiosperm Phylogeny Group). 2016. An update of the Angiosperm Phylogeny Group classification for the orders and families of flowering plants: APG IV. *Bot. J. Linn. Soc.* 181: 1–20.

Appelhans, M. C., Wen, J., Wood, K. R., Allan, G. J., Zimmer, E. A., et al. 2014. Molecular phylogenetic analysis of Hawaiian Rutaceae (*Melicope*, *Platydesma* and *Zanthoxylum*) and their different colonization patterns. *Bot. J. Linn. Soc.* 174: 425–448.

Arana, P. 1987. Perspectivas históricas y proyecciones de la actividad pesquera realizada en el Archipiélago de Juan Fernández, Chile. In J. C. Castilla (ed.), *Islas Oceánicas Chilenas; Conocimiento Científico y Necesidades de Investigaciones*. Santiago: Universidad Católica de Chile.

Arana, P. 2010. *La Isla de Robinson Crusoe*. Valparaíso: Ediciones Universitarias de Valparaíso.

Arana E., P. 1985. Análisis y recomendaciones sobre medidas de regulación en la pesquería de la langosta de Juan Fernández (*Jasus frontalis*). In P. Arana E. (ed.), *Investigaciones Marinas en el Archipiélago de Juan Fernández* (pp. 291–300). Valparaíso: Escuela de Ciencias del Mar, Universidad Católica de Valparaíso.

Archibald, J. K., Crawford, D. J., Santos-Guerra, A., and Mort, M. E. 2006. The utility of automated analysis of inter-simple sequence repeat (ISSR) loci for resolving relationships in the Canary Island species of *Tolpis* (Asteraceae). *Am. J. Bot.* 93: 1154–1162.

Arditti, J., and Ghani, A. K. A. 2000. Numerical and physical properties of orchid seeds and their biological implications. *New Phytol.* 145: 367–421.

Arellano, G. 2012. Evaluación de la dinámica de invasión de *Aristotelia chilensis* (Elaeocarpaceae) y *Rubus ulmifolius* (Rosaceae) en claros de dosel en un bosque de la isla Robinson Crusoe. Tesis Magíster en Áreas Silvestres y Conservación de la Naturaleza, Universidad de Chile (Santiago), Facultad de Ciencias Forestales y de la Conservacion de la Naturaleza.

Armada de Chile. 1965. *Mapa Hidrográfico de Isla Mas a Tierra*. Valparaíso: Instituto Hidrográfico.

Astudillo M. 2014. Geomorfología y evolución geológica de la Isla Robinson Crusoe, Archipiélago Juan Fernández. Thesis, Título de Geología, Depto. de Geología, Universidad de Chile, Santiago.

August-Schmidt, E. M., Haro, G., Borager, A., and D'Antonio, C. M. 2015. Preferential associations of invasive *Lantana camara* (Verbenaceae) in a seasonally dry Hawaiian woodland. *Pacific Sci.* 69: 385–397.

Bacon, C. D., McKenna, M. J., Simmons, M. P., and Wagner, W. H. 2012. Evaluating multiple criteria for species delimitation: an empirical example using Hawaiian palms (Arecaceae: *Pritchardia*). *BMC Evol. Biol.* 2012(12): 23.

Baeza, C., Bastías, C., Stuessy, T., Ruiz, E., Greimler, J., et al. 2013. Cytological investigations on populations of *Taraxacum* (Asteraceae) from the Juan Fernández Archipelago, Chile. *Gayana Bot.* 70: 401–403.

Baeza, C., Vosyka, E., and Stuessy, T. 2004. Recuentos cromosómicos en plantas que crecen en Chile. *Darwiniana* 42: 25–29.

Baeza, C. M., Marticorena, C., Stuessy, T., Ruiz, E., and Negritto, M. 2007. Poaceae en el archipiélago de Juan Fernández (Robinson Crusoe). *Gayana Bot.* 64: 125–174.

Baeza, C. M., Rodríguez, R., Hoeneisen, F., and Stuessy, T. 1999. Consideraciones anatómicas del leño secundario de *Santalum fernandezianum* F. Phil. (Santalaceae), especie extinta del Archipiélago de Juan Fernández, Chile. *Gayana Bot.* 56: 63–65.

Baeza, C. M., Stuessy, T. F., and Marticorena, C. 2002. Notes on the Poaceae of the Robinson Crusoe (Juan Fernández) Islands, Chile. *Brittonia* 54: 154–163.

Bahamonde, N. 1966. Capitulo VIII. El mar y sus recursos. In Anonymous (ed.), *Geografía Económica de Chile, Primer Apéndice* (pp. 81–89). Santiago, Chile: Corporación de Fomento de la Producción (CORFO).

Bailey, C. D., Koch, M. A., Mummenhoff, K., Mayer, M., O'Kane, S. L., Jr., et al. 2006. Toward a global phylogeny of the Brassicaceae. *Mol. Biol. Evol.* 23: 2142–2160.

Bailey, I. W., and Smith, A. C. 1942. Degeneriaceae, a new family of flowering plants from Fiji. *J. Arnold Arbor.* 23: 356–365.

Baker, H. G. 1955. Self-compatibility and establishment after "long-distance" dispersal. *Evolution* 9: 347–348.

Baker, H. G. 1967. Support for Baker's law – as a rule. *Evolution* 21: 853–856.

Baker, H. G., and Cox, P. A. 1984. Further thoughts on dioecism and islands. *Ann. Missouri Bot. Gard.* 71: 244–253.

Baker, P. E., Gledhill, A., Harvey, P. K., and Hawkesworth, C. J. 1987. Geochemical evolution of the Juan Fernandez Islands, SE Pacific. *J. Geol. Soc.* 144: 933–944.

Baker, P. E., and Keyvan-Scocouhi, F. A. 1982. Petrology and geochemistry of the Juan Fernández Islands, south east Pacific. In H.-U. Schmincke, P. E. Baker, V. H. Forjaz (eds.), *Proceedings Symposium on the Activity of Ocean Volcanoes* (pp. 255–264). Ponta Delgada: Universidade do Açores.

Baldwin, B. G. 2003. A phylogenetic perspective on the origin and evolution of Madiinae. In S. Carlquist, B. G. Baldwin, and G. D. Carr (eds.), *Tarweeds & Silverswords: Evolution of the Madiinae (Asteraceae)* (pp. 193–228). St. Louis, MO: Missouri Botanical Garden Press.

Baldwin, B. G., Crawford, D. J., Francisco-Ortega, J., Kim, S.-C., Sang, T., et al. 1998. Molecular phylogenetic insights on the origin and evolution of oceanic island plants. In D. E. Soltis, P. Soltis, and J. J. Doyle (eds.), *Molecular Systematics of Plants,* vol. II*: DNA Sequencing* (pp. 410–441). New York: Kluwer Academic.

Ball, P. W. 1990. Some aspects of the phytogeography of *Carex*. *Canad. J. Bot.* 68: 1462–1472.

Barker, W. R. 1982. Taxonomic studies in *Euphrasia* L. (Scrophulariaceae): a revised infrageneric classification, and a revision of the genus in Australia. *J. Adelaide Bot. Gard.* 5: 1–304.

Barrera, M. E. 1997. Helechos de Juan Fernández. *Mus. Nac. Hist. Nat. Chile Publ. Ocas.* 51: 1–103.

Barrera V. H. 1963. Cambio de nombre a las islas de Juan Fernández. *Rev. Chil. Hist. Geograf.* 131: 425–430.

Barrett, S. C. H. 1995. Mating-system evolution in flowering plants: micro- and macroevolutionary approaches. *Acta Bot. Neerl.* 44: 385–402.

Barría, A. 1996. Incendio penetró bosque nativo en Juan Fernández. *El Mercurio*, February 22.

Barrier, M., Baldwin, B. G., Robichaux, R. H., and Purugganan, M. D. 1999. Interspecific hybrid ancestry of a plant adaptive radiation: allopolyploidy of the Hawaiian silversword alliance (Asteraceae) inferred from floral homeotic gene duplications. *Mol. Biol. Evol.* 16: 1105–1113.

Barrington, D. 2014. A world-level phylogenetic and biogeographic perspective on the fern genus *Polystichum*. Boise, ID: Botany 2014 (Abstr.).

Barros Arana, D. 1904. *El Doctor Don Rodolfo Amando Philippi. Su Vida i Sus Obras (1808–1904).* Santiago: Imprenta Cervantes.

Bate-Smith, E. C. 1958. Plant phenolics as taxonomic guides. *Proc. Linn. Soc.* 169: 198–211.

Bate-Smith, E. C. 1962. The phenolic constituents of plants and their taxonomic significance: I. Dicotyledons. *Bot. J. Linn. Soc.* 58: 73–95.

Beardsley, P. M., and Olmstead, R. G. 2002. Redefining Phrymaceae: the placement of *Mimulus*, tribe Mimuleae, and *Phryma*. *Am. J. Bot.* 89: 1093–1102.

Beardsley, P. M., Schoenig, S. E., Whittall, J. B., and Olmstead, R. G. 2004. Patterns of evolution in western North American *Mimulus* (Phrymaceae). *Am. J. Bot.* 91: 474–489.

Beheregaray, L. B., Gibbs, J. P., Havell, N., Fritts, T. H., Powell, J. R., et al. 2004. Giant tortoises are not so slow: rapid diversification and biogeographic consensus in the Galápagos. *Proc. Natl. Acad. Sci. USA* 101: 6514–6519.

Bellin, J.-N. 1764. *Le Petit Atlas Maritime*, vol. 2 (map 68). Paris.

Bennett, J. R., and Mathews, S. 2006. Phylogeny of the parasitic plant family Orobanchaceae inferred from phytochrome A. *Am. J. Bot.* 93: 1039–1051.

Bernardello, G., Aguilar, R., and Anderson, G. J. 2004. The reproductive biology of *Sophora fernandeziana* (Leguminosae), a vulnerable endemic species from Isla Robinson Crusoe. *Am. J. Bot.* 91: 198–206.

Bernardello, G., Anderson, G., Lopez, P., Cleland, M. A., Stuessy, T. F., et al. 1999. Reproductive biology of *Lactoris fernandeziana* (Lactoridaceae). *Am. J. Bot.* 86: 829–840.

Bernardello, G., Anderson, G. J., Stuessy, T. F., and Crawford, D. J. 2001. A survey of floral traits, breeding systems, floral visitors, and pollination systems of the angiosperms of the Juan Fernández Islands (Chile). *Bot. Rev. (Lancaster)* 67: 255–308.

Bernardello, G., Anderson, G. J., Stuessy, T. F., and Crawford, D. J. 2006. The angiosperm flora of the Archipelago Juan Fernández (Chile): origin and dispersal. *Can. J. Bot.* 84: 1266–1281.

Bernardello, G., Galetto, L., and Anderson, G. J. 2000. Floral nectary structure and nectar chemical composition of some species from Robinson Crusoe Island (Chile). *Can. J. Bot.* 78: 862–872.

Berry, E. W. 1911. A revision of the fossil ferns from the Potomac Group which have been referred to the genera *Cladophlebis* and *Thyrsopteris*. *Proc. US Nat. Mus.* 41: 307–332.

Bertero, C. G. L. 1830. Notice sur l'histoire naturelle de l'ile Juan Fernandez, extraite d'une lettre de M. Bertero. *Ann. Sci. Nat.* 21 : 344–351.

Beurton, C. 1994. Gynoecium and perianth in *Zanthoxylum* s.l. (Rutaceae). *Plant Syst. Evol.* 189: 165–191.

Biffin, E., Lucas, E. J., Craven, L. A., Ribeiro da Costa, I., Harrington, M. G., and Crisp, M. D. 2010. Evolution of exceptional species richness among lineages of fleshy-fruited Myrtaceae. *Ann. Bot.* 106: 79–93.

Biodiversa. 2009a. *Archipiélago Juan Fernández. Sitio Prioritario para la Conservación de la Biodiversidad Global. Sistematización del Estado Actual del Conocimiento*. Santiago: DIPROREN-SAG.

Biodiversa. 2009b. *Priorización y Definición de Objectos de Conservación y Líneas de Acción para la Elaboración del Plan de Conservación de la Biodiversidad Terrestre del Archipiélago Juan Fernández*. Santiago: DIPROREN-SAG.

Bitter, G. 1910. Die Gattung *Acaena*. Vorstudien zu einer Monographie. *Bibliothec. Bot.* 74: 1–336.

Bitter, G. 1921. Ein Gattungsbastard zwischen *Acaena* und *Margyricarpus*: *Margyracaena*. *Feddes Repert. Spec. Nov. Regni Veg.* 17: 239–243.

Bittner, M., Silva, M., Rozas, Z., Jakupovic, J., and Stuessy, T. 1994. Estudio químico del género *Gunnera* en Chile: II. Metabolitos secundários de dos especies continentales y dos especies de las islas de Juan Fernández. *Bol. Soc. Chil. Quim.* 39: 79–83.

Boeger, M. R. T., Cavichiolo, L. E., Pil, M. W., and Labiak, P. H. 2007. Variabilidade fenotípica de *Rumohra adiantiformis* (G.Forst) Ching (Dryopteridaceae). *Hoehnea* 34: 553–561.

Böhle, U. R., Hilger, H. H., and Martin, W. F. 1996. Island colonization and evolution of the insular woody habit in *Echium* L. (Boraginaceae). *Proc. Natl. Acad. Sci. USA* 93: 11740–11745.

Bohm, B. A. 1998a. *Introduction to Flavonoids*. Amsterdam: Harwood Academic.

Bohm, B. A. 1998b. Secondary compounds and evolutionary relationships of island plants. In T. F. Stuessy and M. Ono (eds.), *Evolution and Speciation of Island Plants* (pp. 233–295). Cambridge University Press.

Bohm, B. A., and Fong, C. 1990. Non-polar flavonoids of *Wilkesia* and *Argyroxiphium*. *Phytochemistry* 29: 1175–1177.

Bohm, B. A., and Stuessy, T. F. 2001. *Flavonoids of the Sunflower Family (Asteraceae)*. Wien: Springer-Verlag.

Bohm, B. A., and Yang, J. Y. 2003. Secondary metabolites of Madiinaae and the tarweeds. In S. Carlquist, B. G. Baldwin, and G. D. Carr (eds.), *Tarweeds & Silverswords: Evolution of the Madiinae (Asteraceae)* (pp. 145–170). St. Louis, MO: Missouri Botanical Garden Press.

Bolnick, D. I., and Fitzpatrick, B. M. 2007. Sympatric speciation: models and empirical evidence. *Ann. Rev. Ecol. Evol. Syst.* 38: 459–487.

Bonin, A., Ehrich, D., and Manel, S. 2007. Statistical anaysis of amplified fragment length polymorphism data: a toolbox for molecular ecologists and evolutionists. *Mol. Ecol.* 16: 3737–3758.

Booker, J., Bullard, E. C., and Grasty, R. L. 1967. Palaeomagnetism and age of rocks from Easter Island and Juan Fernandez. *Geophys. J. R. Astronom. Soc.* 12: 469–471.

Borgen, L. 1976. Analysis of a hybrid swarm between *Argyranthemum adauctum* and *A. filifolium* in the Canary Islands. *Norweg. J. Bot.* 23: 121–137.

Borregaard, M. K., Matthews, T. J., and Whittaker, R. J. 2016. The general dynamic model: towards a unified theory of island biogeography? *Global Ecol. Biogeogr.* 25: 805–816.

Bottini, M. C. J., De Bustos, A., Jouve, N., and Poggio, L. 2002. AFLP characterization of natural populations of *Berberis* (Berberidaceae) in Patagonia, Argentina. *Plant Syst. Evol.* 231: 133–142.

Bottini, M. C. J., De Bustos, A., Sanso, A. M., Jouve, N., and Poggio, L. 2007. Relationships in Patagonian species of *Berberis* (Berberidaceae) based on the characterization of rDNA internal transcribed spacer sequences. *Bot. J. Linn. Soc.* 153: 321–328.

Bourne, W. R. P., Brooke, M. de L., Clark, G. S., and Stone, T. 1992. Wildlife conservation problems in the Juan Fernández Archipelago, Chile. *Oryx* 26: 43–51.

Bouzat, J. L. 2010. Conservation genetics of population bottlenecks: the role of chance, selection, and history. *Conserv. Genet.* 11: 463–478.

Bozek, M., Leitch, A. R., Leitch, I. J., Záveská Drábková, L., and Kuta, E. 2012. Chromosome and genome size variation in *Luzula* (Juncaceae), a genus with holocentric chromosomes. *Bot. J. Linn. Soc.* 170: 529–541.

Bramwell, D. and Caujapé-Castells, J. (eds.). 2011. *The Biology of Island Floras*. Cambridge: Cambridge University Press.

Bräuchler, C., Meimberg, H., and Heubl, G. 2010. Molecular phylogeny of Menthinae (Lamiaceae, Nepetoideae, Mentheae): taxonomy, biogeography and conflicts. *Mol. Phylogenet. Evol.* 55: 501–523.

Braun-Blanquet, J. 1928. *Pflanzensoziologie, Grundzüge der Vegetationskunde*. Berlin: Springer.

Braun-Blanquet, J. 1964. *Pflanzensoziologie. Grundzüge der Vegetationskunde* (3rd edn.). Wien: Springer.

Brauner, S., Crawford, D. J., and Stuessy, T. F. 1992. Ribosomal and RAPD variation in the rare plant family Lactoridaceae. *Am. J. Bot.* 79: 1436–1439.

Bremer, K. 1994. *Asteraceae: Cladistics and Classification*. Portland, OR: Timber Press.

Briquet, J. 1935. *International Rules of Botanical Nomenclature . . . Revised by the International Botanical Congress of Cambridge 1930*. Jena: Gustav Fischer.

Britton, N. L. 1901. *Manual of the Flora of the Northern States and Canada*. New York: Henry Holt.

Brizicky, G. K. 1962. Taxonomic and nomenclatural notes on *Zanthoxylum* and *Glycosmis* (Rutaceae). *J. Arnold Arbor.* 43: 80–93.

Brochmann, C., Borgen. L., and Stabbetorp, O. E. 2000. Multiple diploid hybrid speciation of the Canary Island endemic *Argyranthemum sundingii* (Asteraceae). *Plant Syst. Evol.* 220: 77–92.

Brooke, M. de L. 1987a. *The Birds of the Juan Fernández Islands, Chile*. Cambridge: International Council for Bird Preservation, Fauna and Flora Preservation Society, World Wildlife Fund.

Brooke, M. de L. 1987b. Population estimates and breeding biology of the petrels *Pterodroma externa* and *P. longirostris* on Isla Alejandro Selkirk, Juan Fernandez Archipelago. *Condor* 89: 581–586.

Brotherus, V. F. 1924. The musci of the Juan Fernández Islands. In C. Skottsberg (ed.), *The Natural History of the Juan Fernandez and Easter Island, vol. 2: Botany* (pp. 409–448). Uppsala: Almqvist & Wiksells.

Brüggen, J. O. 1950. *Fundamentos de la Geología de Chile*. Santiago: Instituto Geográfico Militar.

Bruhl, J. J. 1995. Sedge genera of the world: relationships and a new classification of the Cyperaceae. *Austral. Syst. Bot.* 8: 125–305.

Brummitt, R. K. 2014. Taxonomy versus cladonomy in the dicot families. *Ann. Missouri Bot. Gard.* 100: 89–99.

Brys, R., Vanden Broeck, A., Mergeay, J., and Jacquemyn, H. 2013. The contribution of mating system variation to reproductive isolation in two closely related *Centaurium* species (Gentianaceae) with a generalized flower morphology. *Evolution* 68: 1281–1293.

Buckley, R. C., and Knedlhans, S. B. 1986. Beachcomber biogeography: interception of dispersing propagules by islands. *J. Biogeogr.* 13: 69–70.

Burrows, C. J. 1994. The seeds always know best. *New Zealand J. Bot.* 32: 349–363.

Bush, C. M., Lu, L., Fritsch, P. W., Li, D.-Z., and Kron, K. A. 2009. Phylogeny of Galutherieae (Ericaceae: Vaccinioideae) based on DNA sequence data from *matK*, *ndhF*, and nrITS. *Int. J. Plant Sci.* 170: 355–364.

Butlin, R. K., Galindo, J., and Grahame, J. W. 2008. Sympatric, parapatric or allopatric: the most important way to classify speciation? *Philos. Trans. R. Soc. Lond. B.* 363: 2997–3007.

Cabrera, A. L. 1949. El género "Senecio" en Chile. *Lilloa* 15: 27–501.

Callander, J. 1768. *Terra Australis Cognita, or Voyages to the Terra Australis, or Southern Hemisphere, during the Sixteenth, Seventeenth, and Eighteenth Centuries*, vol. 2. Edinburgh (see pp. 528–534 for Capt. A. Cowley).

Calviño, C. I., Martínez, S. G., and Downie, S. R. 2008a. The evolutionary history of *Eryngium* (Apiaceae, Saniculoideae): rapid radiations, long distance dispersals, and hybridizations. *Mol. Phylogenet. Evol.* 46: 1129–1150.

Calviño, C. I., Martínez, S. G., and Downie, S. R. 2008b. Morphology and biogeography of Apiaceae subfamily Saniculoideae as inferred by phylogenetic analysis of molecular data. *Am. J. Bot.* 95: 196–214.

Calviño, C. I., Martínez, S. G., and Downie, S. R. 2010. Unraveling the taxonomic complexity of *Eryngium* L. (Apiaceae, Saniculoideae): phylogenetic analysis of 11 non-coding cpDNA loci corroborates rapid radiations. *Plant Div. Evol.* 128: 137–149.

Camus P., Castro, S., and Jaksic, F. 2008. El conejo europeo en Chile: historia de una invasión biológica. *Historia* 41: 305–339.

Candolle, A. P. de. 1833. Genres nouveaux appartenant a la famille des Composées ou Synantherées. *Arch. Bot.* 2: 330–334.

Candolle, A. P. de. 1836. *Prodromus Systematis Naturalis Regni Vegetabilis*, vol. 5. Paris: Treuttel et Würtz.

Candolle, A. P. 1838. *Balbisia* and *Robinsonia*. In *Prodromus Systematis Naturalis Regni Vegetabilis*, Part 6 (pp. 447–448). Paris: Treuttel et Würtz.

Cantley, J. T., Swenson, N. G., Markey, A., and Keeley, S. C. 2014. Biogeographic insights on Pacific *Coprosma* (Rubiaceae) indicate two colonizations to the Hawaiian Islands. *Bot. J. Linn. Soc.* 174: 412–424.

Carlquist, S. 1964. Morphology and relationships of Lactoridaceae. *Aliso* 5: 421–435.

Carlquist, S. 1965. *Island Life*. Garden City, NY: Natural History Press.

Carlquist, S. 1966. The biota of long-distance dispersal: IV. Genetic systems in the floras of oceanic islands. *Evolution*: 20: 433–455.

Carlquist, S. 1967. Anatomy and systematics of *Dendroseris* (sensu lato). *Brittonia* 19: 99–121.

Carlquist, S. 1970. Wood anatomy of insular species of *Plantago* and the problem of raylessness. *Bull. Torrey Bot. Club* 97: 353–361.

Carlquist, S. 1974. *Island Biology*. New York: Columbia University Press.

Carlquist, S. 1980. *Hawaii: A Natural History* (2nd edn.). Honolulu: Pacific Tropical Botanical Garden.

Carlquist, S. 1981. Chance dispersal. *Am. Sci.* 69: 509–516.

Carlquist, S. 1990. Wood anatomy and relationships of Lactoridaceae. *Am. J. Bot.* 77: 1498–1505.

Carlquist, S. 1992. Wood anatomy and stem of *Chloranthus*: summary of wood anatomy of Chloranthaceae, with comments on relationships, vessellessness, and the origin of monocotyledons. *IAWA Bull.* 13: 3–16.

Carlquist, S. 2009. Darwin on island plants. *Bot. J. Linn. Soc.* 161: 20–25.

Carlsen, T., Bleeker, W., Hurka, H., Elven, R., and Brochmann, C. 2009. Biogeography and phylogeny of *Cardamine* (Brassicaceae). *Ann. Missouri Bot. Gard.* 96: 215–236.

Caro, T., Darwin, J., Forrester, T., Ledoux-Bloom, C., and Wells, C. 2011. Conservation in the Anthropocene. *Conserv. Biol.* 26: 185–188.

Carpenter, K. J. 1986. *The History of Scurvy and Vitamin C*. Cambridge University Press.

Carr, G. D. 1985. Monograph of the Hawaiian Madiinae (Asteraceae): *Argyroxiphium*, *Dubautia*, and *Wilkesia*. *Allertonia* 4: 1–123.

Carr, G. D. 1998. Chromosome evolution and speciation in Hawaiian flowering plants. In T. F. Stuessy and M. Ono (eds.), *Evolution and Speciation of Island Plants* (pp. 5–47). Cambridge University Press.

Carr, G. D. 2003. Hybridization in Madiinae. In S. Carlquist, B. G. Baldwin, and G. D. Carr (eds.), *Tarweeds & Silverswords: Evolution of the Madiinae (Asteraceae)* (pp. 79–104). St. Louis: Missouri Botanical Garden Press.

Carr, G. D., and Kyhos, D. W. 1981. Adaptive radiation in the Hawaiian silversword alliance (Compositae-Madiinae): I. Cytogenetics of spontaneous hybrids. *Evolution* 35: 543–556.

Carr, G. D., and Kyhos, D. W. 1986. Adaptive radiation in the Hawaiian silversword alliance (Compositae-Madiinae): II. Cytogenetics of artificial and natural hybrids. *Evolution* 40: 959–976.

Castilla, J. C., and Oliva, D. 1987. Islas oceánicas Chilenas: aspectos descriptivos y potencialidades. In J. C. Castilla (ed.), *Islas Oceánicas Chilenas: Conocimiento Científico y Necesidades de Investigaciones* (pp. 17–35). Santiago: Ediciones Universidad Católica de Chile.

Castro, M., and Rosselló, J. A. 2007. Karyological observations on plant taxa endemic to the Balearic Islands. *Bot. J. Linn. Soc.* 153: 463–476.

Caujapé-Castells, J. 2010. General G_{st} and Ø inflation due to biased intra-population sampling, and its consequences for the conservation of the Canarian Flora. *Conserv. Genet.* 11: 709–720.

Caujapé-Castells, J. 2011. Jesters, red queens, boomerangs and surfers: a molecular outlook on the diversity of the Canarian endemic flora. In D. Bramwell and J. Caujapé-Castells (eds.), *The Biology of Island Floras* (pp. 284–324). Cambridge University Press.

Caujapé-Castells, J., Marrero-Rodríguez, A., Baccarani-Rosas, M., Carrera-García, N., and Vilches-Navarrete, B. 2008a. Population genetics of the endangered Canarian endemic *Atractylis arbuscula* (Asteraceae): implications for taxonomy and conservation. *Plant Syst. Evol.* 273: 99–109.

Caujapé-Castells, J., Naranjo-Suárez, J., Santanna, I., Baccarani-Rosas, M., Carrera-García, N., et al. 2008b. Population genetic suggestions to offset the extinction ratchet in the endangered Canarian endemic *Atractylis preauxiana* (Asteraceae). *Plant Syst. Evol.* 273: 191–199.

Caujapé-Castells, J., Sabbagh, I., Castllano, J. J., Ramos, R., Henriquez, V., et al. 2013. Transformer-4 version 2.0.1, a free multi-platform software to quickly reformat genotype matrices of any marker type, and archive them in the Demiurge information system. *Mol. Ecol. Resour.* 13: 484–493.

Caujapé-Castells, J., Tye, A., Crawford, D. J., Santos-Guerra, A., Sakai, A., et al. 2010. Conservation of oceanic island floras: present and future global challenges. *Perspect. Plant Ecol. Evol. Syst.* 12: 107–129.

Cavieres, A., Mellado, F., Tramón, S., and Rebolledo, S. 2011. *Plan de Conservación. Biodiversidad Terrestre. Archipiélago Juan Fernández*. Santiago: Fundación Biodiversa.

Cereceda, P., Schemenauer, R. S., and Zunino, H. 1994. Distribución de la precipitación en isla Robinson Crusoe. *Rev. Geografía Norte Grande* 21: 33–36.

Chamorro, S., Ruben, H., Olesen, J., McMullen, C., and Traveset, A. 2012. Pollination patterns and plant breeding systems in the Galápagos: a review. *Ann. Bot.* 110: 1489–1501.

Chao, Y.-S., Rouhan, G., Amoroso, V. B., and Chiou, W.-L. 2014. Molecular phylogeny and biogeography of the fern genus *Pteris* (Pteridaceae). *Ann. Bot.* 114: 109–124.

Chase, M. W., Knapp, S., Cox, A. V., Clarkson, J. J., Butsko, Y., et al. 2003. Molecular systematics, GISH and the origin of hybrid taxa in *Nicotiana* (Solanaceae). *Ann. Bot.* 92: 107–127.

Chase, M. W., Morton, C. M., and Kallunki, J. A. 1999. Phylogenetic relationships of Rutaceae: a cladistic analysis of the subfamilies using evidence from *rbcL* and *atpB* sequence variation. *Am. J. Bot.* 86: 1191–1199.

Chase, M. W., Zmarzty, S., Lledo, M. D., Wurdack, K. J., Swensen, S. M., et al. 2002. When in doubt, put it in *Flacourtiaceae*: a molecular phylogenetic analysis based on plastid *rbcL* DNA sequences. *Kew Bull.* 57: 141–181.

Chemisquy, M. A. 2009. Novedades nomenclaturales en el género *Gavilea* (Orchidaceae, Chloraeinae), con especial énfasis en las especies chilenas. *Darwiniana* 47: 315–320.

Chemisquy, M. A., and Morrone, O. 2010. Phylogenetic analysis of the subtribe Chloraeinae (Orchidaceae): a preliminary approach based on three chloroplast markers. *Austral. Syst. Bot.* 23: 38–46.

Chemisquy, M. A., and Morrone, O. 2012. Molecular phylogemy of *Gavilea* (Chloraeinae: Orchidaceae) using plastid and nuclear markers. *Mol. Phylogenet. Evol.* 62: 889–897.

Christenhusz, M. J. M., and Chase, M. W. 2014. Trends and concepts in fern classification. *Ann. Bot.* 113: 571–594.

Christenhusz, M. J. M., Zhang, X.-C., and Schneider, H. 2011. A linear sequence of extant families and genera of lycophytes and ferns. *Phytotaxa* 19: 7–54.

Christensen, C. 1910. On some species of ferns collected by Dr. Carl Skottsberg in temperate South America. *Arkiv för Bot.* 10: 1–32.

Christensen, C., and Skottsberg, C. 1920. The Pteridophyta of the Juan Fernandez Islands. In C. Skottsberg (ed.), *The Natural History of Juan Fernandez and Easter Island*, vol. 2 (pp. 1–46). Uppsala: Almqvist & Wiksells.

Christophersen, E. 1968. Flowering plants from Tristan da Cunha. *Results Norweg. Scient. Exped. Tristan da Cunha 1937–1938*, 55: 1–29.

Cisternas, M. A., Salazar, G. A., Verdugo, G., Novoa, P., Calderon, X., et al. 2012. Phylogenetic analysis of Chloraeinae (Orchidaceae) based on plastid and nuclear DNA sequences. *Bot. J. Linn. Soc.* 168: 258–277.

Clark, L. G. 1989. Systematics of *Chusquea* section *Swallenochloa*, section *Verticillatae*, section *Serpentes*, and section *Longifoliae* (Poaceae-Bambusoideae). *Syst. Bot. Monogr.* 27: 1–127.

Clarke, T. 2001. *Searching for Crusoe: A Journey Among the Last Real Islands.* New York: Ballantine Books.

Clarkson, J. J., Knapp, S., Garcia, V. F., Olmstead, R. G., Leitch, A. R., et al. 2004. Phylogenetic relationships in *Nicotiana* (Solanaceae) inferred from multiple plastid DNA regions. *Mol. Phylogenet. Evol.* 33: 75–90.

Coates, D. J., and Byrne, M. 2005. Genetic variation in plant populations: assessing cause and pattern. In R. J. Henry (ed.), *Plant Diversity and Evolution: Genotypic and Phenotypic Variation in Higher Plants* (pp. 139–164). Wallingford: CAB International.

Cody, M. 2006. *Plants on Islands: Diversity and Dynamics on a Continental Archipelago.* Berkeley: University of California Press.

Cody, M. L., and Overton, J. M. 1996. Short-term evolution of reduced dispersal in island plant populations. *J. Ecol.* 84: 53–61.

Cole, F. R., Medeiros, A. C., Loope, L. L., and Zuehlke, W. 1992. Effects of the Argentine ant on arthropod fauna of Hawaiian high-elevation shrubland. *Ecology* 73: 1313–1322.

Colla, A. 1835. Plantae rariores in regionibus Chilensibus a clarissimo M. D. Bertero nuper detectae. *Mem. Reale Accad. Sci. Torino* 38: 1–42.

Colla, A. 1836. *Herbarium Pedemontanum Justa Methodum Naturalem Dispositum Additis nonnullis Stirpibus Exoticis ad Universos ejusdem Methodi Ordines Exhibendos*, vol. 6. Torino (cf. pp. 215–216).

Colwell, R. K. 1989. Hummingbirds of the Juan Fernández Islands: natural history, evolution and population status. *Ibis* 131: 548–566.

CONAF. 1976. *Plan de Manejo Parque Nacional Juan Fernández.* Santiago: Organización de las Naciones Unidas para la Agricultura y la Alimentación.

CONAF. 2009. *Plan de Manejo Parque Nacional Archipiélago Juan Fernández.* Valparaíso: Gobierno de Chile, Ministerio de Agricultura.

Conner, J. K., and Hartl, D. L. 2004. *A Primer of Ecological Genetics*. Sunderland, MA: Sinauer Associates.

Constance, L. 1964. Systematic botany: an unending synthesis. *Taxon* 13: 257–273.

Contreras-M., A., and Spooner, D. M. 1999. Revision of *Solanum* section *Etuberosum* (subgenus Potatoe). In M. Nee, D. E. Symon, R. N. Lester, and J. P. Jessop (eds.), *Solanaceae IV* (pp. 227–245). Kew: Royal Botanic Gardens.

Cooke, E. 1712. *A Voyage to the South Seas and Round the World Performed in the Years 1708, 1709, 1710 and 1711*. London.

Cordingly, D. 2007. *Cochrane the Dauntless. The Life and Adventures of Admiral Thomas Cochrane, 1775–1860*. London: Bloomsbury.

Correa, M. N. 1956. Las especies argentinas del género *Gavilea*. *Bot. Soc. Argent. Bot*. 6: 73–86.

Correa, M. N. 1966. Una especie nueva y algunas observaciones críticas del género *Gavilea* (Orchidaceae). *Bol. Soc. Argent. Bot*. 11: 60–66.

Correa, M. N. 1968. Una nueva especie del género *Gavilea* (Orchidaceae). *Rev. Mus. de la Plata, secc. Bot*. 11: 75–77.

Cowie, R. H., and Holland, B. S. 2006. Dispersal is fundamental to biogeography and the evolution of biodiversity on oceanic islands. *J. Biogeogr*. 33: 193–198.

Cox, C. B., and Moore, P. D. 1993. *Biogeography: An Ecological and Evolutionary Approach* (5th edn.). Oxford: Blackwell Science.

Coyne, J. A., and Orr, H. A. 2004. *Speciation*. Sunderland, MA: Sinauer Associates.

Crawford, D. J. 1978. Flavonoid chemistry and angiosperm evolution. *Bot. Rev*. 44: 431–456.

Crawford, D. J. 1983. Phylogenetic and systematic inferences from electrophoretic studies. In S. D. Tanksley and T. J. Orton (eds.), *Isozymes in Plant Genetics and Breeding*, Part A (pp. 257–287). Amsterdam: Elsevier.

Crawford, D. J. 1985. Electrophoretic data and plant speciation. *Syst. Bot*. 10: 405–416.

Crawford, D. J. 1989. Enzyme electrophoresis and plant systematics. In D. E. Soltis and P. S. Soltis (eds.), *Isozymes in Plant Biology* (pp. 146–164). Portland, OR: Dioscorides.

Crawford, D. J. 1990. *Plant Molecular Systematics: Macromolecular Approaches*. New York: Wiley.

Crawford, D. J. 2010. Progenitor-derivative species pairs and plant speciation. *Taxon* 59: 1413–1423.

Crawford, D. J. 2014. Genetic diversity in plant populations. In T. F. Stuessy, D. J. Crawford, D. E. Soltis, and P. S. Soltis (eds.), *Plant Systematics: The Origin, Interpretation, and Ordering of Plant Biodiversity* (pp. 85–98). Königstein: Koeltz Scientific Books.

Crawford, D. J., Anderson, G. J., and Bernardello, G. 2011. The reproductive biology of island plants. In D. Bramwell and J. Caujapé-Castells (eds.), *The Biology of Island Floras* (pp. 11–36). Cambridge: Cambridge University Press.

Crawford, D. J., Anderson, G. J., Bernardello, G., and Santos-Guerra, A. 2013. The peproductive biology of island plants: the Juan Fernández and Canary Islands. In J. Caujapé-Castells, G. Niet, and J. M. Fernández-Palacios (eds.), *Proceedings of the Amurga International Conferences on Island Biodiversity 2011: Present and Emerging Knowledge on the Evolution, Diversity and Conservation of the Canarian Flora* (pp. 166–175). Las Palmas de Gran Canaria: Fundación Canaria Amurga-Maspalomas, Jardin Botanico Canario "Viera y Clavijo."

Crawford, D. J., Anderson, G. J., Silva, L. B., Menezes de Sequeira, M., Moura, M., et al. 2015. Breeding systems in *Tolpis* (Asteraceae) in the Macaronesian islands: the Azores, Madeira and the Canaries. *Plant Syst. Evol*. 301: 1191–1206.

Crawford, D. J., Archibald, J. K., Mort, M. E., and Santos-Guerra, A. 2009. Pollen fertility of synthetic intra- and interspecific F_1 hybrids in Canary Island *Tolpis* (Asteraceae). *Plant Syst. Evol.* 279: 103–113.

Crawford, D. J., Brauner, S., Cosner, M. B., and Stuessy, T. F. 1993a. Use of RAPD markers to document the origin of the intergeneric hybrid ×*Margyracaena skottsbergii* (Rosaceae) on the Juan Fernandez Islands. *Am. J. Bot.* 80: 89–92.

Crawford, D. J., Lowrey, T. K., Anderson, G. J., Bernardello, G., Santos-Guerra, A., et al. 2009a. Genetic diversity in the colonizing ancestors of Asteraceae endemic to oceanic islands: Baker's law and polyploidy. In V. A. Funk, A. Susanna, T. F. Stuessy, and R. Bayer (eds.), *Systematics, Evolution, and Biogeography of the Compositae* (pp. 137–151). Vienna: International Association of Plant Taxonomy

Crawford, D. J., Ruiz, E., Stuessy, T. F., Tepe, E., Aqueveque, P., et al. 2001b. Allozyme diversity in the endemic flowering plant species of the Juan Fernández Archipelago, Chile: ecological and historical factors with implications for conservation. *Am. J. Bot.* 88: 2195–2203.

Crawford, D. J., Sang, T., Stuessy, T. F., Kim, S.-C., and Silva O., M. 1998. *Dendroseris* (Asteraceae: Lactuceae) and *Robinsonia* (Asteraceae: Senecioneae) on the Juan Fernández Islands: similarities and differences in biology and phylogeny. In T. F. Stuessy and M. Ono (eds.), *Evolution and Speciation of Island Plants* (pp. 97–119). Cambridge: Cambridge University Press.

Crawford, D. J., and Stuessy, T. F. 1997. Plant speciation on oceanic islands. In K. Iwatsuki and P. H. Raven (eds.), *Evolution and Diversification of Land Plants* (pp. 249–267). Tokyo: Springer-Verlag.

Crawford, D. J., and Stuessy, T. F. 2016. Cryptic variation, molecular data, and the challenge of conserving plant diversity in oceanic archipelagos: the critical role of plant systematics. *Korean J. Plant Taxon.* 46: 129–148.

Crawford, D. J., Stuessy, T. F., Cosner, M. B., Haines, D. W., and Silva O., M. 1993b. Ribosomal and chloroplast DNA restriction site mutations and the radiation of *Robinsonia* (Asteraceae, Senecioneae) on the Juan Fernández Islands. *Plant Syst. Evol.* 184: 233–239.

Crawford, D. J., Stuessy, T. F., Cosner, M. B., Haines, D. W., Silva O., M., et al. 1992a. Evolution of the genus *Dendroseris* (Asteraceae: Lactuceae) on the Juan Fernández Islands: evidence from chloroplast and ribosomal DNA. *Syst. Bot.* 17: 676–682.

Crawford, D. J., Stuessy, T. F., Cosner, M. B., Haines, D., Wiens, D., et al. 1994. *Lactoris fernandeziana* (Lactoridaceae) of the Juan Fernández Islands: allozyme uniformity and field observations. *Conserv. Biol.* 8: 277–280.

Crawford, D. J., Stuessy, T. F., Haines, D. W., Cosner, M. B., Silva O., M., et al. 1992b. Allozyme diversity within and divergence among four species of *Robinsonia* (Asteraceae: Senecioneae), a genus endemic to the Juan Fernandez Islands, Chile. *Am. J. Bot.* 79: 962–966.

Crawford, D. J., Stuessy, T. F., Lammers, T. G., Silva O., M., and Pacheco, P. 1990. Allozyme variation and evolutionary relationships among three species of *Wahlenbergia* (Campanulaceae) in the Juan Fernández Islands. *Bot. Gaz.* 151: 119–124.

Crawford, D. J., Stuessy, T. F., Rodríguez, R., and Rondinelli, M. 1993c. Genetic diversity in *Rhaphithamnus venustus* (Verbenaceae), a species endemic to the Juan Fernández Islands. *Bull. Torrey Bot. Club.* 120: 23–28.

Crawford, D. J., Stuessy, T. F., and Silva, M. 1986. Leaf flavonoid chemistry and the relationships of Lactoridaceae. *Plant Syst. Evol.* 153: 133–139.

Crawford, D. J., Stuessy, T. F., and Silva O., M. 1987a. Allozyme divergence and the evolution of *Dendroseris* (Compositae: Lactuceae) on the Juan Fernández Islands. *Syst. Bot.* 12: 435–443.

Crawford, D. J., Stuessy, T. F., and Silva O., M. 1988. Allozyme variation in *Chenopodium sanctae-clarae*, an endemic species of the Juan Fernández Islands, Chile. *Biochem. Syst. Ecol.* 16: 279–284.

Crawford, D. J., Tago-Nakazawa, M., Stuessy, T. F., Anderson, G. J., Bernardello, G., et al. 2001a. Intersimple sequence repeat (ISSR) variation in *Lactoris fernandziana* (Lactoridaceae), a rare endemic of the Juan Fernández archipelago, Chile. *Plant Species Biol.* 16: 185–192.

Crawford, D. J., Whitkus, R., and Stuessy, T. F. 1987b. Plant evolution and speciation on oceanic islands. In K. Urbanska (ed.), *Differentiation Patterns in Higher Plants* (pp. 184–199). London: Academic Press.

Crisp, M. D., Gilmore, S., and Van Wyk, B.-E. 2000. Molecular phylogeny of the genistoid tribes of papilionoid legumes. In P. S. Herendeen and A. Bruneau (eds.), *Advances in Legume Systematics*, part 9 (pp. 249–276). Kew: Royal Botanic Gardens.

Cronk, Q. C. B. 1997. Islands: stability, diversity, conservation. *Biodiv. Conserv.* 6: 477–493.

Cronk, Q. C. B., and Fuller, J. K. 2001. *Plant Invaders: The Threat to Natural Ecosystems*. London: Earthscan Publications.

Cronquist, A. 1943. The separation of *Erigeron* from *Conyza*. *Bull. Torrey Bot. Club* 70: 629–632.

Cronquist, A. 1947. Revision of the North American species of *Erigeron*, north of Mexico. *Brittonia* 6: 121–302.

Cronquist, A. 1981. *An Integrated System of Classification of Flowering Plants*. New York: Columbia University Press.

Crutzen, P. J., and Stoermer, E. F. 2000. The "Anthropocene." *Global Change Newsl.* 41: 17–18.

Cuddihy, L. W., and Stone, C. P. 1990. *Alteration of Native Hawaiian Vegetation: Effects of Humans, Their Activities and Introductions*. Honolulu: University of Hawaii Cooperative National Park Resources Studies Unit.

Cuevas, J. G. 2002. Informe técnico no. 29. Distribución y abundancia de la flora vascular del Archipiélago de Juan Fernández. Il Parte Final. Unpublished report, Corporación Nacional Forestal V Región, Viña del Mar.

Cuevas, J. G., and Le Quesne, C. 2006. Low vegetation recovery after short-term cattle exclusion on Robinson Crusoe Island. *Plant Ecol.* 183: 105–124.

Cuevas, J. G., Marticorena, A., and Cavieres, L. A. 2004. New additions to the introduced flora of the Juan Fernández islands: origin, distribution, life history traits, and potential of invasion. *Rev. Chil. Hist. Nat.* 77: 523–538.

Cuevas, J. G., and van Leersum, G. 2001. Project "Conservation, restoration, and development of The Juan Fernández islands, Chile." *Rev. Chil. Hist. Nat.* 74: 899–910.

D'Arcy, W. G. 1976. New names and taxa in the Solanaceae. *Ann. Missouri Bot. Gard.* 63: 363–369.

Daly, H., and Magnacca, K. 2003. Hawaiian *Hylaeus* (*Nesoprosopis*) bees (Hymenoptera: Apoidea). In *Insects of Hawaii*, vol. 17. Honolulu: University of Hawaii Press.

Dampier, W. 1697. *A New Voyage Round the World*. London.

Dandy, J. E. (ed.). 1958. The Sloane Herbarium. An annotated list of the Horti Sicci composing it; with biographical accounts of the principal contributors. Based on records compiled by the late James Britten. British Museum: London.

Danton, P. 1998. Redécouverte de l'orchidée des îles de Robinson (Chili): *Gavilea insularis* M. N. Correa. *Bull. Mens. Soc. Linn. Lyon* 67: 226–232.

Danton, P. 2000. Une ascension au Mont Yunque dans l'île de Robinson Crusoe (Chili). *Bull. Mens. Soc. Linn. Lyon* 69: 205–216.

Danton, P. 2004. *Plantas Silvestres de la Isla Robinson Crusoe*. Viña del Mar: CONAF.

Danton, P. 2006a. Contribution à la flore de l'archipel Juan Fernández (Chili): Description de deux taxons nouveaux: *Nicotiana cordifolia* subsp. *sanctaclarae* subsp. nov. (Solanaceae), *Robinsonia saxatilis* sp. nov. (Asteraceae). *Acta Bot. Gallica* 153: 249–255.

Danton P. 2006b. La "myrtisylve" de l'archipel Juan Fernandez (Chili), une forêt en voie de disparation rapide. *Acta Bot. Gallica* 153: 179–199.

Danton, P. 2014. Contribution à la flore de l'archipel Juan Fernández (Chili): description de 4 taxons nouveaux: Angiospermae – *Erigeron corrales-molinensis* sp. nov. (Asteraceae), *Euphrasia formossissima* Skottsb. subsp. *cucharensis* subsp. nov. (Orobanchaceae), *Haloragis masatierrana* var. *applanata* var. nov. et var. *scabrida* var. nov. (Haloragidaceae). *Acta Bot. Gallica* 161: 355–371.

Danton, P., Boudrie, M., Bizot, A., and Viane, R. L. L. 2015. *Pleopeltis ×cerro-altoensis* (Polypodiaceae), a new fern hybrid from Robinson Crusoe Island (Juan Fernández Archipelago, Chile). *Fern Gaz.* 20: 65–78.

Danton, P., Breteau, E., and Baffray, M. 1999. *Les Îles de Robinson: Trésor Vivant des Mers du Sud. Entre Légende et Realité*. Paris: Nathan.

Danton, P., Buord, S., and Stutzin, M. 2013. Las islas de Robinson Crusoe: riqueza biológica amenazada y propuesta de creación de los Jardin y Conservatorio Botánicos. *Chagual* 11: 12–29.

Danton, P., and Perrier, C. 2005. Notes sur la disparition d'une espèce emblématique: *Robinsonia berteroi* (DC.) Sanders, Stuessy & Martic. (Asteraceae), dans l'île Robinson Crusoe, archipel Juan Fernández (Chili). *J. Bot. Soc. Bot. France* 31: 3–8.

Danton, P., and Perrier, C. 2017. *Monografía de las Plantas Vasculares del Archipiélago Juan Fernández, Chile. Ensayo de Valoración para su Preservación*. Gignac: Museo Editions.

Danton, P., Perrier, C., and Martínez, R. G. 2006. Nouveau catalogue de la flore vasculaire de l'archipel Juan Fernández (Chili) [Nuevo catálogo de la flora vascular del Archipiélago Juan Fernández (Chile)]. *Acta Bot. Gallica* 153: 399–587.

Darlington, P. J. 1957. *Zoogeography*. New York: Wiley.

Darwin, C. 1855. Letter to J. D. Hooker, March 7, 1855. In F. Darwin (ed.), *Life and Letters of Charles Darwin*. New York: Basic Books.

Darwin, C. 1859. *On the Origin of Species by Means of Natural Selection, or the Preservation of Favoured Races in the Struggle for Life*. London: John Murray.

Davis, M. A., Chew, M. K., Hobbs, R. J., Lugo, A. E., Ewel, J. J., et al. 2011. Don't judge species on their origins. *Nature* 474: 153–154.

de Joode, D. R., and Wendel, J. F. 1992. Genetic diversity and origin of the Hawaiian Islands cotton, *Gossypium tomentosum*. *Am. J. Bot.* 79: 1311–1319.

de la Sota, E. R. 1972. Notas sobre especies austrosudamericanas del género *Blechnum* L. (Blechnaceae-Pteridophyta). V. *Bol. Soc. Argent. Bot.* 14: 190–197.

de la Sota, E. R., Salino, A., and Costa Asis, F. 2007. Nuevas combinaciones en Polypodiaceae. *Darwiniana* 45: 239.

de Queiroz, A. 2005. The resurrection of oceanic dispersal in historical biogeography. *Trends Ecol. Evol.* 20: 68–73.

de Queiroz, K. 1998. The general lineage concept of species, species criteria, and the process of speciation. In D. J. Howard and S. H. Berlocher (eds.), *Endless Forms: Species and Speciation* (pp. 57–75). New York: Oxford University Press.

de Queiroz, K. 2007. Species concepts and species delimitation. *Syst. Biol.* 56: 879–886.

Decaisne, J. 1834. Monographie des genres *Balbisia* et *Robinsonia*, de la famille des Composées. *Ann. Sc. Natur. Bot. Ser. 2* 1: 1–15.

Defoe, D. 1719. *The Life and Strange Surprising Adventures of Robinson Crusoe*. London.

Delano, A. 1817. *A Narrative of Voyages and Travels in the Northern and Southern Hemispheres: Comprising Three Voyages Round the World; Together with a Voyage of Survey and Discovery, in the Pacific and Oriental Islands*. Boston.

Delprete, P. G., Forneris, G., and Pistarino, A. 2002. Carlo Bertero (1789–1831) in the New World. *Sida* 20: 621–644.

Dempster, L. T. 1980. The genus *Galium* section *Lophogalium* (Rubiaceae) in South America. *Allertonia* 2: 247–279.

Dempster, L. T. 1981. The genus *Galium* (Rubiaceae) in South America. II. *Allertonia* 2: 393–426.

Devey, C. W., Hémond, C., and Stoffers, P. 2000. Metasomatic reactions between carbonate plume melts and mantle harzburgite: the evidence from Friday and Domingo seamounts (Juan Fernandez chain, SE Pacific). *Contrib. Mineral. Petrol.* 139: 68–84.

Diaz, H. F., and Markgraf, V. (eds.). 1992. *El Niño: Historical and Paleoclimatic Aspects of the Southern Oscillation*. Cambridge: Cambridge University Press.

Diem, J., and de Lichtenstein, J. S. 1959. Las Himenofiláceas del área argentino-chilena del sud. *Darwiniana* 11: 611–760.

Dirnböck, T., Greimler, J., López S., P., and Stuessy, T. F. 2003. Predicting future threats to the native vegetation of Robinson Crusoe Island, Juan Fernández Archipelago, Chile. *Conserv. Biol.* 17: 1650–1659.

Dobzhansky, T. 1937. *Genetics and the Origin of Species*. New York: Columbia University Press.

Dohzono, I., and Yokoyama, J. 2010. Impacts of alien bees on native plant-pollinator relationships: a review with special emphasis on plant reproduction. *Appl. Entomol. Zool.* 45: 37–47.

Don, D. 1832. Descriptive catalogue of the Compositae contained in the herbarium of Dr. Gillies; with some additions from other sources. *Philos. Mag. Ann. Chem.* 11: 387–392.

Donoso Zegers, C. 1993. *Bosques Templados de Chile y Argentina*. Santiago: Editorial Universitaria.

Douglas, D. 1914. *Journal Kept by David Douglas during His Travels in North America 1823–1817*. London: Royal Horticultural Society (reprint 1959. New York: Antiquarian Press).

Drábková, L., Kirschner, J., Seberg, O., Petersen, G., and Vlček, Č. 2003. Phylogeny of the Juncaceae based on *rbcL* sequences, with special emphasis on *Luzula* DC. and *Juncus* L. *Plant Syst. Evol.* 240: 133–147.

Drábková, L., Kirschner, J., and Vlček, Č. 2006. Phylogenetic relationships within *Luzula* DC. and *Juncus* L. (Juncaceae): a comparison of phylogenetic signals of *trnL-trnF* intergenic spacer, *trnL* intron and *rbcL* plastome sequence data. *Cladistics* 22: 132–143.

Drábková, L., Kirschner, J., Vlček, Č., and Pačes, V. 2004. *TrnL-trnF* intergenic spacer and *trnL* intron define major clades within *Luzula* and *Juncus* (Juncaceae): importance of structure mutations. *J. Mol. Evol.* 59: 1–10.

Driscoll, H. E., and Barrington, D. S. 2007. Origin of Hawaiian *Polystichum* (Dryopteridaceae) in the context of a world phylogeny. *Am. J. Bot.* 94: 1413–1424.

Dubuisson, J.-Y. 1997a. Systematic relationships within the genus *Trichomanes sensu lato* (Hymenophyllaceae, Filicopsida): cladistic analysis based on anatomical and morphological data. *Bot. J. Linn. Soc.* 123: 265–296.

Dubuisson, J.-Y. 1997b. *rbcL* sequences: a promising tool for the molecular systematics of the fern genus *Trichomanes* (Hymenophyllaceae)? *Mol. Phylogenet. Evol.* 8: 128–138.

Dubuisson, J.-Y., Hennequin, E. J. P., Douzery, R. B., Cranfill, A. R., Smith, A. R., et al. 2003. *rbcL* phylogeny of the fern genus *Trichomanes* (Hymenophyllaceae) with special reference to Neotropical taxa. *Int. J. Plant Sci.* 164: 753–761.

Duek, J. J., and Rodríguez, R. 1972. Lista preliminar de las especies de Pteridophyta en Chile continental e insular. *Bol. Soc. Bol. Concepción* 45: 129–174.

Dunbar-Co, S., Wieczorek, A. M., and Morden, C. W. 2008. Molecular phylogeny and adaptive radiation of the endemic Hawaiian *Plantago* species (Plantaginaceae). *Am. J. Bot.* 95: 1177–1188.

Dupont, Y. L., and Skov, C. 2004. Influence of geographical distribution and floral traits on species richness of bees (Hymenoptera: Apoidea) visiting *Echium* species (Boraginaceae) of the Canary Islands. *Int. J. Plant Sci.* 165: 377–386.

Ebihara, A., Dubuisson, J.-Y., Iwatsuki, K., Hennequin, S., and Ito, M. 2006. A taxonomic revision of Hymenophyllaceae. *Blumea* 51: 221–280.

Ebihara, A., and Iwatsuki, K. 2007. The Hymenophyllaceae of the Pacific Area: 1. *Hymenophyllum* subgenus *Hymenophyllum*. *Bull. Natl. Mus. Nat. Sci. Ser. B* 33: 55–68.

Ebihara, A., Iwatsuki, K., Ito, M., Hennequin, S., and Dubuisson, J.-Y. 2007. A global molecular phylogeny of the fern genus *Trichomanes* (Hymenophyllaceae) with special reference to stem anatomy. *Bot. J. Linn. Soc.* 155: 1–27.

Ebihara, A., Nitta, J. H., and Iwatsuki, K. 2010. The Hymenophyllaceae of the Pacific Area: 2. *Hymenophyllum* (excluding subgen. *Hymenophyllum*). *Bull. Natl. Mus. Nat. Sci. Ser. B* 36: 43–59.

Edmands, S. 2007. Between a rock and a hard place: evaluating the relative risks of inbreeding and outbreeding for conservation and management. *Mol. Ecol.* 16: 463–475.

Edwards, C. E. 2015. Looking to the future of conservation genetics: the case for using quantitative genetic experiments to estimate the ability of rare plants to withstand climate change. *Am. J. Bot.* 102: 1011–1013.

Ehrendorfer, F. 1979. Reproductive biology in island plants. In D. Bramwell (ed.), *Plants and Islands* (pp. 293–306). London: Academic Press.

Eizaguirre, C., and Baltazar-Soares, M. 2014. Evolutionary conservation: evaluating the adaptive potential of species. *Evol. Appl.* 7: 963–967.

Ekblom, R., and Wolf, J. B. W. 2014. A field guide to whole-genome sequencing, assembly and annotation. *Evol. Appl.* 7: 1026–1042.

Eliasson, U. 1995. Patterns of diversity in island plants. In P. M. Vitousek, L. L. Loope, and H. Adsersen (ed.), *Islands: Biological Diversity and Ecosystem Functions* (pp. 35–50). New York: Springer-Verlag.

Ellstrand, N., and Elam, D. 1993. Population genetic consequences of small population size: Implications for plant conservation. *Ann. Rev. Ecol. Syst.* 24: 217–242.

Emerson, B. C. 2002. Evolution on oceanic islands: molecular phylogenetic approaches to understanding pattern and process. *Mol. Ecol.* 11: 951–966.

Emerson, B. C., and Kolm, N. 2005a. Species diversity can drive speciation. *Nature* 434: 1015–1017.

Emerson, B. C., and Kolm, N. 2005b. Emerson and Kolm reply. *Nature* 438: E2.

Engel, M. S. 2000. A new *Lasioglossum* from the Juan Fernández Islands (Hymenoptera: Halictidae). *Rev. Chilena Entomol.* 27: 5–10.

Engler, A. 1882. *Versuch einer Entwicklunsgeschichte der Pflanzenwelt, insbesondere der Florengebiete, seit der Tertiärperiode*. Leipzig: W. Engelman.

Ernst, B., Chamorro, J., Manríquez, P., Orensanz, J. M. L., Parma, A. M., et al. 2013. Sustainability of the Juan Fernández lobster fishery (Chile) and the perils of generic science-based prescriptions. *Global Environ. Change* 23: 1381–1392.

Esquemeling, J. 1685. *The Buccaneers of America, the Second Volume Containing the Dangerous Voyage and Bold Attempts of Captain Bartholomew Sharp and Others Performed upon the Coasts of the South Sea for the Space of Two Years from the Original Journal of the Said Voyage Written by Mr. Basil Ringrose, Gent., Who Was All Along Present at Those Transactions* (2nd edn., English translation). London.

Esselman, E. J., Crawford, D. J., Brauner, S., Stuessy, T. F., Anderson, G. J., et al. 2000. RAPD marker diversity within and divergence among species of *Dendroseris* (Asteraceae: Lactuceae). *Am. J. Bot.* 87: 591–596.

Evans, A. 1930. The thallose Hepaticae of the Juan Fernandez Islands. In C. Skottsberg (ed.), *The Natural History of the Juan Fernandez and Easter Island, vol. 2: Botany* (pp. 551–5860). Uppsala: Almqvist & Wiksells.

Evans, T. M., Jabaily, R. S., Faria, A. P. G. de, Sousa, L. O. F. de, Wendt, T., et al. 2015. Phylogenetic relationships in Bromeliaceae subfamily Bromelioideae based on chloroplast DNA sequence data. *Syst. Bot.* 40: 116–128.

Faith, D. P. 1994. Phylogenetic diversity: a general framework for the prediction of feature diversity. In P. L. Forey, C. J. Humphries, and R. I. Vane-Wright (eds.), *Systematics and Conservation Evaluation* (pp. 251–268). Oxford: Clarendon Press.

Fanning, E. 1833. *Voyages Round the World; With Selected Sketches of Voyages to the South Seas, North and South Pacific Oceans, China, etc.* New York: Collins & Hannay (new edition 1924, under the title of *Voyages & Discoveries in the South Seas, 1792–1832*, edited and published by the Marine Research Society, Salem, MA).

Farías, M., Vargas, G., Tassara, A., Carretier, S., Baize, S., et al. 2010. Land-level changes produced by the M_w 8.8 2010 Chilean earthquake. *Science* 329: 916.

Farley, K. A., Basu, A. R., and Craig, H. 1993. He, Sr and Nd isotopic variations in lavas from the Juan Fernandez Archipelago, SE Pacific. *Contrib. Mineral. Petrol.* 115: 75–87.

Felsenstein, J. 2004. *Inferring Phylogenies*. Sunderland, MA: Sinauer Associates.

Fernández-Palacios, J. M., Kueffer, C., and Drake, D. 2015. A new golden era in island biogeography. *Front. Biogeog.* 7: 14–20.

Ferrara, G., Clarke, W. B., Rama Murthy, V., and Bass, M. N. 1969. K-Ar ages of Juan Fernandez Islands and southeast Pacific dredge hauls. *Eos Transact. Am. Geophys. Union* 50 (4): 329 (Abstr.).

Ferreyra, M. L., Rius, S. P., and Casati, P. 2012. Flavonoids: biosynthesis, biological functions, and biotechnological applications. *Front. Plant Sci.* 3(222): 1–15.

Fior, S., Karis, P. O., Casazza, G., Minuto, L., and Sala, F. 2006. Molecular phylogeny of the Caryophyllaceae (Caryophyllales) inferred from chloroplast *matK* and nuclear rDNA ITS sequences. *Am. J. Bot.* 93: 399–411.

Fisher, A. E., Clark, L. G., and Kelcher, S. A. 2014. Molecular phylogeny estimation of the bamboo genus *Chusquea* (Poaceae: Bambusoideae: Bambuseae) and description of two new subgenera. *Syst. Bot.* 39: 829–844.

Fishman, L., Stathos, A., Beardsley, P. M., Williams, C. F., and Hill, J. P. 2013. Chromosomal rearrangements and the genetics of reproductive barriers in *Mimulus* (monkey flowers). *Evolution* 67: 2547–2560.

Fitzpatrick, B. M., Fordyce, J. A., and Gavrilets, S. 2008. What, if anything, is sympatric speciation? *J. Evol. Biol.* 21: 1452–1459.

Fitzpatrick, B. M., Fordyce, J. A., and Gavrilets, S. 2009. Pattern, process and geographic modes of speciation. *J. Evol. Biol.* 22: 2342–2347.

Fjellheim, S., Jogensen, M. H., Kjos, M., and Borgen, L. 2009. A molecular study of hybridization and homoploid hybrid speciation in *Argyranthemum* (Asteraceae) on Tenerife, Canary Islands. *Bot. J. Linn. Soc.* 159: 19–31.

Fleischer, R. C., and McIntosh, C. E. 2001. Molecular systematics and biogeography of the Hawaiian avifauna. *Stud. Avian Biol.* 22: 51–60.

Forde, M. G. 1964. *Haloragis erecta*: a species complex in evolution. *New Zealand J. Bot.* 2: 425–453.

Fosberg, F. R. 1967. Some ecological effects of wild and semi-wild exotic species of vascular plants. In IUCN (ed.), *Towards a New Relationship of Man and Nature in Temperate Lands,* Part III*: Changes due to Introduced Species* (pp. 98–109). Morges: International Union for Conservation of Nature and Natural Resources.

Fosberg, F. R. 1968. Studies in Pacific Rubiaceae: VI–VIII. *Brittonia* 20: 287–294.

Frakes, L. A. 1979. *Climates through Geologic Time*. Amsterdam: Elsevier.

Francisco-Ortega, J, Crawford, D. J., Santos-Guerra, A., and Carvalho, J. A. 1996. Isozyme variation in the endemic genus *Argyranthemum* (Asteraceae: Anthemideae) in the Macaronesian Islands. *Plant Syst. Evol.* 202: 137–152.

Francisco-Ortega, J., Crawford, D. J., Santos-Guerra, A., and Jansen, R. K. 1997. Origin and evolution of *Argyranthemum* (Asteraceae: Anthemideae) in Macaronesia. In T. J. Givnish and K. J. Sytsma (eds.), *Molecular Evolution and Adaptive Radiation* (pp. 407–431). Cambridge: Cambridge University Press.

Francisco-Ortega, J., Santos-Guerra, A., Kim, S.-C., and Crawford, D. J. 2000. Plant genetic diversity in the Canary Islands: a conservation perspective. *Am. J. Bot.* 87: 909–919.

Frankham, R. 1997. Do island populations have less genetic variation than mainland populations? *Heredity* 78: 311–327.

Frankham, R. 2010. Where are we in conservation genetics and where do we need to go? *Conserv. Genet.* 11: 661–663.

Frankham, R. 2015. Genetic rescue of small inbred populations: metaanalysis reveals large and consistent benefits of gene flow. *Mol. Ecol.* 24: 2610–2618.

Frankham, R., Ballou, J. D., and Briscoe, D. A. 2010. *Introduction to Conservation Genetics* (2nd edn.). Cambridge: Cambridge University Press.

Franzke, A., Pollmann, K., Bleeker, W., Kohrt, R., and Hurka, H. 1998. Molecular systematics of *Cardamine* and allied genera (Brassicaceae): ITS and non-coding chloroplast DNA. *Folia Geobot.* 33: 225–240.

Freire, S. E., Salomón, L., Bayón, N. D., Baeza, C. M., Muñoz-Schick, M., et al. 2016. Taxonomic revision of the genus *Gamochaeta* (Gnaphalieae, Asteraceae) in Chile. *Gayana Bot.* 73: 252–305.

Frenzke, L., Scheiris, E., Pino, G., Symmank, L., Goetghebeur, P., et al. 2015. A revised infrageneric classification of the genus *Peperomia* (Piperaceae). *Taxon* 64: 424–444.

Fries, R. E. 1920. Die Myxomyceten der Juan Fernandez-Inseln. In C. Skottsberg (ed.), *The Natural History of the Juan Fernandez and Easter Island, vol. 2: Botany* (pp. 55–58). Uppsala: Almqvist & Wiksells.

Fries, T. C. E. 1921. Die Gasteromyceten der Juan Fernandez- und Osterinseln. In C. Skottsberg (ed.), *The Natural History of the Juan Fernandez and Easter Island, vol. 2: Botany* (pp. 59–60). Uppsala: Almqvist & Wiksells.

Fritsch, P. W., Lu, L., Bush, C. M., Cruz, B. C., Kron, K. A., et al. 2011. Phylogenetic analysis of the wintergreen group (Ericaceae) based on six genic regions. *Syst. Bot.* 36: 990–1003.

Fritts, T. H., and Rodda, G. H. 1998. The role of introduced species in the degradation of island ecosystems. *Annu. Rev. Ecol. Syst.* 29: 113–140.

Fuentes-Bazan, S., Mansion, G., and Borsch, T. 2012a. Towards a species level tree of the globally diverse genus *Chenopodium* (Chenopodiaceae). *Mol. Phylogenet. Evol.* 62: 359–374.

Fuentes-Bazan, S., Uotila, P., and Borsch, T. 2012b. A novel phylogeny-based generic classification for *Chenopodium* sensu lato, and a tribal rearrangement of Chenopodioideae (Chenopodiaceae). *Willdenowia* 42: 5–24.

Fuenzalida P., H. 1966. Climatología. In Corporación de Fomento de la Producción (CORFO) (ed.), *Geografía Económica de Chile* (2nd edn., pp. 31–44). Santiago de Chile: Editorial Universitaria.

Fuenzalida V., H. 1944. Don Claudio Gay y el Museo Nacional de Historia Natural. *Bol. Mus. Nac. Hist. Nat.* 22: 5–16.

Funk, J. L., and Vitousek, P. M. 2007. Resource-use efficiency and plant invasion in low-resource systems. *Nature* 446: 1079–1081.

Funk, V. A., and Wagner, W. L. 1995. Biogeographic patterns in the Hawaiian Islands. In W. L. Wagner and V. A. Funk (eds.), *Hawaiian Biogeography: Evolution on a Hot Spot Archipelago* (pp. 379–419). Washington, DC: Smithsonian Institution Press.

Funnell, W. 1707. *A Voyage Round the World; Containing an Account of Captain Dampier's expedition into the South-Seas in the Ship St. George, in the Years 1703 and 1704*. London.

Fürstenberg, P. 1906. Dr. Rudolph Amandus Philippi. Sein Leben und seine Werke. *Verhandl. Deutsch. Wissenschafl. Vereins (Santiago)* 5: 1–39.

Gabel, J. D., Les, D. H., Charlton, W. A., and Coombes, A. J. 2000. Phylogeny of *Azara* (Flacourtiaceae). *Am. J. Bot.* 87(6): 128 (Abstr.).

Gabriel y Galán, J. M., Prada, C., Rolleri, C., Ainouche, A., and Vicent, M. 2013. cpDNA supports the identification of the major lineages of American *Blechnum* (Blechnaceae, Polypodiopsida) established by morphology. *Turk. J. Bot.* 37: 769–777.

Galapagos Conservancy. 2015. *Ecosystem Restoration: Project Isabela*. Available at www.galapagos.org/conservation/project-isabela/ (accessed March 11, 2016).

Gallagher, R. E. 1964. *Byron's Journal of His Circumnavigation 1764–1766*. Cambridge: Cambridge University Press (Hakluyt Society).

Gamerro, J. C., and Barreda, V. 2008. New fossil record of Lactoridaceae in southern South America: a palaeobiogeographical approach. *Bot. J. Linn. Soc.* 158: 41–50.

Ganders, F. R., Berbee, M., and Pirseyedi, M. 2000. ITS base sequence phylogeny in *Bidens* (Asteraceae): evidence for the continental relatives of Hawaiian and Marquesan *Bidens*. *Syst. Bot.* 25: 122–133.

Ganders, F. R., Bohm, B. A., and McCormick, S. P. 1990. Flavonoid variation in Hawaiian *Bidens*. *Syst. Bot.* 15: 231–239.

Ganders, F. R., and Nagata, K. M. 1984. The role of hybridization in the evolution of *Bidens* on the Hawaiian Islands. In W. F. Grant (ed.), *Plant Biosystematics* (pp. 179–194). Toronto: Academic Press.

Garcia, J. D. D. 2000. Patterns of insect flower visitation in *Lavandula buchii* Webb (Lamiaceae), an endemic shrub of Tenerife (Canary Islands). *J. Nat. Hist. (Lond.)* 34: 2145–2155.

Garcia-Jacas, N., Susanna, A., Garnatje, T., and Vilatersana, R. 2001. Generic delimitation and phylogeny of the subtribe Centaureinae (Asteraceae): a combined nuclear and chloroplast DNA analysis. *Ann. Bot.* 87: 503–515.

García-Verdugo, C., Friar, E., and Santiago, L. S. 2013. Ecological role of hybridization in adaptive radiations: a case study in the *Dubautia arborea-Dubautia ciliolata* (Asteraceae) complex. *Int. J. Plant Sci.* 174: 749–759.

García-Verdugo, C., Sajeva, M., La Mantia, T., Harrouni, C., Msanda, F., et al. 2015. Do island plant populations really have lower genetic variation than mainland populations? Effects of selection and distribution range on genetic diversity estimates. *Mol. Ecol.* 24: 726–741.

Garnock-Jones, P. J. 2014. Evidence-based review of the taxonomic status of New Zealand's endemic seed plant genera. *New Zealand J. Bot.* 52: 163–212.

Gastony, G. J. 1981. Spore morphology in the Dicksoniaceae: I. The genera *Cystodium*, *Thyrsopteris*, and *Culcita*. *Am. J. Bot.* 68: 808–819.

Gavrilets, S. 2014. Models of speciation: where are we now? *J. Hered.* 105: 743–755.

Gay, C. 1854. *Atlas de la Historia Física y Política de Chile*. Paris: E. Thunot.

Gay, C. 1845–1854. *Historia Física y Política de Chile según Documentos Adquiridos en esta República durante Doce Años de Residencia en ella y Publicada bajo los Auspicios del Supremo Gobierno. Botánica*. Paris.

Geltman, D. V. 1998. New species and new combinations of *Urtica* (Urticaceae) in South America. *Novon* 8: 15–17.

Gerlach, D.C., Hart, S. R., Morales, V. W. J., and Palacios, C. 1986. Mantle heterogeneity beneath the Nazca plate: San Felix and Juan Fernandez islands. *Nature* 322: 165–169.

Giannasi, D. E., and Crawford, D. J. 1986. Biochemical systematics II. A reprise. *Evol. Biol.* 20: 25–248.

Gillespie, R. G. 2004. Community assembly through adaptive radiation in Hawaiian spiders. *Science* 303: 356–359.

Gillespie, R. G., and Clague, D. A. (eds.). 2009. *Encyclopedia of Islands*. Berkeley: University of California Press.

Gillespie, T., Lipkin, B., Sullivan, L., Benowitz, D. R., Pau, S., et al. 2012. The rarest and least protected forests in biodiversity hotspots. *Biodivers. Conserv.* 21: 3597–3611.

Gillett, G. W., and Lim, E. K. S. 1970. An experimental study of the genus *Bidens* in the Hawaiian Islands. *Univ. Calif. Publ. Bot.* 56: 1–63.

Gitaí, J., Horres, R., and Benko-Iseppon, A. M. 2005. Chromosome features and evolution of Bromeliaceae. *Plant Syst. Evol.* 253: 65–80.

Gitaí, J., Paule, J., Zizka, G., Schulte, K., and Benko-Iseppon, A. M. 2014. Chromosome numbers and DNA content in Bromeliaceae: additional data and critical review. *Bot. J. Linn. Soc.* 176: 349–368.

Givnish, T. J. 2010. Ecology of plant speciation. *Taxon* 1326–1366.

Givnish, T. J. 2015. Adaptive radiation versus 'radiation' and 'explosive diversification': why conceptual distinctions are fundamental to understanding evolution. *New Phytol.* 207: 297–303.

Givnish, T. J., Barfuss, M. H. J., Van Ee, B., Riina, R., Schulte, K., et al. 2011. Phylogeny, adaptive radiation, and historical biogeography in Bromeliaceae: insights from an eight-locus plastid phylogeny. *Am. J. Bot.* 98: 872–895.

Givnish, T. J., and Montgomery, R. A. 2014. Common-garden studies on adaptive radiation of photosynthetic physiology among Hawaiian lobeliads. *Proc. R. Soc. Lond. B* 281: 20132944. Available at http://dx.doi.org/10.1098/rspb.2013.2944.

Glenn, T. G., and Schnable, N. A. 2005. Isolating microsatellite DNA loci. *Methods Enzymol.* 395: 202–222.

Global Carex Group. 2015. Making *Carex* monophyletic (Cyperaceae, tribe Cariceae): a new broader circumscription. *Bot. J. Linn. Soc.* 179: 1–42.

Godoy, R. 1989. Espectro biológico de la flora pteridofítica de Chile continental e insular. *Anales Jard. Bot. Madrid* 46: 599–603.

Goldsworthy, S., Francis, J., Boness, D., and Fleischer, R. 2000. Variation in the mitochondrial control region in the Juan Fernández fur seal (*Arctocephalus philippii*). *J. Hered.* 91: 371–377.

Gonzales R., J., and Kessler, M. 2011. A synopsis of the Neotropical species of *Sticherus* (Gleicheniaceae), with descriptions of nine new species. *Phytotaxa* 31: 1–54.

González, A., and Fuertes, J. 2011. Ecología y evolución de las plantas ornitófilas de la Macaronesia. *El Indiferente* 21: 64–75.

González, F., and Rudall, P. 2001. The questionable affinities of *Lactoris*: evidence from branching pattern, inflorescence morphology, and stipule development. *Am. J. Bot.* 88: 2143–2150.

González, F., Rudall, P. J., and Furness, C. A. 2001. Microsporogenesis and systematics of Aristolochiaceae. *Bot. J. Linn. Soc.* 137: 221–242.

Good, R. 1974. *The Geography of Flowering Plants*. London: Longman.

Goodspeed, T. H. 1954. *The Genus* Nicotiana. *Origins, Relationships and Evolution of Its Species in the Light of Their Distribution, Morphology and Cytogenetics*. Waltham, MA: Chronica Botanica.

Gornall, R. J., and Bohm, B. A. 1979. The distribution of flavonoids in the angiosperms. *Bot. Not.* 132: 1–30.

Gotschlich, B. 1904. *Biografía del Dr. Rodulfo Amando Philippi (1803–1904)*. Valdivia: Imprenta Central J. Lampert.

Gottlieb, L. D. 1971. Gel electrophoresis: new approach to the study of evolution. *Bioscience* 21: 939–944.

Gottlieb, L. D. 1973. Genetic differentiation, sympatric speciation, and the origin of a diploid species of *Stephanomeria*. *Am. J. Bot.* 60: 545–553.

Gottlieb, L. D. 1977. Electrophoretic evidence and plant systematics. *Ann. Missouri Bot. Gard.* 64: 161–180.

Gottlieb, L. D. 1981a. Electrophoretic evidence and plant populations. *Prog. Phytochem.* 7: 1–46.

Gottlieb, L. D. 1981b. Gene number in species of Astereae that have different chromosome numbers. *Proc. Natl. Acad. Sci. USA* 78: 3726–3729.

Gottlieb, L. D. 1982. Conservation and duplication of isozymes in plants. *Science* 216: 373–380.

Gottlieb, L. D. 2003. Rethinking classic examples of recent speciation in plants. *New Phytol.* 161: 71–82.

Graham, M. 1824. *Journal of a Residence in Chile during the Year 1822 and a Voyage from Chile to Brazil in 1823*. London.

Grant, P. R. (ed.). 1998. *Evolution on Islands*. Oxford: Oxford University Press.

Grant, P. R., and Grant, B. R. 2008. *How and Why Species Multiply*. Princeton, NJ: Princeton University Press.

Grant, P. R., and Grant, B. R. 2010. Sympatric speciation, immigration, and hybridization in island birds. In J. B. Losos and R. E. Ricklefs (eds.), *The Theory of Island Biogeography Revisited* (pp. 326–357). Princeton, NJ: Princeton University Press.

Grant, V. 1950. The protection of ovules in flowering plants. *Evolution* 3: 179–201.

Grant, V. 1963. *The Origin of Adaptations*. New York: Columbia University Press.

Grant, V. 1981. *Plant Speciation* (2nd edn.). New York: Columbia University Press.

Green, P. T. 1999. Greta's Garbo: stranded seeds and fruits from Greta Beach, Christmas Island, Indian Ocean. *J. Biogeogr.* 26: 937–946.

Greenberg, A. K., and Donoghue, M. J. 2011. Molecular systematics and character evolution in Caryophyllaceae. *Taxon* 60: 1637–1652.

Greimler, J., López S. P., Stuessy, T. F., and Dirnböck, T. 2002a. The vegetation of Robinson Crusoe Island (Isla Masatierra), Juan Fernández Archipelago, Chile. *Pacific Sci.* 56: 263–284 + 1 map.

Greimler, J., López-Sepúlveda, P., Reiter, K., Baeza, C., Peñailillo, P., et al. 2013. Vegetation of Alejandro Selkirk Island (Isla Masafuera), Juan Fernández Archipelago, Chile. *Pacific Sci.* 67: 267–282.

Greimler, J., Stuessy, T. F., Swenson, U., Baeza, C., and Matthei, O. 2002b. Plant invasions on an oceanic archipelago. *Biol. Invasions* 4: 73–85.

Greuter, W., Barrie, F. R., Burdet, H. M., Chaloner, W. G., Demoulin, V., et al. 1994. *International Code of Botanical Nomenclature (Tokyo Code) Adopted by the Fifteenth International Botanical Congress, Yokohama, August–September 1993.* Königstein: Koeltz Sci. Books (*Regnum Veg.* 131).

Grondona, E. 1964. Las especies Argentinas del género *Acaena. Darwiniana* 13: 209–342.

Groppo, M., Pirani, J. R., Salatino, M. L. F., Blanco, S. R., and Kallunki, J. A. 2008. Phylogeny of Rutaceae based on two noncoding regions from cpDNA. *Am. J. Bot.* 95: 985–1005.

Grueber, C. E., Wallis, G. P., and Jamieson, I. G. 2008. Heterozygosity-fitness correlations and their relevance to studies on inbreeding depression in threatened species. *Mol. Ecol.* 17: 3978–3984.

Gunckel L., H. 1971. Las primeras plantas herborizadas en Chile en 1690. *An. Instit. Patagon.* 2(1–2): 134–141.

Gunckel Luër, H. 1968. Nombres vernaculares de plantas endémicas de Juan Fernández. *Rev. Univ. (Univ. Católica de Chile)* 53(31): 1–26.

Gunckel Luër, H. 1984. *Helechos de Chile*. Santiago: Universidad de Chile.

Gunther, B., and Mahalik, P. 1990. *Juania australis* in habitat. *Principes* 34: 79–81.

Guppy, H. B. 1890. The dispersal of plants as illustrated by the flora of the Keeling or Cocos Islands. *Trans. Proc. Philos. Inst. Victoria* 24: 276–306.

Guppy, H. B. 1917. *Plants, Seeds, and Currents in the West Indies and Azores*. London: Williams & Norgate.

Gussarova, G., Popp, M., Vitek, E., and Brochmann, C. 2008. Molecular phylogeny and biogeography of the bipolar *Euphrasia* (Orobanchaceae): recent radiations in an old genus. *Mol. Phylogenet. Evol.* 48: 444–460.

Haberle, S. G. 2003. Late Quaternary vegetation dynamics and human impact on Alexander Selkirk Island, Chile. *J. Biogeogr.* 30: 239–255.

Habicht, J. K. A. 1979. *Paleoclimate, Paleomagnetism and Continental Drift.* Tulsa, OK: American Association of Petroleum Geologists.

Hahn, I. 1996. The birds of the Juan Fernández Islands (Chile) with Special consideration to the Masafuera rayadito (*Aphrastura masafuerae*). Master's Thesis, Institute of Landscape Ecology, Westfälische Wilhelms-Universität, Münster, Germany.

Hahn, I. 1998. Untersuchungen zur Ökologie und zum Lebensraum der Landvogelgemeinschaften des Juan Fernandez-Archipels (Chile). Ph.D. dissertation, Westfälische Wilhelms-Universität, Münster, Germany.

Hahn, I., Römer, U., Vergara, P., and Walter, H. 2009. Biogeography, diversity, and conservation of the birds of the Juan Fernández Islands, Chile. *Vertebrate Zool.* 59: 103–114.

Hahn, I. J., Vergara, P. M., Baumeister, J., Soto, G. E., and Römer, U. 2014. Impact of the 2010 tsunami on an endangered insular soil-plant system. *Nat. Hazards* 71: 987–993.

Hajek, E., and Espinoza, G. A. 1987. Meteorología, climatología y bioclimatología de las islas oceánicas Chilenas. In J. C. Castilla (ed.), *Islas Oceánicas Chilenas: Conocimiento Científico y Necesidades de Investigaciones* (pp. 55–83). Santiago: Ediciones Universidad Católica de Chile.

Hale, R. L. 1923. *The Log of a Forty-Niner: Journal of a Voyage from Newbury-Port to San Francisco on the Brig General Worth Commanded by Capt. Samuel Walton*, ed. by C. H. Russ. Boston, MA: B.J. Brimmer.

Hamrick, J. L. 1989. Isozymes and analyes of genetic structure of plant populations. In D. E. Soltis and P. S. Soltis (eds.), *Isozymes in Plant Biology* (pp. 87–105). Portland, OR: Dioscorides.

Hamrick, J. L., and Godt, M. J. W. 1989. Allozyme diversity in plant species. In A. H. D. Brown, M. T. Clegg, A. L. Kahler, and B. S. Weir (eds.), *Plant Population Genetics, Breeding and Genetic Resources* (pp. 43–63). Sunderland, MA: Sinauer Associates.

Hamrick, J. L., and Godt, M. J. W. 1996. Conservation genetics of endemic plant species. In J. C. Avise and J. L. Hamrick (eds.), *Conservation Genetics: Case Studies from Nature* (pp. 281–304). New York: Chapman & Hall.

Hamrick, J. L., and Godt, M. J. W. 1997. Effects of life history traits on genetic diversity in plant species. In J. Silvertown, M. Franco, and J. L. Harper (eds.), *Plant Life Histories: Ecology, Phylogeny and Evolution* (pp. 102–118). Cambridge: Cambridge University Press.

Hamrick, J. L., Godt, M. J. W., Murawski, D. A., and Loveless, M. D. 1991. Correlations between species traits and allozyme diversity: implications for conservation biology. In D. A. Falk and K. E. Holsinger (eds.), *Genetics and Conservation of Rare Plants* (pp. 75–86). New York: Oxford University Press.

Hamrick, J. L., Linhart, Y. B., and Mitton, J. B. 1979. Relationships between life history characteristics and electrophoretically detectable genetic variation in plants. *Annu. Rev. Ecol. Syst.* 10: 173–200.

Harbaugh, D., and Baldwin, B. G. 2007. Phylogeny and biogeography of the sandalwoods (*Santalum*, Santalaceae): repeated dispersals throughout the Pacific. *Am. J. Bot.* 94: 1028–1040.

Harborne, J. B. 1967. *Comparative Biochemistry of the Flavonoids*. London: Academic Press.

Harborne, J. B., and Turner, B. L. 1984. *Plant Chemosystematics*. London: Academic Press.

Harling, G. 1962. On some Compositae endemic to the Galápagos Islands. *Acta Hort. Berg.* 20: 63–120.

Harley, R. 1986. *Cuminia eriantha*, Labiatae. *Kew Mag.* 3: 151–156.

Harris, H. 1966. Enzyme polymorphisms in man. *Philos. Trans. R. Soc. Lond. B.* 164: 298–310.

Harrisson, K. A., Pavlova A., Telonis-Scott, M., and Sunnucks, P. 2014. Using genomics to characterize evolutionary potential for conservation of wild populations. *Evol. Appl.* 7: 1008–1025.

Harter, D. E. V., Irl, S. D. H., Seo, B., Steinbauer, M. J., Gillespie, R., et al. 2015. Impacts of global climate change on the floras of oceanic islands: projections, implications and current knowledge. *Perspect. Plant Ecol. Evol. Syst.* 17: 160–183.

Hartley, T. G. 1966. A revision of the Malesian species of *Zanthoxylum* (Rutaceae). *J. Arnold Arbor.* 47: 171–221.

Haufler, D. H., Soltis, D. E., and Soltis, P. S. 1995. Phylogeny of the *Polypodium vulgare* complex: insights from chloroplast DNA restriction site data. *Syst. Bot.* 20: 110–119.

Hauk, W. D., Parks, C. R., and Chase, M. W. 2003. Phylogenetic studies of Ophioglossaceae: evidence from *rbcL* and *trnL-F* plastid DNA sequences and morphology. *Mol. Phylogenet. Evol.* 28: 131–151.

Hausdorf, B. 2011. Progress toward a general species concept. *Evolution* 65: 923–931.

Heads, M. J. 1996. Biogeography, taxonomy and evolution in the Pacific genus *Coprosma* (Rubiaceae). *Candollea* 51: 381–405.

Heaney, L. R. 2007. Is a new paradigm emerging for ocean island biogeography? *J. Biogeogr.* 34: 753–757.

Heenan, P. B. 2001. The correct name for Chilean pelu (Fabaceae): the identity of *Edwardsia macnabiana* and the reinstatement of *Sophora cassioides*. *New Zealand J. Bot.* 39: 167–170.

Heenan, P. B., Dawson, M. I., and Wagstaff, S. J. 2004. The relationship of *Sophora* sect. *Edwardsia* (Fabaceae) to *Sophora tomentosa*, the type species of the genus *Sophora*, observed from DNA sequence data and morphological characters. *Bot. J. Linn. Soc.* 146: 439–446.

Helenurm, K., and Ganders, F. R. 1985. Adaptive radiation and genetic differentiation in Hawaiian *Bidens*. *Evolution* 39: 753–765.

Hemsley, W. B. 1884. Report on the botany of Juan Fernandez, the south-eastern Moluccas, and the Admiralty Islands. In C. W. Thomson and J. Murray (superintendents), *Report on the Scientific Results of the Voyage of H.M.S. Challenger during the Years 1873–7, vol. 1: Botany* (pp. 1–96). London: Her Majesty's Stationary Office.

Hendry, A. P., Nosil, P., and Rieseberg, L. H. 2007. The speed of ecological speciation. *Funct. Ecol.* 21: 455–464.

Hennequin, S. 2003. Phylogenetic relationships within the fern genus *Hymenophyllum* s.l. (*Hymenophyllaceae*, Filicopsida): contribution of morphology and cytology. *C. R. Biol.* 326: 599–611.

Hennequin, S., Ebihara, A., Dubuisson, J.-Y., and Schneider, H. 2010. Chromosome number evolution in *Hymenophyllum* (Hymenophyllaceae), with special reference to the subgenus *Hymenophyllum*. *Mol. Phylogenet. Evol.* 55: 47–59.

Hennequin, S., Ebihara, A., Ito, M., Iwatsuki, K., and Dubuisson, J.-Y. 2003. Molecular systematics of the fern genus *Hymenophyllum* s.l. (Hymenophyllaceae) based on chloroplastic coding and non-coding regions. *Mol. Phylogenet. Evol.* 27: 283–301.

Hennequin, S., Ebihara, A., Ito, M., Iwatsuki, K., and Dubuisson, J.-Y. 2006. New insights into the phylogeny of the genus *Hymenophyllum* s.l. (Hymenophyllaceae): revealing the polyphyly of *Mecodium*. *Syst. Bot.* 31: 271–284.

Hennequin, S., Kessler, M., Lindsay, S., and Schneider, H. 2014. Evolutionary patterns in the assembly of fern diversity on the oceanic Mascarene Islands. *J. Biogeogr.* 41: 1651–1663.

Hennig, W. 1966. *Phylogenetic Systematics*, trans. D. D. Davis and R. Zangerl. Urbana: University of Illinois Press.

Henry, R. J. 2013. Evolution of DNA marker technology in plants. In R. J. Henry (ed.), *Molecular Markers in Plants* (pp. 4–19). West Sussex: Wiley-Blackwell.

Hernández, R., and Monleon, J. 1975. La comunidad de pescadores de Juan Fernández. In M. Orellana R. (ed.), *Las Islas de Juan Fernández. Historia, Arqueología y Antropología de la Isla Robinson Crusoe* (pp. 137–153). Santiago: Depto. de Ciéncias Antropológicas y Arqueológicas, Universidad de Chile.

Herzog, T. 1942. Die foliosen Lebermoose der Juan Fernandez.-Inseln und der Osterinsel. In C. Skottsberg (ed.), *The Natural History of the Juan Fernandez and Easter Island, vol. 2: Botany* (pp. 697–752). Uppsala: Almqvist & Wiksells.

Heusser, C. J. 1977. Carlos Muñoz Pizarro (1913–1976). In G. T. Prance and T. S. Elias (eds.), *Extinction Is Forever* (pp. 248–250). New York: NewYork Botanical Garden.

Heywood, V. H. 2011. The hazardous future of island floras. In D. Bramwell and J. Caujapé-Castells (eds.), *The Biology of Island Floras* (pp. 488–510). Cambridge: Cambridge University Press.

Hill, M. O. 1979. *Twinspan: A Fortran Program for Arranging Multivariate Data in an Ordered Two-Way Table by Classification of the Individuals and Attributes*. Ithaca, NY: Cornell University Press.

Hind, N., and Johnson, N. 2006. *Dendroseris litoralis*, Compositae. *Curtis's Bot. Mag.* 23: 314–324.

Hobdy, R. 1993. Lāna'i: a case study: the loss of biodiversity on a small Hawaiian island. *Pacific Sci.* 47: 201–210.

Hoeneisen, M., Becerra, J., Bittner, M., Silva, M., Baeza, C., et al. 1998. Sesquiterpenes of the essential oil of *Santalum fernandezianum* F. Phil (Santalaceae), an extinct species of the Juan Fernández Islands, Chile. *Bol. Soc. Chil. Quim.* 43: 505–508.

Hoggard, R. K., Kores, P. J., Molvray, M., Hoggard, G. D., and Broughton, D. A. 2003. Molecular systematics and biogeography of the amphibious genus *Littorella* (Plantaginaceae). *Am. J. Bot.* 90: 429–435.

Holttum, R. E. 1966. The genus *Arthropteris* J. Sm. in Malesia. *Blumea* 14: 225–229.

Holttum, R. E. 1986. Studies in the fern-genera allied to *Tectaria* Cav: VI. A conspectus of genera in the Old World regarded as related to *Tectaria*, with descriptions of two genera. *Gard. Bull. Straits Settlem.* 39: 153–167.

Hörandl, E. 2014. Nothing in taxonomy makes sense except in the light of evolution: examples from the classification of *Ranunculus*. *Ann. Missouri Bot. Gard.* 100: 14–31.

Hörandl, E., and Appelhaus, M. S. (eds.). 2015. *Next-Generation Sequencing in Plant Systematics*. Königstein: Koeltz Scientific Books.

Hörandl, E., and Emadzade, K. 2012. Evolutionary classification: a case study on the diverse plant genus *Ranunculus* L. (Ranunculaceae). *Perspect. Plant Ecol. Evol. Syst.* 14: 310–324.

Hörandl, E., Paun, O., Johansson, J. T., Lehnebach, C., Armstrong, T., et al. 2005. Phylogenetic relationships and evolutionary traits in *Ranunculus* s.l. (Ranunculaceae) inferred from ITS sequence analysis. *Mol. Phylogenet. Evol.* 36: 305–327.

Horner, H. T., Wanke, S., and Samain, M.-S. 2009. Evolution and systematic value of leaf crystal macropatterns in the genus *Peperomia* (Piperaceae). *Int. J. Plant Sci.* 170: 343–354.

Horres, R., Schulte, K., Weising, K., and Zizka, G. 2007. Systematics of Bromelioideae (Bromeliaceae): evidence from molecular and anatomical studies. *Aliso* 23: 27–43.

Hortal, J., Triantis, K. A., Meiri, S., Thébault, E., and Sfenthourakis, S. 2009. Island species richness increases with habitat diversity. *Am. Nat.* 174(6): E205–17.

Howarth, D. G., and Baum, D. A. 2005. Genealogical evidence of homoploid hybrid speciation in an adaptive radiation of *Scaevola* (Goodeniaceae) in the Hawaiian Islands. *Evolution* 59: 948–961.

Howell, J. 1829. *The Life and Adventures of Alexander Selkirk; Containing the Real Incidents upon which the Romance of Robinson Crusoe Is Founded.* Edinburgh: Oliver & Boyd.

Hu, X-S. 2015. Mating system as a barrier to gene flow. *Evolution* 69: 1158–1177.

Hubbs, C. L., and Norris, K. S. 1971. Original teeming abundance, supposed extinction, and survival of the Juan Fernández fur seal. *Antarctic Res. Ser.* 18: 35–52.

Hubby, J. L., and Lewontin, R. C. 1966. A molecular approach to the study of genic heterozygosity in natural populations: I. The number of alleles at different loci in *Drosophila pseudoobscura*. *Genetics* 54: 577–594.

Huene, R., von, Corvalán, J., Flueh, E. R., Hinz, K., Korstgard, J., et al. and the CONDOR Scientists. 1997. Tectonic control of the subducting Juan Fernández Ridge on the Andean margin near Valparáiso, Chile. *Tectonica* 16: 474–488.

Humphries, C. J., and Richardson, P. M. 1980. Hennig's methods and phytochemistry. In F. A. Bisby, J. G. Vaughan, and C. A. Wright (eds.), *Chemosystematics: Principles and Practice* (pp. 353–3780). London: Academic Press.

Hunter, M. L., Jr., and Gibbs, J. P. 2007. *Fundamentals of Conservation Biology*. Malden, MA: Blackwell.

Hurr, K. A., Lockhar, P. J., Heenan, P. B., and Penny, D. 1999. Evidence for the recent dispersal of *Sophora* (Leguminosae) around the Southern Oceans: molecular data. *J. Biogeogr.* 26: 565–577.

Ingram, K., Bernardello, G., Cover, S., and Wilson, E. 2006. The ants of the Juan Fernández Islands: genesis of an invasive fauna. *Biol. Invas.* 8: 383–387.

IUCN (International Union for the Conservation of Nature). 2001. *Categories and Criteria* (version 3.1). Available at www.redlist.org/info/categories_criteria2001.html.

Iwashina T., Kitajima, J., Shiuchi, T., and Itou, Y. 2005. Chalcones and other flavonoids from *Asarum* sensu lato (Aristolochiaceae). *Biochem. Syst. Ecol.* 33: 571–584.

Iwatsuki, K. 1990. Hymenophyllaceae. In K. U. Kramer and P. S. Green (vol. eds.), *Pteridophytes and Gymnosperms*. In K. Kubitzki (gen. ed.), *The Families and Genera of Vascular Plants*, vol. I (pp. 157–163). Berlin: Springer-Verlag.

Jaksic, F. M., and Fuentes, E. R. 1991. Ecology of a successful invader: the European rabbit in central Chile. In R. H. Groves and F. di Castri (eds.), *Biogeography of Mediterranean Invasions* (pp. 273–283). Cambridge: Cambridge University Press.

Jakubowsky, G., and Stuessy, T. F. 1999. Hybridization and evolution in *Eryngium* (Apiaceae) of the Juan Fernández Islands. *Syst. Assoc. Biennial Conference, Glasgow*, p. 23 (Abstr.)

Johannsen, A. 1937. *A Descriptive Petrography of the Igneous Rocks III*. Chicago: University of Chicago Press.

Johow, F. 1892–93. Los helechos de Juan Fernández. *Anal. Univ. Chile* 82: 741–757, 977–1005 (the two parts were also published together in 1893, 46 pp., perhaps privately, and printed by Imprenta Cervantes: Santiago).

Johow, F. 1893. Las plantas de cultivo en Juan Fernández. *Anal. Univ. Chile* 84: 939–970.

Johow, F. 1896. *Estudios Sobre la Flora de las Islas de Juan Fernández*. Santiago: Gobierno de Chile.

Jones, K. E., Reyes-Betancort, J.A., Hiscock, S. J., and Carine, M. A. 2014. Allopatric diversification, multiple habitat shifts and hybridization in the evolution of *Pericallis* (Asteraceae), a Macaronesian endemic genus. *Am. J. Bot.* 101: 637–651.

Juan, J., and de Ulloa, A. 1748. *Relación Histórica del Viaje a la América Meridional*. Madrid [seen in A. Saumell (ed.). 1990. *Antonio de Ulloa, Viaje a la America Meridional*, Tomo B: Segunda parte del Viage al Reyno del Peru. Madrid: Historia 16].

Juan, J., and de Ulloa, A. 1826. *Noticias Secretas de América, sobre el Estado Naval, Militar, y Político de los Reynos del Perú y Provincias de Quito, Costas de Nueva Granada y Chile*. London.

Judd, W. S., Campbell, C. S., Kellogg, E. A., Stevens, P. F., and Donoghue, M. J. 2007. *Plant Systematics: A Phylogenetic Approach* (3rd edn.). Sunderland, MA: Sinauer Associates.

Judziewicz, E. J., Clark, L. G., Londoño, X., and Stern, M. J. 1999. *American Bamboos*. Washington, DC: Smithsonian Institution Press.

Kadereit, G., Borsch, T., Weising, K., and Freitag, H. 2003. Phylogeny of Amaranthaceae and Chenopodiaceae and the evolution of C_4 photosynthesis. *Int. J. Plant Sci.* 164: 959–986.

Kadereit, G., Mavrodiev, E. V., Zacharias, E. H., and Sukhorukov, A. P. 2010. Molecular phylogeny of Atripliceae (Chenopodioideae, Chenopodiaceae): implications for systematics, biogeography, flower and fruit evolution, and the origin of C_4 photosynthesis. *Am. J. Bot.* 97: 1664–1687.

Kadereit, J. W., Repplinger, M., Schmaltz, N., Uhink, C. H., and Wörz, A. 2008. The phylogeny and biogeography of Apiaceae subf. Saniculoideae tribe Saniculeae: from south to north and south again. *Taxon* 57: 365–382.

Kang, D.-L., Pan, Q.-H., and Yi, Z.-L. 2008. Phylogenetic study on sect. *Duretia* (*Boehmeria*, Urticaceae) based on nrDNA ITS sequences. *Plant Sci. J.* 26: 450–453.

Karrenberg, S., and Widmer, A. 2008. Ecologically relevant genetic variation from a non-*Arabidopsis* perspective. *Curr. Opin. Plant Biol.* 11: 156–162.

Kausel, E. 1949. Comentario sobre las ericáceas y epacridáceas chilenas. *Rev. Universitaria* 34: 155–178 (*Anal. Acad. Chil. Cienc. Nat.* 14).

Kawakami, K., and Okochi, I. 2010. *Restoring the Oceanic Island Ecosystem: Impact and Management of Invasive Alien Species in the Bonin Islands*. Tokyo: Springer.

Keissler, K. 1927. Ascomyceten, Fungi imperfecti und Uredineen von Juan Fernandez. In C. Skottsberg (ed.), *The Natural History of the Juan Fernandez and Easter Island, vol. 2: Botany* (pp. 473–486). Uppsala: Almqvist & Wiksells.

Kelchner, S. A., and Clark, L. G. 1997. Molecular evolution and phylogenetic utility of the chloroplast *rpl16* intron in *Chusquea* and the Bambusoideae (Poaceae). *Mol. Phylogenet. Evol.* 8: 385–397.

Kelchner, S. A., and Bamboo Phylogeny Group. 2013. Higher level phylogenetic relationships within the bamboos (Poaceae: Bambusoideae) based on five plastid markers. *Mol. Phylogenet. Evol.* 67: 404–413.

Kelly, L. M., and González, F. 2003. Phylogenetic relationships in Aristolochiaceae. *Syst. Bot.* 28: 236–249.

Kendrew, W. G. 1961. *The Climates of the Continents*. Oxford: Oxford University Press.

Kerr, M. S. 2004. A phylogenetic and biogeographic analysis of Sanguisorbeae (Rosaceae), with emphasis on the Pleistocene radiation of the high Andean genus *Polylepis*. Ph.D. dissertation, University of Maryland, College Park.

Khan, M. Q., and Narayan, R. K. J. 2007. Phylogenetic diversity and relationships among species of genus *Nicotiana* using RAPDs analysis. *Afr. J. Biotech.* 6: 148–162.

Kiehn, M., Jodl, M., and Jakubowsky, G. 2005. Chromosome numbers of angiosperms from the Juan Fernández Islands, the Tristan da Cunha Archipelago, and from mainland Chile. *Pacific Sci.* 59: 453–460.

Kilian, N., Gemeinholzer, B., and Lack, H. W. 2009. Cichorieae. In V. A. Funk, A. Susanna, T. F. Stuessy and R. J. Bayer (eds.), *Systematics, Evolution and Biogeography of the Compositae* (pp. 341–348). Vienna: International Association for Plant Taxonomy.

Kim, C., Deng, T., Chase, M., Zhang, D.-G., Nie, Z.-L., et al. 2015. Generic phylogeny and character evolution in Urticeae (Urticaceae) inferred from nuclear and plastid DNA regions. *Taxon* 64: 65–78.

Kim, S.-C., Crawford, D. J., Francisco-Ortega, J., and Santos-Guerra, A. 1996a. A common origin for woody *Sonchus* and five related genera in the Macaronesian islands: molecular evidence for extensive radiation. *Proc. Natl. Acad. Sci. USA* 93: 7743–7748.

Kim, S.-C., Crawford, D. J., and Jansen, R. K. 1996b. Phylogenetic relationships among the genera of the subtribe Sonchinae (Asteraceae): evidence from ITS sequences. *Syst. Bot.* 21: 417–432.

Kim, S.-C., Crawford, D. J., Jansen, R. K., and Santos-Guerra, A. 1999. The use of a non-coding region of chloroplast DNA in phylogenetic studies of the subtribe Sonchinae (Asteraceae: Lactuceae). *Plant Syst. Evol.* 215: 85–99.

Kim, S.-C., Lee, C., and Mejías, J. A. 2007. Phylogenetic analysis of chloroplast DNA *matK* gene and ITS of nrDNA sequences reveals polyphyly of the genus *Sonchus* and new relationships among the subtribe Sonchinae (Asteraceae: Cichorieae). *Mol. Phylogenet. Evol.* 44: 578–597.

Kim, Y.-D., Kim, S.-H., and Landrum, L. R. 2004. Taxonomic and phytogeographic implications from ITS phylogeny in *Berberis* (Berberidaceae). *J. Plant Res.* 117: 175–182.

Kingston, N., Waldren, S., and Bradley, U. 2003. The phytogeographical affinities of the Pitcairn Islands: a model for south-eastern Polynesia? *J. Biogeogr.* 30: 1311–1328.

Kirk, H., and Freeland, J. R. 2011. Applications and implications of neutral *versus* non-neutral markers in molecular ecology. *Int. J. Mol. Sci.* 11: 3966–3988.

Kirschner, J. (ed.). 2002. *Rostkovia to Luzula. Species Plantarum: Flora of the World, Part 6: Juncaceae 1*. Canberra: Australian Biological Resources Study.

Knapp, S., Chase, M. W., and Clarkson, J. J. 2004. Nomenclatural changes and a new sectional classification in *Nicotiana* (Solanaceae). *Taxon* 53: 73–82.

Koelling, V. A., Monnahan, P. J., and Kelly, J. K. 2012. A Bayesian method for the joint estimation of outcrossing rate and inbreeding depression. *Heredity* 109: 393–400.

Komarnytskyy, S. I. 2005. Molecular phylogeny of chloroplast DNA of *Nicotiana* species. *Tsitol Genet.* 39(4): 13–19 (in Russian).

Korall, P., Pryer, K. M., Metzgar, J. S., Schneider, H., and Conant, D. S. 2006. Tree ferns: monophyletic groups and their relationships as revealed by four protein-coding plastid loci. *Mol. Phylogenet. Evol.* 39: 830–845.

Kover, P. X., and Mott, R. 2012. Mapping the genetic basis of ecologically and evolutionarily relevant traits in *Arabidopsis thaliana*. *Curr. Opin. Plant Biol.* 15: 212–217.

Koyama, T. 1972. Cyperaceae–Rhynchosporeae and Cladieae. In B. Maguire et al. (eds.), *The Botany of the Guayana Highland,* Part IX. *Mem. New York Bot. Gard.* 23: 23–89. (cf. p. 89).

Kramer, K. U. 1990. Dicksoniaceae. In K. U. Kramer and P. S. Green (vol. eds.), *Pteridophytes and Gymnosperms*, in K. Kubitzki (ed.), *The Families and Genera of Vascular Plants*, vol. 1 (pp. 94–99). Berlin: Springer-Verlag.

Kramer, K. U. 1993. Distribution patterns in major pteridophyte taxa relative to those of angiosperms. *J. Biogeogr.* 20: 287–291.

Kramer, K. U., and Green, P. S. (eds.). 1990. Pteridophytes and gymnosperms. In K. Kubitzki (ed.), *The Families and Genera of Vascular Plants*, vol. 1. Berlin: Springer-Verlag.

Kraus, G. 1882. Ueber das Sandelholz von Juan Fernandez. *Ber. Sitzungen Naturf. Ges. Halle* 1882: 8–9.

Kueffer, C., Daehler, C. C., Torre-Santana, C. W., Lavergne, C., Meyer, J. Y., et al. 2010. A global comparison of plant invasions on oceanic islands. *Perspect. Plant Ecol. Evol. Syst.* 12: 145–161.

Kunkel, G. 1957. Beobachtungen über die Vegetation auf dem Yunque-Massiv. *Bot. Jahrb.* 77: 149–157.

Kunkel, G. 1965. Catalogue of the pteridophytes of the Juan Fernández Islands (Chile). *Nova Hedwigia* 9: 245–284.

Kuntze, O. 1898. *Revisio Generum Plantarum*, pars 3. Leipzig: Arthur Felix. (cf. p. 77).

Kunze, G. 1834. *Thyrsopteris*, eine neue Farngattung. *Linnaea* 9: 506–508.

Kuschel, G. 1952. Los insectos de las islas Juan Fernández. Introducción. *Revista Chil. Entomol.* 2: 3–6.

Labiak, P. H., Sundue, M., Rouhan, G., Hanks, J. G., Mickel, J. T., et al. 2014. Phylogeny and historical biogeography of the lastreopsid ferns (Dryopteridaceae). *Am. J. Bot.* 101: 1207–1228.

Lack, D. 1947. *Darwin's Finches: An Essay on the General Biological Theory of Evolution.* Cambridge: Cambridge University Press.

Lack, H. W., Sack, G., and Bachmann, K. 1978. The genome of *Dendroseris litoralis*, an arboreal insular endemic of the tribe Lactuceae (Asteraceae). *Beitr. Biol. Pflanzen* 54: 425–441.

Lambelet-Haueter, C. 1990. Mauvaises herbes et flore anthropogène: II. Classifications et catégories. *Saussurea* 22: 49–81.

Lambelet-Haueter, C. 1991. Mauvaises herbes et flore anthropogène: I. Définitions, concepts et caractéristiques écologiques. *Saussurea* 21: 47–73.

Lammers, T. G. 1996. Phylogeny, biogeography, and systematics of the *Wahlenbergia fernandeziana* complex (Campanulaceae: Campanuloideae). *Syst. Bot.* 21: 397–415.

Lammers, T. G., Stuessy, T. F., and Silva O., M. 1986. Systematic relationships of the Lactoridaceae, an endemic family of the Juan Fernández Islands, Chile. *Plant Syst. Evol.* 152: 243–266.

Lampert, A., Hastings, A., Grosholz, E. D., Jardine, S. L., and Sanchirico, J. N. 2014. Optimal approaches for balancing invasive species eradication and endangered species management. *Science* 344: 1028–1031.

Landrum, L. R. 1981a. A monograph of the genus *Myrceugenia* (Myrtaceae). *Fl. Neotrop. Monogr.* 29: 1–135.

Landrum, L. R. 1981b. The phylogeny and geography of *Myrceugenia* (Myrtaceae). *Brittonia* 33: 105–129.

Landrum, L. R. 1988. The myrtle family (Myrtaceae) in Chile. *Proc. Calif. Acad. Sci.* 45: 277–317.

Landrum, L. R. 1999. Revision of *Berberis* (Berberidaceae) in Chile and adjacent southern Argentina. *Ann. Missouri Bot. Gard.* 86: 793–834.

Långström, E., and Chase, M. W. 2002. Tribes of Boraginoideae (Boraginaceae) and placement of *Antiphytum*, *Echiochilon*, *Ogastemma* and *Sericostoma*: a phylogenetic analysis based on *atpB* plastid DNA sequence data. *Plant Syst. Evol.* 234: 137–153.

Larsen, C., Ponce, M. M., and Scataglini, M. A. 2013. Revisión de las especies de *Hymenophyllum* (Hymenophyllaceae) del sur de Argentina y Chile. *Gayana Bot.* 70: 274–329.

Laurance, W. F. 2004. The perils of payoff: corruption as a threat to global biodiversity. *Trends Ecol. Evol.* 19: 399–401.

Lee, J., Park, B.-J., Tsunetsugu, Y., Ohira, T., Kagawa, T., et al. 2011. Effect of forest bathing on physiological and psychological responses in young Japanese male subjects. *Publ. Health* 125: 93–100.

Lee, N., and Hauk, W. 2013. Phylogenetic relationships among species of adders tongue ferns (*Ophioglossum* s.s., Ophioglossaceae). *Botany Conference 2013* (Abstr.).

Lee, W. G., Macmillan, B. H., Partridge, T. R., Lister, R., and Lloyd, K. M. 2001. Fruit features in relation to the ecology and distribution of *Acaena* (Rosaceae) species in New Zealand. *New Zealand J. Ecol.* 25: 17–27.

Lehnebach, C. A. 2003. Preliminary checklist of the orchids of Chile. *Bot. J. Linn. Soc.* 143: 449–451.

Leinonen, T., O'Hara, R. B., Cano, J. M., and Merilä, J. 2008. Comparative studies of quantitative trait and neutral marker divergence: a meta-analysis. *J. Evol. Biol.* 21: 1–17.

Leiva, I., Schiller, R., Chamorro, O., López, B., Araya, G., et al. 2013. Nuevos registros sobre especies En Peligro Crítico de la flora de las islas Robinson Crusoe y Santa Clara, Chile. *Gayana Bot.* 70: 377–382.

Leskinen, E., and Alström-Rapaport, C. 1999. Molecular phylogeny of Salicaceae and closely related Flacourtiaceae: evidence from 5.8S, ITS1 and ITS 2 of the rDNA. *Plant. Syst. Evol.* 215: 209–227.

Levin, D. A. 2002. *The Role of Chromosomal Change in Plant Evolution*. New York: Oxford University Press.

Levin, D. A. 1978. The origin of isolating mechanisms in flowering plants. *Evol. Biol.* 11: 185–317.

Levin, D. A. 2000. *The Origin, Expansion, and Demise of Plant Species*. New York: Oxford University Press.

Levin, D. A. 2004. The ecological transition in speciation. *New Phytol.* 161: 91–96.

Levring, T. 1941. Die Meeresalgen der Juan Fernandez-Inseln. In C. Skottsberg (ed.), *The Natural History of the Juan Fernandez and Easter Island, vol. 2: Botany* (pp. 601–670). Uppsala: Almqvist & Wiksells.

Levsen, N. D., Crawford, D. J., Archibald, J. K., Santos-Guerra, A., and Mort, M. E. 2008. Nei's to Bayes': comparing methods for estimating patterns of genetic variation in *Tolpis* (Asteraceae). *Am. J. Bot.* 95: 1466–1474.

Lewis, H. 1966. Speciation in flowering plants. *Science* 152: 167–172.

Lewis, O. 1949. *Sea Routes to the Gold Fields: The Migration by Water to California in 1849–1852*. New York: Knopf.

Li, Q. 2010. Effect of forest bathing trips on human immune function. *Environm. Health Prevent. Med.* 15: 9–17.

Lichtenstein, J. 1944. Las Ophioglosáceas de la Argentina, Chile y Uruguay. *Darwiniana* 6: 380–441.

Linklater, E. 1972. *The Voyage of the Challenger*. London: John Murray.

Linnaeus, C. 1753. *Species Plantarum*, vol. 2. Stockholm.

Linsley, E. 1966. Pollinating insects of the Galápagos Islands. In R. I. Bowman (ed.), *The Galápagos* (pp. 225–232). Berkeley: University of California Press.

Liu, H.-M., Jiang, R.-H., Guo, J., Hovenkamp, P., Perrie, L. R., et al. 2013. Towards a phylogenetic classification of the climbing fern genus *Arthropteris*. *Taxon* 62: 688–700.

Liu, H.-M., Zhang, X.-C., Wang, W., Qiu, Y.-L., and Chen, Z.-D. 2007. Molecular phylogeny of the fern family Dryopteridaceae inferred from chloroplast *rbc*L and *atp*B genes. *Int. J. Plant Sci.* 168: 1311–1323.

Liu, Z., Lu, Z., Wen, X., Otto-Bliesner, B. L., Timmermann, A., et al. 2014. Evolution and forcing mechanisms of El Niño over the past 21,000 years. *Nature* 515: 550–553.

Lloyd, R., and Klekowski, E. 1970. Spore germination and viability in Pteridophyta: evolutionary significance of chlorophyllous spores. *Biotropica* 2: 129–137.

Lomolino, M. V., Riddle, B. R., and Brown, J. H. 2005. *Biogeography* (3rd edn.). Sunderland, MA: Sinauer Associates.

Lomolino. M. V., Brown, J. H., and Sax, D. F. 2010. Island biogeography theory: reticulations and reintegration of "a biogeography of the species." In J. B. Losos and R. E. Ricklefs (eds.), *The Theory of Island Biogeography Revisited* (pp. 13–51). Princeton, NJ: Princeton University Press.

Lönnberg, E. 1921. The birds of the Juan Fernandez Islands. In C. Skottsberg (ed.), *The Natural History of the Juan Fernandez and Easter Islands, vol. 3: Zoology* (pp. 1–17). Uppsala: Almqvist & Wiksells.

Looser, G. 1927. La zarzamora (*Rubus ulmifolius* Schott) en Juan Fernández. *Rev. Chil. Hist. Nat.* 31: 84–85.

Looser, G. 1944. Sinopsis de los *Asplenium* (Filices) de Chile. *Lilloa* 10: 233–264.

Looser, G. 1945. Notas sobre helechos chilenos–III. *Rev. Chil. Hist. Nat.* 48: 110–111.

Looser, G. 1947. Los *Blechnum* (Filices) de Chile. *Rev. Univ. (Univ. Católica de Chile)* 32: 7–104.

López-Sepúlveda, P., Baeza, M., Peñailillo, P., Stuessy, T., Greimler, J., et al. 2013a. Introduced plant species in the Biosphere Reserve Alejandro Selkirk Island, Robinson Crusoe archipelago, Chile, and their threat to the native and endemic vegetation. *Mem. New York Bot. Gard.* 108: 223–243.

López-Sepúlveda, P., Takayama, K., Crawford, D. J., Greimler, J., Peñailillo, P., et al. 2016. Genetic consequences and biogeography of anagenetic speciation of *Rhaphithamnus venustus* (Verbenaceae) in the Juan Fernández archipelago, Chile, based on AFLPs and SSRs. *Plant Spec. Biol* 32: 223–237.

López-Sepúlveda, P., Takayama, K., Greimler, J., Crawford, D. J., Peñailillo, P., et al. 2015a. Speciation and biogeography of *Erigeron* (Asteraceae) in the Juan Fernández Archipelago, Chile, based on AFLPs and SSRs. *Syst. Bot.* 40: 888–899.

López-Sepúlveda, P., Takayama, K., Greimler, J., Crawford, D. J., Peñailillo, P., et al. 2015b. Progressive migration and anagenesis in *Drimys confertifolia* of the Juan Fernández Archipelago, Chile. *J. Plant Res.* 128: 73–90.

López-Sepúlveda, P., Takayama, K., Greimler, J., Peñailillo, P., Crawford, D. J., et al. 2013b. Genetic variation (AFLPs and nuclear microsatellites) in two anagenetically derived endemic species of *Myrceugenia* (Myrtaceae) on the Juan Fernández Islands, Chile. *Am. J. Bot.* 100: 722–734.

Lord, J. 1999. Fleshy-fruitedness in the New Zealand flora. *J. Biogeogr.* 26: 1249–1253.

Losos, J. B., and Parent, C. E. 2010. The speciation-area relationship. In J. B. Losos and R. E. Ricklefs (eds.), *The Theory of Island Biogeography Revisited* (pp. 415–438). Princeton, NJ: Princeton University Press.

Losos, J. B., and Ricklefs, R. E. (eds.). 2010. *The Theory of Island Biogeography Revisted.* Princeton, NJ: Princeton University Press.

Lowrey, T. K. 1986. A biosystematic revision of Hawaiian *Tetramolopium* (Compositae: Astereae). *Allertonia* 4: 1–123.

Lowrey, T. K., and Crawford, D. J. 1985. Allozyme divergence and evolution in *Tetramolopium* (Compositae: Astereae) on the Hawaiian Islands. *Syst. Bot.* 10: 64–72.

Lowry, D. B., and Gould, B. A. 2016. Speciation continuum. In R. Kliman (gen. ed.) and D. Ortiz-Barrientos (section ed.), *Encyclopedia of Evolutionary Biology* (pp. 159–165). New York: Academic Press.

Lowry, D. B., Modliszewski, J. L., Wright, K. M., Wu, C. A., and Willis, J. H. 2008. The strength and genetic basis of reproductive isolating barriers in flowering plants. *Philos. Trans. R. Soc. Lond. B* 363: 3009–3021.

Lowry, P. P., II. 2009. Patterns of species richness, endemism, and diversification in oceanic island floras. In E. Wolanski (ed.), *Oceans and Aquatic Ecosystems*, vol. 2 (pp. 201–220). Paris: Encyclopedia of Life Support Systems (EOLSS).

Lozano, G., and Murillo-A., J. 2001. El género *Dysopsis* (Euphorbiaceae). *Caldasia* 23: 419–426.

Lu, J.-M., Barrington, D. S., and Li, D.-Z. 2007. Molecular phylogeny of the polystichoid ferns in Asia based on *rbcL* sequences. *Syst. Bot.* 32: 26–33.

Luteyn, J. L. 1995. *Pernettya* Gaudichaud. In J. L. Luteyn (ed.), *Ericaceae, Part II: The Superior-Ovaried Genera (Monotropoideae, Pyroloideae, Rhododendroideae, and Vaccinioideae p.p.)* (pp. 365–383). New York: New York Botanical Garden (Flora Neotropica Monograph 66).

Lynch, J. 2009. *San Martín: Argentine Soldier, American Hero.* New Haven, CT: Yale University Press.

Mabberley, D. J. 2008. *Mabberley's Plant-Book: A Portable Dictionary of Plants, Their Classification and Uses* (3rd edn.). Cambridge: Cambridge University Press.

Mabry, T. J., Markham, K. R., and Thomas, M. B. 1970. *The Systematic Identification of Flavonoids*. New York: Springer-Verlag.

MacArthur, R. H., and Wilson, E. O. 1967. *The Theory of Island Biogeography*. Princeton, NJ: Princeton University Press.

Macdonald, G. A., Abbott, A. T., and Peterson, F. L. 1983. *Volcanoes in the Sea: The Geology of Hawaii* (2nd edn.). Honolulu: University of Hawaii Press.

MacPhail, M. K., Partridge, A. D., and Truswell, E. M. 1999. Fossil pollen records of the problematical primitive angiosperm family Lactoridaceae in Australia. *Plant Syst. Evol.* 214: 199–210.

Magee, B. 1795. *Observations on the Islands of Juan Fernandez, Massafuero, and St. Ambrose, in the South Pacific Ocean, and the Coast of Chili, in South-America. Extracted from the Journal of Mr. Bernard Magee, First Officer of the Ship Jefferson, in Her Late Voyage Round the Globe*. Boston: Massachusetts Historical Society (reprint from Johnson Reprint Corporation, 1968).

Magnacca, K. N., Foote, D., and O'Grady, P. M. 2008. A review of the endemic Hawaiian Drosophilidae and their host plants. *Zootaxa* 1728: 1–58.

Malhi, Y., Roberts, J. T., Betts, R. A., Killeen, T. J, Li, W., et al. 2008. Climate change, deforestation, and the fate of the Amazon. *Science* 319: 169–172.

Mallet, J. 2007. Hybrid speciation. *Nature* 446: 279–283.

Mammerickx, J., and Smith, S. M. 1978. *Bathymetry of the Southeast Pacific*. Geol. Soc. Amer. Map and Chart Series MC–26.

Mandáková, T., Heenan, P. B., and Lysak, M. A. 2010 Island species radiation and karyotypic stasis in *Pachycladon* allopolyploids. *BMC Evol. Biol.* 10: 367.

Manen, J.-F. 2000. Relaxation of constraints in promoters of the plastid gene *atpB* in a particular Rubiaceae lineage. *Plant Syst. Evol.* 224: 235–241.

Manen, J.-F., and Natali, A. 1995. Comparison of the evolution of ribulose-1,5-biphosphate carboxylase (*rbcL*) and *atpB-rbcL* noncoding spacer sequences in a recent plant group, the tribe Rubieae (Rubiaceae). *J. Mol. Evol.* 41: 920–927.

Manen, J.-F., Natali, A., and Ehrendorfer, F. 1994. Phylogeny of Rubiaceae-Rubieae inferred from the sequence of a cpDNA intergene region. *Plant Syst. Evol.* 190: 195–211.

Mantese, A., and Medan, D. 1993. Anatomía y arquitectura foliares de *Colletia* y *Adolphia* (Rhamnaceae). *Darwiniana* 32: 91–97.

Markham, K. R. 1982. *Techniques of Flavonoid Identification*. New York: Academic Press.

Marquínez, X., Lohmann, L. G., Faria Salatino, M. L., Salatino, A., and González, F 2009. Generic relationships and dating of lineages in Winteraceae based on nuclear (ITS) and plastid (*rpS16* and *psbA-trnH*) sequence data. *Mol. Phylogenet. Evol.* 53: 435–449.

Marticorena, A. 1996. Revisión del género *Acaena* Mutis ex L. (Rosaceae) en Chile. Master's thesis, Universidad de Concepción, Chile.

Marticorena, A. 2006. Revisión del género *Acaena* (Rosaceae) en Chile. *Ann. Missouri Bot. Gard.* 93: 412–454.

Marticorena, C. 1990. Contribución a la estadística de la flora chilena. *Gayana Bot.* 47: 85–113.

Marticorena, C., and Baeza, C. 2001. Piperaceae. In C. Marticorena and R. Rodríguez (eds.), *Flora de Chile,* vol. 2*: Winteraceae-Ranunculaceae* (pp 26–32). Concepción: Universidad de Concepción, Chile.

Marticorena, C., and Quezada, M. 1985. Catálogo de la flora vascular de Chile. *Gayana Bot.* 42: 1–157.

Marticorena, C., Stuessy, T. F., and Baeza, C. M. 1998. Catalogue of the vascular flora of the Robinson Crusoe or Juan Fernández Islands, Chile. *Gayana Bot.* 55: 187–211.

Martin, N. H., and Willis, J. H. 2007. Ecological divergence associated with mating system causes nearly complete reproductive isolation between sympatric *Mimulus* species. *Evolution* 61: 68–82.

Martin, T. E. 1981. Limitation in small habitat islands: chance or competition? *Auk* 98: 715–734.

Martínez, J. A. 1983. Gualterio Looser Schallemberg. Un naturalista botánico (1898–1982). *Rev. Chil. Hist. Nat.* 56: 91–95.

Marx, H. E., O'Leary, N., Yuan, Y.-W., Lu-Irving, P., Tank, D. C., et al. 2010. A molecular phylogeny and classification of Verbenaceae. *Am. J. Bot.* 97: 1647–1663.

Matthei, O. 1986. El género *Bromus* L. (Poaceae) en Chile. *Gayana Bot.* 43: 47–110.

Matthei, O. 1995. *Manual de las Malezas que Crecen en Chile.* Santiago de Chile: published by author.

Matthei, O., Marticorena, C., and Stuessy, T. F. 1993. La flora adventicia del archipiélago de Juan Fernández. *Gayana Bot.* 50: 69–102.

Matthei, O. R. 1974. El género *Megalachne* Steudel (Gramineae). *Bol. Soc. Biol. Concepción* 48: 165–172.

Mattick, F. 1964. Übersicht über die Florenreiche und Florengebiete der Erde. In H. Melchior (ed.), *A. Engler's Syllabus der Pflanzenfamilien*, Bd. 2, 12te Auflage (pp. 626–629). Berlin: Gebrueder Borntraeger.

Mayr, E. 1942. *Systematics and the Origin of Species from the Viewpoint of a Zoologist.* New York: Columbia University Press.

Mayr, E. 1954. Change of genetic environment and evolution. In J. Huxley, A. C. Hardy and E. B. Ford (eds.), *Evolution as a Process* (pp. 157–180). London: Allen & Unwin.

Mayr, E. 1963. *Animal Species and Evolution.* Cambridge, MA: Belknap Press of Harvard University Press.

McAtee, W. L. 1947. Distribution of seeds by birds. *Am. Midl. Nat.* 38: 214–223.

McCall, G. 2009. Easter Island. In R. G. Gillespie and D. A. Clague (eds.), *Encyclopedia of Islands* (pp. 244–251). Berkeley: University of California Press.

McGlone, M. S., Duncan, R. P., and Heenan, P. B. 2001. Endemism, species selection and the origin and distribution of the vascular plant floras of New Zealand. *J. Biogeogr.* 28: 199–216.

McGlone, M. S., Kershaw, A. P., and Markgraf, V. 1992. El Niño/Southern Oscillation climatic variability in Australasian and South American paleoenvironmental records. In H. F. Diaz and V. Markgraf (eds.), *El Niño: Historical and Paleoclimatic Aspects of the Southern Oscillation* (pp. 435–462). Cambridge: Cambridge University Press.

McHenry, M. A., and Barrington, D. S. 2014. Phylogeny and biogeography of exindusiate Andean *Polystichum* (Dryopteridaceae). *Am. J. Bot.* 101: 365–375.

McKay, J. K., and Latta, R. G. 2002. Adaptive population divergence: markers, QTL and traits. *Trends Ecol. Evol.* 17: 285–291.

McMullen, C. 1987. Breeding systems of selected Galápagos Islands angiosperms. *Am. J. Bot.* 74: 1694–1705.

McMullen, C. 1989. The Galápagos carpenter bee, just how important is it? *Noticias de Galápagos* 48: 16–18.

McMullen, C. 1990. Reproductive biology of Galápagos Islands angiosperms. In J. E. Lawesson, O. Hamann, G. Rogers, G. Reck, and H. Ochoa (eds.), *Botanical Research and Management in Galápagos* (pp. 35–45). St. Louis: Missouri Botanical Garden Press (Monogr. Syst. Bot. Missouri Bot. Gard. 32).

McMullen, C. 1993. Flower-visiting insects of the Galápagos Islands. *Pan Pacific Entomol.* 69: 95–106.

McMullen, C. 2007. Pollination biology of the Galápagos endemic, *Tournefortia rufo-sericea* (Boraginaceae). *Bot. J. Linn. Soc.* 153: 21–31.

McMullen, C. 2009a. Island flora: more than "wretched-looking little weeds." In T. De Roy (ed.), *Galápagos: Preserving Darwin's Legacy* (pp. 62–68). Ontario: Firefly Press.

McMullen, C. 2009b. Pollination biology of a night-flowering Galápagos endemic, *Ipomoea habeliana* (Convolvulaceae). *Bot. J. Linn. Soc.* 160: 11–20.

McMullen, C., and Close, D. 1993. Wind pollination in the Galápagos Islands. *Noticias de Galápagos* 52: 12–17.

McNeill, J., Barrie, F. R., Buck, W. R., Demoulin, V., Greuter, W., et al. (eds.). 2012. *International Code of Nomenclature for Algae, Fungi, and Plants (Melbourne Code)*. Königstein: Koeltz (*Regnum Veg.* 154).

McPhaden, M. J., Zebiak, S. E., and Glantz, M. H. 2006. ENSO as an integrating concept in earth science. *Science* 314: 1740–1745.

Medina, J. T. 1923. Don Juan Francisco de Sobrecasas, autor de la relación de la isla de Juan Fernández. *Rev. Chil. Hist. Geogr.* 45: 461–473 (publication of the manuscript by J. F. de Sobrecasas "Relacion orthographica, physico-médico, matemática de la Isla de San Juan Baptista, álias de Juan Fernández,. . . ." based on his trip to the islands in 1751).

Medina, J. T. 1974. *El Piloto Juan Fernández, Descubridor de las Islas que Llevan su Nombre, y Juan Jufre, Armador de la Expedición que Hizo en Busca de Otras en el Mar del Sur* (2nd edn.). Santiago: Editora Nacional Gabriela Mistral.

Meisner, C. F. 1839. *Plantarum Vascularium Genera*. Leipzig: Libraria Weidmannia.

Mejías, J. A., and Kim, S.-C. 2012. Taxonomic treatment of Cichorieae (Asteraceae) endemic to the Juan Fernández and Desventuradas Islands (SE Pacific). *Ann. Bot. Fenn.* 49: 171–178.

Meloni, M., Reid, A., Caujapé-Castells, J., Sotok M., Fernández-Palacios, J. M., et al. 2015. High genetic diversity and population structure in the endangered Canarian endemic *Ruta oreojasme* (Rutaceae). *Genetica* 143: 571–580.

Mendoza, A., Pérez-García, B., Jaramillo, I. R., and Ricci, M. 1996. Desarrollo del gametofito de *Pteris berteroana* (Pteridaceae: Pterideae). *Rev. Biol. Trop.* 44–45: 51–57.

Menéndez, C. A. 1966. La presencia de *Thyrsopteris* en el Cretácico superior de Cerro Guido, Chile. *Ameghiniana* 4: 299–302.

Merilä, J., and Crnokrak, P. 2001. Comparison of genetic differentiation at marker loci and quantitative traits. *J. Evol. Biol.* 14: 892–903.

Meudt, H. M., and Clark, A. C. 2007. Almost forgotten or latest practice? AFLP applications, analyses, and advances. *Trends Plant Sci.* 12: 106–117.

Meyer, F. G. 1966. Chile-United States botanical expedition to Juan Fernandez Islands, 1965. *Antarctic J.* 1: 238–242.

Meza Torres, E. I., Cerne, B., Ulke, A. G., and Morbelli, M. A. 2015b. Distribution of *Ophioglossum reticulatum* L. in South America. A case of long-distance jump dispersal? *Int. J. Biometeor* 59: 137–150.

Meza Torres, E. I., Macluf, C. C., Morbelli, M. A., and Ferrucci, M. S. 2015a. The circumscription of problematic species of *Ophioglossum* (Ophioglossaceae) from Southern South America: a palynological approach. *Phytotaxa* 205: 145–156.

Middleton, D. J. 1990. *Pernettya* or *Gaultheria*? *Plantsman* 12: 167–177.

Middleton, D. J. 1991. Infrageneric classification of the genus *Gaultheria* L. (Ericaceae). *Bot. J. Linn. Soc.* 106: 229–258.

Middleton, D. J., and Wilcock, C. C. 1990. A critical examination of the status of *Pernettya* as a genus distinct from *Gaultheria*. *Edinb. J. Bot.* 47: 291–301.

Middleton, R. M. 1909. The first Fuegian collection. *J. Bot.* 47: 207–212.

Midway, S. R., and Hodge, A. M. C. 2012. Carlquist revisited: history, success, and applicability of a natural history model. *Biol. Philos.* 27: 497–520.

Miller, J. M. 1988. A new species of *Degeneria* (Degeneriaceae) from the Fiji archipelago. *J. Arnold Arbor.* 69: 275–280.

Miller, J. M. 1989. The archaic flowering plant family Degeneriaceae: its bearing on an old enigma. *Natl. Geogr. Res.* 512: 218–231.

Minster, J. B., and Jordan, T. H. 1978. Present-day plate motions. *J. Geophys. Res.* 83: 5331–5354.

Mione, T., and Anderson, G.J. 1992. Pollen ovule ratios and breeding system evolution in *Solanum* section *Basarthrum*. *Am. J. Bot.* 79: 279–287.

Mitchell, A. D., and Heenan, P.B. 2002. *Sophora* sect. *Edwardsia* (Fabaceae): further evidence from nrDNA sequence data of a recent and rapid radiation around the Southern Oceans. *Bot. J. Linn. Soc.* 140: 435–441.

Mitchell, A. L., and House, S. 1999. *David Douglas: Explorer ad Botanist*. London: Aurum Press.

Monaghan, J. 1973. *Chile, Peru, and the California Gold Rush of 1849*. Berkeley: University of California Press.

Monteiro, R. F., Mantovani, A., and Forzza, R. C. 2015. Morphological phylogenetic analysis of two early-diverging genera of Bromelioideae (Bromeliaceae). *Rodriguesia* 66: 505–521.

Montenegro, G., Teillier, S., Arce, P., and Poblete, V. 1991. Introduction of plants into the Mediterranean-type climate area of Chile. In R. H. Groves and F. di Castri (eds.), *Biogeography of Mediterranean Invasions* (pp. 103–114). Cambridge: Cambridge University Press.

Moody, M. M., and Les, D. H. 2007. Phylogenetic systematics and character evolution in the angiosperm family Haloragaceae. *Am. J. Bot.* 94: 2005–2025.

Moore, D. M. 1983. *Flora of Tierra del Fuego*. Shropshire: Anthony Nelson and Missouri Botanical Garden.

Moore, H. E., Jr. 1969. The genus *Juania* (Palmae-Arecoideae). *Gentes Herb.* 10: 385–393.

Moragues, E., and Traveset, A. 2005. Effect of *Carpobrotus* spp. on the pollination success of native plant species of the Balearic Islands. *Biol. Conserv.* 122: 611–619.

Moran, R. C. 1993. Robinson Crusoe's ferns. *Fiddlehead Forum* 20: 44–47.

Moreira-Muñoz, A. 2011. *Plant Geography of Chile*. Dordrecht: Springer.

Moreira-Muñoz, A., Elortegui F., S., Hobohm, C., and Pinto da Silva Menezes de Sequeira, M. 2014. Endemism on islands: case studies. In C. Hobohm (ed.), *Endemism in Vascular Plants* (pp. 165–204). Dordrecht: Springer.

Morel, Ch. P. 1975. Antecedentes para la restauración de las fortificaciones de la Isla Robinson Crusoe, Archipiélago de Juan Fernández. In M. Orellana R (ed.), *Las Islas de Juan Fernández. Historia, Arqueología y Antropología de la Isla Robinson Crusoe* (pp. 83–95). Santiago: Depto. de Ciéncias Antropológicas y Arqueológicas, Universidad de Chile.

Mort, M. E., Crawford, D. J., Archibald, J. K., O'Leary, T. R., and Santos-Guerra, A. 2010. Plant DNA barcoding: a test using Macaronesian taxa of *Tolpis* (Asteraceae). *Taxon* 59: 581–587.

Mort, M. E., Crawford, D. J., Kelly, J. K., Santos-Guerra, A., Menezes de Sequeira, M., et al. 2015. Mulitplexed-shotgun-genotyping data resolve phylogeny within a very recently derived insular lineage. *Am. J. Bot.* 102: 634–641.

Morton, C. V. 1968. The genera, subgenera, and sections of the Hymenophyllaceae. *Contrib. US Natl. Mus.* 38: 153–214.

Moseley, H. N. 1892. *Notes by a Naturalist: An Account of Observations Made during the Voyage of H.M.S. "Challenger" Round the World in the Years 1872–1876*. London: John Murray.

Muasya, A. M., Vrijdaghs, A., Simpson, D. A., Chase, M. W., Goetghebeur, P., et al. 2009. What is a genus in Cypereae: phylogeny, character homology assessment and generic circumscription in Cypereae. *Bot. Rev.* 75: 52–66.

Mueller-Dombois, D., and Fosberg, F. R. 1998. *Vegetation of the Tropical Pacific Islands*. New York: Springer.

Müller, K., and Borsch, T. 2005. Phylogenetics of Amaranthaceae based on *matK/trnK* sequence data: evidence from parsimony, likelihood, and Bayesian analyses. 2005. *Ann. Missouri Bot. Gard.* 92: 66–102.

Muñoz Pizarro, C. 1944. El itinerario de Don Claudio Gay. *Bol. Mus. Nac. Hist. Nat.* 22: 27–44.

Muñoz Pizarro, C. 1960. *Las Especies de Plantas Descritas por R. A. Philippi en el Siglo XIX*. Santiago: Universidad de Chile.

Muñoz Pizarro, C. 1974. El Archipiélago de Juan Fernández y la conservación de sus recursos naturales renovables. Santiago: Museo Nacional de Historia Natural (Serie Educativa No. 9).

Munroe, E. G. 1948. The geographical distribution of butterflies in the West Indies. Ph.D. dissertation, Cornell University, Ithaca, NY.

Munroe, E. G. 1963. Perspectives in biogeography. *Can. Entomol.* 95: 299–308.

Murakami, N., and Schaal, B. A. 1994. Chloroplast DNA variation and the phylogeny of *Asplenium* sect. *Hymenasplenium* (Aspleniaceae) in the New World Tropics. *J. Plant Res.* 107: 245–251.

Murillo-A., J., Ruiz, E., Landrum, L. R., Stuessy, T. F., and Barfuss, M. H. J. 2012. Phylogenetic relationships of *Myrceugenia* (Myrtaceae) based on plastid and nuclear DNA sequences. *Mol. Phylogenet. Evol.* 62: 764–776.

Murillo-A., J., Stuessy, T. F., and Ruiz, E. 2013. Phylogenetic relationships among *Myrceugenia, Blepharocalyx*, and *Luma* (Myrtaceae) based on paired-sites models and the secondary structures of ITS and ETS sequences. *Plant Syst. Evol.* 299: 713–729.

Murillo-A., J. Stuessy, T. F., and Ruiz, E. 2016. Explaining disjunct distributions in the flora of southern South America: evolutionary history and biogeography of *Myrceugenia* (Myrtaceae). *J. Biogeogr.* 43: 979–990.

Murillo-Aldana, J., and Ruiz, E. 2011. Revalidación de *Nothomyrcia* (Myrtaceae), un género endémico del Archipiélago de Juan Fernández. *Gayana Bot.* 68: 129–134.

Murphy, B. P., Johnson, J. P. L., Gasparini, N. M., and Sklar, L. S. 2016. Chemical weathering as a mechanism for the climatic control of bedrock river incision. *Nature* 532: 223–227.

Murray, D. R. 1986. Seed dispersal by water. In D. R. Murray (ed.), *Seed Dispersal* (pp. 49–85). Sydney: Academic Press.

Nakai, T. 1950. A new classification of the Gleicheniales. *Bull. Nat. Sci. Mus. Tokyo* 29: 1–71.

Narayan, R. 1987. Nuclear DNA changes, genome differentiation and evolution in *Nicotiana* (Solanaceae). *Plant Syst. Evol.* 157: 161–180.

Natali, A., and Manen, J.-F. 1995. The phylogenetic reconstruction inferred from the sequence of a cpDNA intergene region as a framework for the taxonomic revision of a systematic group: the example of the tribe Rubieae (Rubiaceae). *Giorn. Bot. Ital.* 129: 57–66.

Natland, J. H. 2003. Capture of helium and other volatiles during the growth of olivine phenocrysts in picritic basalts from the Juan Fernandez Islands. *J. Petrol.* 44: 421–456.

Naumann, J., Salomo, K., Der, J. P., Wafula, E. K., Bolin, J. F., et al. 2013. Single-copy nuclear genes place haustorial Hydnoraceae within Piperales and reveal a Cretaceous origin of multiple parasitic angiosperm lineages. *PLoS One* 8(11):1–12.

Naumann, J., Symmank, L., Samain, M.-S., Müller, K. F., Neinhuis, C., et al. 2011. Chasing the hare: evaluating the phylogenetic utility of a nuclear single copy gene region at and below species level within the species rich group *Peperomia* (Piperaceae). *BMC Evol. Biol.* 11: 357. Available at www.biomedcentral.com/1471–2148/11/357.

Navas B., E. 1961. El género *Urtica* en Chile. *Bol. Soc. Argent. Bot.* 9: 395–413.

Nazaire, M., and Hufford, L. 2012. A broad phylogenetic analysis of Boraginaceae: implications for the relationships of *Mertensia*. *Syst. Bot.* 37: 758–783.

Neal, P. R., and Anderson, G. J. 2005. Are 'mating systems' 'breeding systems' of inconsistent and confusing terminology in plant reproductive biology? Or is it the other way around? *Plant Syst. Evol.* 250: 173–185.

Nee, M. 1996. A new species of *Acanthosyris* (Santalaceae) from Bolivia and a key to the woody South American Santalaceae. *Brittonia* 48: 574–579.

Nei, M. 1973. Analysis of gene diversity in subdivided populations. *Proc. Natl. Acad. Sci. USA* 70: 3321–3323.

Nei, M. 1978. Estimation of average heterozygosity and genetic distance from a small number of individuals. *Genetics* 89: 583–590.

Nei, M., Maruyama, T., and Chakraborty, R. 1975. The bottleneck effect and genetic variability in populations. *Evolution* 29: 1–10.

Neshyba, S., and Silva S., N. 1985. Algunos aspectos climatológicos de las islas Robinson Crusoe y Santa Clara (Pacífico Sud Oriental). In P. Arana (ed.), *Investigaciones Marinas en el Archipiélago de Juan Fernández* (pp. 43–54). Santiago: Editorial Universitaria.

Nesom, G. L. 2008. Classification of subtribe Conyzinae (Asteraceae: Astereae). *Lundellia* 11: 8–38.

Newell, R. E., Kidson, J. W., Vincent, D. G., and Boer, G. J. 1972. *The General Circulation of the Tropical Atmosphere*, vol. 1. Cambridge, MA: MIT Press.

Newman, D., and Pilson, D. 1997. Increased probability of extinction due to decreased genetic effective population size: experimental populations of *Clarkia pulchella*. *Evolution* 51: 354–362.

Nielsen, L. R., Cowan, R. S., Siegismund, H. R., Adsersen, H., Philipp, M., et al. 2003. Morphometric, AFLP and plastid microsatellite variation in populations of *Scalesia divisa* and *S. incisa* (Asteraceae) from the Galápagos Islands. *Bot. J. Linn. Soc.* 143: 243–254.

Nielsen, L. R., Philipp, M., Adsersen, H., and Siegismund, H. R. 2000. Breeding system of *Scalesia divisa* Andersson – an endemic Asteraceae from the Galápagos Islands. *Norske Vidensk.-Akad., Mat.-Naturvidensk. Kl., Skrift. N. Ser.* 39: 127–138.

Nishida, H., and Nishida, M. 1979. *Thyrsopterorachis*, gen. nov., a tree fern rachis from the Upper Cretaceous of Hokkaido, Japan. *Bot. Mag. Tokyo* 92: 187–195.

Nishida, H., and Nishida, M. 1981. The vegetation of Mas a Tierra (Robinson Crusoe) Island, Juan Fernandez. *Fac. Sci. Chiba Univ.* 41–48.

Noben, S., and Lehnert, M. 2013. The genus *Dicksonia* (Dicksoniaceae) in the western Pacific. *Phytotaxa* 155: 23–34.

North Dakota Agricultural Experiment Station (NDAES). 1975. *Recommended Chemical Soil Test Procedures for the North Central Region*. Fargo: NDAES (Bulletin No. 499).

Nosil, P., Harmon, L. J., and Seehausen, O. 2009. Ecological explanations for (incomplete) speciation. *Trends Ecol. Evol.* 24: 145–156.

Novoa, P. 2015. *Expedición Botánica a la Isla Alejandro Selkirk*. Viña del Mar: Jardin Botánico Nacional.

Novoa, R., and Villaseca, C. (eds.). 1989. *Mapa Agroclimático de Chile*. Santiago de Chile: Instituto de Investigaciones Agropecuarias (INIA), Ministerio de Agricultura.

Noyes, R. D. 2000. Biogeographical and evolutionary insights on *Erigeron* and allies (Asteraceae) from ITS sequence data. *Plant Syst. Evol.* 220: 93–114.

Oakley, C. G. 2015. The influence of natural variation in population size on the ecological and quantitative genetics of the endangered endemic plant *Hypericum cumulicola*. *Int. J. Plant Sci.* 176: 11–19.

Oehrens, E., and Garrido, N. 1986. Posibilidad del control biológico de la zarzamora en el archipiélago de Juan Fernández. *Bol. Soc. Biol. Concepción Chile* 57: 205–206.

Oldfield, S. 2011. Botanic gardens and the conservation of island floras. In D. Bramwell and J. Caujapé-Castells (eds.), *The Biology of Island Floras* (pp. 474–487). Cambridge: Cambridge University Press.

Olesen, J. M., Valido, A., and Dupont, Y. L. 2002. Polinización de las plantas Canarias. *El Indiferente* 13: 18–29.

Oliver, W. R. B. 1935. The genus *Coprosma*. *B.P. Bishop Mus. Bull.* 132: 1–207.

Ollerton, J., Cranmer, L., Stelzer, R., Sullivan, S., and Chittka, L. 2009. Bird pollination of Canary Island endemic plants. *Naturwissenschaften* 96: 221–232.

Olmstead, R. G., dePamphilis, C. W., Wolfe, A. D., Young, N. D., Elisens, W. J., et al. 2001. Disintegration of the Scrophulariaceae. *Am. J. Bot.* 88: 348–361.

Olmstead, R. G., and Reeves, P. A. 1995. Evidence for the polyphyly of the Scrophulariaceae based on chloroplast *rbcL* and *ndhF* sequences. *Ann. Missouri Bot. Gard.* 82: 176–193.

Ono, M. 1975. Chromosome numbers of some endemic species of the Bonin Islands. I. *Bot. Mag. (Tokyo)* 88: 323–328.

Ono, M. 1991. The flora of the Bonin (Ogasawara) Islands: endemism and dispersal modes. *Aliso* 13: 95–105.

Ono, M. 1998. Conservation of the endemic vascular plant species of the Bonin (Ogasawara) Islands. In T. F. Stuessy and M. Ono (eds.), *Evolution and Speciation in Island Plants* (pp. 169–180). Cambridge: Cambridge University Press.

Ono, M., and Kobayashi, S. 1985. Flowering plants endemic to the Bonin Islands. In M. Ono and K. Okutomi (eds.), *Endemic Plant Species and Vegetation of the Bonin (Ogasawara) Islands* (pp. 1–96). Kamakura: Aboc-sha.

Ono, M., and Masuda, Y. 1981. Chromosome numbers of some endemic species of the Bonin Islands, part II. *Ogasawara Res.* 4: 1–24.

Orchard, A. E. 1975. Taxonomic revisions in the family Haloragaceae: I. The genera *Haloragis*, *Haloragodendron*, *Glischrocaryon*, *Meziella* and *Gonocarpus*. *Bull. Auckland Instit. & Mus.* 10: 1–299.

Orellana R., M. 1975. Historia de los primeros poblamientos de la Isla de Robinson Crusoe. In M. Orellana R. (ed.), *Las Islas de Juan Fernandez Historia, Arqueología y Antropología de la Isla Robinson Crusoe* (pp. 11–22). Santiago: Dpto. Ciéncias Antropológicas y Arqueológicas, Universidad de Chile.

Ortiz R., A. (ed.). 1982. *Estudio de los Recursos Físicos: Archipiélago Juan Fernández*. Santiago: Instituto Nacional de Investigación de Recursos Naturales – CORFO.

Otto, E. M., Janßen, T., Kreier, H.-P., and Schneider, H. 2009. New insights into the phylogeny of *Pleopeltis* and related Neotropical genera (Polypodiaceae, Polypodiopsida). *Mol. Phylogenet. Evol.* 53: 190–201.

Ovalle, A. de. 1649. *An Historical Relation of the Kingdom of Chile, of the Islands of the Kingdom of Chile*. Rome (translated from Spanish and included in Pinkerton, J. 1813. *A General Collection of the Best and Most Interesting Voyages and Travels in All Parts of the World*, London, pp. 90–91).

Oxelman, B., Kornhall, P., Olmstead, R. G., and Bremer, B. 2005. Further disintegration of Scrophulariaceae. *Taxon* 54: 411–425.

Pacheco, P., Crawford, D. J., Stuessy, T. F., and Silva O., M. 1985. Flavonoid evolution in *Robinsonia* (Compositae) of the Juan Fernandez Islands. *Am. J. Bot.* 72: 989–998.

Pacheco, P., Crawford, D. J., Stuessy, T. F., and Silva O., M. 1991b. Flavonoid evolution in *Dendroseris* (Compositae) in the Juan Fernández Islands, Chile. *Am. J. Bot.* 78: 534–543.

Pacheco, P., Crawford, D. J., Stuessy, T. F., and Silva O., M. 1993. Flavonoid chemistry and evolution of *Gunnera* (Guneraceae) in the Juan Fernández Islands, Chile. *Gayana Bot.* 50: 17–30.

Pacheco, P., Stuessy, T. F., and Crawford, D. J. 1991a. Natural interspecific hybridization in *Gunnera* (Gunneraceae) of the Juan Fernández Islands, Chile. *Pacific Sci.* 45: 389–399.

Pannell, J. R. 2015. Evolution of the mating system in colonizing plants. *Mol. Ecol.* 24: 2018–2037.

Paradis, E. 2006. *Analysis of Phylogenetics and Evolution with R*. New York: Springer.

Parker de Bassi, M. T. 1987. *Tras la Estela del Dresden*. Santiago: Ediciones Tusitala.

Parodi, L. R. 1945. Sinopsis de las gramíneas chilenas del género *Chusquea*. *Rev. Univ. (Univ. Católica de Chile)* 30(1): 61–71.

Parra, M. J., Rodríguez, R., Cavieres, L., Muñoz-Tapia, L., and Atala, C. 2012. Latitudinal patterns in Pteridophyte distribution of continental Chile. *Gayana Bot.* 72: 58–69.

Patiño, J, Carine, M., Fernández-Palacios, J. M., Otto, R., Schaefer, H., et al. 2014. The anagenetic world of spore-producing land plants. *New Phytol.* 201: 305–311.

Peck, S. B. 1990. Eyeless arthropods of the Galapagos Islands, Ecuador: composition and origin of the cryptozoic fauna of a young, tropical, oceanic archipelago. *Biotropica* 22: 366–381.

Peck, S. B., and Kukalova-Peck, J. 1990. Origin and biogeography of the beetles (Coleoptera) of the Galápagos Archipelago, Ecuador. *Can. J. Zool.* 68: 1617–1638.

Pedley, L. 1986. Derivation and dispersal of *Acacia* (Leguminosae), with particular reference to Australia, and the recognition of *Senegalia* and *Racosperma*. *Bot. J. Linn. Soc.* 92: 219–284.

Pedley, L. 1987. *Racosperma* Martius (Leguminosae: Mimosoideae) in Queensland: a checklist. *Austrobaileya* 2: 344–357.

Pelser, P. B., Kennedy, A. H., Tepe, E. J., Shidler, J. B., Nordenstam, B., et al. 2010a. Patterns and causes of incongruence between plastid and nuclear Senecioneae (Asteraceae) phylogenies. *Am. J. Bot.* 97: 856–873.

Pelser, P. B., Nordenstam, B., Kadereit, J. W., and Watson, L. E. 2007. An ITS phylogeny of tribe Senecioneae (Asteraceae) and a new delimitation of *Senecio* L. *Taxon* 56: 1077–1104.

Pelser, P. B., Tepe, E. J., Kennedy, A. H., and Watson, L. E. 2010b. The fate of *Robinsonia* (Asteraceae): sunk in *Senecio*, but still monophyletic? *Phytotaxa* 5: 31–46.

Peña, C., Negritto, M. A., Ruiz, E., Baeza, C. M., and Finot, V. L. 2017. Revisión de *Megalachne* Steud. (Poaceae: Pooideae: Poeae), género endémico del Archipiélago de Juan Fernández, Chile. *Gayana Bot.* 74: 189–199.

Peña, R. C., and Cassels, B. K. 1996. Phylogenetic relationships among Chilean *Sophora* species. *Biochem. Syst. Evol.* 24: 725–733.

Peña, R. C., Iturriaga, L., Montenegro, G., and Cassels, B. K. 2000. Phylogenetic and biogeographic aspects of *Sophora* sect. *Edwardsia* (Papilionaceae). *Pacific Sci.* 54: 159–167.

Peña, R. C., Iturriaga, L., Mujica, A. M., and Montenegro, G. 1993. Análisis micromorfológico de polen de *Sophora* (Papilionaceae). Hipótesis filogenética sobre el origin de la sección *Edwardsia*. *Gayana Bot.* 50: 57–65.

Pereira S., E. 1971. *Los Primeros Contactos entre Chile y los Estados Unidos 1778–1809*. Santiago: Andres Bello.

Pérez, M. E. 2010. Niña de 12 años alertó de la ola gigante en Juan Fernández y evitó una tragedia mayor. *La Tercera* 1: 18–19.

Pérez de Paz, J., and Caujapé-Castells, J. 2013. A review of the allozyme data set for the Canarian endemic flora: causes of the high genetic diversity levels and implications for conservation. *Ann. Bot.* 111: 1059–1073.

Perrie, L. R., Bayly, M. J., Lehnebach, C. A., and Brownsey, P. J. 2007. Molecular phylogenetics and molecular dating of the New Zealand Gleicheniaceae. *Brittonia* 59: 129–141.

Perrie, L. R., and Brownsey, P. J. 2005. Insights into the biogeography and polyploid evolution of New Zealand *Asplenium* from chloroplast DNA sequence data. *Am. Fern J.* 95: 1–21.

Perrie, L. R., Brownsey, P. J., Lockhart, P. J., Brown, E. A., and Large, M. F. 2003. Biogeography of temperate Australasian *Polystichum* ferns as inferred from chloroplast sequence and AFLP. *J. Biogeogr.* 30: 1729–1736.

Perrie, L. R., Wilson, R. K., Shepherd, L. D., Ohlsen, D. J., Batty, E. L., et al. 2014. Molecular phylogenetics and generic taxonomy of Blechnaceae ferns. *Taxon* 63: 745–758.

Peterson, B. 1964. Carl Skottsberg 1880–1963. *Taxon* 13: 1–7.

Pfosser, M. F., Guzy-Wrobelska, J., Sun, B. Y., Stuessy, T. F., Sugawara, T., et al. 2002. The origin of species of *Acer* (Sapindaceae) endemic to Ullung Island, Korea. *Syst. Bot.* 27: 351–367.

Pfosser, M., Jakubowsky, G., Schluter, P. M., Fer, T., Kato, H., et al. 2005. Evolution of *Dystaenia takesimana* (Apiaceae), endemic to Ullung Island, Korea. *Plant Syst. Evol.* 256: 159–170.

Pfrender, M. E., Spitze, K., Hicks, J., Morgan, K., Latta, L., et al. 2000. Lack of concordance between genetic diversity estimates at the molecular and quantitative-trait levels. *Conserv. Genet.* 1: 263–269.

Philbrick, C. T., and Anderson, G. J. 1987. Implications of pollen/ovule ratios and pollen size for the reproductive biology of *Potamogeton* and autogamy in aquatic angiosperms. *Syst. Bot.* 12: 98–105.

Philippi, F. 1892. El sándalo de Juan Fernández. *Anal. Mus. Nac. Chile, Segunda Sección, Bot.* 5–7.

Philippi, R. 1856a. Observaciones sobre la flora de Juan Fernández. *Anales Univ. Chile* 1856: 157–169.

Philippi, R. A. 1856b. Bemerkungen über die Flora der Insel Juan Fernandez. *Bot. Zeitung (Berlin)* 14: 641–650.

Pickard, J. 1983. Vegetation of Lord Howe Island. *Cunninghamia* 1(2): 133–266.

Pilger, R. 1920. Ueber einige Gramineae der Skottsbergschen Sammlung von Juan Fernández. *Repert. Spec. Nov. Regni. Veg.* 16: 385–388.

Porter, D. M. 1976. Geography and dispersal of Galápagos islands vascular plants. *Nature* 264: 745–746.

Porter, D. M. 1983. Vascular plants of the Galapagos: origins and dispersal. In R. I. Bowman, M. Berson, and A. E. Leviton (eds.), *Patterns of Evolution in Galápagos Organisms* (pp. 33–96). San Francisco: Pacific Division, AAAS.

Potter, D., Eriksson, T., Evans, R. C., Oh, S., Smedmark, J. E. E., et al. 2007. Phylogeny and classification of Rosaceae. *Plant Syst. Evol.* 266: 5–43.

Powell, E. A., and Kron, K. A. 2001. An analysis of the phylogenetic relationships in the Wintergreen Group (*Gaultheria*, *Diplycosia*, *Pernettya*, *Tepuia*; Ericaceae). *Syst. Bot.* 26: 808–817.

Prado, J., Del Nero Rodrigues, C., Salatino, A., and Salatino, M. L. F. 2007. Phylogenetic relationships among Pteridaceae, including Brazilian species, inferred from *rbcL* sequences. *Taxon* 56: 355–364.

Prince, R. A., Schweller, W. J., Ness, G. E., Coulbourn, W. T., Shepherd, G. L., et al. 1980. *Bathymetry of the Peru-Chile Trench and Continental Margin* (Geol. Soc. Am. Map and Chart Series MC-26), Pt. I.

Prohens, J., Anderson, G. J., Herraiz, J., Bernardello, G., Santos Guerra, A., et al. 2007. Genetic diversity and conservation of two endangered eggplant relatives (*Solanum vespertilio* Aiton and *S. lidii* Sunding) endemic to the Canary Islands. *Genet. Res. Crop Evol.* 54: 451–464.

Pryer, K. M., Smith, A. R., Hunt, J. S., and Dubuisson, J.-Y. 2001. *rbcL* data reveal two monophyletic groups of filmy ferns (Filicopsida: Hymenophyllaceae). *Am. J. Bot.* 88: 1118–1130.

Pugsley, H. W. 1936. Enumeration of the species of *Euphrasia* L. Sect. *Semicalcaratae* Benth. *J. Bot. (Lond.)* 74: 273–288.

Puppo, P., Curto, M., Velo-Antón, G., Pérez de Paz, P. L., and Meimberg, H. 2014. The influence of geological history on diversification in insular species: genetic and morphological patterns of *Micromeria* Benth. (Lamiaceae) in Tenerife (Canary archipelago). *J. Biogeogr.* 41: 1871–1882.

Purvis, A., Gittleman, J. L., and Brooks, T. (eds.). 2005. *Phylogeny and Conservation*. Cambridge: Cambridge University Press.

Pyšek, P., Hulme, P. E., Meyerson, L. A., Smth, G. F., Boatwright, J. S., et al. 2013. Hitting the right target: taxonomic challenges for, and of, plant invasions. *AoB Plants* 5: plate 042; doi:10.1093/aobpla/plt042

Pyšek, P., Richardson, D. M., Rejmánek, M., Webster, G., Williamson, M., et al. 2004. Alien plants in checklists and floras: towards better communication between taxonomists and ecologists. *Taxon* 53: 131–143.

Qiu, Y.-L., Chase, M. W., Les, D. H., and Parks, C. R. 1993. Molecular phylogenetics of the Magnoliidae: cladistic analyses of nucleotide sequences of the plastid gene *rbc*L. *Ann. Missouri Bot. Gard.* 80: 587–606.

Quensel, P. 1954. Additional comments on the geology of the Juan Fernandez Islands. In C. Skottsberg (ed.), *The Natural History of Juan Fernandez and Easter Island, vol. 1* (pp. 37–87). Uppsala: Almqvist & Wiksells.

Rahmstorf, S. 2002. Ocean circulation and climate during the past 120,000 years. *Nature* 419: 207–214.

Rahn, K. 1984. *Plantago* sect. *Oliganthos* in southern South America, a taxonomic revision. *Nord. J. Bot.* 4: 601–627.

Rahn, K. 1996. A phylogenetic study of the Plantaginaceae. *Bot. J. Linn. Soc.* 120: 145–198.

Ramalho, R. S., Quartau, R., Trenhaile, A. S., Mitchell, N. C., Woodroffe, C. D., et al. 2013. Coastal evolution on volcanic oceanic islands: a complex interplay between volcanism, erosion, sedimentation, sea-level change and biogenic production. *Earth Sci. Rev.* 127: 140–170.

Ramsey, J., Bradshaw, H. D., and Schemske, D. W. 2003. Components of reproductive isolation between the monkeyflowers *Mimulus lewisii* and *M. cardinalis* (Phrymaceae). *Evolution* 57: 1520–1534.

Raven, P. H., Kyhos, D. W., and Cave, M. S. 1971. Chromosome numbers and relationships in Annoniflorae. *Taxon* 20: 479–483.

Reed, D. H., and Frankham, R. 2001. How closely correlated are molecular and quantitative measures of genetic variation? A meta-analysis. *Evolution* 55: 1095–1103.

Reiche, C. 1896–1911. *Flora de Chile*. Santiago: Imprenta Cervantes.

Renner, S. 2014. The relative and absolute frequencies of angiosperm sexual systems: dioecy, monoecy, gynodioecy, and an updated online database. *Am. J. Bot.* 101: 1588–1596.

Renner, S., and Ricklefs, R. 1995. Dioecy and its correlates in the flowering plants. *Am. J. Bot.* 82: 596–606.

Retamales, H. A., and Scharaschkin, T. 2015. Comparative leaf anatomy and micromorphology of the Chilean Myrtaceae: taxonomic and ecological implications. *Flora* 217: 138–154.

Reynolds, S. C. P. 2002. *A Catalogue of Alien Plants in Ireland*. Glasnevin: National Botanic Gardens.

Reznicek, A. A. 1990. Evolution in sedges (*Carex*, Cyperaceae). *Can. J. Bot.* 68: 1409–1432.

Ricci, M. 1996. Variation in distribution and abundance of the endemic flora of Juan Fernández Islands, Chile: Pteridophyta. *Biodivers. Conserv.* 5: 1521–1532.

Ricci, M. 2001. Evaluation of conservation status of *Lactoris fernandeziana* Philippi (Lactoridaceae) in Chile. *Biodivers. Conserv.* 10: 2129–2138.

Ricci, M. 2006. Conservation status and *ex situ* cultivation efforts of endemic flora of the Juan Fernández Archipelago. *Biodivers. Conserv.* 15: 3111–3130.

Ricci, M. and Eaton, L. 1994. The rescue of *Wahlenbergia larrainii* in Robinson Crusoe Island, Chile. *Biol. Conserv.* 68: 89–93.

Richardson, J. E., Fay, M. F., Cronk, Q. C. B., Bowman, D., and Chase, M. W. 2000. A phylogenetic analysis of Rhamnaceae using *rbcL* and *trnL-F* plastid DNA sequences. *Am. J. Bot.* 87: 1309–1324.

Rick, C. 1966. Some plant-animal relations on the Galápagos Islands. In R. I. Bowman (ed.), *The Galápagos* (pp. 215–224). Berkeley: University of California Press.

Ridley, H. N. 1930. *The Dispersal of Plants Throughout the World*. Ashford, Kent: L. Reeve.

Rieseberg, L. H. 1997. Hybrid origins of plant species. *Annu. Rev. Ecol. Syst.* 28: 359–389.

Rieseberg, L. H., and Brouillet, L. 1994. Are many plant species paraphyletic? *Taxon* 43: 21–32.

Rieseberg, L. H., and Willis, J. H. 2007. Plant speciation. *Science* 317: 910–914.

Ritland K. 2002. Extensions of models for the estimation of mating systems using *n* independent loci. *Heredity* 88: 221–228.

Roalson, E. H., Columbus, J. T., and Friar, E. A. 2001. Phylogenetic relationships in Cariceae (Cyperaceae) based on ITS (nrDNA) and *trnT-L-F* (cpDNA) region sequences: assessment of subgeneric and sectional relationships in *Carex* with emphasis on section *Acrocystis*. *Syst. Bot.* 26: 318–341.

Rodríguez, R. 1995. Pteridophyta. In C. Marticorena and R. Rodríguez (eds.), *Flora de Chile, vol. I: Pteridophyta: Gymnospermae* (pp. 119–309). Concepción, Chile: Universidad de Concepción.

Rodríguez, R. 1990. Comentarios fitogeográficos y taxonómicos de Pteridophyta chilenos. *Gayana Bot.* 46: 199–208.

Rodríguez, R. 2011. Rutaceae. In C. Marticorena and R. Rodríguez (eds.), *Flora de Chile, vol. 3(1): Misodendraceae-Zygophyllaceae* (pp. 109–114). Concepción, Chile: Universidad de Concepción.

Rodríguez, R. 2015. Notas taxonómicas sobre Pteridófitos chilenos. *Gayana Bot.* 72: 94–100.

Rodríguez, R., and Ponce, M. 2007. Nuevas combinaciones en Gleicheniaceae. *Darwiniana* 45: 236–241.

Rodríguez-Ríos, R. 1990. *Gleichenia lepidota* n. sp. y la familia Gleicheniaceae del archipiélago de Juan Fernández, Chile. *Gayana Bot.* 47: 37–45.

Roff, D. A. 1990. The evolution of flightlessness in insects. *Ecol. Monogr.* 60: 389–421.

Rogers, W. 1712. *A Cruising Voyage Round the World: First to the South-Seas, Thence to the East-Indies, and Homewards by the Cape of Good Hope. Begun in 1708, and Finish'd in 1711.* London.

Roggeveen, J. 1838. *Daagverhaal der Ontdekkings-reis von Mr. Jacob Roggeveen in de Jaren 1721 en 1722*. Middelburg (in the English translation by A. Sharp (ed.), *The Journal of Jacob Roggeveen*. Oxford University Press, 1970).

Romell, L. 1926. Basidiomycetes from Juan Fernandez, collected by C. Skottsberg. In C. Skottsberg (ed.), *The Natural History of the Juan Fernandez and Easter Island, vol. 2: Botany* (pp. 465–471). Uppsala: Almqvist & Wiksells.

Romero A., H. 1985. *Geografía de los Climas*. In E. López S. (dir.), *Geografia de Chile*, vol. 11. Santiago: Instituto Geográfico Militar.

Romero García, A. T., Blanca López, G., and Morales Torres, C. 1988. Revisión del género *Agrostis* L. (Poaceae) en la Peninsula Ibérica. *Ruizia* 7: 1–160.

Rønsted, N., Chase, M. W., Albach, D. C., and Bello M. A. 2002. Phylogenetic relationships within *Plantago* (Plantaginaceae): evidence from nuclear ribosomal ITS and plastid *trnL-F* sequence data. *Bot. J. Linn. Soc.* 139: 323–338.

Rossbach, R. P. 1940. *Spergularia* in North and South America. *Rhodora* 42: 57–83, 105–143, 158–193, 203–213.

Rossbach, R. P. 1943. El género *Spergularia* (Caryophyllaceae) en Chile. *Darwiniana* 6: 211–256.

Rothfels, C. J., Sundue, M. A., Kuo, L.-Y., Larsson, A., Kato, M., et al. 2012. A revised family-level classification for eupolypod II ferns (Polypodiidae: Polypodiales). *Taxon* 61: 515–533.

Rothfels, C. J., Windham, M. D., Gruszi, A. L., Gastony, G. J., and Pryer, K. M. 2008. Toward a monophyletic *Notholaena* (Pteridaceae): resolving patterns of evolutionary convergence in xeric-adapted ferns. *Taxon* 57: 712–724.

Rothfels, C. J., Windham, M. D., and Pryer, K. M. 2013. A plastid phylogeny of the cosmopolitan fern family Cystopteridaceae (Polypodiopsida). *Syst. Bot.* 38: 295–306.

Roy, M. S., Torres-Mura, J. C., and Hertel, F. 1998. Evolution and history of hummingbirds (Aves: Trochilidae) from the Juan Fernandez Islands, Chile. *Ibis* 140: 265–273.

Rúgolo de Agrasar, Z., and De Paula, M. E. 1978. *Agrostis* L. In E. G. Nicora (ed.), *Flora Patagonica, Parte III: Gramineae* (pp. 369–394). Buenos Aires: INTA.

Rúgolo de Agrasar, Z. E., and Molina, A. M. 1997. Las especies del género *Agrostis* L. (Gramineae: Agrostideae) de Chile. *Gayana Bot.* 54: 91–156.

Ruh, M. 1975. Alfredo de Rodt, subdelegado en Juan Fernández 1877–1905. In M. Orellana R. (ed.), *Las Islas de Juan Fernández. Historia, Arqueología y Antropología de la Isla Robinson Crusoe* (pp. 97–136). Santiago: Depto. Ciéncias Antropológicas y Arqueológicas, Universidad de Chile.

Ruiz, E. 2001. Ranunculaceae. In C. Marticorena and R. Rodríguez (eds.), *Flora de Chile, vol. 2: Winteraceae-Ranunculaceae* (pp. 40–84). Concepción, Chile: Universidad de Concepción.

Ruiz, E., Becerra, J., Silva O., M., Crawford, D. J., and Stuessy, T. F. 1994. Flavonoid chemistry of the endemic species of *Myrceugenia* (Myrtaceae) of the Juan Fernández Islands and relatives in continental South America. *Brittonia* 46: 187–193.

Ruiz, E., Crawford, D. J., González, F., Samuel, R., Becerra, J., et al. 2004. Phylogenetic relationships and genetic divergence among endemic species of *Berberis*, *Gunnera*, *Myrceugenia* and *Sophora* of the Juan Fernández Islands (Chile) and their continental progenitors based on isozymes and nrITS sequences. *Taxon* 53: 321–332.

Ruiz, E., Donoso, C., González, F., Becerra, J., Marticorena, C., et al. 1999. Phenetic relationships between Juan Fernández and continental Chilean species of *Sophora* (Fabaceae) based on flavonoid patterns. *Bol. Chil. Quim.* 44: 351–356.

Ruiz, E., Marticorena, C., Crawford, D. J., Stuessy, T. F., González, F., et al. 2000. Morphological and ITS sequence divergence between taxa of *Cuminia* (Lamiaceae), an endemic genus of the Juan Fernández Islands, Chile. *Brittonia* 52: 341–350.

Ruiz, E., Toro, O., Crawford, D. J., Stuessy, T. F., Negritto, M. A., et al. 2008. Phylogenetic relationships among Chilean species of *Drimys* (Winteraceae) based on ITS sequences and insertion/deletion events. *Gayana Bot.* 65: 220–228.

Rundle, H. D., and Nosil, P. 2005. Ecological speciation. *Ecol. Lett.* 8: 336–352.

Russ, C. H. 1923. *The Log of a Forty-Niner [Richard L. Hale]: Being the Record of Adventures by Sea and Shore to the California Gold-Fields and the Pacific Northwest, 1849–1854.* Boston, MA: B. J. Brimmer.

Ryves, T. B., Clement, E. J., and Foster, M. C. 1996. *Alien Grasses of the British Isles*. London: Botanical Society of the British Isles.

Saarela, J. M., Liu, Q., Peterson, P. M., Soreng, R. J., and Paszko, B. 2010. Phylogenetics of the grass "Aveneae-type plastid DNA clade" (Poaceae: Pooideae, Poeae) based on plastid and nuclear ribosomal DNA sequence data. In O. Seberg, G. Petersen, A. S. Barfod, and J. I. Davis (eds.), *Diversity, Phylogeny, and Evolution in the Monocotyledons* (pp. 557–586). Aarhus, Denmark: Aarhus University Press.

Saarela, J. M., Peterson, P. M., Keane, R. M., Cayouette, J., and Graham, S. W. 2007. Molecular phylogenetics of *Bromus* (Poaceae: Pooideae) based on chloroplast and nuclear DNA sequence data. *Aliso* 23: 450–467.

Saiz, F., and Ojeda, P. 1988. *Oryctolagus cuniculus* L. en Juan Fernández. Problema y control. *Ann. Mus. Hist. Nat. Valparaíso* 19: 91–98.

Saiz, G. F., De la Hoz, E., Toro, H., Zuniga, L., Vasquez, E., et al. 1982. *Proposición de un Método de Control Integrado del Conejo en el Archipiélago de Juan Fernández.* Valparaíso: Universidad Católica de Valparaíso.

Sakai, A., Wagner, W., Ferguson, D., and Herbst, D. 1995. Origins of dioecy in the Hawaiian flora. *Ecology* 76: 2517–2529.

Salisbury, E. J. 1964. Carl Johan Fredrik Skottsberg 1880–1963. *Biograph. Mem. Fellows R. Soc.* 10: 244–256.

Samain, M.-S., Mathieu, G., Pino, G., Symmank, L., Cieza, N., et al. 2011. The geophytic *Peperomia* subgenus *Tildenia* (Piperaceae) in the Andes with the description of new species in a phylogenetic framework. *Plant Ecol. Evol.* 144: 148–176.

Samain, M.-S., Vanderschaeve, L., Chaerle, P., Goetghebeur, P., Neinhuis, C., et al. 2009. Is morphology telling the truth about the evolution of the species rich genus *Peperomia* (Piperaceae)? *Plant Syst. Evol.* 278: 1–21.

Sampson, F. B. 1995. Pollen morphology of Lactoridaceae: a re-examination. *Grana* 34: 100–107.

Sanders, R., Stuessy, T. F., and Marticorena, C. 1982. Recent changes in the flora of the Juan Fernández Islands, Chile. *Taxon* 31: 284–289.

Sanders, R. W., Stuessy, T. F., Marticorena, C., and Silva O., M. 1987. Phytogeography and evolution of *Dendroseris* and *Robinsonia*, tree-Compositae of the Juan Fernández Islands. *Opera Bot.* 92: 195–215.

Sanders, R. T., Stuessy, T. F., and Rodríguez, R. 1983. Chromosome numbers from the flora of the Juan Fernandez Islands. *Am. J. Bot.* 70: 799–810.

Sang, T., Crawford, D. J., Kim, S.-C., and Stuessy, T. F. 1994. Radiation of the endemic genus *Dendroseris* (Asteraceae) on the Juan Fernández Islands: evidence from sequences of the ITS regions of nuclear ribosomal DNA. *Am. J. Bot.* 81: 1494–1501.

Sang, T., Crawford, D. J., Stuessy, T. F., and Silva O., M. 1995. ITS sequences and the phylogeny of the genus *Robinsonia* (Asteraceae). *Syst. Bot.* 20: 55–64.

Santibañez E., J. 1945. La corriente de Humboldt en las costas de Chile. *Liga Marítima de Chile Publ.* 28: 1–37.

Saunders, A., Glen, A., Campbell, K., Atkinson, R., Sawyer, J., et al. 2011. *Estudio Sobre la Factibilidad del Manejo de Especies Invasoras en el Archipiélago de Juan Fernández, Chile.* Lincoln: Landcare Research.

Saville, D. B. O., and Calder, J. A. 1953. Phylogeny of *Carex* in the light of parasitism by the smut fungi. *Can. J. Bot.* 31: 164–174.

Sax D. F., and Gaines S. D. 2008. Species invasions and extinction: the future of native biodiversity on islands. *Proc. Natl. Acad. Sci. USA* 105: 11490–11497.

Schneider, H., Russell, S. J., Cox, C. J., Bakker, F., Henderson, S., et al. 2004. Chloroplast phylogeny of asplenioid ferns based on *rbcL* and *trnL-F* spacer sequences (Polypodiidae, Aspleniaceae) and its implications for biogeography. *Syst. Bot.* 29: 260–274.

Schneider, J., Winterfeld, G., Hoffmann, M. H., and Röser, M. 2011. Duthieeae, a new tribe of grasses (Poaceae) identified among the early diverging lineages of subfamily Pooideae: molecular phylogenetics, morphological delineation, cytogenetics and biogeography. *Syst. Biodivers.* 9: 27–44.

Schoener, T. W. 2010. The MacArthur-Wilson equilibrium model: a chronicle of what it said and how it was tested. In J. B. Losos and Ricklefs (eds.), *The Theory of Island Biogeography Revisited* (pp. 52–87). Princeton, NJ: Princeton University Press.

Schönswetter, P., and Tribsch, A. 2005. Vicariance and dispersal in the alpine perennial *Bupleurum stellatum* L. (Apiaceae). *Taxon* 54: 725–732.

Schopf, T. J. M. 1980. *Paleoceanography.* Cambridge, MA: Harvard University Press.

Schouten, W. C. 1619. *The Relation of a Wonderfull Voiage.* Amsterdam.

Schuettpelz, E., and Pryer, K. M. 2007. Fern phylogeny inferred from 400 leptosporangiate species and three plastid genes. *Taxon* 56: 1037–1050.

Schulte, K., Barfuss, M. H. J., and Zizka, G. 2009. Phylogeny of Bromelioideae (Bromeliaceae) inferred from nuclear and plastid DNA loci reveals the evolution of the tank habit within the subfamily. *Mol. Phylogenet. Evol.* 51: 327–339.

Schulte, K., Horres, R., and Zizka, G. 2005. Molecular phylogeny of Bromelioideae and its implications on biogeography and the evolution of CAM in the family (Poales, Bromeliaceae). *Senckenberg. Biol.* 85: 1–13.

Schulz, O. E. 1903. Monographie der Gattung *Cardamine. Engler Bot. Jahrb.* 32: 280–623.

Schumer, M., Rosenthal, G. G., and Andolfatto, P. 2014. How common is homoploid hybrid speciation? *Evolution* 68: 1553–1560.

Schwaar, J. 1979. Feuchtwälder auf Juan Fernandez. *Phytocoenologia* 6: 514–523.

Scott, A. J. 1978. A review of the classification of *Chenopodium* L. and related genera (Chenopodiaceae). *Bot. Jahrb. Syst.* 100: 205–220.

Seaman, F. C., and Funk, V. A. 1983. Cladistic analysis of complex natural products: developing transformation series from sesquiterpene lactone data. *Taxon* 32: 1–27.

Sede, S. M., Dürnhöfer, M. S., and Zapata, F. 2013. Phylogenetics of *Escallonia* (Escalloniaceae) based on plastid DNA sequence data. *Bot. J. Linn. Soc.* 173: 442–451.

Seigler, D. S. 1998. *Plant Secondary Metabolism.* Boston, MA: Kluwer.

Seltzer, G. O., Rodbell, D. T., and Wright, H. E. 2003. Late-quaternary paleoclimates of the southern tropical Andes and adjacent regions. *Palaeogeogr. Palaeoclim. Palaeoecol.* 194: 1–3 (introduction to symposium, pp. 5–338).

Setchell, W. A. 1937. The Codiums of the Juan Fernandez Islands. In C. Skottsberg (ed.), *The Natural History of the Juan Fernandez and Easter Island, vol. 2: Botany* (pp. 537–600). Uppsala: Almqvist & Wiksells.

Seward, A. C. 1910. *Fossil Plants: A Text-Book for Students of Botany and Geology.* Cambridge: Cambridge University Press.

Shapiro, B. J., Leducq, J.-P., and Mallet, J. 2016. What is speciation? *PLoS Genet.* 12: e1005860 doi: 10.1371/journal.pgen.1005860.

Sharp, A. (ed.). 1970. *The Journal of Jacob Roggeveen.* Oxford: Clarendon Press.

Shelvocke, G. 1726. *A Voyage Round the World.* London (in the edited version by W. G. Perrin (ed.), London: Cassell, 1928).

Shepherd, L. D., Perrie, L. R., Parris, B. S., and Brownsey, P. J. 2007. A molecular phylogeny for the New Zealand Blechnaceae ferns from analyses of chloroplast *trnL-trnF* DNA sequences. *New Zealand J. Bot.* 45: 67–80.

Shinohara, W., Nakato, N., Yatabe-Kakugawa, Y., Oka, T., Kim, J. K., et al. 2013. The use of *matK* in Ophioglossaceae phylogeny and determination of *Mankyua* chromosome number shed light on chromosome number evolution in Ophioglossaceae. *Syst. Bot.* 38: 564–570.

Sigel, E. M., Windham, M. D., Haufler, C. H., and Pryer, K. M. 2014. Phylogeny, divergence time estimates, and phylogeography of the diploid species of the *Polypodium vulgare* complex (Polypodiaceae). *Syst. Bot.* 39: 1042–1055.

Silva, M. 2016. Guardaparques hallan plantas extintas en archipiélago Juan Fernández. *Economía y Negocios.* Available online at www.economiaynegocios.cl/noticias (accessed April 3, 2016).

Simberloff, D. 2001. Eradication of island invasives: practical actions and results achieved. *Trends Ecol. Evol.* 16: 273–274.

Simberloff, D., Martin, J.-L., Genovesi, P., Maris, V., Wardle, D. A., et al. 2013. Impacts of biological invasions: what's what and the way forward. *Trends Ecol. Evol.* 28: 58–66.

Simberloff, D. S., and Wilson, E. O. 1970. Experimental zoogeography of islands. A two-year record of colonization. *Ecology* 51: 934–937.

Simpson, B. B. 1975. Pleistocene changes in the flora of the high tropical Andes. *Paleobiology* 1: 273–294.

Simpson, G. G. 1953. *The Major Features of Evolution.* New York: Columbia University Press.

Sjöstedt, B. 1975. Revision of the genus *Cardamine* L. (Cruciferae) in South and Central America. *Bot. Notiser* 128: 8–19.

Skottsberg, C. 1910. Juan Fernandez-öarnas sandeltrad. *Svensk Bot. Tidskr.* 4: 167–173.

Skottsberg, C. 1914. Botanische Ergebnisse der Schwedischen Expedition nach Patagonien und dem Feuerlande 1907–1909. IV. Studien über die Vegetation der Juan Fernandez-Inseln. *Kungl. Svenska Vetenskapsakad. Handl.* 51: 1–73.

Skottsberg, C. 1918. *Til Robinson-ön och Världens ände.* Stockholm: Albert Bonniers.

Skottsberg, C. (ed.). 1920–1956. *The Natural History of Juan Fernandez and Easter Island, vol. 1: Geography, Geology, Origin of Island Life*; vol. 2: *Botany.* Uppsala: Almqvist & Wiksells.

Skottsberg, C. (ed.). 1921–1940. *The Natural History of Juan Fernandez and Easter Island, vol. 3: Zoology.* Uppsala: Almqvist & Wiksells.

Skottsberg, C. 1921. The phanerogams of the Juan Fernandez Islands. In C. Skottsberg (ed.), *The Natural History of Juan Fernandez and Easter Island, vol. 2* (pp. 95–240). Uppsala: Almqvist & Wiksells.

Skottsberg, C. 1925. Juan Fernandez and Hawaii. A phytogeographical discussion. *Bernice P. Bishop Mus. Bull.* 16: 1–47.

Skottsberg, C. 1928. Pollinationsbiologie und Samenverbreitung auf den Juan Fernandez-Inseln. In C. Skottsberg (ed.), *The Natural History of Juan Fernandez and Easter Island, vol. 2: Botany* (pp. 503–547). Uppsala: Almqvist & Wiksells.

Skottsberg, C. 1934. Le peuplement des îles pacifiques du Chili. In L. Berland et al. (eds.), *Contribution à l'étude du Peuplement Zoologique et Botanique des îles du Pacifique* (pp. 271–280). Paris: Paul Lechevalier et fils.

Skottsberg, C. 1936. *Greigia berteroi* Skottsb. and its systematic position. *Meddel. Göteb. Botan. Trädg.* 11: 209–222.

Skottsberg, C. 1941. Marine algal communities of the Juan Fernandez Islands, with remarks on the composition of the flora. In C. Skottsberg (ed.), *The Natural History of the Juan Fernandez and Easter Island, vol. 2: Botany* (pp. 671–696). Uppsala: Almqvist & Wiksells.

Skottsberg, C. 1946. *Peperomia berteroana* Miq. and *P. tristanensis* Christoph., an interesting case of disjunction. *Acta Horti Gothob.* 16: 251–288.

Skottsberg, C. 1953a. The vegetation of Juan Fernandez and Easter Islands. In C. Skottsberg (ed.), *The Natural History of Juan Fernandez and Easter Island, vol. 2: Botany*, Part 6 (pp. 793–960). Uppsala: Almqvist and Wiksells.

Skottsberg, C. 1953b. A supplement to the phanerogams of the Juan Fernandez Islands. In C. Skottsberg (ed.), *The Natural History of Juan Fernandez and Easter Island, vol. 2: Botany*, Part 6 (pp. 763–792). Uppsala: Almqvist & Wiksells.

Skottsberg, C. 1954. A geographical sketch of the Juan Fernandez Islands. In C. Skottsberg (ed.), *The Natural History of Juan Fernandez and Easter Island, vol. 1* (pp. 89–192). Uppsala: Almqvist & Wiksells.

Skottsberg, C. 1956. Derivation of the flora and fauna of Juan Fernandez and Easter Island. In C. Skottsberg (ed.), *The Natural History of Juan Fernandez and Easter Island, vol. 1: Geography, Geology, Origin of Island Life* (pp. 193–438). Uppsala: Almqvist & Wiksells.

Skottsberg, C. 1958. Über *Yunquea Tenzii* Skottsb. *Ber. Deutsch. Bot. Ges.* 71: 45–52.

Sleumer, H. 1936. Revision der Gattung *Pernettya* Gaud. *Notiz. Bot. Gard. Mus. Berlin-Dahlem* 12: 626–655.

Sleumer, H. 1977. Revision der Gattung *Azara* R. et P. (Flacourtiaceae). *Bot. Jahrb. Syst.* 98: 151–175.

Sleumer, H. 1985. Taxonomy of the genus *Pernettya* Gaud. (Ericaceae). *Bot. Jahrb. Syst.* 105: 449–480.

Smith, A. R. 1993. Phytogeographic principles and their use in understanding fern relationships. *J. Biogeogr.* 20: 255–264.

Smith, A. R., and Moran, R. C. 1987. New combinations in *Megalastrum* (Dryopteridaceae). *Am. Fern J.* 77: 124–130.

Smith, A. R., Pryer, K. M., Schuettpelz, E., Korall, P., Schneider, H., et al. 2006. A classification for extant ferns. *Taxon* 55: 705–731.

Smith, A. R., and Tejero-Díez, J. D. 2014. *Pleopeltis* (Polypodiaceae), a redefinition of the genus and nomenclatural novelties. *Bot. Sci.* 92: 43–58.

Smith, C. W. 1989. Non-native plants. In C. P. Stone and D. B. Stone (eds.), *Conservation Biology in Hawaii* (pp. 60–69). Honolulu: University of Hawaii Cooperative National Park Resources Studies Unit.

Smith, L. B., and Downs, R. J. 1979. Bromelioideae (Bromeliaceae). *Fl. Neotrop. Monogr.* 14: 1493–2142.

Smith, R. J., Muir, R. D. J., Walpole, M. J., Balmford, A., and Leader-Williams, N. 2003. Governance and the loss of biodiversity. *Nature* 426: 67–70.

Smith-Ramírez, C., and Arellano-Cataldo, G. 2013. Necesidad de que las Fuerzas Armadas realicen un control de las plantas invasoras en el Archipiélago de Juan Fernández, Chile. *Bosque (Valdivia)* 34 (1): 3–6.

Smith-Ramírez, C., Arellano, G., Hagen, E., Vargas, R., Castillo, J., et al. 2013. El rol de *Turdus falcklandii* (Aves: Passeriforme) como dispersor de plantas invasoras en el archipiélago de Juan Fernández. *Rev. Chil. Hist. Nat.* 86: 33–48.

Sobel, J. M. 2014. Ecogeographic isolation and speciation in the genus *Mimulus*. *Am. Nat.* 184: 565–579.

Sobel, J. M., and Chen, G. F. 2014. Unification of methods for estimating the strength of reproductive isolation. *Evolution* 68: 1511–1523.

Sobel, J. M., Chen, G. F., Watt, L. R., and Schemske, D. W. 2010. The biology of speciation. *Evolution* 64: 295–315.

Sobel, J. M., and Streisfeld, M. A. 2015. Strong premating reproductive isolation drives incipient speciation in *Mimulus aurantiacus*. *Evolution* 69: 447–461.

Solbrig, O. T. 1962. The South American species of *Erigeron*. *Contrib. Gray Herb.* 191: 3–79.

Solbrig, O. T. 1971. The population biology of dandelions. *Am. Sci.* 59: 686–694.

Solbrig, O. T., Anderson, L. C., Kyhos, D. W., and Raven, P. H. 1969. Chromosome numbers in Compositae VII: Astereae III. *Am. J. Bot.* 56: 348–353.

Soltis, D. E., Albert, V. A., Leebens-Mack, J., Bell, C. D., Paterson, A., et al. 2009. Polyploidy and angiosperm diversification. *Am. J. Bot.* 96: 336–348.

Soltis, D. E., Gitzendanner, M. A., Stull, G., Chester, M., Chanderbali, A., et al. 2013. The potential of genomics in plant systematics. *Taxon* 62: 886–898.

Soltis, D. E., Smith, S. A., Cellinese, N., Wurdack, K. J., Tank, D. C., et al. 2011. Angiosperm phylogeny: 17 genes, 640 taxa. *Am. J. Bot.* 98: 704–730.

Soltis, D. E., Soltis, P. S., Endress, P. K., and Chase, M. W. 2005. *Phylogeny and Evolution of Angiosperms*. Sunderland, MA: Sinauer Associates.

Soltis, D. E., Soltis, P. S., Nikrent, D. L., Johnson, L. A., Hahn, W. J., et al. 1997. Angiosperm phylogeny inferred from 18S ribosomal DNA sequences. *Ann. Missouri Bot. Gard.* 84: 1–49.

Soltis, D. E., Visger, C. J., and Soltis, P. S. 2014. The polyploidy revolution then . . . and now: Stebbins revisited. *Am. J. Bot.* 101: 1057–1078.

Soraru, S. B. 1972. Revisión de las Urticaceae argentinas. *Darwiniana* 17: 246–325.

Soreng, R. J., Davidse, G., Peterson, P. M., Zuloaga, F. O., Judziewicz, E. J., et al. 2000. *Catalogue of New World grasses (Poaceae)*. Available at www.tropicos.org/project/cnwg (accessed November 29, 2015).

Soza, V. L., and Olmstead, R. G. 2010. Evolution of breeding systems and fruits in New World *Galium* and relatives (Rubiaceae). *Am. J. Bot.* 97: 1630–1646.

Spalik, K., Piwczyński, M., Danderson, C. A., Kurzyna-Młynik, R., Bone, T. S., et al. 2010. Amphitropic amphiantarctic disjunctions in Apiaceae subfamily Apioideae. *J. Biogeogr.* 37: 1977–1994.

Spooner, D., Stuessy, T. F., Crawford, D. J., and Silva O., M. 1987. Chromosome numbers from the flora of the Juan Fernandez Islands. II. *Rhodora* 89: 351–356.

Spooner, D. M., Anderson, G. J., and Jansen, R. K. 1993. Chloroplast DNA evidence for the interrelationships of tomatoes, potatoes, and pepinos (Solanaceae). *Am. J. Bot.* 80: 676–688.

Spooner, D. M., Douches, D. S., and Contreras M., A. 1992. Allozyme variation within *Solanum* sect. *Petota*, ser. *Etuberosa* (Solanaceae). *Am. J. Bot.* 79: 467–471.

Spooner, D. M., Tivang, J., Nienhuis, J., Miller, J. T., Douches, D. S., et al. 1996. Comparison of four molecular markers in measuring relationships among the wild potato relatives *Solanum* section *Etuberosum* (subgenus *Potatoe*). *Theor. Appl. Genet.* 92: 532–540.

Spry, W. J. J. 1877. *The Cruise of Her Majesty's Ship "Challenger": Voyages over Many Seas, Scenes in Many Lands*. New York: Harper & Brothers.

St. John, H., and Takeuchi, W. 1988. Enlargement of O'ahu *Cyrtandra* (Gesneriaceae). Hawaiian plant studies 48. *Phytologia* 65: 207–215.

Stafleu, F. A., and Cowan, R. S. 1976. *Taxonomic Literature,* vol. I*: A-G*. Utrecht: Bohn, Scheltema & Holkema.

Stafleu, F. A., Demoulin, V., Greuter, W., Hiepko, P., Linczevski, I. A., et al. 1978. *International Code of Botanical Nomenclature, Adopted by the Twelfth International Botanical Congress, Leningrad, July 1975*. Utrecht: Bohn, Scheltema & Holkema (*Regnum Veg*. 97).

Staples, G. W., and Cowie, R. H. (eds.). 2001. *Hawaii's Invasive Species: A Guide to the Invasive Plants and Animals in the Hawaiian Islands*. Honolulu: Mutual Publishing and Bishop Museum Press.

Starr, J. R., and Ford, B. A. 2009. Phylogeny and evolution in Cariceae (Cyperaceae): current knowledge and future directions. *Bot. Rev.* 75: 110–137.

Starr, J. R., Harris, S. A., and Simpson, D. A. 2004. Phylogeny of the unispicate taxa in Cyperaceae tribe Cariceae I: generic relationships and evolutionary scenarios. *Syst. Bot*. 29: 528–544.

Starr, J. R., Harris, S. A., and Simpson, D. A. 2008. Phylogeny of the unispicate taxa in Cyperaceae tribe Cariceae II: the limits of *Uncinia*. In B. A. Ford and R. C. F. Naczi (eds.), *Sedges: Uses, Diversity, and Systematics of the Cyperaceae* (pp. 243–262). St. Louis, MO: Missouri Botanical Garden Press.

Stathos, A., and Fishman, L. 2014. Chromosomal rearrangements directly cause underdominant F_1 pollen sterility in *Mimulus lewisii-Mimulus cardinalis* hybrids. *Evolution* 68: 3109–3119.

Stearns, H. T. 1966. *Geology of the State of Hawaii*. Palo Alto, CA: Pacific Books.

Stebbins, G. L., Jr. 1950. *Variation and Evolution in Plants*. New York: Columbia University Press.

Stebbins, G. L. 1957. Self-fertilization and population variability in the higher plants. *Am. Nat.* 41: 337–354.

Stebbins, G. L. 1958. The inviability, weakness, and sterility of intespecific hybrids. *Adv. Genet.* 9: 147–215.

Stebbins, G. L. 1971. *Chromosomal Evolution in Higher Plants*. London: Edward Arnold.

Stebbins, G. L., Jenkins, J. A., and Walters, M. S. 1953. Chromosomes and phylogeny in the Compositae, tribe Cichorieae. *Univ. Calif. Publ. Bot*. 26: 401–430.

Steele, P. R., and Pires, J. C. 2011. Biodiversity assessment: state-of-the-art techniques in phylogenomics and species identification. *Am. J. Bot.* 98: 415–425.

Steinbauer, M. J., et al. 2016. Topography-driven isolation, speciation and a global increase of endemism with elevation. *Global Ecol. Biogeogr.* 25: 1097–1107.

Stevens, P. F. 1971. A classification of the Ericaceae: subfamilies and tribes. *Bot. J. Linn. Soc.* 64: 1–53.

Stiefkens, L., Bernardello, G., and Anderson, G. J. 2001. The somatic chromosomes of *Sophora fernandeziana* (Fabaceae), an endemic tree from Robinson Crusoe Island. *Pacific Sci.* 55: 71–75.

Stone, C. P., Smith, C. W., and Tunison, J. T. (eds.). 1992. *Alien Plant Invasions in Native Ecosystems of Hawaii: Management and Research*. Honolulu: University of Hawaii Cooperative National Park Resource Studies Unit.

Ström, K. M. 1921. Freshwater algae from Juan Fernandez and Easter Island. In C. Skottsberg (ed.), *The Natural History of the Juan Fernandez and Easter Island, vol. 2: Botany* (pp. 35–93). Uppsala: Almqvist & Wiksells.

Strong, M. T. 1997. *Machaerina* (Cyperaceae) in South America. *Novon* 7: 308–319.

Strother, J. L. 1976. Chromosome studies in Compositae. *Am. J. Bot.* 63: 247–250.

Stuessy, T. F. 1995. Fernandezian region: CPD site PO8. Juan Fernandez Islands, Chile. In S. D. Davis, V. H. Heywood, and A. C. Hamilton (eds.), *Centres of Plant Diversity*, vol. 2 (pp. 565–568). Cambridge: WWF & IUCN.

Stuessy, T. F. 2007. Evolution of specific and genetic diversity during ontogeny of island floras: the importance of understanding process for interpreting island biogeographic patterns. In M. C. Ebach and R. S. Tangney (eds.), *Biogeography in a Changing World* (pp. 117–133). Boca Raton, FL: CRC Press.

Stuessy, T. F. 2009. *Plant Taxonomy: The Systematic Evaluation of Comparative Data* (2nd edn.). New York: Columbia University Press.

Stuessy, T. F. 2010. Paraphyly and the origin and classification of angiosperms. *Taxon* 59: 689–693.

Stuessy, T. F., and Crawford, D. J. 1983. Flavonoids and phylogenetic reconstruction. *Plant Syst. Evol.* 143: 83–107.

Stuessy, T. F., and Crawford, D. J. 1998. Chromosomal stasis during speciation in angiosperms of oceanic islands. In T. F. Stuessy and M. Ono (eds.), *Evolution and Speciation of Island Plants* (pp. 307–324). Cambridge: Cambridge University Press.

Stuessy, T. F., Crawford, D. J., Anderson, G. J., and Jensen, R. J. 1998c. Systematics, biogeography, and conservation of Lactoridaceae. *Perspect. Plant Ecol. Evol. Syst.* 1: 267–290.

Stuessy, T. F., Crawford, D. J., and Marticorena, C. 1990. Patterns of phylogeny in the endemic vascular flora of the Juan Fernández Islands, Chile. *Syst. Bot.* 15: 338–346.

Stuessy, T. F., Crawford, D. J., Marticorena, C., and Rodríguez, R. 1998a. Island biogeography of angiosperms of the Juan Fernández archipelago. In T. F. Stuessy and M. Ono (eds.), *Evolution and Speciation of Island Plants* (pp. 121–138). Cambridge: Cambridge University Press.

Stuessy, T. F., Crawford, D. J., Marticorena, C., and Silva O., M. 1998e. Isolating mechanisms and modes of speciation in endemic angiosperms of Juan Fernández Islands. In T. F. Stuessy and M. Ono (eds.) *Evolution and Speciation of Island Plants* (pp. 79–96). Cambridge: Cambridge University Press.

Stuessy, T. F., Crawford, D. J., Soltis, D. E., and Soltis, P. S. 2014a. *Plant Systematics: The Origin, Interpretation, and Ordering of Plant Biodiversity*. Königstein: Koeltz Scientific Books.

Stuessy, T. F., Foland, K. A., Sutter, J. F., Sanders, R. W., and Silva O., M. 1984. Botanical and geological significance of potassium-argon dates from the Juan Fernandez Islands. *Science* 255: 49–51.

Stuessy, T. F., Greimler, J., and Dirnböck, T. 2005a. Landscape modification and impact on specific and genetic diversity in oceanic islands. In H. Balslev and I. Friis (eds.), *Plant Diversity and Complexity Patterns: Local, Regional and Global Dimensions* (pp. 89–101). Copenhagen: Danish Academy of Sciences (*Biol. Skr.* 55).

Stuessy, T. F., Jakubowsky, G., Gómez, R. S., Pfosser, M., Schluter, P. M., et al. 2006. Anagenetic evolution in island plants. *J. Biogeogr.* 33: 1259–1265.

Stuessy, T. F., König, C., and López Sepúlveda, P. 2014b. Paraphyly and endemic genera of oceanic islands: implications for conservation. *Ann. Missouri Bot. Gard.* 100: 50–78.

Stuessy, T. F., Marticorena, C., Rodríguez, R., Crawford, D. J., and Silva O., M. 1992. Endemism in the vascular flora of the Juan Fernández Islands. *Aliso* 13: 297–307.

Stuessy, T. F., and Ono, M. (eds.). 1998. *Evolution and Speciation of Island Plants*. Cambridge: Cambridge University Press.

Stuessy, T. F., Ruiz, E., Crawford, D. J., and Tremetsberger, K. 2005b. Testing degrees of genetic divergence and populational variation in oceanic island archipelagos: Juan Fernández as a model system. *Nova Acta Leopoldina* 92: 147–165.

Stuessy, T. F., Sanders, R. W., and Matthei, O. R. 1983. *Juania australis* revisited in the Juan Fernández Islands, Chile. *Principes* 27: 71–74.

Stuessy, T. F., Sanders, R. W., and Silva O., M. 1984. Phytogeography and evolution of the flora of the Juan Fernández Islands: a progress report. In F. J. Radovsky, P. H. Raven, and S. H. Sohmer (eds.), *Biogeography of the Tropical Pacific* (pp. 55–69). Lawrence, KS: Association of Systematics Collections and the Bishop Museum.

Stuessy, T. F., Sang, T., and DeVore, M. 1996. Phylogeny and biogeography of the subfamily Barnadesioideae with implications for early evolution of the Compositae. In D. J. Hind and H. J. Beentje (eds.), *Proceeding of the International Compositae Conference (1994),* vol. 1*: Compositae: Systematics* (pp. 463–490). Kew: Royal Botanic Gardens.

Stuessy, T. F., Swenson, U., Crawford, D. J., Anderson, G., and Silva O., M. 1998d. Plant conservation in the Juan Fernández archipelago, Chile. *Aliso* 16: 89–101.

Stuessy, T. F., Swenson, U., Marticorena, C., Matthei, O., and Crawford, D. J. 1998b. Loss of plant diversity and extinction on Robinson Crusoe Islands, Chile. In C.-I. Peng and P. P. Lowry, II (eds.), *Rare, Threatened, and Endangered Floras of Asia and the Pacific Rim* (pp. 243–257). Taipei: Institute of Botany, Academica Sinica (Monograph Series No. 16).

Stuessy, T. F., Takayama, K., and López-Sepúlveda, P. 2012. Founder effects are invisible in endemic species of oceanic islands. *J. Biogeog.* 39: 1565–1566.

Stuessy, T. F., Takayama, K., López-Sepúlveda, P., and Crawford, D. J. 2014c. Interpretation of patterns of genetic variation in endemic plant species of oceanic islands. *Bot. J. Linn. Soc.* 174: 276–288.

Sugiura, S. 2008. Male territorial behaviour of the endemic large carpenter bee, *Xylocopa (Koptortosoma) ogasawarensis* (Hymenoptera: Apidae), on the oceanic Ogasawara Islands. *Eur. J. Entomol.* 105: 153–157.

Sukhorukov, A. P., and Zhang, M. 2013. Fruit and seed anatomy of *Chenopodium* and related genera (Chenopodioideae, Chenopodiaceae/Amarathaceae): implications for evolution and taxonomy. *PLoS One* 8(4): e61906.

Sun, B.-Y., and Stuessy, T. F. 1998. Preliminary observations on the evolution of endemic angiosperms of Ullung Island, Korea. In T. F. Stuessy and M. Ono (eds.), *Evolution and Speciation of Island Plants* (pp. 181–202). Cambridge: Cambridge University Press.

Sun, B.-Y., Stuessy, T., and Crawford, D. 1990. Chromosome counts from the flora of the Juan Fernández Islands, Chile. III. *Pacific Sci.* 44: 258–264.

Sun, B.-Y., Stuessy, T. F., Humaña, A. M., Riveros, M., and Crawford, D. J. 1996. Evolution of *Rhaphithamnus venustus* (Verbenaceae), a gynodioecious hummingbird-pollinated endemic of the Juan Fernández Islands, Chile. *Pacific Sci.* 50: 55–65.

Sun, M., and Ganders, F. R. 1988. Mixed mating sytems in Hawaiian *Bidens*. *Evolution* 42: 516–527.

Sundue, M., Rouhan, G., and Moran, R. 2010. *Megalastrum* (Dryopteridaceae) of the Circumaustral region: Chile, Argentina, and Southern Islands of the Atlantic, Pacific, and Indian Oceans. *Syst. Bot.* 35: 461–475.

Sungkaew, S., Stapleton, C. M. A., Salamin, N., and Hodkinson, T. R. 2009. Non-monophyly of the woody bamboos (Bambuseae; Poaceae): a multi-gene region phylogenetic analysis of Bambusoideae s.s. *J. Plant Res.* 122: 95–108.

Susanna, A., Galbany-Casals, M., Romaschenko, K., Barres, L., Martín, J., et al. 2011. Lessons from *Plectocephalus* (Compositae, Cardueae-Centaureinae): ITS disorientation in annuals and Beringian dispersal as revealed by molecular analyses. *Ann. Bot.* 108: 263–277.

Susanna, A., and Garcia-Jacas, N. 2001. Tribe Cardueae. In Anderberg, A. et al. (eds.), *Compositae*. In J. W. Kadereit and C. Jeffrey (eds.), *Flowering Plants, vol. VIII: Eudicots, Asterales. Compositae* (pp. 123–147). In K. Kubitzki (series ed.), *The Families and Genera of Vascular Plants*. Berlin: Springer.

Sutcliffe, T. ["T.S."]. 1839. *The Earthquake of Juan Fernandez, as It Occurred in the Year 1835, Authenticated by the Retired Governor of that Island, to Which is Added a Refutation of Several Misstatements that Have Been Published in the "Nautical Magazine" of 1837, and the Public Papers*. Manchester.

Sutcliffe, T. 1841. *Sixteen Years in Chile and Peru from 1822 to 1839*. London: Fisher (on p. x, end of the Preface, June 1,1841, is given, but the book may have appeared later).

Swain, T. (ed.). 1963. *Chemical Plant Taxonomy*. New York: Academic Press.

Swenson, U., Stuessy, T. F., Baeza, M., and Crawford, D. J. 1997. New and historical plant introductions, and potential pests, in the Juan Fernandez Islands, Chile. *Pacific Sci.* 51: 233–253.

Swenson, U., Ulfsson, V., and Havran, C. 2016. Hawaiiöarna: en flora på fallrepet. *Svensk Bot. Tidskr.* 110: 80–107.

Symmank, L., Samain, M.-S., Smith, J. F., Pino, G., Stoll, A., et al. 2011. The extraordinary journey of *Peperomia* subgenus *Tildenia* (Piperaceae): insights into diversification and colonization patterns from its cradle in Peru to the Trans-Mexican Volcanic Belt. *J. Biogeogr.* 38: 2337–2349.

Takayama, K., Crawford, D. J., López-Sepúlveda, P., Greimler, J., and Stuessy, T. F. Submitted. Factors driving adaptive radiation in plants of oceanic islands: a case study from the Juan Fernández Archipelago. *J. Plant Res.*

Takayama, K., López, P., König, C., Kohl, G., Novak, J., et al. 2011. A simple and cost-effective approach for microsatellite isolation in non-model plant species using small-scale 454 pyrosequencing. *Taxon* 60: 1442–1449.

Takayama, K., López-Sepúlveda, P., Greimler, J., Crawford, D. J., Peñailillo, P., et al. 2015a. Relationships and genetic consequences of contrasting modes of speciation among endemic species of *Robinsonia* (Asteraceae, Senecioneae) of the Juan Fernández Archipelago, Chile, based on AFLPs and SSRs. *New Phytol.* 205: 415–428.

Takayama, K., López-Sepúlveda, P., Greimler, J., Crawford, D. J., Peñailillo, P., et al. 2015b. Genetic consequences of cladogenetic vs. anagenetic speciation in endemic plants of oceanic islands. *AoB Plants* 7: plv102. doi: 10.1093/aobpla/plv102

Takayama, K., López-Sepúlveda, P., Kohl, G., Novak, J., and Stuessy, T. F. 2012a. Development of microsatellite markers in species of *Erigeron* (Asteraceae) endemic to the Juan Fernández Archipelago, Chile. *Am. J. Bot. Primer Notes* e1–e3. Doi: 10.3732/ajb.1200218.

Takayama, K., Sun, B.-Y., and Stuessy, T. F. 2012b. Genetic consequences of anagenetic speciation in *Acer okamotoanum* (Sapindaceae) on Ullung Island, Korea. *Ann. Bot.* 109: 321–330.

Takayama, K., Sun, B.-Y., and Stuessy, T. 2013. Anagenetic speciation in Ullung Island, Korea: genetic diversity and structure in the island endemic species, *Acer takesimense* (Sapindaceae). *J. Plant Res.* 126: 323–333.

Takhtajan, A. 1969. *Flowering Plants: Origin and Dispersal*. Edinburgh: Oliver & Boyd.

Takhtajan, A. 1986. *Floristic Regions of the World*. Berkeley: University of California Press.

Tank, D. C., Beardsley, P. M., Kelchner, S. A., and Olmstead, R. G. 2006. Review of the systematics of Scrophulariaceae s.l. and their current disposition. *Austral. Syst. Bot.* 19: 289–307.

Tatemichi, Y. (ed.). 1990. *Illustrated Book of the Genus* Nicotiana. Toyoda: Japan Tobacco.

Tay, M. L., Meudt, H. M., Garnock-Jones, P. J., and Ritchie, P. A. 2010. DNA sequences from three genomes reveal multiple long-distance dispersals and non-monophyly of sections in Australasian *Plantago* (Plantaginaceae). *Austral. Syst. Bot.* 23: 47–68.

Taylor, C. 2003. Urticaceae. In C. Marticorena and R. Rodríguez (eds.), *Flora de Chile, vol. 3(1): Misodendraceae-Zygophyllaceae* (pp. 44–61). Concepción, Chile: Universidad de Concepción.

Taylor, C. M., and Muñoz-Schick, M. 1994. The botanical works of Philippi, father and son, in Chile. *Ann. Missouri Bot. Gard.* 81: 743–748.

Taylor, R. L., and Mulligan, G. A. 1968. *Flora of the Queen Charlotte Islands, part 1: Systematics of the Vascular Plants; part 2: Cytological Aspects of the Vascular Plants*. Ottawa: Canada Department of Agriculture (Research Branch Monograph no. 4).

Taylor, S., and Kumar, L. 2014. Climate change and weed impacts on small island ecosystems: *Lantana camara* L. (Magnoliopsida: Verbenaceae) distribution in Fiji. *Pacific Sci.* 68: 117–133.

Teillier, S., and Escobar, F. 2013. Revisión del género *Gaultheria* L. (Ericaceae) en Chile. *Gayana Bot.* 70: 136–153.

Templeton, A. R. 1981. Mechanisms of speciation: a population genetic approach. *Annu. Rev. Ecol. Evol.* 12: 23–48.

Templeton, A. R. 1982. Genetic architecture of speciation. In C. Barigozzi (ed.), *Mechanisms of Speciation* (pp. 105–121). New York: Alan R. Liss.

Templeton, A. R. 2008. The reality and importance of founder speciation in evolution. *BioEssays* 30: 470–479.

Těšitel, J., Řiha, P., Svobodová, Š., Malinová, T., and Štech, M. 2010. Phylogeny, life history evolution and biogeography of the rhinanthoid Orobanchaceae. *Folia Geobot.* 45: 347–367.

The Marie Currie SPECIATION Network. 2012. What do we need to know about speciation? *Trends Ecol. Evol.* 27: 27–39.

Thompson, P. M., and Ollason, J. C. 2001. Lagged effects of ocean climate change on fulmar population dynamics. *Nature* 413: 417–420.

Thorne, R. F. 1963. Biotic distribution patterns in the tropical Pacific. In J. L. Gressitt (ed.), *Pacific Basin Biogeography* (pp. 311–354). Honolulu: Bishop Museum Press.

Thornton, I. 2007. *Island Colonization: The Origin and Development of Island Communities*, ed. T. New. Cambridge: Cambridge University Press.

Tichy, L. 2002. JUICE, software for vegetation classification. *J. Veg. Sci.* 13: 451–453.

Tiffin, P., and Ross-Ibarra, J. 2014. Advances and limits of using population genetics to understand local adaptation. *Trends Ecol. Evol.* 29: 673–680.

Tobe, H., Stuessy, T. F., Raven, P. H., and Oginuma, K. 1993. Embryology and karyomorphology of Lactoridaceae. *Am. J. Bot.* 80: 933–946.

Toft, C. A., and Schoener, T. W. 1983. Abundance and diversity of orb spiders on 106 Bahamian Islands: biogeography at an intermediate trophic level. *Oikos* 41: 411–426.

Tomb, A., Chambers, K., Kyhos, D., Powell, M., and Raven, P. 1978. Chromosome numbers in the Compositae: XIV. Lactuceae. *Am. J. Bot.* 65: 717–721.

Tomlinson, P. B. 1969. The anatomy of the vegetative organs of *Juania australis* (Palmae). *Gentes Herb.* 10: 412–424.

Torres, D. 1987. Antecedentes sobre el lobo fino de Juan Fernández *Arctocephalus philippii* y proyecciones para su estudio. In J. C. Castilla (ed.), *Islas Oceánicas Chilenas: Conocimiento Científico y Necesidades de Investigaciones* (pp. 289–317). Santiago: Universidad Católica de Chile.

Tortosa, R. D. 1989. El género *Colletia* (Rhamnaceae). *Parodiana* 5: 279–332.

Traveset, A., and Richardson, D. 2006. Biological invasions as disruptors of plant reproductive mutualisms. *Trends Ecol. Evol.* 21: 208–216.

Trénel, P., Gustafsson, M. H. G., Baker, W. J., Asmussen-Lange, C. B., Dransfield, J., et al. 2007. Mid-Tertiary dispersal, not Gondwanan vicariance explains distribution patterns in the wax palm subfamily (Ceroxyloideae: Arecaceae). *Mol. Phylogenet. Evol.* 45: 272–288.

Trusty, J. L., Kesler, H. C., Rodríguez, J., and Francisco-Ortega, J. 2011. Conservation status of endemic plants on Isla del Coco, Costa Rica: applying IUCN Red List criteria on a small island. In D. Bramwell and J. Caujapé-Castells (eds.), *The Biology of Island Floras* (pp. 452–473). Cambridge: Cambridge University Press.

Trusty, J. L., Olmstead, R. G., Bogler, D. J., Santos-Guerra, A., and Francisco-Ortega, J. 2004. Using molecular data to test a biogeographic connection of the Macaronesian genus *Bystropogon* (Lamiaceae) to the New World: A case of conflicting phylogenies. *Syst. Bot.* 29: 702–715.

Tryon, R. 1971. Development and evolution of fern floras of oceanic islands. *Biotropica* 2: 76–84 (also published as pp. 54–62 in W. L. Stern (ed.), *Adaptive Aspects of Insular Evolution*. Pullman, WA: Washington State University Press, 1971).

Tryon, R. M. 1956. A revision of the American species of *Notholaena*. *Contr. Gray Herb.* 179: 1–106.

Tryon, R. M., and Tryon, A. F. 1982. *Ferns and Allied Plants, with Special Reference to Tropical America*. New York: Springer-Verlag.

Tryon, R. M., Tryon, A. F., and Kramer, K. U. 1990. Pteridaceae. In K. U. Kramer and P. S. Green (vol. ed.), *Pteridophytes and Gymnosperms*. In K. Kubitzki (gen. ed.), *The Families and Genera of Vascular Plants*, vol. 1 (pp. 230–256). Berlin: Springer-Verlag.

Tucker, S. C., and Douglas, A. W. 1996. Floral structure, development, and relationships of paleoherbs: *Saruma*, *Cabomba*, *Lactoris*, and selected Piperales. In D. W. Taylor and L. J. Hickey (eds.), *Flowering Plant Origin, Evolution and Phylogeny* (pp. 141–175). New York: Chapman & Hall.

Turner, M. G., Baker, W. L., Peterson, C. J., and Peet, R. K. 1998. Factors influencing succession: lessons from large, infrequent natural disturbances. *Ecosystems* 1: 511–523.

Turrill, W. B. 1959. Plant geography. In R. C. Rollins and G. Taylor (eds.), *Vistas in Botany*, vol. 2 (pp. 172–229). London: Pergamon.

Uhl, N. W. 1969. Floral anatomy of *Juania*, *Ravenea*, and *Ceroxylon* (Palmae-Arecoideae). *Gentes Herb.* 10: 394–411.

UNEP. 2003. Pilot assessmants: ecological and socio-economic impacts of invasive alien species on islands. CBD/SBSTTA/9/ INF/33.

Valdebenito, H., Stuessy, T. F., and Crawford, D. J. 1990a. A new biogeographic connection between islands in the Atlantic and Pacific Oceans. *Nature* 347: 549–550.

Valdebenito, H., Stuessy, T. F., and Crawford, D. J. 1990b. Synonymy in *Peperomia berteroana* (Piperaceae) results in biological disjunction between Pacific and Atlantic Oceans. *Brittonia* 42: 121–124.

Valdebenito, H., Stuessy, T. F., Crawford, D. J., and Silva O., M. 1992a. Evolution of *Erigeron* (Compositae) in the Juan Fernández Islands, Chile. *Syst. Bot.* 17: 470–480.

Valdebenito, H., Stuessy, T. F., Crawford, D. J., and Silva O., M. 1992b. Evolution of *Peperomia* (Piperaceae) in the Juan Fernández Islands, Chile. *Plant Syst. Evol.* 182: 107–119.

Valdebenito, H. A. 1989. Evolution of *Erigeron* (Compositae) and *Peperomia* (Piperaceae) in the Juan Fernández Islands, Chile. Ph.D. dissertation, The Ohio State University, Columbus.

Valenzuela F., R. 1978. *La Protección Jurídica del Patrimonio Ambiental de las Islas Oceánicas Chilenas*. Valparaíso: Ediciones Universitarias.

van Balgooy, M. M. J. 1960. Preliminary plant-geographical analysis of the Pacific. *Blumea* 10: 385–430.

van Balgooy, M. M. J. 1971. Plant-geography of the Pacific. *Blumea Suppl.* 6: 1–222.

Van den Heede, C. J., Viane, R. L. L., and Chase, M. W. 2003. Phylogenetic analysis of *Asplenium* subgenus *Ceterach* (Pteridophyta: Aspleniaceae) based on plastid and nuclear ribosomal ITS DNA sequences. *Am. J. Bot.* 90: 481–495.

Van der Aart, P. J. M., and Vulto, J. C. 1992. Evolutionary status, biogeography and human effects, and general ecology. In P. J. C. Kuiper and M. Bos (eds.), *Plantago: A Multidisciplinary Study* (pp. 4–6). Berlin: Springer-Verlag.

van der Pijl, L. 1982. *Principles of Dispersal in Higher Plants* (3rd edn.). New York: Springer-Verlag.

van Hengstum, T., Lachmuth, S., Oostermeijer, J. G. B., den Nijs, H. C. M., Meirmans, P. G., et al. 2012. Human-induced hybridization among congeneric endemic plants on Tenerife, Canary Islands. *Plant Syst. Evol.* 298 : 1119–1131.

Van Loon, H. 1972. Pressure in the southern hemisphere. In C. W. Newton (ed.), *Meteorology of the Southern Hemisphere* (pp. 59–86). Boston: American Meteorological Society.

Vanderpoorten, A., Devos, N., Goffinet, B., Hardy, O. J., and Shaw, A. J. 2008. The barriers to oceanic island radiation in bryophytes: insights from the phylogeography of the moss *Grimmia montana*. *J. Biogeogr.* 35: 654–663.

Vargas, R., Cuevas, J. G., Le Quesne, C., Reif, A., and Bannister, J. 2010. Spatial distribution and regeneration strategies of the main forest trees on Robinson Crusoe Island. *Rev. Chil. Hist. Nat.* 83: 349–363.

Vargas, R., Reif, A., and Faúndez, M. J. 2011. The forests of Robinson Crusoe Island, Chile: an endemism hotspot in danger. *Bosque* 32: 155–164.

Vaux, F., Trewick, S. A., and Morgan-Richards, M. 2016. Lineages, splits and divergence challenge whether the terms anagenesis and cladogenesis are necessary. *Biol. J. Linn. Soc.* 117: 165–176.

Vazačová, K., and Münzbergová, Z. 2014. Dispersal ability of island endemic plants: what can we learn using multiple dispersal traits? *Flora* 209: 530–539.

Verboom, G. A. 2006. A phylogeny of the schoenoid sedges (Cyperaceae: Schoeneae) based on plastid DNA sequences, with special reference to the genera found in Africa. *Mol. Phylogenet. Evol.* 38: 79–89.

Vergara C., H., and Morales G., E. 1985. Morfología submarina del segmento central del Cordón Asísmico Juan Fernández, Pacífico Suroriental. In P. Arana E. (ed.), *Investigaciones Marinas en el Archipiélago de Juan Fernández* (pp. 25–34). Santiago: Editorial Universitaria.

Vergara C., H., and Valenzuela A., E. 1982. Morfología submarina del guyot O'Higgins, estremo oriental del cordón asísmico Juan Fernández. *Actas III Congr. Geol. Chil.* 1: 132–145.

Vicuña Mackenna, B. 1883. *Juan Fernández. Historia Verdadera de la Isla de Robinson Crusoe.* Santiago: Rafael Jover (facsimile ed., 1974, Ediciones Universitarias de Valparaíso, Chile).

Vignolo-Lutati, F. 1955. L'opera botanica del dottore Carlo Bertero di S. Vittoria d'Alba (1789–1831) nelle Antille e Sud-America (1816–1821 e 1827–1831), quale risulta dalle collezioni dell'Instituto del Orto Botanico della Universita di Torino. *Mem. Acad. Sci. Torino, ser. 3a*, t. II (1): 1–266.

Viljoen, J.-A., Muasya, A. M., Barrett, R. L., Bruhl, J. J., Gibbs, A. K., et al. 2013. Radiation and repeated transoceanic dispersal of Schoeneae (Cyperaceae) through the southern hemisphere. *Am. J. Bot.* 100: 2494–2508.

Vives Moreno, A. 2008. In memoriam Doctor Eugene Gordon Munroe (1919–2008). *SHILAP Rev. Lepidopterol.* 36: 449–455.

von Bohlen V., C. 1995. El género *Mimulus* L. (Scrophulariaceae) en Chile. *Gayana Bot.* 52: 7–28.

von Rodt, C. 1907. *Aus Central- und Südamerica.* Bern: W. Wälchli (see pp. 120 and 219 for comments on her cousin, K. A. von Rodt).

von Wettstein, R. 1896. *Monographie der Gattung* Euphrasia. Leipzig: Wilhelm Engelmann.

Vuilleumier, B. S. 1971. Pleistocene changes in the fauna and flora of South America. *Science* 173: 771–780.

Wace, N. M. 1960. The botany of southern oceanic islands. *Proc. R. Soc. Lond. Series B Biol. Sci.* 152: 475–490.

Wagner, S. T., Hesse, L., Isnard, S., Samain, M.-S., Bolin, J., et al. 2014. Major trends in stem anatomy and growth forms in the perianth-bearing Piperales, with species focus on *Aristolochia*. *Ann. Bot.* 113: 1139–1154.

Wagner, W. L., and Funk, V. A. (eds.). 1995. *Hawaiian Biogeography: Evolution on a Hot Spot Archipelago.* Washington, DC: Smithsonian Institution Press.

Wagner, W. L., Bruegmann, M. M., Herbst, D. M., and Lau, J. Q. C. 1999. Hawaiian vascular plants at risk: 1999. *Bishop Mus. Occas. Pap.* 60: 1–58.

Wagner, W., Weller, S., and Sakai, A. 2005. Monograph of *Schiedea*. *Syst. Bot. Monogr.* 72: 1–169.

Wagner, W. L., Herbst, D. R., and Sohmer, S. H. 1990. *Manual of the Flowering Plants of Hawaii*, vols. 1 and 2. Honolulu: University of Hawai'i Press and Bishop Museum Press.

Walker, L. R., and Bellingham, P. 2011. *Island Environments in a Changing World.* Cambridge: Cambridge University Press.

Wallace, A. R. 1880. *Island Life.* London: Macmillan.

Wallace, A. R. 1895. *Natural Selection and Tropical Nature: Essays on Descriptive and Theoretical Biology.* London: Macmillan.

Walter, R., and Robins, B. 1748. *A Voyage Round the World in the Years MDCCXL, I, II, III, IV by George Anson.* London (edited with an introduction by Glyndwr Williams, Oxford University Press, 1974).

Walter, H. 1973. *Vegetation of the Earth, in Relation to Climate and the Eco-physiological Conditions.* London: English Universities Press.

Wang, J. 2015. Does G_{st} underestimate differentiation from marker data? *Mol. Ecol.* 24: 3546–3558.

Wanke, S., Jaramillo, M. A., Borsch, T., Samain, M.-S., Quandt, D., et al. 2007. Evolution of Piperales–*matK* gene and *trnK* intron sequence data reveal lineage specific resolution contrast. *Mol. Phylogenet. Evol.* 42: 477–497.

Wanke, S., Samain, M.-S., Vanderschaeve, L., Mathieu, G., Goetghebeur, P., et al. 2006. Phylogeny of the genus *Peperomia* (Piperaceae) inferred from the *trn*K/*mat*K region (cpDNA). *Plant Biol.* 8: 93–102.

Wanntorp, L., and Wanntorp, H.-E. 2003. The biogeography of *Gunnera* L.: vicariance and dispersal. *J. Biogeogr.* 30: 979–987.

Wanntorp, L., Wanntorp, H.-E., and Källersjö, M. 2002. Phylogenetic relationships of *Gunnera* based on nuclear ribosomal DNA ITS region, *rbcL* and *rps16* intron sequences. *Syst. Bot.* 27: 512–521.

Wanntorp, L., Wanntorp, H.-E., Källersjö, M., and Oxelman, B. 2001. Phylogeny of *Gunnera*. *Plant Syst. Evol.* 226: 85–107.

Wanntorp, L., Wanntorp, H.-E., and Rutishauser, R. 2003. On the homology of the scales in *Gunnera* (Gunneraceae). *Bot. J. Linn. Soc.* 142: 301–308.

Warren, B. H., et al. 2014. Islands as model systems in ecology and evolution: prospects fifty years after MacArthur-Wilson. *Ecol. Lett.* 18: 200–217.

Warwick, S. I., Mummenhoff, K., Sauder, C. A., Koch, M. A., and Al-Shehbaz, I. A. 2010. Closing the gaps: phylogenetic relationships in the Brassicaceae based on DNA sequence data of nuclear ribosomal ITS region. *Plant Syst. Evol.* 285: 209–232.

Waterman, P. G. 1975. New combinations in *Zanthoxylum* L. (1753). *Taxon* 24: 361–366.

Waterway, M. J., and Starr, J. R. 2007. Phylogenetic relationships in tribe Cariceae (Cyperaceae) based on nested analyses of four molecular data sets. *Aliso* 23: 165–192.

Watson, L., and Dallwitz, M. J. 1992. *The Grass Genera of the World.* Cambridge: Cambridge University Press.

Weeks, A. R., Sgro, C. M., Young, A. G., Frankham, R., Mitchell, N. J., et al. 2011. Assessing the benefits and risks of translocations in changing environments: a genetic perspective. *Evol. Appl.* 4: 709–725

Weigelt, P., Steinbauer, M. J., Cabral, J. S., and Kreft, H. 2016. Late quaternary climate change shapes island biodiversity. *Nature* 532: 99–102.

Weigend, M., and Luebert, F. 2009. Weeding the nettles: I. Clarifying species limits in perennial, rhizomatous *Urtica* (Urticaceae) from southern and central Chile and Argentina. *Phytotaxa* 2: 1–12.

Wessinger, C. A., Freeman, C. C., Mort, M. E., Rausher, M. D., and Hileman, L. C. 2016. Multiplexed shotgun genotyping resolves species relationships within the North American genus *Penstemon*. Am. *J. Bot.* 103: 9121–922.

Wester, L. 1991. Invasions and extinctions on Masatierra (Juan Fernández Islands): a review of early historical evidence. *J. Hist. Geogr.* 17: 18–34.

Wettstein, R. 1922. *Euphrasia formosissima*. Addendum in C. Skottsberg (ed.), *The Natural History of Juan Fernandez and Easter Island, vol. 2* (pp. 209–210). Uppsala: Almqvist & Wiksells.

Wettstein, R. von. 1896. *Monographie der Gattung* Euphrasia. Leipzig: Wilhelm Engelmann.

Wheeler, G. A. 1988. Taxonomic notes on *Carex* (Cyperaceae) of austral South America. *Aliso* 12: 97–102.

Wheeler, G. A. 2007. *Carex* and *Uncinia* (Cyperaceae, Cariceae) from the Juan Fernández Archipelago, Chile. *Darwiniana* 45: 120–141.

Wheeler, G. A., and Goetghebeur, P. 1995. Four new species of *Uncinia* (Cyperaceae) from northern South America. *Aliso* 14: 141–146.

Whelan, C. J., Wenny, D. G., and Marquis, R. J. 2008. Ecosystem services provided by birds. *Ann. NY Acad. Sci.* 1134: 25–60.

Whittaker, R. J. 1998. *Island Biogeography: Ecology, Evolution, and Conservation.* Oxford University Press.

Whittaker, R. J., and Fernández-Palacios, J. M. 2007. *Island Biogeography: Ecology, Evolution, and Conservation* (2nd edn.). New York: Oxford University Press.

Whittaker, R. J., Ladle, R. J., Araújo, M. B., Fernández-Palacios, J. M., Delgado, J. D., et al. 2007. The island immaturity-speciation pulse model of island evolution: an alternative to the "diversity begets diversity" model. *Ecography* 30: 321–327.

Whittaker, R. J., Triantis, K. A., and Ladle, R. J. 2008. A general dynamic theory of oceanic island biogeography. *J. Biogeogr.* 35: 977–994.

Whittaker, R. J., Triantis, K. A., and Ladle, R. J. 2010. A general dynamic theory of oceanic island biogeography: extending the MacArthur-Wilson theory to accommodate the rise and fall of volcanic islands. In J. B. Losos and R. E. Ricklefs (eds.), *The Theory of Island Biogeography Revisited* (pp. 88–115). Princeton, NJ: Princeton University Press.

Wichman, S. R., Wright, S. D., Cameron, E. K., Keeling, D. J., and Gardner, R. C. 2002. Elevated genetic heterogeneity and Pleistocene climatic instability: inferences from nrDNA in New Zealand *Coprosma* (Rubiaceae). *J. Biogeogr.* 29: 943–954.

Wiggins, I., and Porter, D. M. 1971. *Flora of the Galapagos Islands.* Stanford, CA: Stanford University Press.

Will, B., and Zizka, G. 1999. A review of the genus *Greigia* Regel (Bromeliaceae) in Chile. *Harvard Papers Bot.* 4: 205–239.

Williamson, M. 1996. *Biological Invasions.* London: Chapman & Hall.

Willmer, P. 2011. *Pollination and Floral Ecology.* Princeton, NJ: Princeton University Press.

Willson, M. F., Irvine, A. K., and Walsh, N. G. 1989. Vertebrate dispersal syndromes in some Australian and New Zealand plant communities with geographic comparisons. *Biotropica* 21: 133–147.

Wilmot-Dear, C. M., and Friis, I. 1996. The New World species of *Boehmeria* and *Pouzolzia* (Urticaceae, tribus Boehmerieae). A taxonomic revision. *Opera Bot.* 129: 1–103.

Wilson, E. O. 1973. The ants of Easter Island and Juan Fernández. *Pacific Insects* 15: 285–287.

Wilson, K. A. 1996. Alien ferns in Hawaii. *Pacific Sci.* 50: 127–141.

Windham, M. D. 1986. Reassessment of the phylogenetic relationships of *Notholaena*. *Am. J. Bot.* 73: 742 (Abstr.).

Windham, M. D. 1987. *Argyrochosma*, a new genus of cheilanthoid ferns. *Am. Fern J.* 77: 37–41.

Windham, M. D., Wolf, P. G., and Ranker, T. A. 1986. Factors affecting prolonged spore viability in herbarium collections of three species of *Pellaea*. *Am. Fern J.* 76: 141–148.

Wolf, P. G., Sipes, S. D., White, M. R., Martines, M. L., Pryer, K. M., et al. 1999. Phylogenetic relationships of the enigmatic fern families Hymenophyllopsidaceae and Lophosoriaceae: evidence from *rbc*L nucleotide sequences. *Plant Syst. Evol.* 219: 263–270.

Woodward, R. L., Jr. 1969. *Robinson Crusoe's Island.* Chapel Hill: University of North Carolina Press.

Wörz, A. 2005. A new subgeneric classification of the genus *Eryngium* L. (Apiaceae, Saniculoideae). *Bot. Jahrb. Syst.* 126: 253–259.

Wörz, A. 2007. The "Botanische Reiseverein" – a 19th-century joint stock company for the collecting of herbarium specimens. *Huntia (Pittsburgh)* 13: 121–141.

Wörz, A. 2011. Revision of *Eryngium* L. (Apiaceae-Saniculoideae): general part and Palaearctic species. *Biblioth. Bot.* 159: 1–498.

Wright, S. 1951. The genetic structure of populations. *Ann. Eugenics* 15: 323–354.

Wu, M.-J., Huang, T.-C., and Huang, S.-F. 2009. Phylogenetic biogeography of *Euphrasia* section *Malesianae* (Orobanchaceae) in Taiwan and Malesia. *Blumea* 54: 242–247.

Wysocki, W. P., Clark, L. G., Attigala, L., Ruiz-Sánchez, E., and Duvall, M. R. 2015. Evolution of the bamboos (Bambusoideae; Poaceae): a full plastome phylogenomic analysis. *BMC Evol. Biol.* 15: 50. Doi 10.1186/s12862-015-0321-5.

Yakimowski, S. B., and Rieseberg, L. H. 2014. The role of homoploid hybridization in evolution: a century of studies synthesizing genetics and ecology. *Am. J. Bot.* 101: 1–12.

Yañez, E., Silva, C., Vega, R., Espíndola, F., Álvarez, L., et al. 2009. Seamounts in the south-eastern Pacific Ocean and biodiversity on Juan Fernández seamounts, Chile. *Latin Am. J. Aquat. Res.* 37: 555–570.

Yañez R., E., Pizarro G., L., Barbieri B., M. A., and Barra A., O. 1985. Dinámica del stock de langosta (*Jasus frontalis* H. Milne Edwards, 1837) explotado en el archipiélago de Juan Fernández (33°40'S – 80°W). In P. Arana (ed.), *Investigaciones Marinas en el Archipiélago de Juan Fernández* (pp. 251–271). Valparaíso: Escuela de Ciencias del Mar, Universidad Católica de Valparaíso.

Yatskievych, G., and Arbeláez A., A. L. 2008. A new species and three generic transfers in the fern genus *Notholaena* (Pteridaceae). *Novon* 18: 120–124.

Yelles-Chaouche, A., Francheteau, J., and Patriat, P. 1987. Evolution of the Juan Fernandez microplate during the last three million years. *Earth Planet. Sci. Lett.* 86: 269–286.

Yen, A. C., and Olmstead, R. G. 2000a. Molecular systematics of Cyperaceae tribe Cariceae based on two chloroplast DNA regions: *ndhF* and *trnL* intron-intergenic spacer. *Syst. Bot.* 25: 479–494.

Yen, A. C., and Olmstead, R. G. 2000b. Phylogenetic analysis of *Carex* (Cyperaceae): generic and subgeneric relationships based on chloroplast DNA. In K. L. Wilson and D. A. Morrison (eds.), *Monocots: Systematics and Evolution* (pp. 602–609). Collingwood, Victoria, Austrialia: CSIRO Publishing.

Yu, R., Baloch, S. U., Liu, L., Pan, Q., Gong, S., et al. 2015. The phylogenetic relationships among germplasm resources of wild ramie (*Boehmeria nivea* L. Gaud) in China based on *trn*L-F and ITS sequences. *Pakistan J. Bot.* 47: 1451–1457.

Yuncker, T. 1953. The Piperaceae of Argentina, Bolivia and Chile. *Lilloa* 27: 97–303.

Zahlbruckner, A. 1924. Die Flechten der Juan Fernandez-Inseln. In C. Skottsberg (ed.), *The Natural History of the Juan Fernandez and Easter Island, vol. 2: Botany.* Uppsala: Almqvist & Wiksells.

Zapata, F. 2013. A multilocus phylogenetic analysis of *Escallonia* (Escalloniaceae): diversification in montane South America. *Am. J. Bot.* 100: 526–545.

Zavada, M. S., and Benson, J. M. 1987. First fossil evidence for the primitive angiosperm family Lactoridaceae. *Am. J. Bot.* 74: 1590–1594.

Zavada, M. S., and Taylor, T. N. 1986. Pollen morphology of Lactoridaceae. *Plant Syst. Evol.* 154: 31–39.

Záveská Drábková, L., and Vlček, Č. 2010. Molecular phylogeny of the genus *Luzula* DC. (Juncaceae, Monocotyledones) based on plastome and nuclear ribosomal regions: a case of incongruence, incomplete lineage sorting and hybridization. *Mol. Phylogenet. Evol.* 57: 536–551.

Zhang, L., Rothfels, C. J., Ebihara, A., Schuettpelz, E., Le Péchon, T., et al. 2015. A global plastid phylogeny of the brake fern genus *Pteris* (Pteridaceae) and related genera in the Pteridoideae. *Cladistics* 31: 406–423.

Zhang, L., Schuettpelz, E., Rothfels, C. J., Zhou, X.-M., Gao, X.-F., and Zhang, L.-B. 2016. Circumscription and phylogeny of the fern family Tectariaceae based on plastid and nuclear markers, with the description of two new genera: *Draconopteris* and *Malaifilix* (Tectariaceae). *Taxon* 65: 723–738.

Zhang, L.-B., Zhang, L., Dong, S.-Y., Sessa, E. B., Gao, X.-F., et al. 2012. Molecular circumscription and major evolutionary lineages of the fern genus *Dryopteris* (Dryopteridaceae). *BMC Evol. Biol.* 12: 180; www.biomedcentral.com/1471–2148/12/180.

Ziegler, A. C. 2002. *Hawaiian Natural History, Ecology, and Evolution*. Honolulu: University of Hawaii Press.

Zimmerman, E. C. 1948. *Insects of Hawaii*. Honolulu: University of Hawaii Press.

Zizka, G. 1991. Die Chonta-Palme (*Juania australis*) der Juan Fernández Inseln (Chile). *Der Palmengarten* 2: 21–24.

Zizka, G., Trumpler, K., and Zöllner, O. 2002. Revision of the genus *Ochagavia* (Bromeliaceae, Bromelioideae). *Willdenowia* 32: 331–350.

Index of Scientific and Vernacular Names

All illustrations are indicated by **boldface**; colored figures (e.g., **C1**) appear on the unnumbered pages following p. 108.

Animals

Plants

Index of Topics

All illustrations are indicated by **boldface**; colored figures (e.g., **C1**) appear on the unnumbered pages following p. 108.